1608-1674 John Milton, English poet and statesman

1642-1727 Sir Isaac Newton, English mathematician and philosopher

1646-1716 Baron Gottfried Wilhelm von Leibniz, German mathematician and philosopher

1685-1750 Johann Sebastian Bach, German composer and organist

1685-1759 George Frederick Handel, English composer

1706-1790 Benjamin Franklin, American scientist and statesman

1707-1778 Carolus Linnaeus, Swedish botanist

1723-1790 Adam Smith, Scottish economist

1726-1797 James Hutton, Scottish geologist and discoverer of time in the evolutionary sense

1736-1806 Charles Augustin de Coulomb, French physicist

1736-1819 James Watt, Scottish inventor: the steam engine

1743-1794 Antoine Laurent Lavoisier, French founder of modern chemistry

1745-1827 Count Alessandro Volta, Italian physicist

1749-1832 Johann Wolfgang von Goethe, German poet and amateur scientist

1753-1814 Count Rumford (Benjamin Thompson), scientist and statesman, American in origin

1766-1844 John Dalton, English chemist and father of atomic theory

1775-1836 André Marie Ampère, French mathematician and physicist

1791-1867 Michael Faraday, English physicist and chemist

1809-1882 Charles Robert Darwin, English biologist

1822-1884 Gregor Johann Mendel, Austrian monk, founder of genetics

1831-1879 James C. Maxwell, Scottish physicist

1834-1907 Dmitri Ivanovich Mendeleev, Russian chemist

1845-1923 Wilhelm Konrad Roentgen, German physicist

1856-1940 J. J. Thomson, English physicist

1858-1947 Max Planck, German physicist, father of quantum theory

1867-1934 Marie Sklodowska Curie, Polish physicist and chemist

1871-1937 Baron Ernest Rutherford, New Zealand-born English physicist

1879-1955 Albert Einstein, German physicist, founder of relativity theory

1885-1962 Niels Bohr, Danish physicist

SCIENCE RESTATED

SCIENCE RESTATED

Physics and Chemistry for the Non-Scientist

HAROLD G. CASSIDY

Professor in Chemistry, Yale University

Freeman, Cooper & Company

1736 Stockton Street, San Francisco, California 94133

Printed in the United States of America
Library of Congress Catalogue Card Number 72-119371
SBN 87735-007-8

This text is dedicated to the memory of

PROFESSOR HERBERT SPENCER HARNED

and

DEAN WILLIAM CLYDE DEVANE

Preface

This text is the outcome of events set in motion over fifteen years ago. Then, in 1956 a committee of chairmen from Yale science departments met under the chairmanship of Professor Herbert Harned of the Chemistry Department. This committee gave official recognition to the fact that the educational needs in science of students who are not going to be professional scientists are different in significant ways from the needs of students in the regular pre-professional courses. They laid down in broad outline the features that seemed desirable in a course suitable for persons not going on in physical science. Essentially, it was intended that the values of science as a mode of thought, as a substantive underpinning to personal philosophy, as a method for controlling the phenomenal world through intellectual concepts, should be brought home insofar as possible to these students.

It was not glossed over that the students would have to learn a vocabulary of technical terms (as I wrote later in the Yale Alumni Magazine), for it was realized that a major barrier between the non-scientist and the sciences is the specialized vocabulary needed to comprehend a science. Also it was felt that it would be an important objective of the course to bring home to the student the beauty of scientific language—a language of precision and subtlety that can be appreciated by an intelligent literary person.

Dr. Arthur R. Quinton of the Physics Department and I were invited to give this course. I had providently prepared for this invitation, foreseeing through service on a number of ground-preparing committees that this action would eventuate. A Faculty Fellowship from the Fund for the Advancement of Education allowed me to spend a year renewing my acquaintance with humanities subjects, attitudes, and goals. There was at this time a "science" requirement at Yale for non-science students. Dean William C. DeVane made available to me, with the help of a grant from Carnegie Corporation of New York, classes consisting of small groups—twenty or so in size—of selected students, many of whom were antagonistic to, or indifferent to, "science."

Professor Quinton and the other physicists who taught sections of this course over the years—Drs. John M. Bailey, William A. Blanpied, William F. Hoffmann, McAllister H. Hull, and Jens C. Zorn—taught me what physics I know, and much more than physics.

As I reconstruct what happened I can see why it is so difficult to alter a curriculum. I had worked with groups of these students for several years. Many of them were genuinely concerned to learn what I had to offer, and were generous and constructive in critical appraisals, making of these experiences teaching of the happiest kind. At the same time it was apparent to me that sensitive communication was far from being certain. For one thing there was no suitable text. We tried out some seven different ones in sequence, supplementing as needed with notes and exercises. But the trouble was more fundamental, it seemed to me, being a matter of both content and conceptual approach. Also issues were complicated by the changing student climate. Then, at the end of one year I suddenly realized that I need not follow accepted fashion in this introductory course. So obvious an insight it seems now. Then it had the effect of a revelation.

With the full and generous support of Dean DeVane and Professor Harned, to the memory of whom this text is gratefully dedicated, I discarded *all* conventional content. I think it took three or four years of teaching the course to develop the confidence to take this step. It is difficult to analyze the obvious. The following summer was spent rewriting the course, and since then it has been fully experimental and has evolved to its present state. I scribbled down one of Professor Harned's comments when I told him of this step and of my plans. He said, "You can't go wrong if you go back to the roots. You can always teach the roots to anyone interested, and it's no more difficult than some passages from Shakespeare, or Eliot."

It is clearly not proper to depart from classical and tested paths unless on the basis of a carefully thought-out, rational philosophy. I have discussed this philosophy in some detail in my *Knowledge, Experience and Action, an Essay on Education* (Teachers College Press, 1969), and will not repeat here except as a very brief summary. The course is designed to speak to the condition of the individual student, recognizing that he is unique, important, and that his growth is the justification of education. I am concerned to connect science to his life through his perceptions, his ability to transmute experience into mental constructs by acts of symbolic transformation; that he be able to understand philosophically what he is doing as he connects the world of experience (an existential and contingent world, indeed) to the beautifully ordered cognitive world of scientific constructs, where essential laws rule, and then returns by his own volition to put into practice—to bring to action, and so test—the results of that enterprise. No intelligent, alert, imaginative student has yet failed to be moved by this vision. This at least is the first step in enlisting his attention; in helping him to recover lost curiosity, and a sense of his ability to gain the upper hand over the subject.

The rest of the course follows, and is described in the introductions to the several Parts. Basically, I have picked out those aspects of Physics and Chemistry that I think will still have important influences upon world culture twenty or more years from now. I include only the necessary classical background. I have avoided historical interludes not because I do not think history important—far from it—but because of the exigencies of time and space.

I hope that the teacher will not be disappointed because I have failed to touch some of his favorite topics. This is a teaching text, not a compendium, and there should be plenty of time for each teacher to introduce, modify, omit—whatever he wishes. I urge this. What we want is to teach that science contributes to a way of life; that it combats alienation by offering doors open to the depth that is everywhere, at any point, below the surface; that it can give the student that "raw hedonistic pleasure," as John Weichsel put it, that comes from his own conceptual grasp of some facet of reality. We want to help the student view himself and his work in the context of his contemporaries, and our contribution is to that part of the contemporary scene that is the product of the scientific technology that rests on physical science. We want to help him to live as a constructive, functioning part of the whole system of humanity—past, present, and future—to the best of our ability.

At the same time we do not wish to baffle and confuse the student. When very little is known about a subject it can happen that ritual behavior, in-group language, and other devices are imported to lend seeming depth and apparent content. We know so much in physical science that we can be secure enough to acknowledge limits and ignorance, and actually practice humility (traits of the utmost value to society, and presently highly regarded by perceptive students). Simultaneously we must consciously import the utmost clarity and simplicity into our teaching, we must eschew unessential jargon; we must seek simplicity while distrusting it; we must try for the utmost clarity, openness, and communion with our students.

I need hardly say that I have a great many people to thank for help in bringing this text into existence. I have already mentioned the support given by the Ford Foundation and by Carnegie Corporation.

Research Corporation at this time gave me their first Venture Grant, and enabled me to open up the new field of Redox Polymers, and to carry on active chemical research, which then was generously supported by the Public Health Service. This was important to a teacher who has never suffered from any teaching/research dichotomy. A sabbatical year spent at the Center for Advanced Studies at Wesleyan University under the direction of Paul Horgan enabled me to prepare the first draft of this text. During that period Professor Richard W. Lindquist was very helpful, especially with one chapter that responded to his expert eye. Professor Willard V. Quine and Professor José Gomez-Ibañez both receive my thanks.

Kathryn Childs Cassidy has been a source of strength and support during the times of painful academic frustration that seem inevitably to face an innovator. So also, Dr. Orville F. Rogers, who provided a room with a view on Casco Bay, for summer writing.

I acknowledge with thanks the labors of Mrs. Barbara Satton, Secretary at the Center for Advanced Studies. Particularly, I must thank Mrs. Mildred Ross Bray at Yale who, with the greatest patience and understanding through repeated revision, imperturbably turned out her precise and elegant typescripts.

The staff at Freeman, Cooper & Company come in for thanks because of their insistence on perfection (within human limits): Margaret C. Freeman, James Levorsen, and James Mennick—the latter two converted my "drawings" to understandable form.

I am indebted to the following publishers for permission to quote, or use figures from, their publications (all credited, I hope, at the point of use): Addison-Wesley, American Institute of Physics, *American Journal of Science*, Johann Ambrosius Barth, Cambridge University Press, Cornell University Press, Dover Publications, Inc., Education Development Center, The Franklin Institute, Gauthier-Villars, Harper & Row, Holt, Rinehart & Winston, Inc., Houghton Mifflin Co., The Johns Hopkins Press, *Journal of Chemical Education*, McGraw-Hill Book Company, The Macmillan Company, Prentice-Hall, Inc., Routledge & Kegan Paul, Ltd., The Royal Society, Springer Verlag, Taylor & Francis, Ltd., John Wiley & Sons, Inc.

HAROLD CASSIDY

New Haven
Winter, 1970

Contents

The Intellectual Context — **Part I**

1. Perspective 4
2. To Set the Scene 34

The Classical Background — **Part II**

3. The Pervasive Electron 60
4. Magnetism 96
5. Light 126

The Fruit of Discontent — **Part III**

6. Relativity 168
7. Electromagnetic Radiation 202
8. Quanta and Photoelectric Theory 220

The Modern View of Matter — **Part IV**

9. Constraint and Variety. An Overall View of the Nature of Matter 254
10. Atomic Structure 268
11. Atomic Properties 290
12. Nuclear Investigations. The Great Machines 308
13. Nuclear Properties and Processes 334
14. Molecular Properties and Reactions 358
15. Molecular Structure 382

Larger Issues and Unifying Principles — Part V

16. Chance, Probability, and Statistics 422
17. Process. An Introduction to Cybernetics 438
18. Bodies, Particles, and Fields 468
19. Directional Arrows and Universal Laws 476
20. Toward a Natural Philosophy 494

Appendixes

Appendix 1. Trigonometry 505
Appendix 2. Some Physical Constants 515
Appendix 3. Some Relative Nuclidic Masses 517
Appendix 4. Alphabetical List of the Elements 519

SCIENCE RESTATED

Part I

Chapter 1. Perspective

Chapter 2. To Set the Scene

The Intellectual Context ~ PART I

As we look out upon the world external to ourselves we cannot fail to observe the prevalence of change: material objects change their spatial positions; matter of one kind is transformed into something else. The motions and transformations of matter have always intrigued people, and many have sought to understand and to control them. Those who worked through scientific methods developed Physics and Chemistry. We are concerned in this book with some of their work. And in this first Part we are especially concerned with the intellectual aspect of their enterprises, because by paying attention to this aspect we can economize time and effort. If we can understand the principles, then we needn't learn every fact.

It is important to know what modern scientists think about the motions and transformations of matter, and to understand how they go about getting and using this knowledge. This is because science and scientific thinking are dominant in this era in which we live. Through scientific technology, which is applied science, the material bases of our lives are undergoing fantastic changes; and the conceptual bases are likewise changing. At the same time, as we shall show throughout this book, it is quite feasible for the layman to understand modern science (and outside of his field **everyone** is a layman). The methods of thinking and of expression in science are not of some special order inaccessible to the ordinary person. After all, science is the work of human beings who range over as wide a gamut of intelligence and creativity, and ignorance and stupidity, as may be found in any other endeavor bar none. That scientific thinking is dominant in our culture is a reflection of the tremendous material successes of scientific technology. It is difficult to escape the conclusion that scientists must be doing something right. But what is it, and what are its limitations? It may take the humanist, looking on and concerned with intellectual history, to evaluate what is right and what the limits are. This humanist could be you, reading these words.

But one has to understand in order to criticize, and it is just for this reason that we offer Chapters 1 and 2. These chapters are intended to place the remainder of the book in context in the whole sphere of knowledge and experience. The two chapters that comprise this Part are to be read with imagination. The matters touched on are relevant to the entire College experience of the reader, not to say to his whole life. If this Part is approached with a prosaic attitude it may well seem elementary. Many of our most important insights speak with a still small voice, and are drowned out for the imperceptive by the raucous cries of novelty-mongers, image-formers, and intellectual masseurs. But if it is read with perception, and with continued effort to relate it to personal experiences; if it is re-read from time to time, and supplemented by the Student's developing insights, this Part can perform its real function of tying the rest of the book, especially the purely physical part, to the other parts of his life.

In the first chapter we analyze and present a metaphor of the structure of a college or university with respect to knowledge, experience, and action. What we are after, here, is a broad view of the academic context within which the reader of this book finds himself. It is a premise of this text that meaning arises out of organized connectedness. An isolated fact, if it could exist, would be meaningless, and we have been very careful in connecting all facts and theories in this book. But further, we are concerned to make connections to the rest of the Student's experience and knowledge, which is the reason for our analysis of college activity.

In this chapter, too, we present the framework of a theory of knowledge which we use to organize the entire book. We say **a** theory of knowledge, recognizing that we are being arbitrary (for this is neither a history of philosophy nor a text on epistemology). We are concerned to formulate the way that a scientist deals with the age-old problem of experience, knowledge, and action; how experience may become knowledge, and how this may be put into practice in action. Again we construct a metaphor, hoping in this way to alert the Student to the fact that things are not as simple and straightforward as they may seem; to help him become aware that scientists are far from knowing everything, yet that what they know comprises a vast network of logically validated facts and theories which is strengthened and made credible by its agreement with the way the world is.

In Chapter 2 we attack the problem of how to conceptualize change. Since we shall be dealing throughout the rest of the book, with motion of things and transformations of matter it is necessary to show how change can be grasped intellectually. Moreover, since the Student lives in the midst of change, the fundamental principles dealt with in this chapter may help him in many ways that are not explicitly treated. The discussion of motion—of change of position in space—makes use of the framework developed in Chapter 1 and leads directly to Part II.

Chapter 1

1.1 The literate citizen
1.2 On Science and the non-Scientist
1.3 The Context of knowledge, experience, and action
1.4 On meaning
1.5 The nature of Science
1.6 Toward a life view
1.7 Practical application of the foregoing ideas

Perspective ~ 1

Our purpose is to help non-scientists become literate in science.

The literate citizen 1.1

Our society is so complicated that specialization is inevitable for most of us. It is almost the only way in which a person, inevitably subject to limitations on his time and talents, can achieve a degree of individuality and excellence. Whether he plays the piano, does chemistry or law, plumbing or baseball, a citizen to be good at something must specialize in it. At the same time he lives with others whose interests and needs differ from his. To be able to understand them and communicate with them in a meaningful way he must acquire a degree of breadth in their areas of interest. The world of knowledge and experience is so diverse and so vast that no one alone can encompass it. Thus each of us lives in this dilemma: because we are finite human beings we must specialize if we are to be good at anything; and at the same time, our needs as social beings to communicate together call for widening our range of knowledge and experience in areas where we will not be specialists. The intricate pattern of our society is woven of specialties held together by the overlapping nonspecialized knowledge of literate people. It is strengthened by mutual trust, when it is guarded by mutual understanding, and by experience. It rests in part on the memory-bank of our heritage, housed in the library.

A literate person is one who has learned to live understandingly in the face of this dilemma. He has learned through formal and informal education enough of the history and traditions of his and other cultures to be alert to the consequences of his acts; he has some grasp of the great areas of knowledge, experience, and the applications of these, as reflected in history and in the current scene. He has learned how to evaluate current affairs rationally, and is fortified against false doctrines whether these come in the name of Science or of the Humanities. He is a specialist in some area of knowledge, experience, action, and he is aware of the accepted truths of other areas, of the regions of uncertainty in them, and of the reasons for acceptance and for uncertainty.

1.2 On science and the non-scientist

It is obvious that science and technology (which is the art of applied science) are changing our culture and our world. No amount of wishful thinking, no amount of intellectual revolt, can stop the changes. Through them people are getting material things that they want. What seems to be the case, however, is that many of the material, mental, and spiritual prices that are being paid for these changes may bankrupt or even destroy us.

For example, at present the population of the world is increasing at a rate of over two persons per second which, conservatively estimated, is equivalent to a new city of 170,000 people each day. Improved public health, lessened infant mortality, and increased longevity, brought about in considerable part as the result of discoveries in science and medical technology, have contributed to this dangerous state of affairs. These additional people require food and shelter, which "use up" resources and room, both of which are in finite supply on this Spaceship Earth. Animals and plants thus have pressure put upon them, and it is said that many species, products of aeons of Nature's inventiveness, are becoming extinct. At the same time material wastes and pollution increase and crowding and noise cause mental pollutions. Unless this sequence of processes is brought under control the predictable result will be overall degradation of our culture.

Our recourse is not to try to stop all change, but to guide it into scientifically and humanistically sound channels. It follows, then, that the man—whether scientist or non-scientist in training and temperament—who would understand what is happening in this, our Spaceship (which in a sense is all we have), must understand Science. He does not have to become a producing scientist, but he should know enough, and have the judgment, to evaluate movements and ideas and to estimate consequences. He should be able to place his estimates and evaluations in the context of his culture for, as we shall see, only if he makes such connections between knowledge, experience, and action does he give fullest meaning to his own life and that of his society.

The context of knowledge, experience, and action 1.3

What we wish to do first is to develop the context in which this course is studied. It should be thought of as containing a core which makes contact at as many points as possible with other courses.

The intellectual structure of a college or university is a pattern or reflection of the culture which supports it. We must first grasp this intellectual pattern; then it can be seen where this course fits in and how connections can be made to other courses. For example, you cannot encompass all the hundreds of courses offered in your college catalogue. How do you achieve a good balance?

The fields of *knowledge* which are the product of rational mental activity, and of *experience* which is subjective; the results of perceptions of all kinds, and of *action* which may be thought of as the application of knowledge and experience to practice, are so vast that they have conventionally been divided into segments: the Sciences, the Humanities, the Philosophies, the Technologies. The dividing lines are fairly logical ones, but not hard and fast. A degree of overlapping facilitates communication between people specializing within different segments. Each segment comprises disciplines that are housed in departments in the college or university; departments are further subdivided.

We begin by defining and putting together the segments. The Humanities and Sciences may be delineated by means of Figure 1-1.

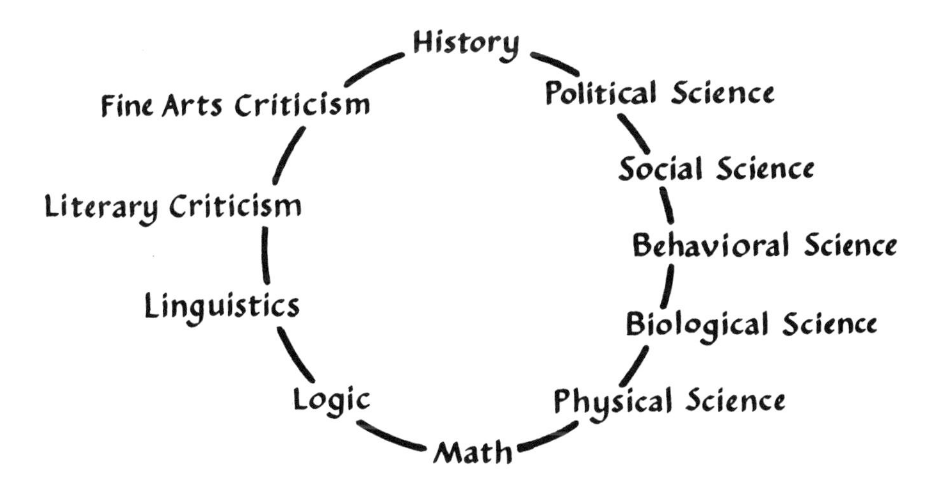

Fig. 1-1 *A definition of the Sciences and the Humanities.*

This is produced by arranging the disciplines listed in a college catalogue into groups, and placing these—somewhat logically—around the circle shown. On the Science side there is logic in moving from physical sciences (astronomy, physics, chemistry, geology) to policy sciences (political science, economics). We move from phenomena produced by the interactions of myriads of *identical* entities such as atoms or molecules in a gas to phenomena determined by the actions of highly individual people. The arrangement is logical in terms of complexity of the systems being studied. Physics and chemistry are concerned with relatively simple interactions between particles; biology with the more complex form and functions of living organisms; the behavioral, social, and policy sciences with still more complicated interactions that involve physical, mental, and spiritual dimensions of individuals and groups of people; of families, states, and nations.

The logic of our arrangement is not so compelling on the humanistic side. Here the disciplines are much older than most of the Sciences. However, the arrangement is based upon the kinds of data that are the subjects of critical study. At the lower left of the diagram the subjects are largely discursive and written with well-defined grammar and syntax. At the upper part they may be discursive (music, dance) or presentational (painting, architecture) but without well-defined grammar or syntax.

This diagram is meant to imply that all these disciplines are related; though in some way the Sciences differ from the Humanities. But the diagram is not intended to be taken too literally: literary criticism is not in any sense "opposite" to psychology. Below, we shall look more closely at the differences between the Sciences and the Humanities.

If we now place this circle as an equatorial belt about a sphere, Figure 1-2, we may thereby expand the Sciences and Humanities to entire hemispheres. Now we may locate at one pole of our "Sphere of Knowledge and Experience" the Philosophies, and at the other the Technologies. We mean to indicate by this figure that there are scientific philosophies as well as humanistic philosophies and that there are humanistic as well as scientific technologies. There is room on this sphere for any discipline that belongs in a college or university.

In all of the disciplines on our sphere, three kinds of activity are carried on, usually simultaneously and always with different degrees of emphasis. In literary criticism, or in chemistry, we obtain information about a state of affairs. We read the works themselves; we learn in chemistry how the substances look and react with each other. This is "data-gathering." But the human brain is a pattern-forming organ, and so whenever several pieces of data are communicated to a person's

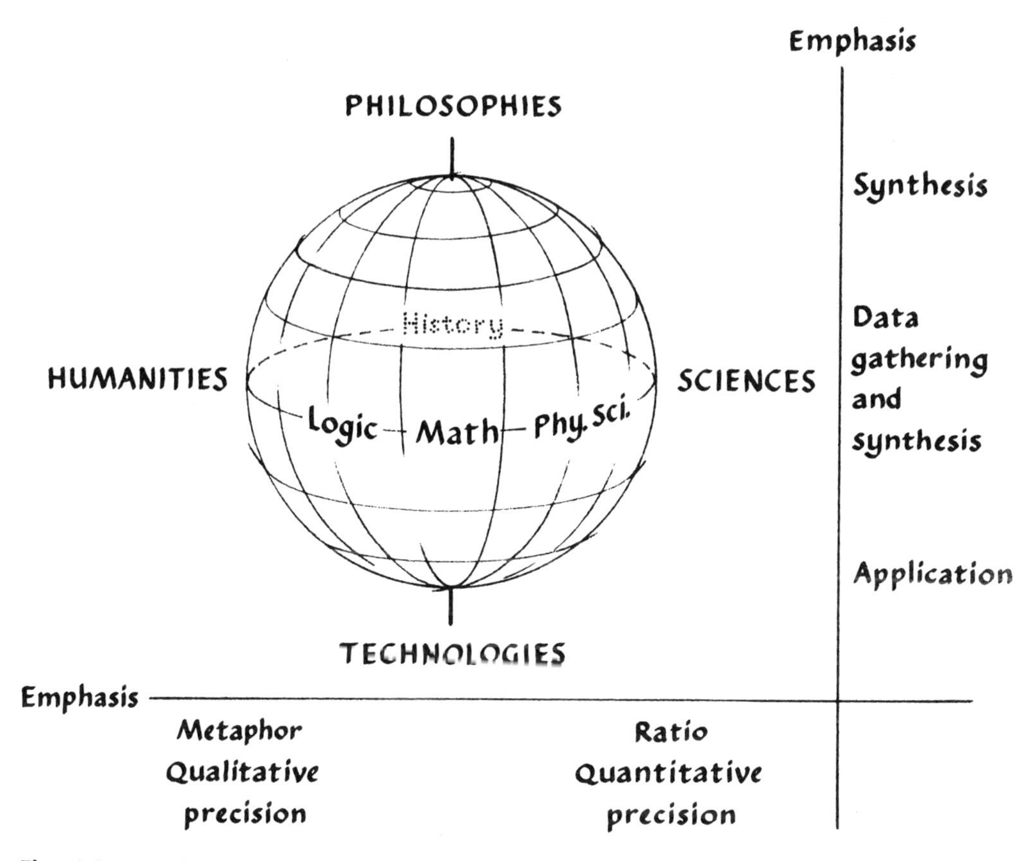

Fig. 1-2 *The intellectual structure of the University: the "Sphere of Knowledge, Experience, and Action."*

brain they are almost inevitably compared. Out of such comparisons of many data we tend to build patterns, to suppose relationships, to devise conceptual models. These may grow into theories and perhaps laws. This pattern seeking and forming activity is "synthesis." At the same time, caution warns us that it is easy to be misled by perceptions and to be in error with respect to hypotheses, so we tend to check our hypothetical suppositions against the facts themselves, looking to see how well the latter fit the facts and how well the hypotheses predict what we shall find in new factual situations. In other words, we seek to validate our knowledge against experience. This is "application to practice." The three, as we said, go on together, and we separate them by names only in order to discuss them. Figure 1-3 puts them back together.

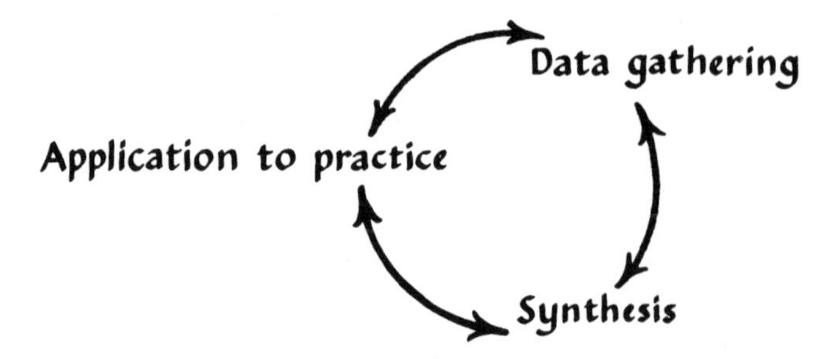

Fig. 1-3 *The principal activities of the persons engaged in the disciplines implied in Figure 1-2.*

We said, also, that emphases differ. *This* fact is used in Figure 1-2 to clarify the differences between the groups of disciplines along a vertical axis of our Sphere. The Philosophies emphasize synthesis. One should have facts on which to base a philosophy, and any philosophy, even one concerned with possible worlds—concerned, that is, not only with existing Man but with potential Man—should be tested against the way the World actually is unless it is meant as a purely intellectual exercise. The equatorial belt of disciplines, the most familiar ones in a liberal arts curriculum, emphasizes data-gathering *and* synthesis described previously. It must be evident why we put fine arts and literary *criticism* here. This is what most departments of English, of Literature, of Art, do emphasize: they usually have not taught creative writing or painting as their primary emphasis. Nowadays some of these departments are broadening their interests.

The Technologies emphasize application to practice. It is characteristic of a technology that it is concerned with the application of many disciplines to each given unique problem. For example, a civil engineer (a scientific technologist) applies a knowledge of metallurgy, chemistry of soils, properties of concrete, and so forth, to his particular unique problem, say of building a bridge. He may also bring to bear on his *engineering* decisions political and aesthetic considerations; economic ones he has always with him. Or a physician (a humanistic technologist) treating an ill person, focuses his art and science on *this* patient. It may well be that what he has learned about the human heart in his literature courses is as useful in one particular case as what he learned of the heart in his anatomy course. Medicine, law, theology, and rhetoric are the oldest disciplines in the university. They grew out of life. They are still among the humanistic technologies concerned with the application of many theoretical disciplines to the solution of particular problems. Every discipline, whether Science or Humanities, has its technology, and each emphasizes application to practice in particular cases. Virgil I. Grissom, Astronaut, put the concept of application to practice succinctly. He said that when a test pilot takes up a new plane and puts it through its paces, "he is really flying somebody's theory."

We have, then, ranges of emphases that help us to see the differences between Philosophy, "pure" Science and Arts, and "applied" Science and Arts. Now, looking from left to right in Figure 1-2 we can set up ranges of properties that distinguish the Humanities from the Sciences. One such useful distinction may be made in terms of methods of communication. *The scientist is concerned with phenomena that are publicly verifiable.* Although he carries out individual observations and experiments, the data that he works with are preferably repeatable and open to verification by anyone who is suitably trained. He aims for a quantitative precision of statement which, because it must depend on mathematics if possible, and if not (as is usually the case) then on carefully defined words, may seem cold to an outsider. The humanist, concerned to communicate or discuss individual, unique, subjective, private experience, usually relies on language that is connotatively rich—often speaking a different personal message to each listener. He aims at qualitative precision, and his language may be warm and personal, though it is not invariably so, for some literary criticism is impersonal and cold. If he uses other means of communication, as in painting, dance, and music, the same argument holds. *His* facts, too, *may* be publicly verifiable.

Any discipline in college can be located on this sphere. It describes the intellectual structure of a college or university. It implies that in principle all disciplines are connected and form a whole. It is a formal device designed to enable one to grasp the pattern of knowledge and experience, so that although he can "take" only a few of the courses offered by the college he can know in principle the types of thinking going on in others. The sphere is a device to show the context of this Course so that it gains in meaning.

On meaning 1.4

Meaning, we say, inheres in connectedness. If anything—a thought, a person, a particle—were completely isolated (assuming for the sake of the point that this were possible) it would be meaningless. Likewise, because this Course is connected along the surface of our Sphere to all other areas of knowledge, experience, and action, it receives and confers meaning. This is a matter of some importance to a student when he is trying to build for himself a "life view." For as we shall see, it is a wonderful characteristic of these intellectual activities that they weave ever more richly interconnected patterns and validate each other. Thus they may give increasing meaning to the whole, *and to the literate person who contributes to and develops these patterns.*

We can now consider the question "Why study science?" Modern psychiatry has joined with ancient wisdom in asserting that:

> Man lives in three dimensions: the somatic, the mental, and the spiritual. The spiritual dimension cannot be ignored, for it is what makes us human. . . . But where do we hear of that which most deeply inspires man; where is the innate desire to give as much meaning as possible to one's life, to actualize as many values as possible—what I should like to call the will-to-meaning?
>
> This will-to-meaning is the most human phenomenon of all, since an animal certainly never worries about the meaning of its existence. . . .[1]

This is a modern statement of the ancient wisdom of body, mind, and spirit. Now the somatic dimension is limited. As Frankl points out it is simply an animal dimension. It is also limiting, for without adequate nourishment the body cannot function and consequently neither can the mental and spiritual dimensions be at their best. Similarly, the mental dimension exerts some controls over the somatic and spiritual, and the spiritual, as much evidence shows us, can definitively influence both mind and body. This is to say that all are interlinked.

The studies of the sciences, humanities, philosophies, and technologies are approached by way of the mental dimension, but clearly not exclusively so: the study of literature, or of molecules, is made with body, mind, and spirit; we suggest that the excitement and even ecstasy of educating ourselves rests on the integral linkage of all three dimensions. What we often see, on the other hand, is a kind of spiritual distress, an existential frustration. This may arise from the conflict between a person's will-to-meaning and his spiritual values unsupported by knowledge. Without scientific and other knowledge of his world a person is incapable of establishing his own scale of values. He loses his capacity for intelligent decision and action.

This is not the place for an essay on existential freedom, the values inherent in science, and the personal aspects of knowledge. What we wish to suggest here is that inherent in the nature of man the pattern-former, as a specifically human attribute, is his *will-to-meaning*. It is to actualize, to make real, the values therein that in our Culture we insist upon formal education. It is specifically the values inherent in the field of scientific knowledge and experience that lead us to believe that every literate person should study and comprehend some science. It

---

[1] From the Introduction to *The Doctor and the Soul*. Viktor E. Frankl, Alfred A. Knopf, New York, 1966, p. *x*. Viktor Frankl survived two concentration camps and knows whereof he writes.

is the basic reason for this book—to show what meaning certain ideas from physics and chemistry can have for a modern literate man. We have treated physical science as we have here as part of an holistic view of knowledge and experience to oppose in an explicit way nihilistic philosophies and intellectual and spiritual alienation.

The nature of science 1.5

Science is an orderly body of knowledge—a corpus of beliefs about our physical environment derived from our experiences.

Physical science, as we restrict it in this book, is derived from experiences with matter in motion or at rest (physics) and with transformations of matter (chemistry). When it is difficult to say whether physics *or* chemistry is what one is doing, it means that it is not important to make the distinction—one is doing physical chemistry or chemical physics.

The knowledge called science originates in *perception:* one experiences something through his senses. The processes of converting perceptions to knowledge and of ordering this knowledge are not well understood, though we perform these acts every day. We can, however, make certain general statements that bear on the subject. Suppose I go into my laboratory to do chemistry. Say I want to make a certain new compound, never before made, for which I have imagined some use. I have thought about it and decided upon a procedure. I set up apparatus consisting of suitable flasks and heating and cooling devices; I measure into the flask the necessary solid and liquid chemical reagents, taking them from particular labeled bottles. I then heat the mixture (if heat is needed) and watch what goes on: changes in color, changes in state from solid to liquid or to gas, and so on. After a time I let the mixture cool, having decided from what I observed that the reaction has been completed. I observe that some material separates (say) in crystalline form. This is what I expected to happen. I separate the crystals, redissolve them in some suitable solvent, and allow them to crystallize again upon slow cooling. I repeat this process of purification until no further change in the properties of the material occurs, and now I feel that it is pure. A completely untutored person watching all this would be baffled, and probably bored. He would have missed much of what was going on. All he would have seen would have been operations carried out on material objects. Science has a great deal to do with the other things that were going on: things associated with "thinking about it," with choosing "suitable apparatus," "necessary chemical reagents," with "measuring," "expecting certain things to happen," and so on. Look now at these operations in a more general way.

Both the onlooker and I perceive a number of things going on. In addition I act physically—picking up flasks, and so forth, connecting them, weighing reagents, and so on. Physically, I act on and thus change the material objects—or some of them. But in the act of perception alone we two are both coupled to the same objects by the same reflected light that we see. We might symbolize all this as follows:

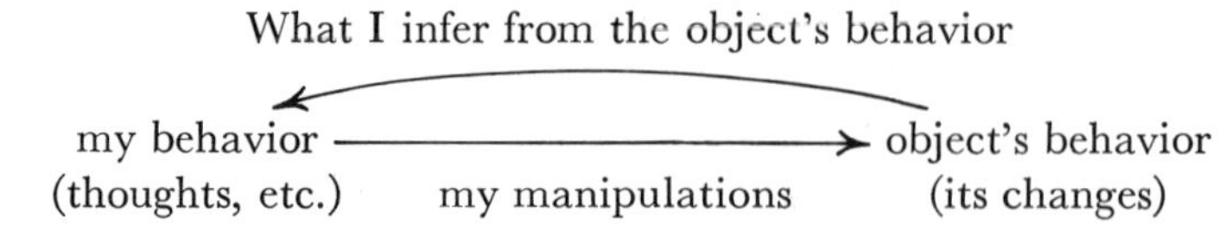

We mean by this that observer and object are linked: my behavior changes that of the object (perhaps) and its behavior has an effect on me (perhaps). In physical science it is not very common that my act of *perception* changes the behavior of the object, but in the life and social sciences it is very common that the act of observing changes the behavior of an observed, responsive creature. That the behavior of the object may "change" my behavior is most common: I watch what is going on in the flask, and if it seems unusual, or unexpected, if the reaction seems to be going too fast for safety, for example, I take appropriate action. Of course, in this case I change my behavior in the light of the object's behavior. The onlooker, not knowing what is going on, may behave quite differently in response to what he sees happening. Thus both the object and the observer may be changed in some way by this interaction.

At the same time, we know that the act of perception may produce false information about the thing perceived. As a simple example, I might be color-blind, and so miss some information, or I might obtain erroneous information; if there is a thermometer in the flask I might read it incorrectly because of some imperfection in the glass or in my way of looking at it. And so on. An objective of much of laboratory teaching is to learn how to avoid errors of perception.

Another objective of ours is to learn what to see. This is a complicated matter. The amount of information that impinges on our eyes is usually a great deal more than can be handled by them or taken in by our minds. Even that which passes through the iris to the retina is usually more than can be handled by that organ. Thus some sort of selection must take place. This selected pattern is further selected and rearranged when the impulses (electrical and chemical) that travel along the optic nerve reach the brain. It is not surprising, then, that our untutored onlooker fails to see things that I have learned to see, and sees things that I (possibly to my loss) have learned to ignore or not see. One of the major aims of science is to handle experiences in such a way that any person who is suitably trained in careful, comprehensive, and alert perception, and in the avoidance of erroneous perception, will agree with any other about some observation—each person thus validating the work of the other. The fruits to us of such scientific efforts amply testify to the considerable degree of success in the use of the methods.

So much for the experience itself, the perception that may lead to knowledge. What can we say about the critical step from experience to knowledge? Knowledge is what can be written down in books or on the labels of bottles: it requires words, or symbols, drawings, and so forth, together with concepts. The process of going from the experience to the written or verbal or other communication is "symbolic transformation." This step, when it is made for the first time, is a creative act of an intuitive kind. After it has been made and fitted into the body of knowledge and experience it can be taught, and made available to everyone.

For example, I reach and take down from the shelf a bottle of white crystalline material labeled 'sodium chloride, NaCl.' That name conveys knowledge which was gained as the end product of a sequence of perceptions, operations, and symbolic transformations. A logical reconstruction of what happened might go as follows. There is a familiar substance known as table salt. A chemist has recrystallized it from a solution many times until at last he feels certain that it is pure. He then examines it chemically (by means we need not detail) and concludes on the basis of chains of other knowledge that it is composed of the chemical elements sodium (Na) and chlorine (Cl) in a one-to-one ratio (1:1). That is, in the compound, the ratio of sodium to chlorine atoms (the atomic ratio) is the same number of each. He has thus found out experimentally what this stuff is, and he can label it NaCl and give it the more specific name 'sodium chloride,' based on chemical syntax and grammar. And so it came into the bottle on my shelf.

To the chemist the name denotes many properties. One is that the formula weight of this compound can be calculated by adding the atomic weights of Na (22.99) and Cl (35.46). It is 58.45. He has learned that one formula weight of any substance, measured in grams, contains 6×10^{23} molecules of the compound.[1] This allows him to weigh out an amount of the material that will give him the desired number of molecules of NaCl needed in his reaction. Thus he may proceed from the

---

[1] *Reminder:* 6×10^{23} is a shorthand way of writing 6 with 23 zeros after it, then a decimal point. Thus 1×10 is the same as 1×10^1 (the exponent '1' is usually omitted), which is 10. Similarly, 3×10^4 is 30,000. When the exponent is preceded by a minus sign it signifies the position of the number to the right of the decimal point. This derives from the rule of exponents that says $1/10^3$ may be written 10^{-3}. Thus 4×10^{-1} is 0.4: the number 4 is in the first place to the right of the decimal. Then 5×10^{-5} is 0.00005. Similarly, 4.75×10^2 is 475; 4.75×10^{-2} is 0.0475.

crystalline material he experienced, through the knowledge conveyed by the label on the bottle, back to a quantitative measure of the reactive entities in a weighed amount of the material. (He learns many shortcuts.)

What has been going on here? From one point of view, I have been carrying out certain visible operations. We might name these existential, or phenomenological. Although they might give me a certain pleasurable sense, they are analytically rather shallow: over two thousand years ago Heraclitus told us that one can never step into the same river twice, and he has been quoted ever since because what he said is true. So, I did the experiment and that is that.

But it is not enough, of course. Suppose we make use of certain insights and formulations of the physicist-philosophers, F. S. C. Northrop and Henry Margenau, and imagine these manipulations, these experienced acts, plotted on a surface which will be named the "plane of perceptions." We agree that analytically it is a rather shallow region, which is why we call the whole surface a plane. A cross section through it, labeled the 'P-plane,' is shown in Figure 1-4. If you need to visualize something you might think of the beakers and flasks on a laboratory desk, and imagine that what we are doing physically is weaving a network of perceptions and other subjective experiences over this surface. (Some people have difficulty with these notions. If you are one of them, read on, then re-read. Did you "understand" *Burnt Norton* the first time you read it?)

Now on the basis of these perceptions which are in some ways unique to him (and which can be carried out by others who also undergo their own unique experiences) the scientist conceives a 'construct' or a set of linked constructs. When we say that an experience is unique to the person we mean this literally. It has ineffable components, perhaps. In other areas, as in art or literature, there may arise out of experience a 'work'—something unique to the man. In science, the emphasis is more on reproducibility and public agreement. This is achieved through the constructs. These are invented, abstract 'concepts by intellection' as they are called by Northrop to distinguish them from the mental results of the experiences themselves, which he describes as 'concepts by intuition.' When I pick up the bottle of salt, I am operating at or close to the plane of perceptions; when I calculate the number of molecules present in a given weight of the salt I am operating in the 'field of constructs,' shown as the 'C-field' in Figure 1-4.

A construct is a kind of invention. It is in the first place thought up by someone in a creative, imaginative act, as the idealization of some behavior or property of nature. For example, of our own *experience* we know several kinds of space: visual, tactile, dreamed, and other private kinds. But the physicist's kind of space, the *constructed* space defined in equations of physics, is a kind of absolute (with a small *a*). It is uniform,

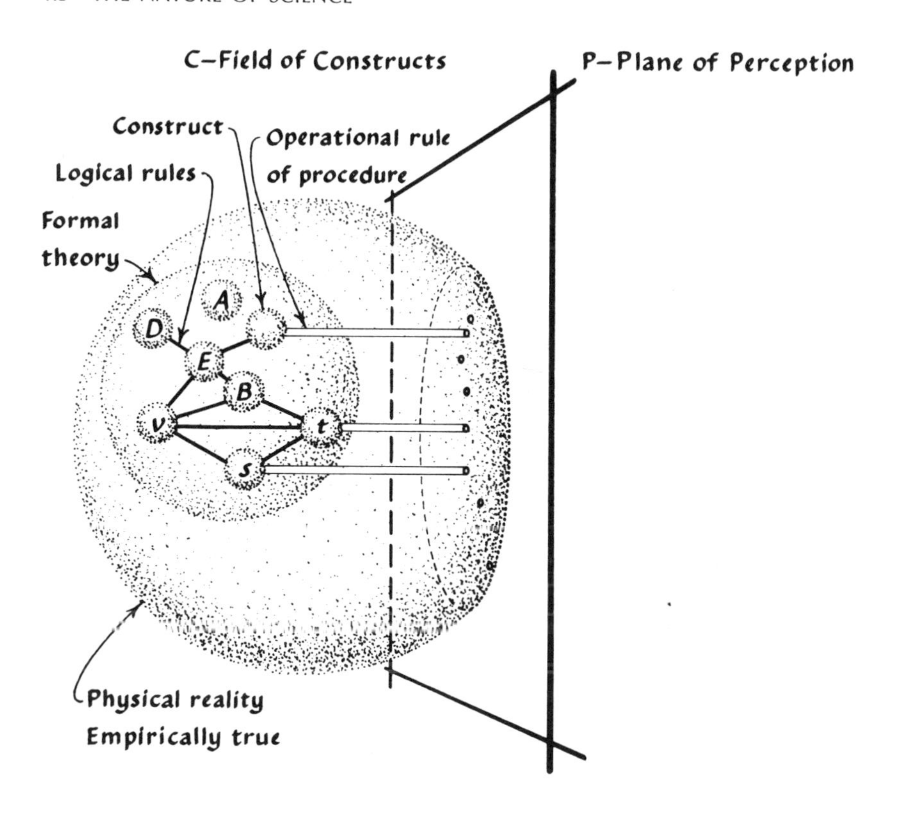

C–Field of Constructs	P–Plane of Perception
Regions of	
Knowledge	Experience
The cognitive	The existential
"Truth about"	"Truth to"
Public world of science	Private world of sense perception
Concepts by intellection	Concepts by intuition

Constructs

Theory-making ←
(Intuitive step to concepts which are related by essential laws which are invariant relationships).

Application to practice →
(Deductive step to existential laws).

Fig. 1-4 *A metaphor of the structure of physical (scientific) reality. [Adapted from H. Margenau,* The Nature of Physical Reality. *McGraw-Hill Book Co., New York, 1950, by permission.]*

homogeneous, not visualizable. It is arrived at by following definite 'operational' rules of procedure which are symbolized by the double lines in Figure 1-4. Take for example a distance between your dormitory room and the classroom. Some mornings the distance is much too great as you struggle along to class; other days you hardly notice it. Your roommate tells you that he has similar experiences. We now agree to follow an operational recipe which says you take a measuring tape; you put one end down at your door-sill, and the other in a direct line to the classroom . . . and so on. You measure the straight-line distance, and give it a value 's' in feet or yards or meters depending on how your tape is calibrated. Now *anyone* who follows the rule with that or any other good tape will arrive at the same value no matter how he feels about it. This is the value of the distance and it is called the 'displacement' when it is distance-in-a-specified-direction. Displacement is a construct; it is independent of how you or anyone else feels; it is public, and publicly testable.

Now you can see what we are getting at. Out of the welter of private experiences the scientist invents 'constructs' such as distance (s) time (t) velocity (v) in Figure 1-4. He develops a body of connected, defined constructs. They are symbolized by the circles in the figure; the double lines are the pathway of symbolic transformation, so to say. The construct in some way transcends the experience.

Constructs may be connected or related to each other by equations. For example, distance (s) and time (t) may be set in the relation $v = s/t$, where v is now the construct 'average velocity.' This relationship says that when s and t are found to have certain values (postulated, or measured in an actual experience in which the two are linked) the object traveling a distance s (meters) in a time t (seconds) exhibits a velocity v (meters-per-second). Constructs are what they are in order that they be acceptable to any scientist. Thus from the private experiences that each of us may have we may come to a common meeting ground: a public world of scientific knowledge comprised of a network of linked constructs. This is, indeed, one embodiment of our will-to-meaning. It is a meta-language; a language that transcends (meta) the existential experiences and thus may be used to talk *about* them. It is not one of them.

It must be evident to the perceptive student that concepts that function like constructs are not unique to science. The critic of an art, or of literature, must have in his mind small-*a* absolutes of an analogous kind. These would comprise his theory of beauty, or form, or style, and so on. For example, analyze in these terms *The Life of Forms in Art* by Henri Focillon.[1] One cannot escape the recognition of an analogous type of thinking as it is carried on by the creative humanist. As an

[1] Wittenborn, Schultz, New York, 1948. See also, Harold G. Cassidy, *The Sciences and The Arts: A New Alliance*, Harper & Brothers, New York, 1962.

exclamation point I might add that this art critic, writing of Gothic architecture, gives one of the best definitions of experiment I have read. He says (p. 9):

> By experiment I mean an investigation that is supported by prior knowledge, based upon a hypothesis, conducted with intelligent reason, and carried out in the realm of technique.

For a construct to have meaning it must be connected to at least two others, thus being part of the network that leads to something new. For example, in Figure 1-4, B is a construct with meaning, *e.g.*, it connects v with t and so may represent 'acceleration.' On the other hand D is redundant; it may be merely some other way of saying E; and A is irrelevant, and completely meaningless in this context because it is completely isolated. A group of constructs connected among themselves by logical rules (the single lines) comprise a 'formal theory.' The constructs validate each other. When the constructs are also connected to perceptions and are empirically verified they become 'physical reality,' our subject.

Since the constructs are in a sense absolutes (we made them so) the relations between them are 'invariant-relations.' Simple ones are called relationships; far-ranging ones are called *laws*. Moreover, these invariant-relations are in a sense *essential*, in contrast to the *existential* at the P-plane. For example: I pick up a book. I may or may not then drop it—I may not even have picked it up in the first place, but suppose I did. These acts are in the realm of the existential, where I have a certain freedom (and responsibility). But when I picked up the book I lifted it against gravity; if I release it it will ineluctably fall; and the laws that govern these behaviors and properties are essential laws; that is, they are inherent in the way things are; we cannot legislate them. The laws that predict what will happen are then by arrangement absolute: what actually happens is relative to them.

Recall that Ptolemy taught that the Earth, the throne of Man, was the center of the universe, a fixed point around which the heavenly bodies moved. Copernicus dethroned Man, consigning him to a place on a minor planet, swinging with others around the sun, which is one of a myriad of stars. Now Man is back at the center of the universe once more, but this time in a conceptual sense only. He conceives constructs and their relationships that are absolute. 'Reality' is relative to *them*. The implications of this change in viewpoint for our stewardship of this spaceship Earth are incalculable.

Toward a life view 1.6

We conceive, then, that the goal of a liberal education is to assist a student to develop for himself a life view, that is, a personal philosophy constructed from what knowledge, experience, and reason tell him about the fit between facts, theories, hopes, ideals, and reality. He should become able to interpret the heritage of time-tested knowledge, and to go beyond this as necessary to new meanings suitable to the uses of his own life and time. For him to do this his teachers and his own independent reading should acquaint him with the best, most relevant theories, hopes, and dreams of his heritage and that of other cultures. They should assist him (or her) to become truly self-aware: aware of the unique gifts he alone holds in *his* hands. A college may only provide an intellectual scaffolding to sustain, challenge, and encourage him.

This course is designed to contribute toward this goal. We have chosen relatively few aspects of physical science, and treated each in some depth. We have chosen those subjects that have been affecting our culture most strongly and that we think will continue to be relevant for several decades. Since it is impossible to imagine what will turn up of importance in this area during the student's lifetime, we endeavor to treat these so as to show their scientific anatomy, so to speak, and to place them in context in intellectual history where this is possible without undue digression. Thus we endeavor to prepare the student to deal with new ideas.

We return, finally, to the fact that all areas of knowledge and experience are connected—even though tenuously in some cases. When Galileo discovered crags, crevices, and excrescences on the surface of the Moon, and spots which moved about on the Sun, he destroyed the effectiveness of many literary devices of the day. Incautious indeed would thereafter be the poet who compared his Love's complexion to the "satiny" surface of fair Diana; or his constancy to the immutable perfection of the Sun. The scientific results, discussed by Marjorie Nicolson,[1] did not destroy poetry. They eventually enriched literature and all the arts, as modern science is now in process of doing.

This is, further, not a one-way street. Humanists certainly do affect Science through their influences on the cultural context of Science. One need only suggest, here, what has been discussed explicitly by Max Jammer,[2] namely, that the writings of some early existentialists set the intellectual context for and "contributed decisively to" the development of new constructs of modern quantum theory.

In sum, then, we conceive it to be the Student's primary aim in

---

[1] Marjorie Hope Nicolson, "The Telescope and Imagination," *Modern Philology*, *32*, 233 (1935). See her *The Breaking of the Circle*, Columbia University Press, New York, revised ed. (paperback), 1965 (Northwestern University Press, 1960), and *Newton Demands the Muse*, Princeton University Press, 1946, paperback edition 1966.

[2] Max Jammer, *The Conceptual Development of Quantum Mechanics*, McGraw-Hill Book Co., New York, 1966, pp. 166 ff. For other examples see my *The Sciences and The Arts*, *loc. cit.*, Chap. 9.

College to build for himself a life view. To do this he must be aware of his own abilities, and aware that apparent limitations can be transcended by him given a sufficiently strong will-to-meaning. He must specialize, of course, yet he must also develop the breadth to understand and communicate with others. This is why a person not specializing in science, yet needs to know enough science to be literate in this area. This conception, built upon an holistic view of knowledge and experience, has been the basis for the choice of subjects and for their treatment in the following chapters.

Practical application of the foregoing ideas 1.7

The ability to transform experience (at the P-plane) into the symbols of speech and writing makes it possible to communicate knowledge; to store and study it, and to use it. Once the vital transformation has been made, the way is opened to apply the tools of logic, of mathematics, and of the various sciences, as this book exemplifies. It is a matter of great moment that not only do we convert experiences of phenomena 'out there,' outside ourselves, into symbolic form, moving thereby away from the P-plane, but also we are able to make the return journey. As we have seen, the totality of constructs and their connecting relations comprise theory. The return from theory to 'practice' is the essence of technology. This circuit which we illustrate in this section may be thought of as an encoding and decoding chain: perceptions encoded from stimuli received by our eyes, ears, and other receptor organs are re-coded into images and ideas which, by further re-coding processes, emerge through guided muscular action as written or spoken words or other symbols. (We must not suppose, of course, that by putting names —coding, decoding, and so forth—on these complicated processes we have explained them. Rather, the purpose is to make it easier to recall them in other connections.) These symbols may be manipulated, as we said, and then issue in action of some kind in which they are applied to practice. What we wish to do now is to illustrate this kind of circuit, or chain. Because of our many limitations we shall deal very simply and baldly with these complex matters, and rely on the bibliographic references at the end of this chapter to expand and qualify our treatment.

A *set* is a group the members of which show some particular similarity to each other. This similarity is used to define the set. The members of the set, which may be objects, or attributes, or relations, are usually called 'elements,' and written in braces, thus: the set of all even numbers between 3 and eleven is $\{4, 6, 8, 10\}$ and no others. This is a finite set. As an example of an infinite set we may take the set of all the points at a distance r from a given origin point O. These points comprise the surface of a sphere with center at O and radius r, and cannot be listed because infinite.

It will be observed that a description of a set is a rule which states the *property* that determines the set. That is, it states something characteristic of the members of the set which makes it reasonable to gather them together under one description, though in general each member differs in some way from every other. (Every object differs from every other if only in not occupying the same space at the given time.) Moreover, every object has almost an infinity of properties—macroscopic, microscopic, qualitative, quantitative, relational, and so on—which would baffle any attempt at complete description. It is characteristic of a science that it does not attempt to concern itself in detail with all the infinity of properties of an individual object: only those that are relevant to the particular end in view. In Figure 1-4 the rules of correlation between constructs are in effect rules designating particular relevant properties. They leave behind the individuality of the 'thing out there,' the observed phenomenon, in connecting the concept by intuition to the construct; in connecting, to put it another way, the aesthetic component of reality (at the P-plane) with the theoretic (in the C-field).

Since a set may be a collection of objects which have some particular property in common, and at the same time the members or elements of the set have a large number of properties other than the particular property (or properties) that defines the set, it is evident that a member of one set may also belong to other sets. For example, the Author is at the same time an element of the sets 'people named Cassidy,' 'persons weighing between 150 and 180 pounds,' 'adults,' 'teachers of physical science,' 'male people,' and so on. In designating a set, then, we ignore the properties of the elements in which we are not interested. For example, in connection with the even numbers between 3 and 11, we ignored the properties 'no primes are included,' and 'none is divisible by 7,' and so on.

It is convenient sometimes to symbolize a set by means of a Venn diagram. We let the area in a circle or other figure stand for the set, say of persons weighing between 150 and 180 pounds, inclusive. We can write this definition inside the circle (Figure 1-5) or name the set 'P' (which has nothing to do with the P-plane, being merely conventional) and we write a definition: "P is the set of all persons weighing from 150 to 180 pounds, inclusive." The *size* of the circle is irrelevant: we just have an enclosure which is correlated to the set of elements, P. Space outside of the enclosure represents all the entities that do not belong to the set. These are not-P, symbolized by $\tilde{P}$. P and $\tilde{P}$ make up a 'universe of discourse,' 'U.' This is what we are concerned with, rather than all the other things we might discuss. Suppose, now, we have another set, which we designate 'Q' and define as "the set of all

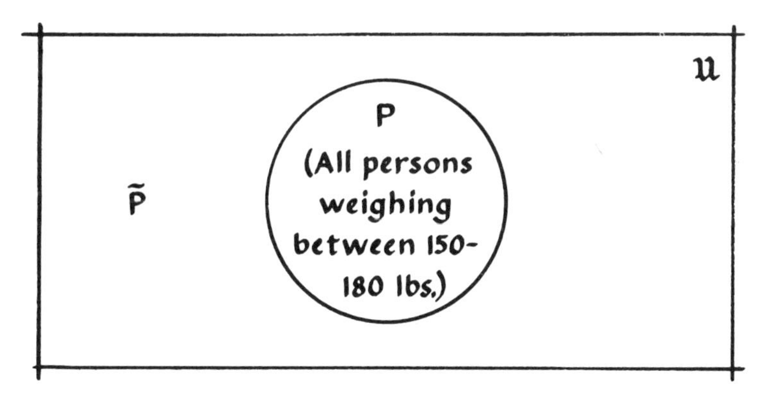

Fig. 1-5 *Venn diagram of a set* P *and its complement* P *in the universe of discourse* u.

male people." Clearly, there are some male people belonging to set P as well. We show this in the Venn diagram symbolism in Figure 1-6. The overlapping, hatched area *b* includes all those persons who weigh between 150 and 180 pounds (set P) *and* who are male (set Q). Those who are male and weigh other than 150 to 180 pounds are represented by the area *c*, in Q yet not in P; those who weigh 150 to 180 pounds but are not male are represented by the area *a*.

It is important to notice that so far we have been dealing abstractly, using the names of entities, which we put in single quotation marks when we introduce them. If we were concerned about one or more actual people, we should have to *look and see* whether or not they belonged in this and/or that set: we would have to examine the data. The set has a name 'P' or 'Q' but this does not name some new entity: it is the name of the *relationship* that unites the elements of the set and is true of any or all the elements of the set. The names P and Q, and so forth, are used as convenient shorthand for the longer statements.

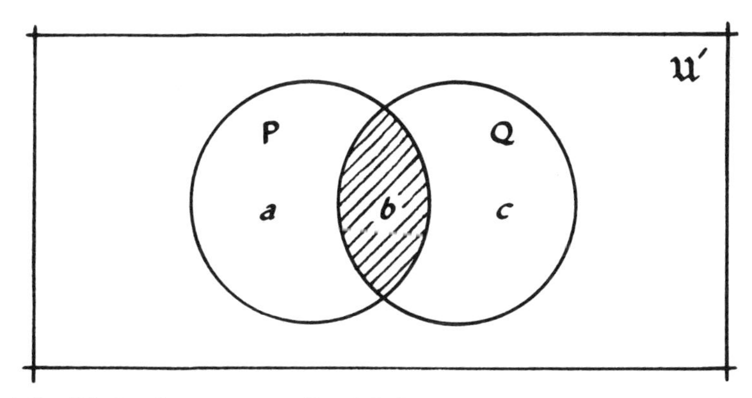

Fig. 1-6 *Relations between two sets* P *and* Q: b *represents the subset* $P \cap Q$; a *plus* b *plus* c *represents* $P \cup Q$.

Further, a number of connectives are used to designate compound relationships between P and Q and in general any sets. We shall concern ourselves with three only of these: 'complement' $\sim$, 'intersection' $\cap$, and 'union' $\cup$. In the universe of discourse, $\tilde{P}$ is said to be the complement of P; it complements P in the sense of filling out the rest of the possibilities we are discussing, *e.g.*, people of different weights (Figure 1-5). The area of overlap, or intersection, of P and Q in Figure 1-6, namely the hatched area *b*, represents that subset of P and Q in which the elements belong to both sets. It would be designated $P \cap Q$, read "the intersection of P and Q." The area *a* plus *b* plus *c* is written $P \cup Q$, read "the union of P with Q." It represents the set of all those persons who *either* weigh between 150 and 180 pounds *or* are male, *or both.*

Many persons who don't care for abstractions enjoy manipulating Venn diagrams. The reason may be that the diagram presents a 'Gestalt,' or whole comprehensive picture at a glance. It is easy to see that the area *a* may be designated $P - (P \cap Q)$, where '$-$' means minus.

Suppose, now, that we look at the *statements* we have been making about these sets. Notice that we conceived of certain *things* as belonging to a set: we do this kind of thinking all the time as part of the pattern-forming drive of our minds. We can then make categorical statements about these things, or about the sets. For example, we can say of a person "this is a male;" or "this person is a member of the set 'Q';" or "Q is the set of all males," and so forth. These statements are neither the male person, nor the set of male persons. They are *about* them, and so we signalize this fact by giving the statements lower-case symbols. Let '*p*' be the statement 'this person weighs within 150 to 180 pounds,' and let '*q*' represent the statement 'this person is male.' Each of these is a simple statement which may refer to some person. A statement may or may not have meaning; if it has meaning it is called a 'proposition' and may or may not be true. *Whether it is in fact true or false is its truth value. In fact means empirically.* We look and see.

Simple propositions are made into compound ones by connectives, and again we shall concern ourselves with three only: 'and' '$\wedge$'; 'either/or or both' '$\vee$'; and 'it is not the case that' '$\sim$'. Then if we wish to say "this person weighs within 150 to 180 pounds and is male" we write $p \wedge q$; or "either this person weighs within 150 to 180 pounds or is not male, or both" we write $p \vee \sim q$.

The connection between the proposition and the set of things or other entities may now be shown as a table 1-1. We have moved conceptually away from the things 'out there' to verbal or written propositions. We have also made compound statements.

In designating our propositions *p* and *q* we saved ourselves con-

Table 1-1

Table of Connectives

Propositions		*Translation*	*Sets*	
Name	Symbol		Name	Symbol
Conjunction	$\wedge$	. . . and . . .	Intersection	$\cap$
Disjunction	$\vee$	. . or . . or both . .	Union	$\cup$
Negation	$\sim$	. . . not . . .	Complement	$\sim$

siderable repetitive writing of the whole sentences. These letters were used as a kind of shorthand. But a great deal more than this is implied in their logical use. For p and q may be *any* statements of relations such that one may say of each that it is true or false. Furthermore, given any two propositions, p and q, either of which may be true or false, there are four possible combinations of them. These can readily be found by systematic cross-tabulation. The *compound statement* may be true or false, depending on the *truth value* of the simple propositions *and* on the meaning of the connective. The possibilities are generally shown in the form of a 'truth table.' Thus, given any p and any q, in the compound statement $p \wedge q$, only if both are true is the compound statement true. This is the meaning of 'and,' or '$\wedge$.' This is summarized conveniently in a truth table for the relation $p \wedge q$, Table 1-2.

Table 1-2

Truth Table for Conjunction

Possible combinations		Truth value of the compound statement
p	q	$p \wedge q$
T	T	T
T	F	F
F	T	F
F	F	F

The truth table for negation must show that if p is false then $\sim p$ must be true, and if p is true $\sim p$ must be false, as shown in Table 1-3.

Table 1-3

Truth Table for Negation

p	$\sim p$
T .	. F
F .	. T

The connective '$\vee$' means "either . . . or . . . or both," and its truth table is shown in Table 1-4.

Table 1-4

Truth Table for Inclusive Disjunction

p	q	$p \vee q$
T .	. T	T
T .	. F	T
F .	. T	T
F .	. F	F

This truth table clearly follows the definition of the connective, because if either p or q or both are true, then the compound statement is true. If both are false, then the statement must be false.

The three connectives serve to generate the whole of symbolic logic. We cannot show this in detail, of course. We are dealing with highly general statements, so that we can agree that if two compound statements have the same truth table, then they are equivalent. (We use parentheses, brackets, braces, to set off expressions.) Suppose, then, that we have the statement:

$$\sim[(p \wedge q) \vee (\sim p \wedge \sim q)]$$

What is its truth table? We proceed in a systematic manner, first writing down the truth tables of $(p \wedge q)$ (from Table 1-2) and from this $(\sim p \wedge \sim q)$, and then applying the relation '$\vee$' to them, thus (Tables 1-5, 6, 7).

Table 1-5

p	q	$(p \wedge q)$
T .	. T . .	. T
T .	. F . .	. F
F .	. T . .	. F
F .	. F . .	. F

Table 1-6

$\sim p$	$\sim q$	$\sim p \wedge \sim q$
F	F	F
F	T	F
T	F	F
T	T	T

Table 1-7

$(p \wedge q)$	$(\sim p \wedge \sim q)$	$[(p \wedge q) \vee (\sim p \wedge \sim q)]$
T	F	T
F	F	F
F	F	F
F	T	T

Then, negating the result '$\sim$[]' we can write the truth table of the expression:

Table 1-8

p	q	$\sim[(p \wedge q) \vee (\sim p \wedge \sim q)]$
T	T	F
T	F	T
F	T	T
F	F	F

This statement may be put verbally as "either p or q but not both." We say that we have analyzed the original complicated statement to find its truth table.

We now are in a position to connect the manipulations of compound statements and simple statements with those of sets and subsets. The connections are made in the following way. Statements, which may be simple or compound, are as usual designated p, q, r, ... These statements relate to a *set of logical possibilities*, u. We recognize among the logical possibilities in u, sub-sets for which the statements p, q, r, . . . are respectively true. These sub-sets are designated, respectively, P, Q, R, . . . and are the 'truth sets' of the statements p, q, r, ... Since a set may be a collection of physical objects (in the sense previously discussed) we have here a mode of symbolic transformation between the physical set and statements related to the set. We designate the universe set as U.

There is an interesting kind of statement called an 'open statement,' in which one or more 'variables' occur. When the variable is given a definite value, the statement becomes a definite one, which may be true or false, depending on the 'value' taken on by the variable. For example, the open statement might be 'x is a Senior.' Here the universe of discourse is the entire student body at a given time in a given college. If the name that is substituted for x is that of a Freshman, or Sophomore, or Junior, the definite statement turns out to be false. It is true if the person is authentically a Senior. Or, again, the statement might be 'the integer x is a prime.' Here the universe of discourse is all the real numbers (an infinite set). The statement is true when, for example, 3, or 5, or 2, or any other prime number, is substituted for x; it is false in other cases, as for example if 4 or 6 is substituted for x. Such sentences that may sometimes be true, sometimes false, are called 'propositional functions.'

The variable is a symbol—v, w, x, y, z are commonly used—which refers to an element, or elements, of a set. Some of the elements—those which on substitution in the propositional function yield true propositions—form a true sub-set of the universe of discourse. For example, the propositional function:

$$x^2 - x - 6 = 0$$

may be *asserted*, in the universe of discourse of all real numbers. It is true only for certain numbers, namely -2 and $+3$. Thus the truth set of the statement is $\{-2, +3\}$. We designate this P. Another statement might be:

$$x^2 - 4 = 0$$

The truth set of this statement is 'Q' $= \{2, -2\}$. We solve two equations simultaneously when we find one or more elements that are common to the truth sets of the two equations. In this case, we want $P \cap Q$; the intersection of their truth sets. The Venn diagram for this is shown in Figure 1-7. If the truth sets turned out to have no element in common, the equations would be inconsistent.

It may be added that in practice Venn diagrams are not of much use in solving simultaneous equations! However, we have introduced them because many people who initially have trouble understanding abstract discursive formulations, such as those of statements in terms of p, q, and so forth, and those of sets, P, Q, and so forth, or of algebra, do not seem to be troubled by geometry—or at least not nearly as much. Furthermore, *all of these can be converted* (laboriously) *into words.* To see the connections between all these modes of expression is perhaps to see that for some purposes one will be better than another, and to under-

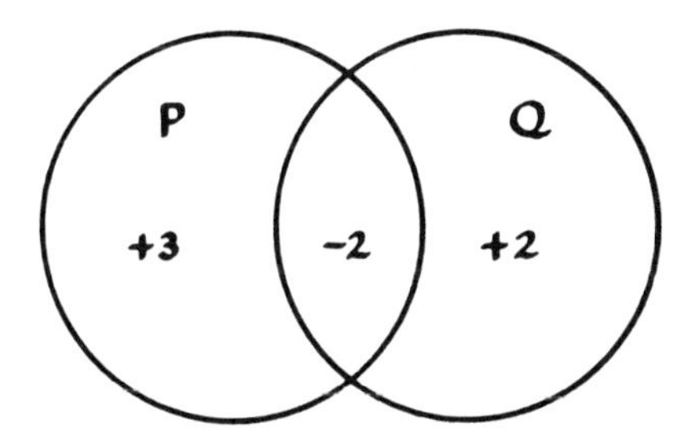

Fig. 1-7 *Simultaneous solution of two equations with solutions* $P = \{+3, -2\}$ *and* $Q = \{+2, -2\}$ *as represented by* $P \cap Q$ *in the Venn diagram.*

stand them all more readily. Our final step, then, will be to show in principle how these may be embodied in machinery. By machine we mean a material object constructed by man for a definite purpose.

We have seen, in principle at least, that logic is basic to mathematics, and we know that mathematics is basic to science and technology. We have a chain of connections, here, which is itself logical. Suppose we adopt the following conventions. This is a code that we shall use. In Figure 1-8 is shown a flashlight circuit, with the switch open. If the switch is closed, and the battery is charged, current flows

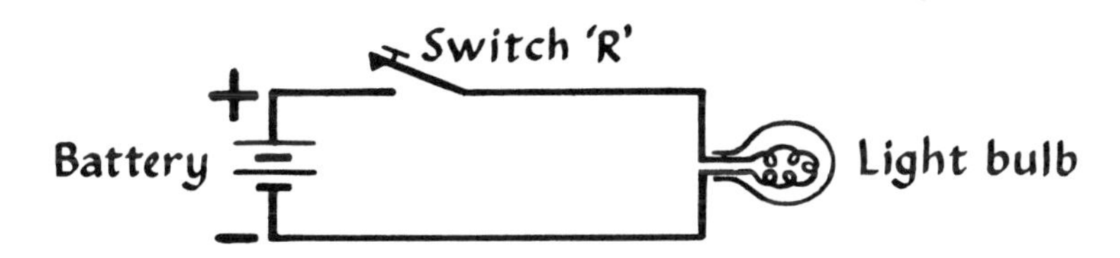

Fig. 1-8 *Simple circuit, with battery, light bulb, and on-off switch shown in the 'off' position.*

through the circuit from one pole of the battery to the other and in the process heats the filament of the bulb to a white-hot, light-giving glow. We shall designate the switch 'R' (it is a member of a set containing this particular switch). We shall associate with it a statement p, "the switch R is closed," and we shall know when the switch is truly closed because the light will come on. So if we see the light on, we know that current is flowing, and we designate the state "current is flowing" 'T.' The flashlight may therefore be thought of as *embodying* in hardware a logical proposition: p can be true or false, and which it is can be recognized empirically by whether or not the light is on, given that the flashlight functions.

The flashlight could, of course, be hitched up backwards, as in Figure 1-9. The proposition $\sim p$ is true of this circuit; that is, the light is on when the switch R is open. This is a 'not' circuit for "it is not the case that the light comes on when the switch is closed."

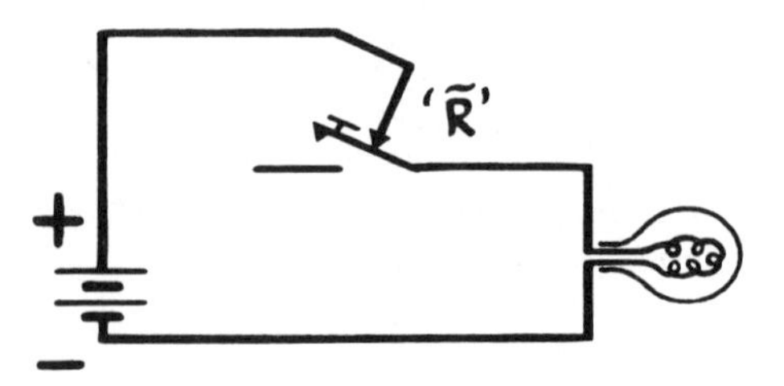

Fig. 1-9 *The circuit of Fig. 1-8 is altered so that when the switch is 'off' the current flows and the light is on.*

Now connect two switches "in a series" as shown in Figure 1-10. It is apparent that this circuit represents the logical relationship of conjunction, for only when R and S are both closed will the light come on. (See Table 1-2, and let p refer to R and q to S.) Correspondingly, if the switches are arranged in parallel, as in Figure 1-11, the circuit embodies the logical relationship of disjunction, for the light will come on if *either* one *or* the other or *both* of the switches are closed.

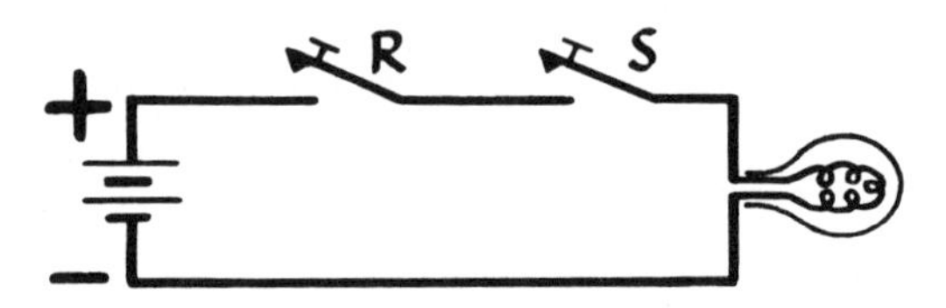

Fig. 1-10 *Two switches 'in series.' This embodies the relation* R $\cap$ S *in hardware, given the stated conventions.*

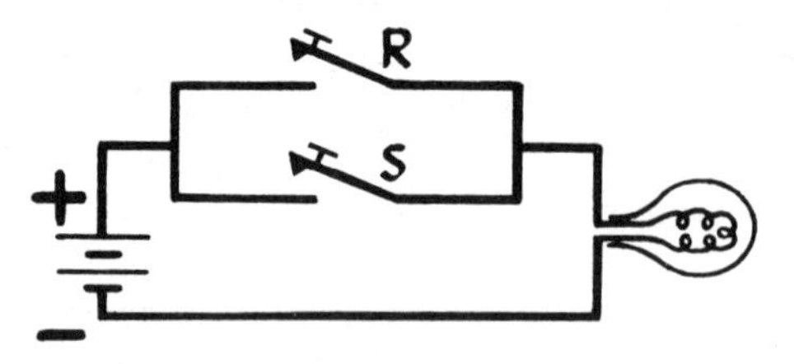

Fig. 1-11 *Two switches 'in parallel.' This embodies the relation* R $\cup$ S *in hardware, given the stated conventions.*

On-off switches made from electronic components rather than the mechanical ones illustrated but employing the same logical principles are used in computers. Modules—compact building-blocks that function as components of larger assemblages—embodying in hardware these and other logical relationships can be combined according to need (*i.e.*, programmed) to carry out logical manipulations of an either/or, true/false, type. Because these logical relationships are at the basis of mathematics, and because true/false may be coded as 1/0, which are the two numbers of binary arithmetic, one can imagine in a general way that these machines can do mathematical problems when properly programmed.

Exercises

1.1 Place the courses you are taking on the Sphere of Knowledge and Experience, Figure 1-2. Are you satisfied with the distribution?

1.2 What is meant by the expression "An idea exists in my mind"? In what respects, if any, does it make sense?

1.3 Interpret these lines from the poem "Einstein" by Archibald MacLeish. He is speaking of Einstein:

> . . . he lies upon his bed
> Exerting on Arcturus and the moon
> Forces proportional inversely to
> The squares of their remoteness and conceives
> The universe. . . .

1.4 What, do you suppose, might happen if a group of scientists found a set of constructs (A) to be true of nature, and another group found a set (B) true, and it appeared that paradoxes were involved?

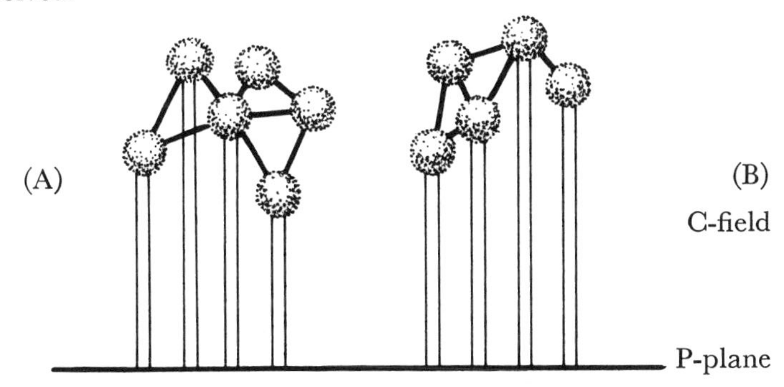

Bibliography

M. L. J. Abercrombie. *The Anatomy of Judgment. An investigation into the processes of Perception and Reasoning*. Basic Books, New York, 1960.

Henry Margenau, *The Nature of Physical Reality*. McGraw-Hill Book Company, Inc., New York, 1950. Most highly recommended to those reading this text. It has served the Author as a philosophical framework.

Harold G. Cassidy, *Knowledge, Experience and Action. An Essay on Education.* Teachers College Press, Columbia University, New York, 1969. This is a statement of much of the philosophy underlying the construction and teaching of this textbook.

Susanne Langer, *Philosophy in a New Key: A Study of the Symbolism of Reason, Rite, and Art*. Harvard University Press, Cambridge, Mass., 1942, 1951. This is a delightfully written book, full of insight.

Colin Cherry, *On Human Communication. A Review, a Survey, and a Criticism.* M.I.T. Press, Cambridge, Mass., 1957. This is a comprehensive, well-written book not, says the Author, for experts: indeed he dedicates the book "To my dog, Pym."

Irving Adler, *The New Mathematics*. Signet Science Library, The New American Library, New York, 1958. This paperback is highly recommended to anyone who is uneasy about mathematics. The Author writes with beautiful simplicity and authenticity.

John G. Kemeny, H. Mirkil, J. L. Snell, and G. L. Thompson. *Finite Mathematical Structures*. Prentice-Hall, Englewood Cliffs, N. J., 1959. This book is for the more sophisticated reader. It could conveniently follow Adler's book. It, too, is clear and well written.

J. Eldon Whitesitt, *Boolean Algebra and its Applications*. Addison-Wesley Publishing Company, Inc., Reading, Mass., 1961. For the person intrigued by sets and circuits this small volume can make fascinating reading.

Alice Mary Hilton, *Logic, Computing Machines, and Automation*. Meridian Books, The World Publishing Co., Cleveland, 1964. This comprehensive and difficult book is recommended to those who wish an introduction to the subjects listed in its title.

Chapter 2

2.1 Toward a physical view
2.2 Form and function
2.3 Motion and its conceptualization
2.4 Constant velocity
2.5 Function
2.6 Constant velocity continued
2.7 Accelerated motion
2.8 Recapitulation

To Set the Scene ~ 2

Toward a physical view 2.1

How few types of fundamental physical entities give rise to how fantastic a variety of material things! There are only some ninety-two naturally-occurring atoms to account for material diversity—diversity of all substances in our environment. These atoms are all constructed on the same basic plan from three fundamental particles: *Protons* and *neutrons* combine to produce a nucleus. Each kind of atom has its own nucleus. Outside of and around the nucleus are *electrons*, arranged in from one to six shells. That the simple idea of a nucleus comprised of protons and neutrons hides deeper complexity is taught us by the results of bombardment of nuclei with atomic projectiles fired by 'atom-smashers.' When it is struck by a high-energy particle the nucleus may be smashed into smaller pieces that splash out from it. But even here the variety found has been relatively small: somewhere between one and two hundred subatomic pieces, among which a pattern of relationships appears. The wonderful variety of the material, macroscopic world arises from these relatively few fundamental particles. How?

For one thing, the three particles combine to form atoms, atoms combine to form molecules, and molecules aggregate to form larger structures which, becoming large enough, reach visible size. And at each step possible variety increases. We live in the middle range of things with respect to size. We can see tiny particles of matter, such as a pinhead, and can look out at the Moon and the Stars. But to explore the far reaches of the heavens, or the inner depths of molecules and atoms, requires instruments, intellectual constructs, and mathematics: and *comprehension* is difficult—just about as difficult as comprehending the statement that "In 1950, the United States consumed 2.7

billion tons of materials of all kinds . . ."—when we calculate this number to 36,000 pounds per person it remains hard to visualize.

The diameter of the head of a small pin might be one millimeter, 10^{-3} m;[1] of an atom in the metal of that head, 10^{-10} m. Thus 10^{7} atoms might be lined up across the diameter of that pin-head. Magnify each atom to the size of a pinhead, and the diameter of the head goes up to 10^{4} m—a hundred football fields on end. At the center of the atom is a nucleus, some 10^{-15} m across. Magnify this to the size of a pinhead and the atomic diameter is now the length of a football field. These gee-whiz-type calculations will tell us very little unless we ponder them with the insights that come with further knowledge. The function of this text is to provide the modern knowledge, and to aid the attendant insights. The especial function of this chapter is to review certain basic notions that will start us on the way to the goals stated in Chapter 1.

2.2 Form and function

It is not surprising that we find ourselves in the middle range of things in respect of size (or, for that matter, of any other quality); but it is fortuitous. It means that we can look in two directions from where we stand. In respect of size we find that our native senses set limits to what we can know by direct perception. The smallest drop of water that we can easily see might have a diameter of 0.1 millimeter, or thereabouts. For convenience let us imagine its volume to be 0.01 centimeters cubed, or (0.01 cm^{3}). This is a number taken for ease of calculation. It has the right order of magnitude: persons with very sharp eyesight might be able to see smaller drops; those with poorer would not perceive this little drop. We shall find that this little drop contains some 3×10^{16} water molecules. Instruments tell us that these molecules are jittering about in vigorous motion, colliding with each other and escaping from the surface of the drop—as well as falling back in. We learn that each molecule has the structure H—O—H, where the formula means that two atoms of hydrogen are attached or bonded to each atom of oxygen. Other instruments tell us that the three atoms

---

[1] *Recall:* One millimeter is 1/1000 meter. We shall use as our standard unit of length the meter (m). This has internationally accepted embodiment in a bar of platinum-iridium alloy carefully preserved by the French Bureau of Standards. Common prefixes that indicate powers of ten are:

Prefix	mega-	kilo-	deci-	centi-	milli-	micro-
Abbreviation	M	k	d	c	m	μ
Power of ten	10^{6}	10^{3}	10^{-1}	10^{-2}	10^{-3}	10^{-6}

Coming into use are terms like 'megabuck' for one million dollars. Larger and smaller numbers may be designated tera-, 10^{12}; giga-, 10^{9}, nano-, 10^{-9}; pico-, 10^{-12}.

jitter with respect to each other under the blows from neighboring molecules. Further, the atoms are held together in the molecule by shared electrons which, however, are in rapid movement about the three nuclei. And, indeed, there is evidence for motion of the subnuclear entities (whatever they are) that give rise to the form of the nucleus.

We therefore may describe the hierarchy of material structure in terms of the interplay of *functions* and *forms*, where function implies something dynamic, and form a structure that has some constant interrelation of its parts in time. The form of the nucleus is a result of the functioning of subnuclear particles; the form of the atom is a result of the functioning of nucleus and electrons; the form of the molecule results from the functioning of its atoms; the form of the macroscopic droplet is the result of the functioning of its component molecules. And so on, to rivers and oceans, rain and snow and hail. All material objects whatsoever may be described in these terms. Fortunately for us, at our macroscopic level of size behaviors of things do not reflect what the individual atoms or molecules in them are doing. This is partly because the random motions of atoms and molecules tend to cancel out overall, and partly because of constraints between them (which we shall have to discuss elsewhere). But imagine the plight of a virus particle. Being about the same size as many molecules in its environment it must be battered about unmercifully. If we were so battered, thinking and most rational actions would be impossible. Life would be much worse than the worst psychedelic assaults.

This well-established knowledge supports certain ancient intuitions with new insights and facts: all material objects are the sites of continual motion; things move, and undergo transformations; they change to other things and move to other places; they come from other things and arrive from other places. No thing remains constant forever.

Motion and its conceptualization 2.3

Change is of many kinds. Changing position is named motion. How do we conceptualize this? In the first place, motion is always relative to something. A tree or a building, rooted to the Earth, is not moving in that its motion is zero relative to the Earth. But the Earth is hurtling through space both in its orbit around the Sun, and with the Sun in its flight through space relative to certain stars. Thus relative to an observer in outer space (whom we are at liberty to imagine) the tree and building are in all probability moving. They are moving relative to the Sun and the Moon; this we ourselves can see as their shadows move. Thus our first act in discussing motion or anything else is to specify the system in which we are interested. The trees and building are motionless relative to the Earth. Let us imagine a ball resting motionless on level ground beneath the tree, near a building.

Experience has taught us that tree, ball, and building will remain motionless indefinitely unless something disturbs them. If the ball is given a push it starts to roll. Experience tells us that it will slow down and eventually stop: that, we know, is the result of frictional forces. How far it rolls depends on the 'force' of the push. These relationships were formulated by Isaac Newton in his famous First Law:

> *Every body continues in its state of rest, or of uniform motion in a straight line, unless it is compelled to change that state by forces impressed upon it.*

We were brought up in a culture the language and daily actions of which imply implicit and widely held belief in this Law.

But changing position can take place in many ways: it may be rectilinear motion—that is, motion in a straight line; it may be curvilinear; or it may be oscillatory. It may change from one to the other. Further, it may occur with constant velocity; or with changing velocity, in which case any change is said to be 'accelerated' motion. When the ball at rest is given a push, it accelerates from 'zero velocity' to some state of motion under the force of the push. This change occurs over a short period of time, and we say that the ball received an 'impulse' imparted by the force during this short period of time. The ball continues to roll for a while, slows down, and stops. These are the phenomena we observe—the phenomenological (in the scientific sense of the word) or existential aspects of the behavior of the system: ball-on-level-ground.

Motion of an object, we saw, is relative to some other object, which may itself be moving when some other system is considered. Here we conceptualize the motion of the ball by setting up (in imagination, or by marks on the ground, and so forth) a Cartesian representation of the local space (Figure 2-1). (This technique was invented by the French philosopher and mathematician René Descartes (1596–1650); it has been highly developed since his lifetime. The technique is familiar in the construction of graphs.) The Cartesian coordinate system consists of three mutually perpendicular lines, the X, Y, and Z axes, conventionally labeled as shown in the Figure. Since we are interested in the motion of the ball on the flat (plane) surface, we shall neglect the Z direction, and draw only the two coordinate axes X and Y as in Figure 2-2. Notice that, in thinking of lines and exact angles and planes, we are well into the field of constructs. At the same time, we have a tangible situation, and so we may give tangible body to these constructs. This we might do by drawing lines on the ground

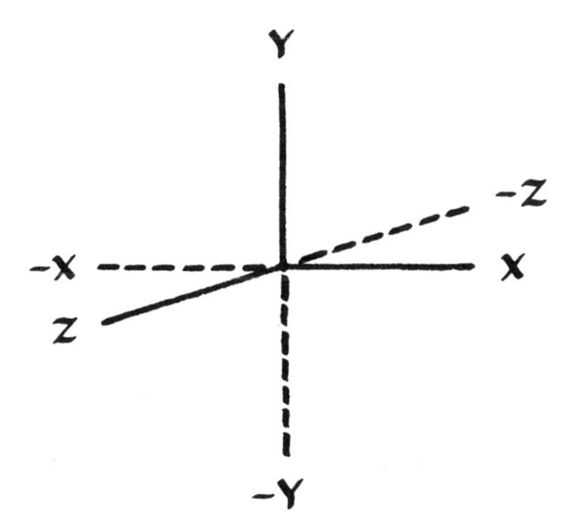

Fig. 2-1 *Conventional labeling of a Cartesian coordinate system.*

to represent the axes, and by calibrating these axes by means of a meter stick, as in Figure 2-2. Now, by algebra and geometry, any point on the plane can be identified by two numbers: its distance parallel to the X axis from the Y, a number we shall represent by x, and its

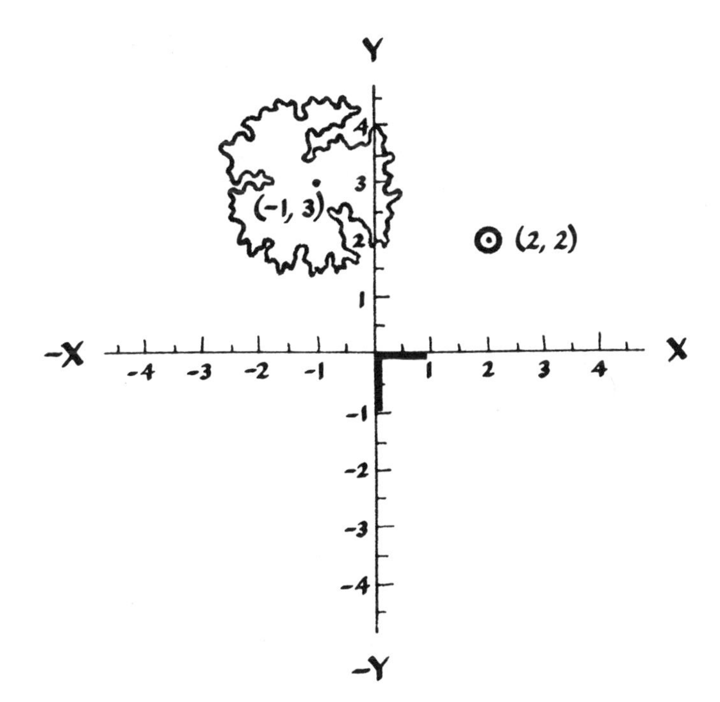

Fig. 2-2 *Location of points in a Cartesian plane.*

distance parallel to the Y axis from the X, represented by y. We use lower-case x and y to represent measured *values*, that is numbers of meters, along the *axes* X and Y, and may speak of x and y as the coordinates of the point. The ball might be found to rest at the point (2, 2) (by actual measurement) if we should set up these Cartesian coordinates so that the origin, the point of intersection of X and Y, lies at the corner of the building, two sides of which lie along the +X and −Y axes. The convention is that the two numbers form an 'ordered pair' with x always written first: (x, y).[1] The tree might then be located at (−1, 3). Our embodiment falls short of the niceties of geometrical points: the ball rests at (2, 2) as a center of the region of contact with the ground; the 'position' of the tree is that of its center if we imagine a section through it at ground level.

These Cartesian axes constitute the frame of reference which gives meaning to statements about motions of objects in this system. *We* have some control over where we set up this frame: it might be convenient to place the origin at the point where the ball rests, with the Y axis parallel to the building, as in Figure 2-3. We would then identify the positions of other objects with different numbers, *but they have not moved relative to eath other*. (Convince yourself that in rotating the axes and placing the origin at the ball we have not changed relative positions. The ball is still 2 m away from the wall of the building at its nearest point, and the tree is still 'three meters to the left and one up' from the ball.) The geometry of the system preserves their relationships to each other. This can be seen by constructing a right-angle triangle with sides parallel to the axes and the hypotenuse reaching from ball to tree. By Pythagoras' theorem, the distance from ball to tree, s, is equal to $\sqrt{3^2 + 1^2}$; $s = \pm\sqrt{10}$. On squaring, the signs become plus, and we lose them; thus s is ± and we have to tell whether + or − *by inspecting the situation* (Figure 2-4).

2.4 Constant velocity

We give the ball a gentle push so that it accelerates from zero speed to, say, four meters per minute. We conceptualize what is happening in the following way, ignoring for the present the initial push in which act we 'coupled' ourselves to the system for a moment and then withdrew (except that we are still coupled optically which allows us to see what is going on!). For simplicity, imagine that the first arrangement of the axes is used. The ball starts at (2, 2); when we stop pushing it it has moved only a negligible distance. We pick up a

---

[1] Convince yourself, if you have any lingering doubts, that only in special cases are (x, y) and (y, x) identical points. Do this by picking some values for x, and y and plotting them.

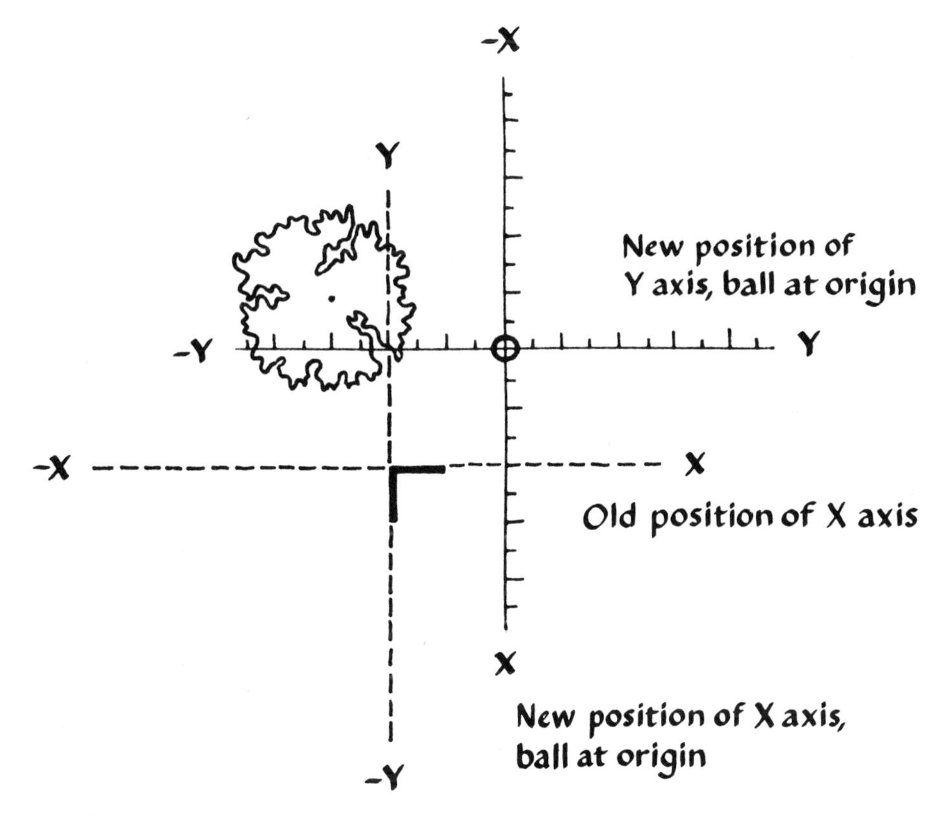

Fig. 2-3 *Effect of displacement and rotation of axes.*

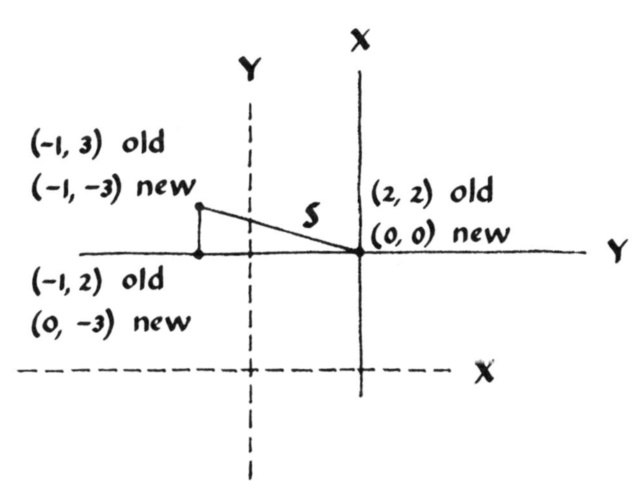

Fig. 2-4 *The displaced and rotated axes give new numbers to the points but do not change their relative positions.*

stopwatch and time it, starting with time zero when it just crosses the (3, 2) point. We have arranged for it to move parallel to the X axis at constant velocity. We then obtain data of the following kind, the units being given at the head of the column (Table 2-1). The speed of

Table 2-1

An Idealized Velocity Measurement

Position of ball (in meters)	Measured at time (in seconds)	Calc. distance (s) traveled from beginning (in meters)	Time of travel (t) (in min)	Speed of travel s/t (m/min)
(3, 2)	0	0	0	—
(4, 2)	15	1	0.25	4
(5, 2)	30	2	0.50	4
(6, 2)	45	3	0.75	4
(7, 2)	60	4	1.00	4

the ball is distance traveled divided by time taken. This is average speed, and has the same value over any interval since the speed is constant. As we can state the direction, we can speak of this as a 'velocity': the ball has a velocity 4 m/min parallel to the X axis. We see from our data sheet, which was derived from observing the ball, that the value of y remained constant. (Such is the condition in this system for motion parallel to the X axis.)

These data are made to be of the utmost simplicity in order to display several additional points. We have here two related changes: in position and in time. We say that distance traveled is a 'function' of elapsed time or, depending on our point of view, that time is a function of distance.

2.5 Function

'Function' is a relationship basic to mathematics and science. It is the name of any mathematical expression that describes a relation between *variables*, between the linked values of things the values of which can change. In Table 2-1 the two sets of values—position measured and time measured—are *correlated* because each value of position was measured at a certain time. Here are two sets of 'events' both derived from observation of the system. We are not *necessarily* concerned[1]

[1] The most unconnected things could be 'correlated' in this sense. For example, the death-rate of dogs in Kansas City might be correlated to rainfall in New Haven, but very few persons except devotees of the amazing would find the correlation interesting. We surely would not say that one phenomenon is related to the other.

with a cause-effect relation, but it is clear that the events in the column labeled 's' have a unique relation to those in the column labeled 't'. Let us, to give the data a more general flavor, label each entity, as shown in Table 2-2. The set, or group, S comprises all these s's, and

Table 2-2

Re-Labeled Measurements from Table 2-1

Distance (s) (in meters)	Time of travel (t) (in minutes)	Average speed s_i/t_i (m/min)
s_0 0	t_0 0	—
s_1 1	t_1 0.25	4
s_2 2	t_2 0.50	4
s_3 3	t_3 0.75	4
s_4 4	t_4 1.00	4

the set T comprises all the t's. 'Function' means a rule that relates each element of one set, for example each t in the set T, to an element of another set, for example, and s in the set S, as shown in Figure 2-5. In technical terms, T is the 'domain' of the function and S is said to be the 'range' of the function.

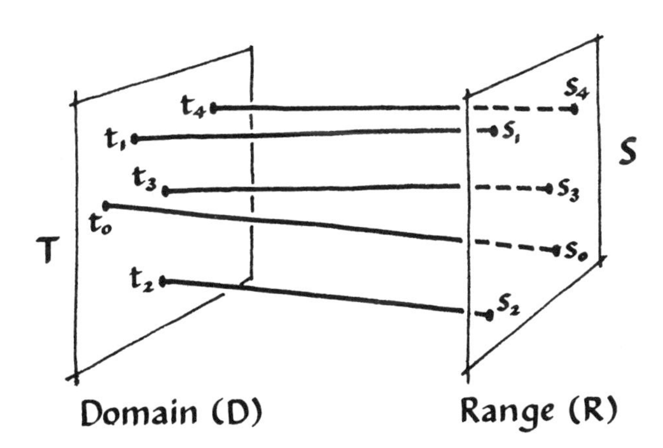

Fig. 2-5 *Diagram of a functional relation.*

If, as in the tables, t is any element in the set T, then the rule '**f**' assigns it an 'object' **f**(t) which is an element in S. The term '**f**(t)' is read "the value taken on at t under the rule **f**," or more concisely "the value of **f** at t." **f** is said to be a function on the domain (here, T) and the range of **f**, as noted above, is the range of the function (here, S).

In the most general sense, the object $\mathbf{f}(x)$ may also be an element of the domain D as well as of the range R, and the various values of $\mathbf{f}(x)$ may all be different or all the same or mixed. For example, let D consist of the members of your class at college, and $\mathbf{f}(1)$ be the first name of the student (1), $\mathbf{f}(\mathrm{m})$ the first name of the m'th student, and so on. Notice that m is a variable in a propositional function of the kind described in Section 1-7. The range R will consist of a set of many names, and probably several of the students listed in the domain D will have the same first name, listed in R.

The essential act of a scientific investigation after collecting the data (the first two columns of Table 2-1) is to calculate them in some suitable way (the third and fourth columns) so that the rule, the functional relation between them (column five), may be displayed in the simplest terms. Here we have been excruciatingly obvious, perhaps. In scientific research the data often appear chaotic, and it is often moot what aspects of it are related functionally. Notice that in the terms of Figure 1-4 the rolling ball is experienced by you at the P-plane. x and y are values of constructs, and would be plotted in the C-field. These are arrived at along double-line pathways that represent the operations of measurement with a meter-stick. The values of the constructs are connected by a functional relation symbolized by a single line.

Many students have some difficulty with these notions through misunderstanding what is being done here. In mathematics an equation may be written $y = \mathbf{f}(x)$; it is read, "y is a function of x." The letters y and x stand for variables, and in the statement $y = \mathbf{f}(x)$ the rule $\mathbf{f}$ assigns to every value of the 'independent variable x' a value of the 'dependent variable y.' In other words, each value of x in the domain of this independent variable is connected by the rule $\mathbf{f}$ to a value of y in the range of the dependent variable. Now, looking at Table 2-1, let x be replaced by t, and y be replaced by s; then we look for the rule $\mathbf{f}$ such that $s = \mathbf{f}(t)$. This rule is '4 m/min.' That is, $s = 4\,t$, when s is in meters and t in minutes. But we might be interested in the relation $t = \mathbf{g}(s)$, where $\mathbf{g}$ is *this* functional relation. It is 0.25 min/meter.

From our practical point of view, mathematics is a device for keeping our reasoning straight. This is one reason for the care in defining terms and for the rigid rules.

2.6 Constant velocity continued

One way of looking for a functional relation between data is to plot them on a graph. Most of us did this kind of exercise in primary or secondary school. The procedure is to set up Cartesian axes, in this case labeling them 'distance' and 'time,' and plot the values of the calculated (or raw) data on them, connecting the points with a line, as in Figure 2-6. We see that this relationship is a straight-line one. The

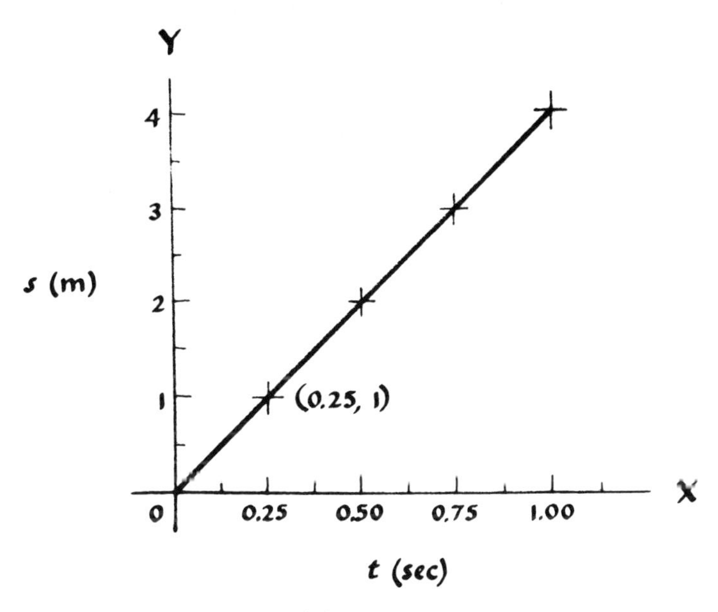

Fig. 2-6 *Plot of* s *versus* t *in constant velocity.*

'slope' of this line is the distance up (along the Y axis) that the line rises for a given distance out along the X: that is, 4 m/sec. It is a visualization of the functional relation. It should be noted that these axes *are not a frame of reference.* A frame of reference is correlated to the source of data, and is thus associated with phenomena close to the P-plane. These axes are mathematical and geometrical tools, like numbers. They belong in a separate language called a 'metalanguage' to suggest that it is higher, or transcending, or at a higher level of abstraction than some other language. It is used to *talk about* data.

Accelerated motion 2.7

It is found experimentally that all accelerations are the result of force applied to the accelerated object. It is a little difficult to speak in detail about the change in velocity *during* the impulse imparted to the ball to start it rolling. Clearly the velocity must have changed, since the ball went from a state of rest to one of motion. If the duration of time that the push was applied to the ball is represented by Δt, where the Δ means 'an interval of' and is measured as the interval between starting time t_0 and end of push t_1, that is,

$$\Delta t = t_1 - t_0 \text{ (in sec)}$$

and if the force is represented by **F**, then the impulse is defined as

$$\text{impulse} = \mathbf{F}\Delta t$$

We shall assume that the force is constant and that the time interval is short. (**F** is measured in units called 'newtons' (N), so impulse is measured in newton seconds.) Observe that we have not done anything occult here. We have replaced words by algebraic symbols. These are easy to manipulate logically by rules learned in secondary school, and they save a lot of writing effort.

The relation between force and acceleration (a) was also enunciated by Newton. This is his Second Law of motion. It was translated by Lord Kelvin as

> *Change of motion is proportional to force applied, and takes place in the direction of the straight line in which the force acts.*

Since we have undertaken to see how the physicist takes hold, with "conceptual hands," of perceived behaviors, and are here especially concerned with the kind of change called motion, it is desirable to see how acceleration and force, both constructs, are conceptualized and applied to practice. Once more, we shall deliberately choose the familiar example of a falling object. Here, the object falls under the constant force of gravitation. By Newton's law, change of rate of motion (represented by a) is proportional to applied force (represented by **F**). That is:

$$\mathbf{F} \propto a$$

where $\propto$ *means* 'is proportional to.' It is a property of a proportionality that it may be restated as an equality by introducing a constant in the following way:

$$\mathbf{F} = ma$$

We have replaced the proportionality sign $\propto$ by '$= m$', an equality sign and a constant, m. This makes our statement more precise. "Is proportional to" is still somewhat vague. "Is equal to" is about as firm as you can get. Here m is the constant of proportionality. It is named the 'mass' of the body that is accelerated, and is measured in kilograms.

We study the free fall of a heavy ball as we drop it from a height. We do things as conveniently as possible. For example, we might set up a calibrated Y axis in a stair-well, drop a ball close to the axis and arrange five cameras to take flash shots of the ball at specified times: an electric clock actuates camera '0' at the origin just as the ball is released, the shutter of camera '1' is released exactly one second later, '2' one second after this, and so on. We set each camera up by trial until it is in position to catch the ball which passes in front of each as the one-second, two-second, and so on, intervals come up. The axis is photographed simultaneously so that we obtain a reading of how far the ball has fallen in 0, 1, 2, 3, 4 seconds, measuring at the position of the center of the ball, if this was at zero on the axis initially. We arrange the axis so that the ball starts at the origin. Data for a number of such runs, averaged, are simulated in Table 2-3 and Figure 2-7. The usual convention is that perpendicular distances 'up' with respect to the surface of the Earth shall be deemed positive. Since the ball falls, all

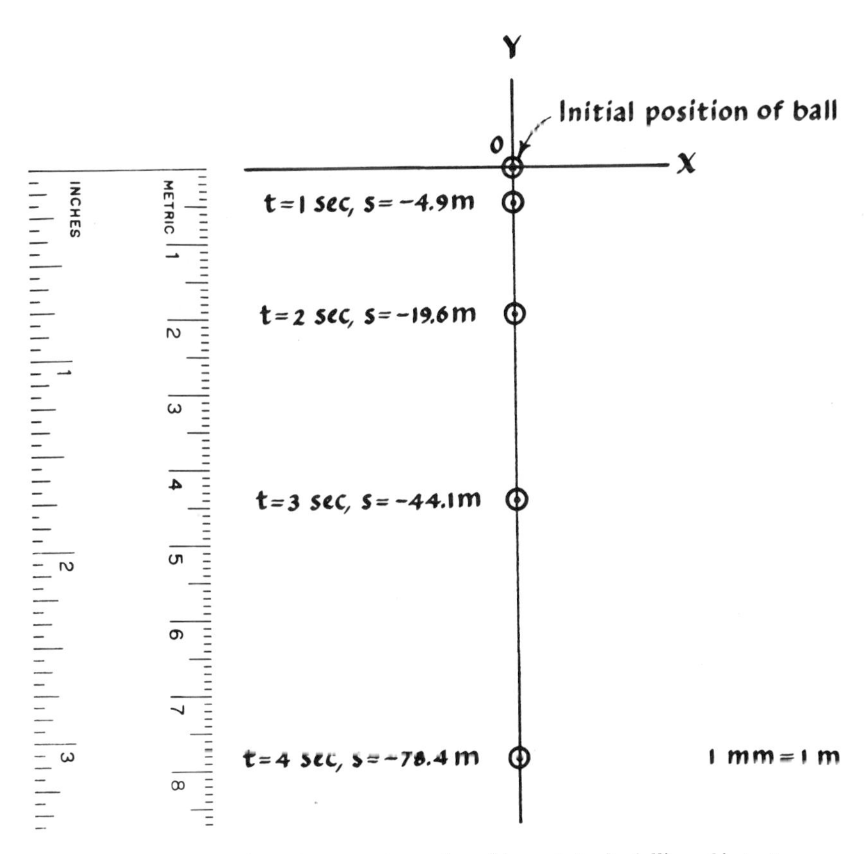

Fig. 2-7 *Scale drawing, 1 mm = 1 m, of positions of freely falling object at successive seconds.*

Table 2-3 *(Figures 2-7, 2-8)*

Idealized Measurements on a Freely Falling Object

Distance fallen, s, in meters	Time elapsed, t, in seconds
0	0
−4.9	1
−19.6	2
−44.1	3
−78.4	4

distances in Table 2-3 are given negative signs. (Zero may be written + or −, or ±, or just left as 0.) Our problem is to find a functional relation between displacement and elapsed time, s and t, hidden in these data. Experiments with different freely falling objects heavy enough that air-friction is not a seriously disturbing factor yield substantially the same tables of data.

To begin at the simplest, the relation is not linear: s/t does not calculate to be constant. Confirming this visually, if s is plotted versus t in a graph (Figure 2-8) the resulting line is not straight, but curved. (Notice that we have plotted increasing negative numbers up along a Y axis. This is quite permissible because as we insisted above this is not a frame of reference but a mathematical tool. The curve is an expression in our metalanguage of a relation between two constructs. Please be sure that this is not a cause of sublimal confusion.) However, the relation is certainly a regular one. Though we did not measure positions of the ball in the intermediate times we assume that they can be calculated by interpolation from the curve since we watched the ball fall and it did not behave erratically; thus at 2.5 sec the ball must have fallen close to 31 m, as shown by the dashed line in the Figure. (It is not safe, however, to extrapolate the curve beyond 4 sec unless we are sure that the ball continued to fall. It might just then have landed on the ground!)

One might reason thus: if the velocity *had* been constant, then $\Delta s/\Delta t$ = constant (k), whence $\Delta s = k\Delta t$. Here we can see that the velocity is continually changing. The average velocity ($\bar{v}$) *in terms of s and t* from the beginning to any time t by definition is $\bar{v} = (s_i - s_0)/(t_i - t_0) = \Delta s/\Delta t$, where i in this case may refer to readings 0, 1, 2, 3, or 4. The definition of average velocity *in terms of the velocity* at the beginning v_0 and that at the end, v_i, of a time interval t_0 to t_i is:

$$\bar{v} = (v_0 + v_i)/2$$

Since we know that v_0 is 0, and can calculate the value of $\bar{v}$ as $\Delta s/\Delta t$ (Table 2-4), we can calculate from these relations the actual velocity

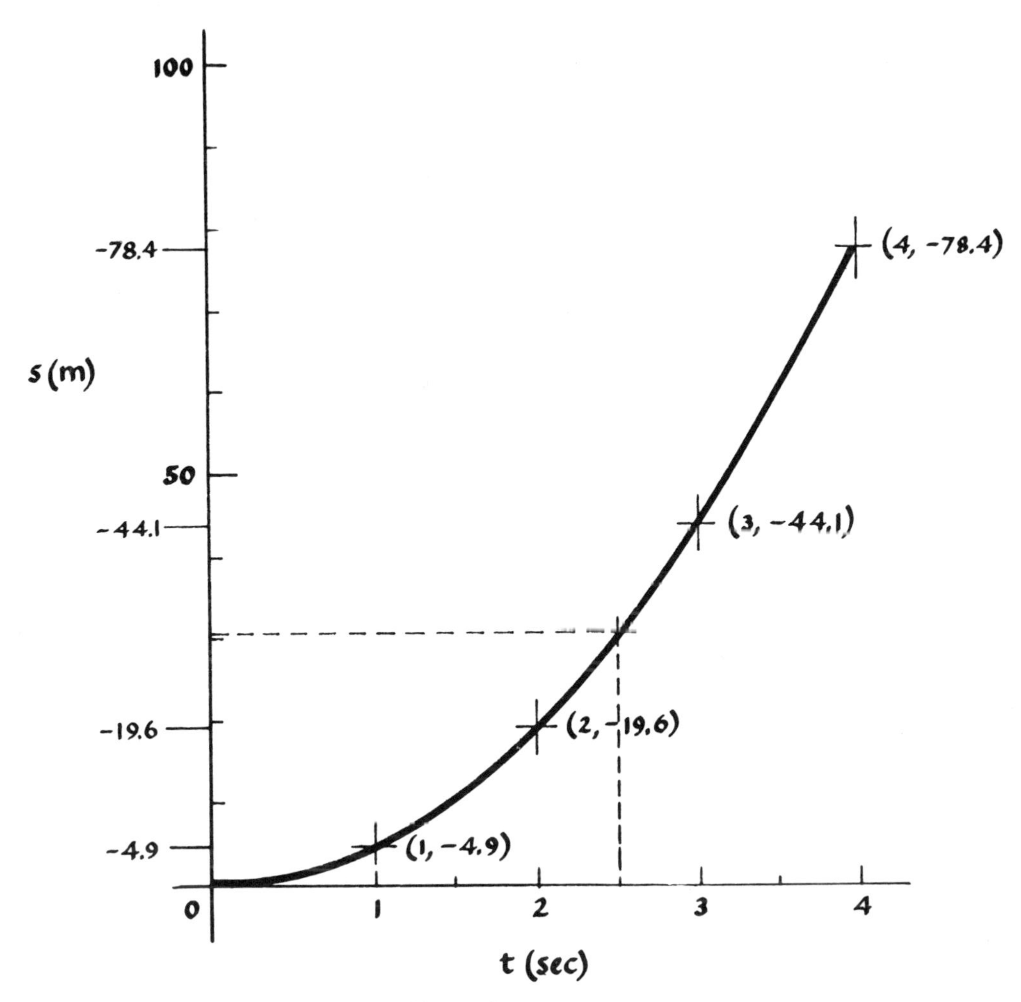

Fig. 2-8 *Plot of* s *versus* t *for object falling freely.*

Table 2-4 *(Figure 2-9)*

Change of Velocity with Time in Free Fall

Δs in meters	Δt in sec	$\Delta s/\Delta t = \bar{v}$ m/sec	$v_i = 2\bar{v}$ m/sec	v_i/t (m/sec)/sec
0	0	0	0	0
4.9	1	4.9	9.8	9.8
19.6	2	9.8	19.6	9.8
44.1	3	14.7	29.4	9.8
78.4	4	19.6	39.2	9.8

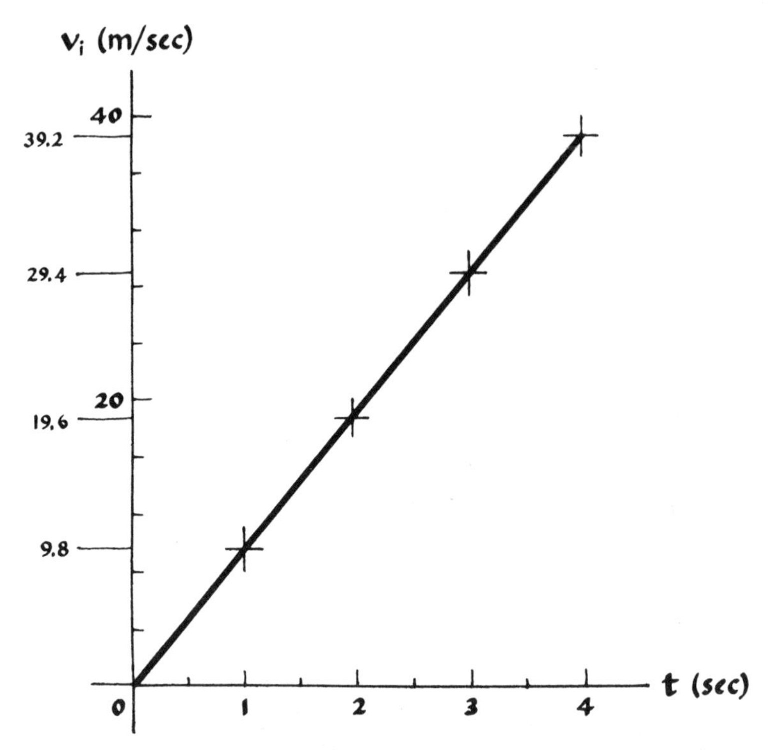

Fig. 2-9 *Plot of* v_i *in m/sec versus* t *in sec for constant acceleration.*

at the *end* of 1, 2, 3, 4 seconds. That is, in this case:

$$v_i = 2\bar{v}$$

The results are shown in Table 2-4. These data are plotted in Figure 2-9. Here we can see the regularity: $v_i = \mathbf{f}(t)$, where **f** is 9.8 m/sec². Average acceleration is defined as:

$$\bar{a} = (v_i - v_0)/(t_i - t_0) = \Delta v/\Delta t$$

Since in Table 2-4 $v_0 = 0$, and $t_0 = 0$, $\bar{a} = v_i/t$; it is found to be constant. The units of acceleration are (m/sec)/sec; that is, m/(sec × sec), or m/sec². This may seem odd: whoever heard of a 'square second'! But this is a metalanguage, and we are discussing constructs: length, time, velocity, acceleration, and their connecting invariant-relations. These expressions are legitimate tools of thought. Having discovered this constant relation, which allows all the data to be compressed into the simple equation, we can put numbers into the construct-system connecting a, v, s, and t for this experienced falling object. We have solved the problem. We have reduced the mass of raw data to an orderly relation.

Recapitulation 2.8

We spoke in Chapter 1 of symbolic transformation. This is probably the most difficult step that humanists and scientists have to take in bringing order out of raw data. Fruitful constructs are hard to come by in the first place. After we understand them we use them quite freely. It becomes a matter of getting used to them and then using them correctly. Things experienced are turned into mental constructs symbolically represented. These constructs may be manipulated by the rules of logic, or mathematics, or geometry, or by other rules such as symmetry considerations. The consequences of these manipulations may then be tested in practice—in ways that we can experience—so as to see whether we are on the right track: whether, that is, the consequences (deductions) fit the way things are.

A famous example of this procedure was Einstein's prediction that a beam of light passing close to the Sun would be bent out of its otherwise straight path. He derived this conclusion from constructs in his theory of relativity which made him postulate that light would be affected by gravity. From mathematical calculations he concluded that the magnitude of the angle of bending would be 1.73 seconds of arc. So convincing was his argument that two expensive expeditions were fitted out and dispatched, one to Brazil and one to Africa, to set up telescopes in the path of an eclipse of the Sun to make the necessary test of the prediction. (The preparations were carried out although the First World War was in progress, so important was the test considered.) Einstein's prediction was verified, and this validated the whole network of constructs and logical relations that led to it. We shall look into this in some detail later.

We have demonstrated similar thinking here, in this chapter. The ideas 'velocity' and 'acceleration' are constructs (developed by Galileo in his investigations of motion). The construct 'average acceleration', $\bar{v}$, was connected to distance traveled $(s_i - s_0)$ during the time interval $(t_i - t_0)$ by the relation $\bar{v} = (s_i - s_0)/(t_i - t_0)$. This is a general relation because, for example, t_0 is the starting point in time (you may begin the demonstration whenever you wish) and t_i is a particular time later when you decide to measure s_i. Given the functional relationship for any given physical situation, we can calculate one of the terms if the others are known. Actually, we do this kind of thing all the time, if we drive a car. We know how far we want to go (Δs), and we estimate our average speed $(\bar{v})$, thus being able to calculate how long it will take to get there (Δt). The discussion given here exposes the inner nature of this common calculation, thereby relating it to all scientific thinking.

The same applies to acceleration, and to Newton's law relating force and acceleration to mass. Force, acceleration, and mass are constructs, with very precise meanings. They are connected by the *law* (a name for an important and general invariant-relation):

$$\mathbf{F} = ma$$

This law describes with great accuracy the acceleration and deceleration of a car, the behaviors of moving objects of all kinds. It has been thoroughly checked out in experience. We rest our lives on it every day. The power of correct constructs to enable prediction and control of experienced motions is what we have demonstrated in this chapter. We have also demonstrated how they may be arrived at, and how they may be manipulated. What we have presented here is applicable in some form throughout the entire field of knowledge, experience, and action; indeed, to all the physical uses of life.

Exercises

2.1 Refer to Figure 2-1. Transpose two labels, *e.g.*, name the Y axis of the coordinate system (*a*) in Figure 2-1 the X axis, and the X the Y. Label this coordinate system (*tr*). Imagine that these axes are made of wooden rods, so that you can pick them up and turn them around. A test for identity would be whether you could superimpose (in your imagination, of course) the one on the other. If each part of the one corresponded with the same part of the other, the two would be identical . . . When we suggest transposing labels, we mean that, if the axes were color coded (say X is red and Y is green), you now paint the axis that was Y red, making it an X axis, and the former X axis green, so that it becomes a Y in the transposed-label system. a) Are the two coordinate systems (*a*) and (*tr*) the same? b) How might this be demonstrated? c) What is the relationship between them? d) Now, transpose in the system (*tr*) any two labels, for example, Y and Z. What has happened?

2.2 Refer to Table 2-1. What is the speed in m/sec?

2.3 Refer to Figure 2-3. Suppose that the coordinate axes are arranged with the origin still at the position of the ball, but the axes rotated slightly clock-wise so that the Y axis passes through the center of the tree. What, now, are the coordinates of ball and tree?

2.4 On a piece of graph paper, draw Cartesian axes X and Y at right angles to each other. Locate the three points (1, 2), (5, 2), (5, 5) and connect them by straight lines. Let the side parallel to the X axis be A, and that parallel to the Y axis be B, and the remaining side be R. a) Demonstrate algebraically and geometrically that the length of R is 5 units. b) What is the slope of R? c) Leaving the triangular figure as it is on the paper, displace the coordinates so that their origin is at the point that was (1, 2) and the X axis lies along the side A. What are the coordinates of the points in this system? d) What is the slope of R? e) Rotate the pair of axes, now, so that instead of X lying along A, Y lies along R. What are the coordinates of the three points of the triangle now? (Cut a strip of the graph paper and use it as a ruler along the new axes.) f) What do you suppose is the slope of R in this system of coordinates? (See Appendix on Trigonometry if you have difficulties.) g) What is the slope of A? h) Generalize the effects of displacement and/or rotation of coordinates upon the lengths of the sides, and upon the slope of the lines.

2.5 If you start from in front of your house and drive 22 miles out into the country and 22 back to the same spot in front of your house, taking 2 hours and 12 minutes for the round trip, a) What is your average speed? b) What is your average velocity for the whole trip? (Careful, now!) Explain your answer.

2.6 In respect to the illustration in which D is the set of members of your class, and R the set of first names, state the functional relation verbally: what is the meaning of **f**(m)?

2.7 In the following tables are listed certain functional relations. In each case describe the domain and the range of the function.

(a) x	**f**(x)	(b) x	**g**(x)	(c) x	**h**(x)
0	2	a	*	Galileo	1564
1	2	b	?	Newton	1642
2	1	c	@	Kepler	1571
3	0	d	*	Dante	1265
4	0	e	!	Descartes	1596
5	7	f	#	Shakespeare	1564

2.8 Calculate data for several points, and plot curves for the functions (a) $y = x^3$; (b) $y = x^{1/3}$; (c) $y = x^{-2}$; (d) $y = x^{-1/2}$. The first two are parabolic; the second two, hyperbolic curves. (If you have forgotten about exponents, see footnote in Chapter 1.) You can sketch in a curve from only a few points, *e.g.*, $x = 0, 1, 2, 3, -1, -2, -3$. Set these up in a table, as the first column; then the values of y can be put in as other columns. Then sketch the graph.

Answers to Exercises

2.1 a) No. b) They cannot be superimposed so as to have each contiguous axis with the same label and sign; hence they are not identical (Figure 2-10). c) The relation is that of image to mirror image (Figure 2-11). d) We are back where we began, with (a).

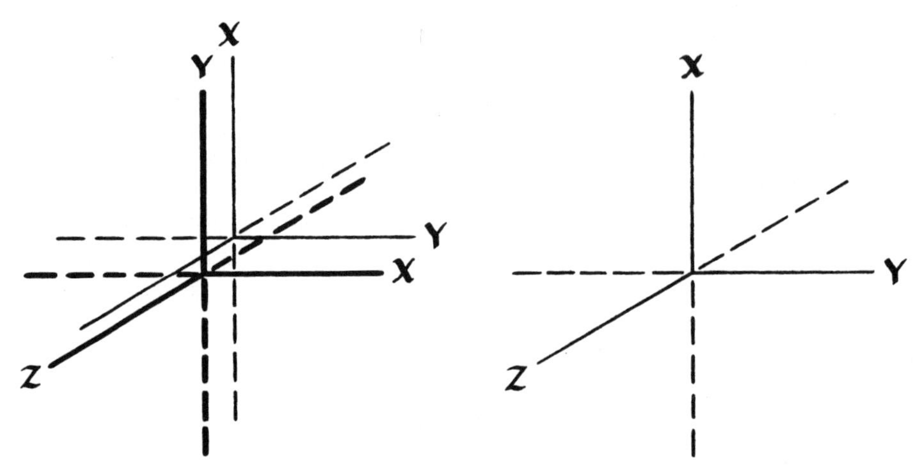

Fig. 2-10 *Attempt to superimpose exactly two coordinate systems with two transposed labels.*

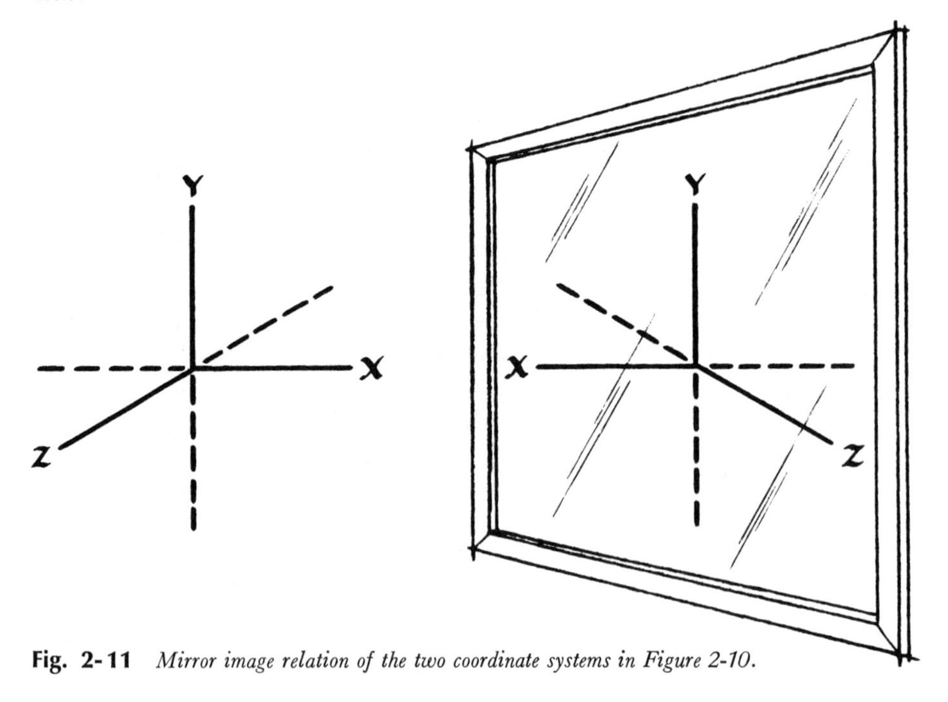

Fig. 2-11 *Mirror image relation of the two coordinate systems in Figure 2-10.*

2.2 1/15 m/sec.

2.3 Ball (0, 0); tree $(0, -\sqrt{10})$.

2.4 a) $R = \sqrt{A^2 + B^2} = \sqrt{16 + 9} = \sqrt{25} = 5$. b) 3/4; c) (0, 0), (4, 0), (4, 3); d) 3/4; e) (0, 0), (~2.4, ~3.3), (0, 5); f) undefined because the denominator is 0; g) approx. 3.3/2.4.

2.5 a) 20 mph; b) Zero. By definition. Velocity involves direction, and whatever the net direction out it was exactly canceled by the net return direction.

2.6 The function **f** is the value of the first name; **f**(m) is the value of **f** at m, namely m's first name.

2.7 a) Domain {1 to 5}; Range {0, 1, 2, 7}
b) D = {a to f}; R = {*, ?, @, !, #}
c) D = {list of the names}; R = {list of the dates}

Part II

Chapter 3. The Pervasive Electron

Chapter 4. Magnetism

Chapter 5. Light

The Classical Background ~ PART II

The edifice of science is like a Gothic cathedral: heavily buttressed, with tiers of solid masonry, and above all a few adventuresome spires reaching further upward. Elegant design and meaningful symbolism are built in—and there abound whimsical-seeming conceits and gargoyles. One can spend a lifetime studying such an edifice, interpreting it and contributing to it, and still not exhaust the possibilities. But we do not have a lifetime to spend; only one strand of an academic year, and we wish to comprehend the general plan, especially to understand the more recently constructed parts. We therefore must study only a few aspects of the edifice, leading in a direct line from the well-laid foundations to the two or three most interesting spires. This Part deals with some of the buttressing.

It is impossible to introduce modern physical science with any sense of its validity without providing some classical background. Modern science grew out of this background, and as we shall see in Part III it emerged because of the appearance of imperfections in the classical theories. It is an important part of our intention to show how these imperfections became apparent and what was done about them. Thus we must provide some classical background.

In this Part we employ a certain amount of algebra. The Student should look upon this as a language with its own grammar and syntax. Mastery is essential to understanding modern science. This is normal. After all one would expect a book on German or French literature to contain passages in the original language. At the same time we have been careful to derive and explain everything in detail so that the Student willing to put in the necessary effort can emerge from this Part with a real sense of mastery of the language. And we have put in nothing that will not be used later.

Electricity (Chapter 3) and Light (Chapter 5) are the two themes of this Part, with light taking precedence. We review elementary aspects of electricity so we can talk about the behavior of the charged particles that make up all matter. Because these charged particles are in motion we had to show in Chapter 2 how motion could be grasped intellectually. In Chapter 3 we discuss chiefly the properties of essentially stationary charged bodies: static electrical effects. There are two cogent reasons for this. One is that it enables us to develop the law of force between charged particles, and the other is that we can then develop the notion of a force field—a region in space about a charged object wherein another charged particle experiences force. The construct 'field' is important to modern science as the conjugate of 'particle.' The particle is a small discrete object; the field an extended continuous region. We then discuss briefly the flow of electricity.

Flowing charges generate in their neighborhood a magnetic field, and this is the subject of Chapter 4. We show how the magnetic field can be conceptualized relative to the flowing current that generates it, and we use the information collected in Chapters 3 and 4 to describe the cyclotron particle accelerator.

A classical introduction to certain properties of light is the burden of Chapter 5. Our object in presenting this material becomes clear when we discuss the problems—the seeds of discontent—which grew to bear fruit in the evolution of modern physics.

By the end of this Part we will have reviewed the minimum classical physics and chemistry needed to comprehend modern physical science. Much of this material may have been presented to you in primary and secondary school, so it may sound familiar. If it does, read on, and think of the epistemological implications. Think in terms of the observations, the constructs invented to make sense of them, and the operational rules and ingenious devices used quantitatively to connect observations at the P-plane with the numerical values of constructs. Apply these techniques to problems that face you in other studies. And in any case, read with imagination.

Chapter 3

3.1 Context
3.2 Static electricity
3.3 Electrical behavior of solid matter
3.4 Electrical behavior in liquids
3.5 Gaseous conduction
3.6 Conceptual approach: lines of force
3.7 Conceptual approach: electrical field
3.8 Coulomb's law
3.9 Vector representation
3.10 Conceptual approach: field intensity
3.11 Conceptual approach: potential
3.12 Current electricity
3.13 Charge on the electron
3.14 Caveat
3.15 Summary

The Pervasive Electron ~ 3

Context 3.1

Although in the usual sense of 'seeing' no one has ever seen an electron, and no one ever will, we believe that they exist; that they are real entities. The reasons for this belief, and the sources of meaning of the name 'electron,' illustrate clearly what we have been discussing in the preceding chapters. Since no one can see an electron, or feel one, clearly there is no direct sensory intuition for 'an electron' at the *P*-plane. The term must therefore refer to a construct. We first look at the connections of this construct to the *P*-plane, which support our belief in the entity 'electron,' and then at the meaning of the construct. Here we have to consider problems of conceptualization and measurement. Having gone through this discussion in detail we will have presented an example applicable to other questions of 'existence' of entities and the 'meaning' of the sticky terms that are bound to appear in later chapters.

In order to facilitate discussion we shall accept for the present—subject of course to later proof—that, as we said in Chapter 2, atoms, the constituents of all matter, are constructed of three fundamental particles: neutrons which bear no charge, electrons which are negatively charged, and protons which are positively charged.

When we make a statement like this the student may understandably enquire how we know that the charge on the electron is negative. (The reifying word 'is' must be handled carefully. Recall the student looking through a telescope at a bright star. "The name of that star is 'Venus'," says the astronomer. "Goodness, how extraordinarily clever of you to find it out," says the viewer, "How in the world did you do it? ") The answer leads us to the observations of classical natural philosophers. The French scientist Charles Du Fay discovered that there are two kinds of electricity, which he named 'vitreous' and 'resinous.' When a vitreous substance such as glass is rubbed with silk the glass becomes charged and is repelled by another piece of rubbed glass. On the other hand when a resinous material such as amber is rubbed with fur it becomes charged and is repelled by another piece of rubbed amber. However, rubbed glass and rubbed amber attract each other. Silk behaves as a resinous substance, hair as a vitreous. Evidently there are two kinds of electrification—let us call them 'v' and 'r'—such that v repels v and attracts r, and r repels r and attracts v: like kinds repel, unlike attract. The name 'positive' electrification was given by Benjamin Franklin, who conceived that a glass rod rubbed on silk gained 'electrical fluid' from the silk. It became positively electrified, and left the silk which had lost some of its normal quota of fluid negatively electrified. Thus '$+$ charged' became associated with vitreous electricity and '$-$' with resinous. The symbols were assigned in the course of talking and writing about observations, as part of a process we have previously discussed. Later evidence led to assigning the symbol '$-$', referring to charge, to electrons, and '$+$' to protons. Along the way a development common to our language and ways of thought occurred: one says "the electron is negative, or is negatively charged," and lo and behold! a construct becomes reified into a thing, to the confusion of beginning students. We could code the two kinds of charge by means of any convenient pair of complementary or contrasting terms. In the universe of discourse about 'charged particles', if 'P' is the set of particles exhibiting vitreous electrification, then '$\tilde{P}$' is the set exhibiting resinous (see Section 1-7).

3.2 Static electricity

The behaviors described as vitreous and resinous electrification are manifestations of static, that is nonmoving, electricity. The behaviors, the properties, shown by this type of electricity may be demonstrated conveniently by means of an electroscope, shown in Figure 3-1. The electroscope shown consists of a cylindrical chamber—the shape is not important. This chamber has glass windows front and rear. Its function is to protect the delicate apparatus inside from drafts, dust, and so forth. Through an insulated stopper in the top of the chamber there projects a metal rod A. This rod has soldered to the top a metal plate upon which objects may be placed. At its other end inside the chamber there is fastened a flap of gold leaf, a, or two pieces of gold leaf, b. Gold leaf is metallic gold hammered out to such thinness that it is almost translucent. It is very flexible and fragile. The case of the electroscope is 'grounded', that is, it is attached to a water-pipe or some metallic conductor that goes into the Earth. We shall accept that electrons may

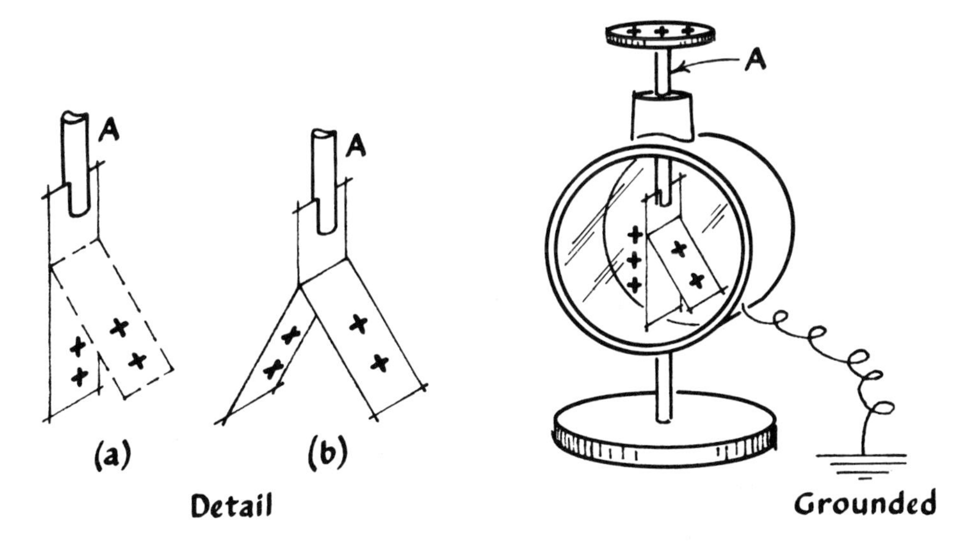

Fig. 3-1 *A charged electroscope. The rod A is insulated from the case, which is usually grounded. The rod, inside the case, has one* (a) *or two* (b) *strips of gold leaf fastened to it. When even a small charge is given to the rod the repulsive forces are sufficient to lift and thus separate the strips. This is the objective evidence that the electroscope is charged.*

flow through metals, which conduct them as an electric current. Thus, no charge can accumulate on the case, for it is connected directly by the metallic conductor to the vast neutral reservoir Earth. The principle of the electroscope is this: if it becomes charged so that, say, the gold leaves are (+) as in Figure 3-1, they repel. Since they are hinged at one end, they can only open, like the leaves of a book. The angle of opening is proportional to the charge: the greater the charge, the greater the force of repulsion and thus the greater the force available to lift the leaves against gravity, that is to open them.

Consider what occurs when a glass rod, rubbed with silk and thus electrified (+), is brought up toward the disk of the uncharged electroscope. In order to avoid excessive circumlocution we shall use modern terms and theory: when the glass, originally uncharged, is rubbed with silk—a process that ensures close contact between the two substances—electrons flow onto the silk *from* the glass, leaving it deficient in electrons, and thus (+). The silk becomes charged (−). This may be shown by the tendency of the silk cloth to adhere to the glass. It is found that plus charges in solid substances are firmly fixed in the matrix of the substance while *some* of the electrons present may be mobile. It is these that flow. The charges on silk and glass tend to remain static. These materials do not conduct charges. (The explanation in terms of material structures will appear in later chapters.) Therefore one can hold one end of the glass rod without appreciably affecting the charge on the other end. The (+) charged rod is brought up toward the disk, but without touching it. The leaves are seen to open, Figure 3-2. The reason

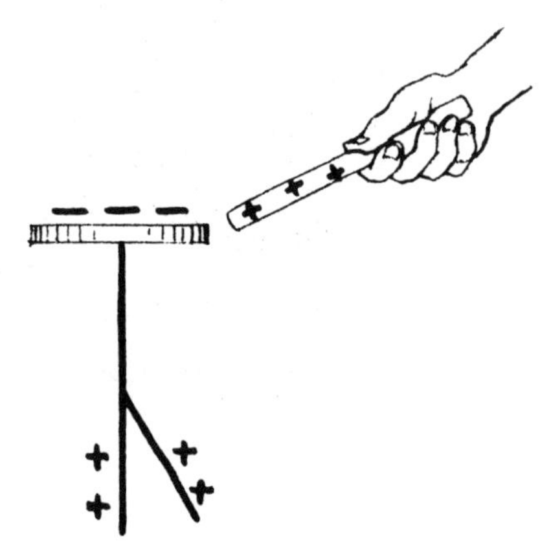

Fig. 3-2 *A plus-charged glass rod is brought near the plate of an electroscope. Negative charge is induced on the plate, and as a consequence, positive charge on the other end of the rod. The leaves separate in response.*

is that opposite charges attract. The (+) rod attracts electrons (−) which flow on to the disk as diagrammed, leaving the far end of the metal (+) and thus causing the leaves to diverge. If the rod is touched to the disk, electrons flow onto it from the disk, for it is always more strongly charged than the disk. Then when it is removed the movable electrons remaining on the metal of the electroscope redistribute so as to make the residual (+) charges as far apart from each other as possible. This is understandable since the electrons repel also. The result is that illustrated in Figure 3-1: the electroscope is charged (+).

Suppose that an electroscope is charged a little (+), so that the leaves could spread further with increased charge. It now is an instrument for detecting a charged body and for determining the nature of its charge (Figure 3-3a). You bring up to this test-instrument a (+) charged rod: the leaves diverge still more for the reason explained in connection with Figure 3-2 (see Figure 3-3b). Suppose the rod is (−) charged. It repels electrons (there are always some movable ones in the metal) toward the leaves, neutralizing some or all of their charge, and the leaves close (Figure 3-3c). If the rod were without any charge it would have no effect on the leaves. We have in the slightly charged electroscope [by convention always charged (+)] a handy device for detecting and identifying static charges. If we calibrate the electro-

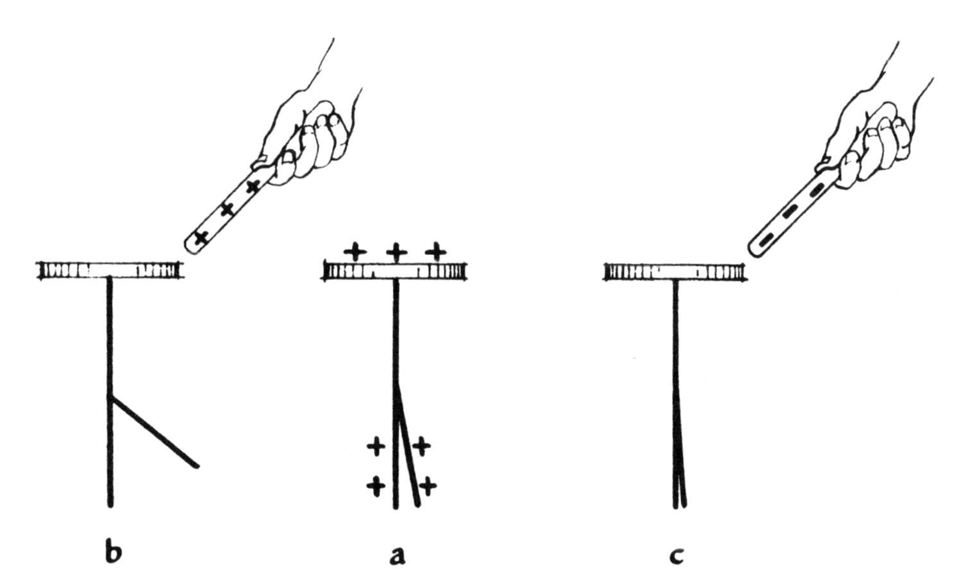

Fig. 3-3 *A lightly plus-charged electroscope* (a) *serves as a test instrument to detect the presence and quality of charge on an object.* (b) *A plus-charged glass rod, brought close to the plate, by induction causes the leaves to diverge further.* (c) *A minus-charged rubber rod, brought close to the plate, by induction causes the leaves to collapse.*

scope in terms of angle-of-leaves to amount-of-charge, it becomes a delicate electric meter. All these observations and the hardware are close to the *P*-plane. The explanation carries us into the *C*-field.

That charges may be separated, as shown in Figure 3-2, can be demonstrated in an elegant classical way. The neutral hollow metal objects, *A* and *B* (Figure 3-4), one spherical and the other cigar-shaped, standing on insulating bases, are made to touch. They are hollow only to make them lightweight, and the two shapes are convenient for dis-

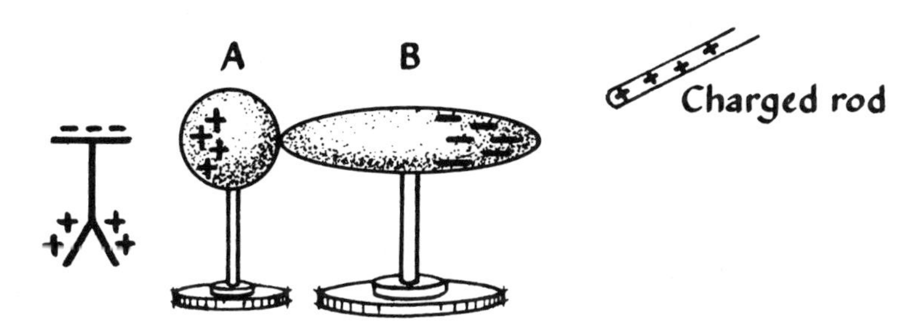

Fig. 3-4 *Separation of charge, first step. Two touching metal objects A and B are charged by induction. An electroscope may be used to show that the far side of A carries a plus charge.*

tinguishing which is which on a demonstration table. A glass rod charged (+) is brought close to one end and an electroscope is used to show that the far end is now charged (+). With the glass rod close to *B*, *A* is moved away, out of contact with *B*, Figure 3-5. Now the electro-

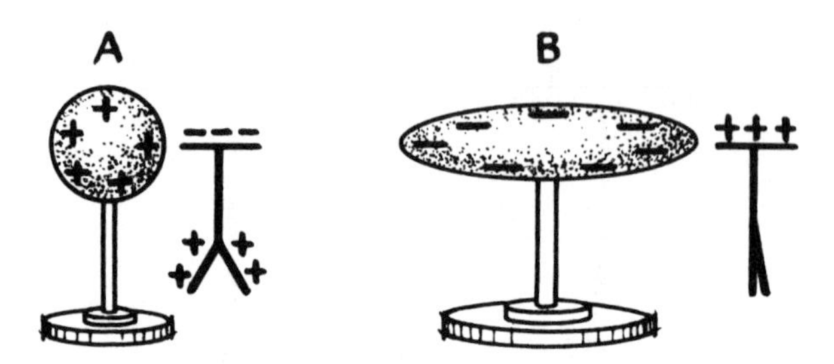

Fig. 3-5 *Separation of charge, second step. Keeping the rod near to B, A is moved out of touch by means of the insulated stand. An electroscope would now show that A is net plus-charged and B is net minus-charged. This shows that the 'charges' are mobile and separable.*

scope shows that *A* is (+) with the charges distributed all over it, and that *B* is minus: some charges in the system of two touching objects were separated from each other in this simple way. The electroscope is our tool for showing what has occurred. Another, in some ways simpler, tool consists of a light ball of pith suspended by a silk thread from an insulating stand. When it is charged (+) with a charged glass rod, it is an instrument for detecting charge: it is repelled by (+)-charged objects (Figure 3-6a) and attracted by (−)-charged objects (Figure 3-6b). Uncharged objects do not affect it.

3.3 Electrical behavior of solid matter

Materials of all kinds may be classified with respect to their ability to conduct a current of electricity. On the one hand there are the excellent conductors silver and copper; on the other the essentially non-conducting materials such as porcelain, paraffin, rubber, and polyethylene plastic, which are used as insulators. These extremes define a range: from excellent to good to poor conductors to better and better insulators. In general, metals are likely to conduct electricity while non-metals do not. This ability of metals is due to their atomic constitution and material structure. It comes from the fact that, while all the positively charged nuclei of the constituent atoms as well as most of the electrons are relatively fixed in the structure of any material, there are in metals some loosely held electrons which can be pushed or pulled,

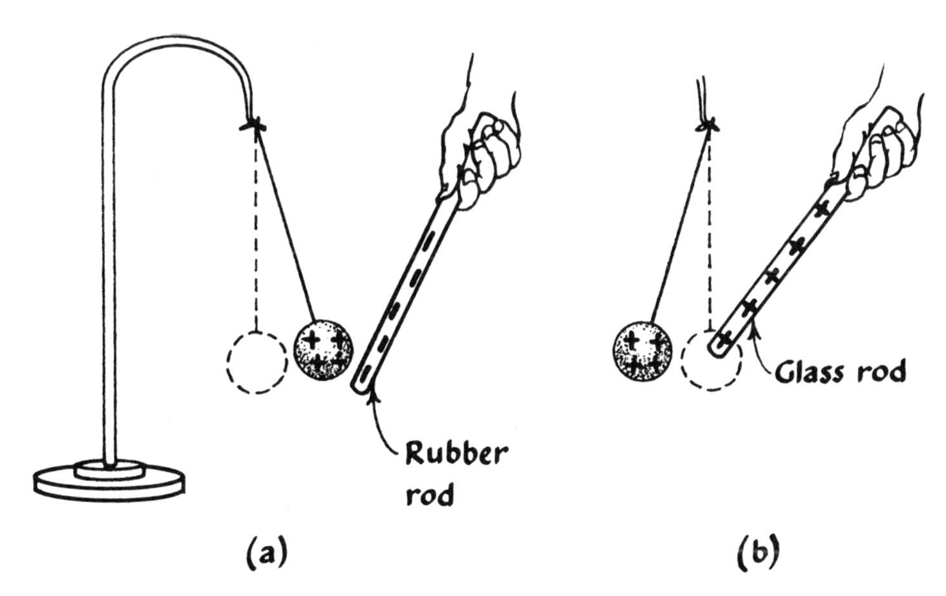

Fig. 3-6 *Behavior of a charged pith-ball test instrument. The ball, charged plus by touching it with a charged glass rod, is repelled by a plus charge* a *and swings away. It is attracted by a minus charge* b *and swings toward it.*

or otherwise moved, through the metallic structure. These comprise an electric current when they flow through a wire. The metal may be visualized as a fixed network of atoms through which can flow, as water through porous material, electrons. When no current is flowing, and the metal is in a neutral state, there are as many (+) as (−) charges present. When a current is flowing the (−) charges continually change place. As many enter a given section of wire as come out of it, and the wire remains neutral.

In a good insulator (the material of which is sometimes called a dielectric) virtually all the electrons present are firmly held to their respective plus-charged nuclei: they resist efforts to move them along. A measure of the insulating effectiveness of a dielectric is this resistance of its electrons to flow.

Electrical behavior in liquids 3.4

We saw in Section 3.3 that some solid materials conduct electricity while others do not. The same may be said of liquids. It is not surprising to find that mercury, a liquid metal, conducts electrons, while mineral oil, a liquid paraffin, does not. Some solutions conduct by a different mechanism. We describe it here for the purpose of showing that electrons may leave the surface of a metal or enter it. Accept that zinc (symbol Zn) forms a compound with chlorine which has the formula $ZnCl_2$, meaning that there are two chlorine atoms attached to one zinc atom. Also, copper (Cu) forms a chloride $CuCl_2$. Each is dissolved in water. In water solution the $ZnCl_2$ separates into Zn^{++} ions and 2 Cl^- ions; similarly, the copper separates into Cu^{++} and 2 Cl^- ions. One solution (say the $CuCl_2$) is poured into a cup of porous ceramic (unglazed porcelain) resting in a beaker; the other is poured into the beaker outside of the cup. The porous cup is a device to prevent rapid mixing of the two solutions, and we can confirm that rapid mixing is prevented because we can see that the blue copper chloride solution diffuses only very slowly through the walls of the cup into the colorless zinc chloride solution. Now we stick a strip of zinc into the zinc chloride solution, and copper into the copper chloride solution. Finally, we connect the zinc to the copper strip by wires that go to an electric current-measuring device, an ammeter. The apparatus is shown in Figure 3-7. Immediately the connection is made, the ammeter shows that current

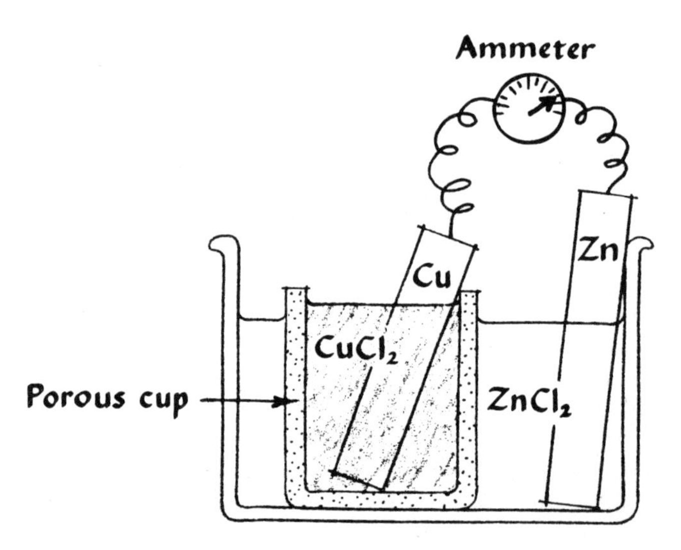

Fig. 3-7 *A simple type of battery. A zinc strip dips into a zinc chloride solution which is separated from a copper chloride solution by a porous wall. Into the copper chloride solution a copper strip dips. The metal strips are connected by a wire through an ammeter, which indicates that current is passing. The current consists of electrons flowing from the zinc metal to the copper plate. At the copper plate the electrons discharge copper ions, and copper plates out.* (Cu^{++} *plus* $2e^- \rightarrow Cu^0$ *metal.*) *At the zinc plate zinc metal loses electrons and goes into solution as zinc ions.* (Zn^0 *metal* $\rightarrow Zn^{++} + 2e^-$.) *Chloride ions* ($Cl^-$) *migrate through the porous wall to neutralize electrically the accumulating positively charged zinc ions.*

is flowing through the wire. As we watch, shiny copper plates out on the copper strip. At the same time the surface of the zinc becomes etched. (If we had weighed the metal strips before, and weighed them from time to time during the demonstration, we would find that the copper was gaining weight; the zinc was losing weight.)

What is happening is interpreted in the following way, to be explained in some detail when we study atomic structure. That a current flows must mean that in some way electrons move through the wire. They are found to move from the zinc to the copper. At the surface of the zinc, atoms of zinc give up two electrons each and pass into the solution as zinc ions (Zn^{++}). At the surface of the copper, copper ions (Cu^{++}) in the copper chloride solution plate out by accepting from the metal strip two electrons per ion and become metallic copper. Electrons thus enter at the zinc side, pass through the wire, are captured at the copper side. But as the zinc goes into solution as Zn^{++} ions it must be neutralized, otherwise a (+) charge would build up, and then electrons would not tend to leave that place. Also, as the copper plates out there are left behind in the solution two negative chloride ions (Cl^-) per copper ion, and if they were not removed the tendency of Cu^{++} to leave the solution, which would be negative, would decrease. What happens is that Cl^- ions diffuse from the copper chloride solution across through the solution in the holes of the porous barrier into the zinc chloride, just neutralizing the newly formed zinc ions. The concentration of the copper chloride solution falls; that of the zinc chloride rises. The whole reaction is driven by the respective tendencies of zinc to become an ion, and copper to plate out. Under suitable conditions, then, electrons may leave or enter a metal surface in contact with a solution.

Gaseous conduction 3.5

An ordinary gas, such as air, consists of atoms or groups of atoms in rapid, random motion. These collide with each other and with any material objects present. For example at the surface of a metal plate in the air there is rapid bombardment by gas atoms, and any electron that might tend to escape would very likely be prevented from getting far. Certainly, it would take a random path, and would be difficult to follow and identify. When the gas is largely removed, however, electrons can be made to stream off a metal surface, particularly if it is hot. The resulting stream of electrons is a cathode ray. Such a stream is produced in a television tube.

A glass bulb is prepared into which are sealed a filament and a section of metal rod with a central hole along its axis (Figure 3-8). These are electrically connected outside of the bulb. The bulb is then evacu-

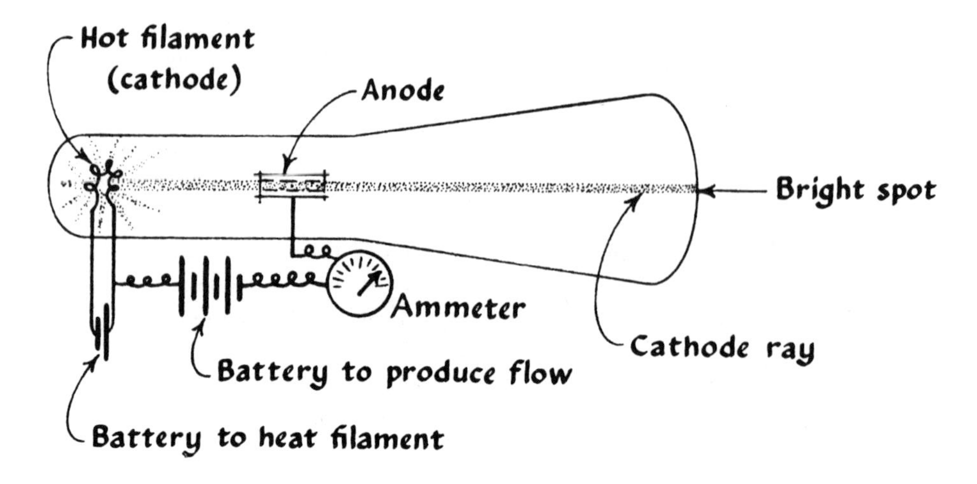

Fig. 3-8 *A simple cathode-ray tube. Electrons boil off the hot cathode and are accelerated toward the anode, which has a hole in it. Some pass through the hole and travel down the tube. Where they strike the glass end of the tube a patch of light appears.*

ated until the gas pressure inside it is low. This removes most of the gas particles so that an electron may travel far without colliding with more than a few of them. By means of a small battery, the filament may be heated red-hot or white-hot. If, now, a strong battery is connected so that the filament is (−) to the (+) metal rod, electrons boiling off the filament accelerate toward the (+) electrode (named an anode) and many of them pass into it, as shown by the ammeter A, which indicates that a current is flowing. The only way a current can flow is by the passage of electrons across the gap from the filament (the negative electrode, or cathode) to the anode unless they flow along the surface of the glass. Very few take this latter path. The anode has an axial hole, and some electrons are moving so fast by the time they reach the anode that they pass through it and issue as a stream on the other side. They strike the end of the bulb and make a bright spot. If before the bulb was sealed a little mercury was put in, this stream becomes visible as a purplish ray. The reasons for the bright spot and the purple ray are explained in a later chapter and need not distract us from the visible features. That this ray contains electrons was demonstrated by J. J. Thomson (Section 4.7). He showed that the ray was repelled (bent away) by a (−)-charged plate and was attracted toward a (+)-charged plate. Moreover, as we shall see later, he was able to obtain numerical information about the electrons. He used different materials for cathodes and found that rays with the same properties were produced. He was thus forced to conclude that electrons were constituents of all these materials. This same conclusion had been arrived at through conduction in solutions, using many different electrodes.

Conceptual approach: lines of force 3.6

The phenomena we have described are but a minute portion of the available data. We now move into the C-field to show how these phenomena, with their multiplicity of appearances, were grasped conceptually. We first take up static charges, and as is the usual procedure idealize the situation to simplify it.

Consider a charged particle fixed in space, as on an insulated stand. Let it be charged (+). We explore the neighborhood of this charged particle by means of a (+) charged electroscope, which is our standard test instrument. At the same time we imagine what would be the case if the charge on the electroscope were so small (and the electroscope itself so small) that it does not cause distortion of the forces about the (+) charged particle.

From whatever side we approach, the electroscope senses (+) character which falls off in strength in a direction radial to the center of the fixed (+) particle. The force is directed away from the (+) center. This is shown two-dimensionally in Figure 3-9, where the arrows represent lines of force imagined to take their source in the (+) charge as recognized by a standard (+) charged electroscope. These lines of force were invented by Michael Faraday for purposes of visualization. Figure 3-9 is a cross-section through a sphere of such lines. If the fixed particles is charged (−), the response of the standard (+) charged

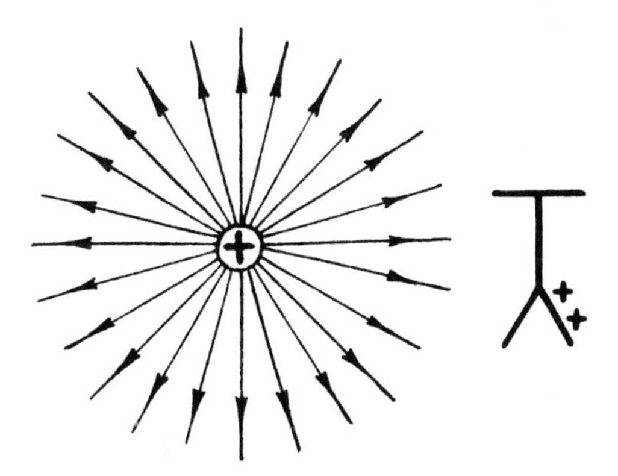

Fig. 3-9 *A test instrument shows that in the neighborhood of a plus charge there is plus character in space. This is conceptualized as lines of force, and symbolized by vector arrows.*

electroscope is as in Figure 3-10, that is, attraction toward the (−) object.

Fig. 3-10 *Lines of force about a negatively charged particle.*

Suppose, now, two fixed particles, both charged (+). Exploration of their neighborhood with the standard detecting tool would lead to the map of lines of force shown in cross-section in the Figure 3-11. The

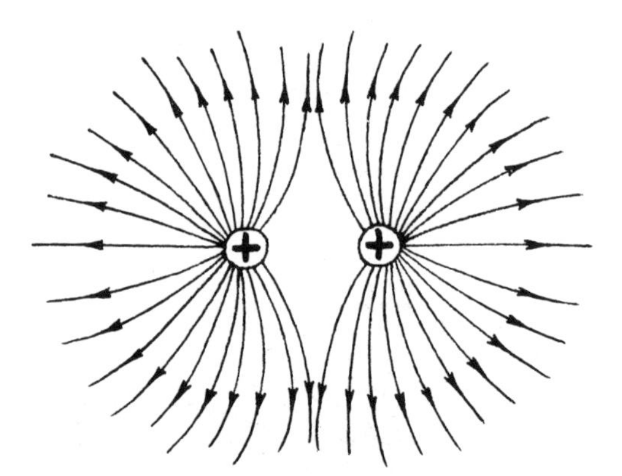

Fig. 3-11 *Lines of force in the neighborhood of two like-charged particles.*

curvature of the lines implies in a graphic way the repulsion between the two (+) charges. Were one fixed particle (+) and the other (−), the map of lines of force would be like that in Figure 3-12 (you will remember the force is actually three-dimensional). If two plates are arranged close together, parallel to each other, and connected to the poles of a battery, so that one plate is charged (+) and the other (−),

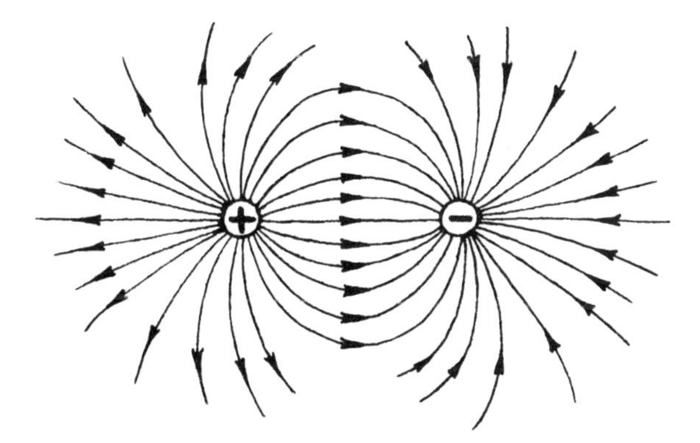

Fig. 3-12 *Lines of force in the neighborhood of two unlike-charged particles.*

the map of lines of force between them is found upon exploration to be that in Figure 3-13 (in cross-section). It is as though at the surfaces of each plate there are closely arranged point charges, all interacting as in Figure 3-12. Between the plates (this will become important) the distribution of lines of force is uniform over the whole area; only at the edges do the lines diverge in a non-uniform way.

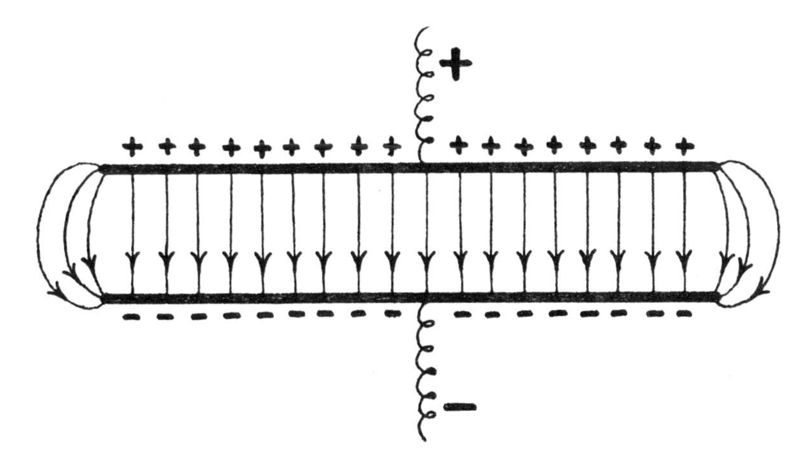

Fig. 3-13 *Lines of force in the neighborhood of two unlike-charged plates.*

Conceptual approach: electrical field 3.7

Here are two entities obviously affecting each other—after all, the force involved in the interaction between a charged rod and an electroscope is enough to lift the gold leaf of an electroscope against gravity, thus doing work on the leaf. So "something" is there, yet it is not visible in the space between rod and electroscope. One invents invisible lines of force, comprising an electrical field about the charged entities. We must be clear about this: the field is a construct invented so as to gain conceptual grasp of a cluster, or class, of phenomena. A field is a region of space in which at any point a *suitable* test-object or tool will detect a force, and therefore at any point there is a potential for doing work. We shall explore these behaviors of fields in some detail later. The points to be made here are that the field is an invention—so be careful about reifying it. Further, it is recognized only by a suitable tool *which has the same kind of field*. Charge-fields are recognized only by charged particles. These two points will continue to be important throughout the text.

It is a characteristic of a field that its force-giving and potential properties can be measured, and it is characteristic of scientists that they are not satisfied until adequate quantification has been achieved. There is a reason for this. If the single lines between constructs (Figure 1-4) can be converted to proportions and these stated mathematically, then one can go experimentally to any field and map it with numerical values in a *convincing* way. The first breakthrough in this enterprise was made by Charles Coulomb in the eighteenth century. It is worth describing because of its simple elegance.

3.8 Coulomb's law

By the time Coulomb began his beautiful work with charged objects it was known, or suspected, that the forces of attraction and repulsion between charged bodies vary inversely with the distance between them. Coulomb set up a torsion balance (which he had invented) consisting of a fine wire suspended from a chuck that could be turned. It carried at the lower end a light rod, or straw, with a tiny light ball of pith on one end and a paper disk to counterbalance it and to damp motion on the other end (Figure 3-14). This just touched a fixed pith-ball of identical size. When the pith-balls were charged by being touched by another charged ball, the suspended one would swing away in an arc, twisting the wire until the force resisting further twist just balanced the force of repulsion between the balls. This force could be measured from previous knowledge of the force needed to cause the wire to twist a certain amount: the wire is a spring. The balls could be forced together by applying more twist to the wire through the calibrated arm at its upper end. By forcing the balls toward each other, by twisting the torsion wire, Coulomb demonstrated quantitatively that the force of repulsion or attraction between two charges varies inversely with the square of the distance between them. This is written in mathematical shorthand as follows. We let **F** represent, or stand for,

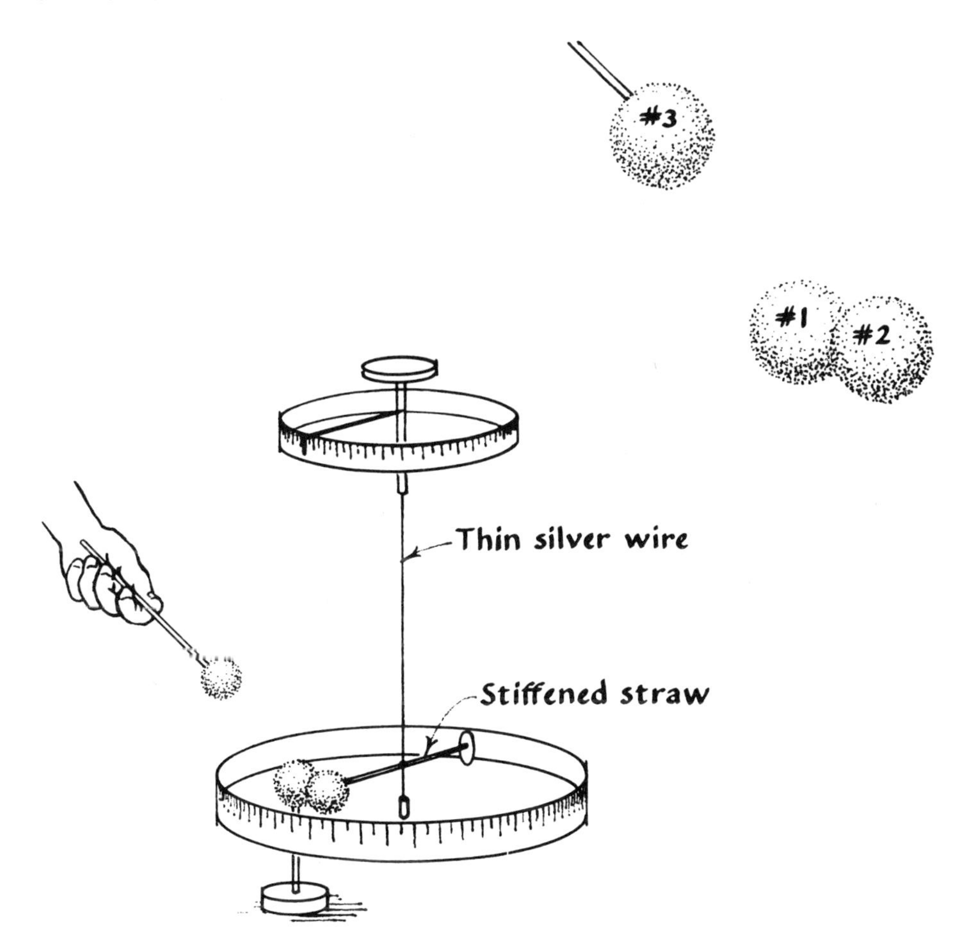

Fig. 3-14 *Coulomb's apparatus. [Redrawn from an illustration in Morris H. Shamos,* Great Experiments in Physics, *Holt, Rinehart & Winston, Inc., New York, 1960, p. 172]. See text.*

'force.' This is measured by the amount (the angle through which) the wire is twisted. The sign of proportionality, as before, is '$\propto$.' The 'distance between the balls' is represented by s, and can be measured by means of a good ruler. When Coulomb did the experiment he found that whatever the charges on the balls, the closer they were brought together, that is, the smaller the value of s, the larger the **F**. Moreover he could find an exact proportionality not between **F** and s, but between the value of **F** and that calculated by squaring s, *i.e.*, s^2. To symbolize this, he would write:

$$\mathbf{F} \propto 1/s^2 \qquad (3\text{-}1)$$

This is an elegant way of saying that the force gets larger as s^2 gets smaller, since it is apparent that the smaller s^2 is, the larger $1/s^2$ becomes, and hence in a proportional way, **F**.

As an example, suppose that in the apparatus in Figure 3-14 the two balls #1 and #2 have been charged equally by touching them simultaneously with #3, which was charged up outside the apparatus. Ball #1 is fixed; #2 swings away, twisting the wire, until the twisting force just balances the repulsion force. Suppose the distance between the two balls is now 4 cm, and the angle of twist, a measure of the force on the wire, is 20°. You bring the balls closer to each other by turning the knob at the top of the wire, twisting it, and 'forcing' the balls closer. You find that to bring the balls to 3 cm you need a certain added twist, then to bring them to 2.5 cm, still more, and so on. You construct a table of data, as observed. This might be named a 'protocol'; in all experiments the protocol takes the form: "I did this" (*i.e.*, applied a force **F**); "I observed that" (*i.e.*, the new distance apart of the balls, *s*). The relation is inverse: as *s* goes down, **F** goes up. Now you wish to find the relationship (if any) between the two sets of numbers (of course we are 'supposing,' for you know quite well that there is a relationship: Coulomb found it through a procedure somewhat like this; we are 'acting out' the original drama). Suppose, then, you plot **F** against *s*. You obtain a curve, Figure 3-15. This looks very much like the accelera-

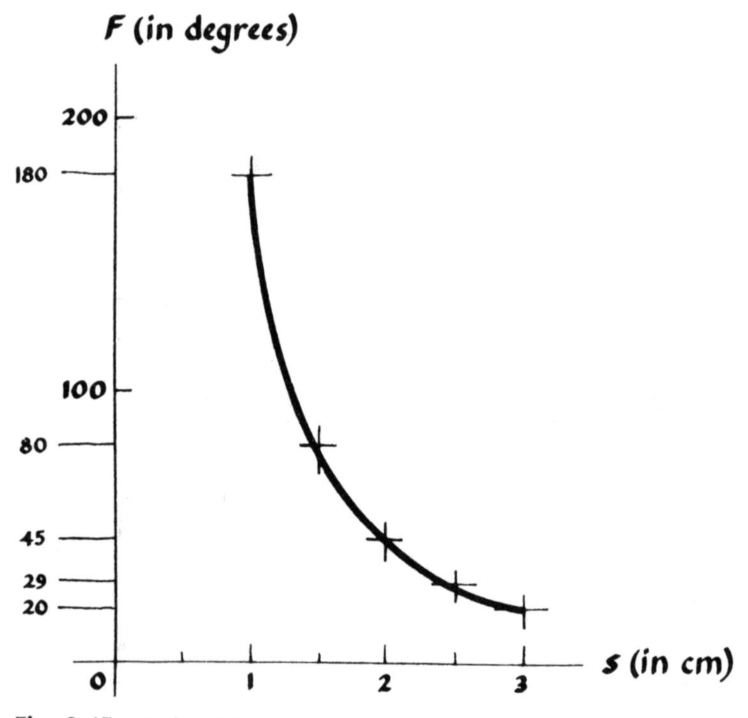

Fig. 3-15 *A plot of the raw data.*

tion problem of the previous chapter. You try plotting **F** against $1/s^2$, and you get a straight line (Figure 3-16). Now you can confirm that:

$$\mathbf{F} \propto 1/s^2.$$

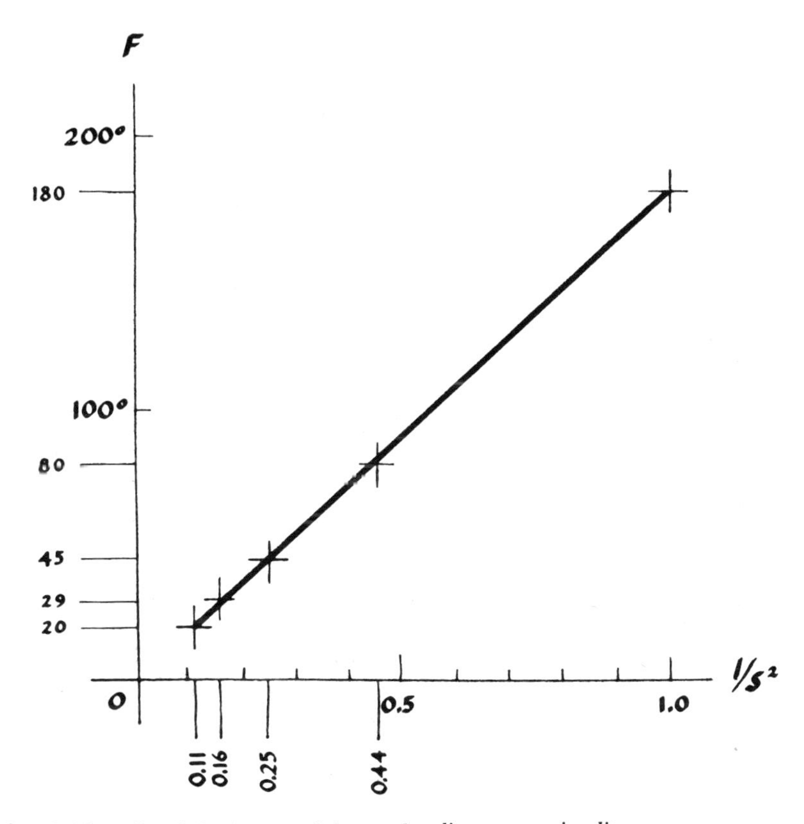

Fig. 3-16 *Plot of the data re-coded to produce linear proportionality.*

Coulomb did not have any way of measuring a magnitude for the charge on a pith-ball. He was able to get the relation between **F** and the size of the charge by the following type of reasoning and experiment.

Since charge spreads out over a surface, if you have a charged ball and touch it with an uncharged one of exactly the same surface area, the charge will spread out over the doubled area with half the charge on each ball. Thus without actually knowing the size of a charge to begin with, it is possible to change it to 1/2, and this to 1/4; or by touching a 1/2 and 1/4, one can get two balls with the total 3/4 charge distributed equally between them. If we let a ball be numbered 1 and another 2, for convenience of distinguishing them (they could be called '*a*' and '*b*', or 'left' and 'right', or whatever), and we let q represent 'charge,' so that q_1 represents the size of the charge on ball 1, then by experiments such as already described, putting charges of different

Table 3-1

Data—Coulomb's Law

F (in degrees)	s (in cm apart in a straight line)	s^2	$1/s^2$
20	3	9	0.11
29	2.5	6.25	0.16
45	2	4	0.25
80	1.5	2.25	0.44
180	1	1	1.0

relative size on the balls, Coulomb could show that the force (**F**) between the balls was proportional to the product of the sizes of the charges, that is, in symbols,

$$\mathbf{F} \propto q_1 q_2 \tag{3-2}$$

Moreover, if the charge is given a sign (+) or (−), depending on its nature, then if q_1q_2 is (+), which happens when a (+) value is multiplied by a (+) value or a (−) by a (−), the sign indicates that the force was repulsive, for like charges repel. If **F** comes out (−), as would happen if q_1 is (+) and q_2 is (−), or vice versa, the sign indicates attraction force.

Putting the two proportionalities together we see 'by inspection' (intuitively) that

$$\mathbf{F} \propto q_1 q_2 / s^2 \tag{3-3}$$

We saw that an equation of this type can be converted to an equality by introducing a 'constant of proportionality.' We let this constant be represented by 'K' and obtain a statement in equation form of Coulomb's law:

$$\mathbf{F} = K q_1 q_2 / s^2 \tag{3-4}$$

Here the force **F** is measured in newtons, q_1 and q_2 in coulombs (C), and the distance s in meters. (The proportionality constant K is taken for ordinary calculations as 9×10^9 Nm^2/C^2). This **F** is referred to as a 'Coulomb force,' meaning that it is a force between charged particles.

A handy way of checking equations is in terms of their 'dimensions,' that is, the units in which they are given.

The units must be the same on both sides of the equals sign, otherwise things are not equal: you can't equate words with shoes, for example. In the case of equation (3-4) *force is measured in dimensions called newtons* (N), q *in coulombs, and distance in meters. The constant* K *is in the dimensions* N m^2/C^2. *Therefore, writing the equation out in terms of its dimensions:*

$$N = \frac{N\,m^2}{C^2} \times \frac{C \times C}{m^2}$$

The equation is consistent in its dimensions, as can be seen by cancellation. We are concerned about this because units are important: if you receive 16 moneys it is of some consequence whether the units are dollars, cents, or cowrie shells. The modern student who thinks nothing of dealing in rupees, dollars (U.S.), dollars (Canadian), dollars (Australian), or francs, as he flies about the World, should accept in the same spirit dealing with newtons, coulombs, meters.

As an example of the use of this law, suppose a small ball carrying a charge of 2×10^{-7} C. Another ball carrying a charge 1×10^{-8} C is brought to 10 cm (0.1 meter) from the first. What is the force experienced by each? The force, equal and opposite on each ball, is calculated to be:

$$\mathbf{F} = 9 \times 10^9 \times 2 \times 10^{-7} \times 1 \times 10^{-8}/(0.1)^2 = +18 \times 10^{-4} \text{ N}.$$

If one of the two balls had had a charge of the same magnitude but negative in sign, the force would have been -18×10^{-4} N, the minus sign signalling attraction. If both had been negative the force would have been $+18 \times 10^{-4}$ N.

Vector representation 3.9

Suppose we have two small charged spheres fixed at a distance of 10 cm from each other. For physical flavor we might use the example in 3.8 and imagine one charged $+2 \times 10^{-7}$ coul and the other $+1 \times 10^{-8}$ coul. Then by Coulomb's law, the force of repulsion between them is 18×10^{-4} N. The repulsion force can be represented by arrows directed away, the one from the other, in the direction they would be displaced if one or both were released. To complete the representation we let the length of the arrow represent the strength of the force: we might let 1×10^{-4} N be represented by 1 mm, when each arrow would be 1.8 cm long, as in Figure 3-17. These arrows are *vector* arrows. A

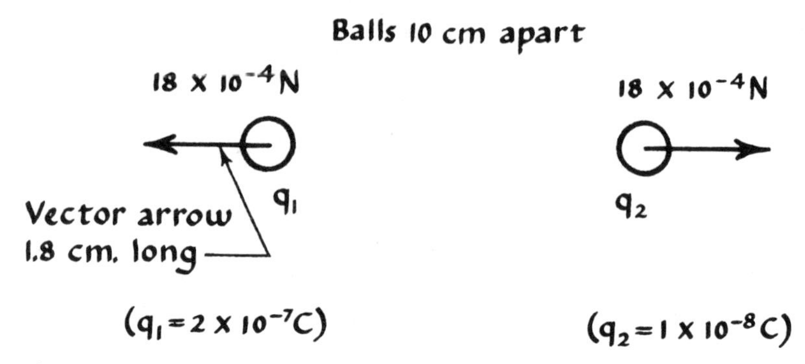

Fig. 3-17 *The forces on two electrons 10 cm apart.*

vector quantity is defined by both a magnitude *and* a direction. Thus displacement (*e.g.*, as on a map, one mile due east); velocity (*e.g.*, 100 m/sec straight up as of a bullet); acceleration (*e.g.*, 32 ft. per second per second straight down as in a freely falling object); force (*e.g.*, 18×10^{-4} N in a +X direction as in something pushed) are vector quantities, and can be represented by arrows. (Quantities that have only magnitude and no particular direction are called *scalar* quantities. Thus 3 sec, for example, or 1 lb., or one quart, or 32°F are scalar quantities. We digress now to discuss vectors because they will turn out to be useful throughout this text. At the same time they represent an elegant form of mathematics combined with geometry. They lend a pleasing *Gestalt* to any theory.)

The convenient thing about a vector arrow is that is can be manipulated by geometry or trigonometry. This is felicitous. It can save a great deal of labor and also yield a clear picture of what is described. For example, we have a vector arrow *PQ*, Figure 3-18. The Cartesian

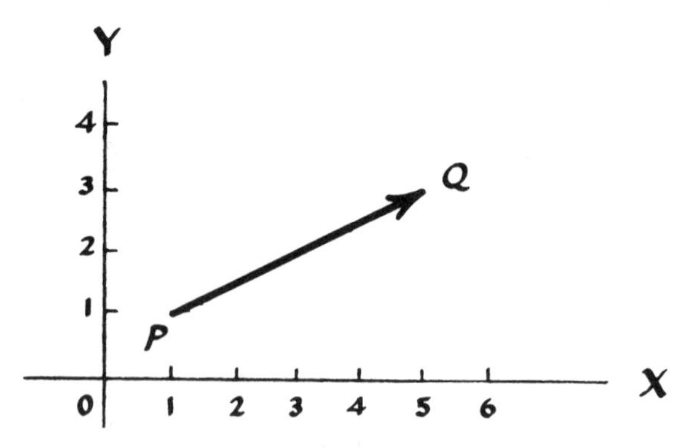

Fig. 3-18 *A vector arrow* PQ.

coordinates of P in this diagram are (1, 1) and of $Q(5, 3)$. Putting this into general terms, we say that the coordinates of P are (x_p, y_p) and of $Q(x_q, y_q)$. The vector is written $[(x_q - x_p), (y_q - y_p)]$ or, in the example, Figure 3-18 $[(5 - 1), (3 - 1)] = [4, 2]$. This notation, [4, 2], is equivalent to moving the axes without changing their directions so that the origin is at P, as in Figure 3-19, for now the vector is $[(4 - 0),$

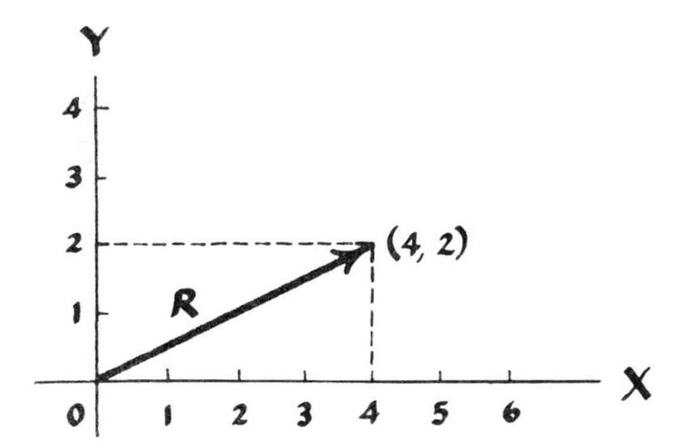

Fig. 3-19 *The vector arrow of Figure 3-18 moved so that its origin* (P) *is at the origin of the coordinate system.*

$2 - 0]$. Clearly, $(5 - 1) = (4 - 0)$; and $(3 - 1) = (2 - 0)$. *The arrow has the same length R and direction relative to the X and Y axes as before.* Using Pythagoras' theorem, $R = \sqrt{x^2 + y^2} = \sqrt{20}$. Vector arrows can be moved around in a coordinate system without changing their values *provided only* that their length and direction are not altered. For example, all the vector quantities in Figure 3-20 are identical.

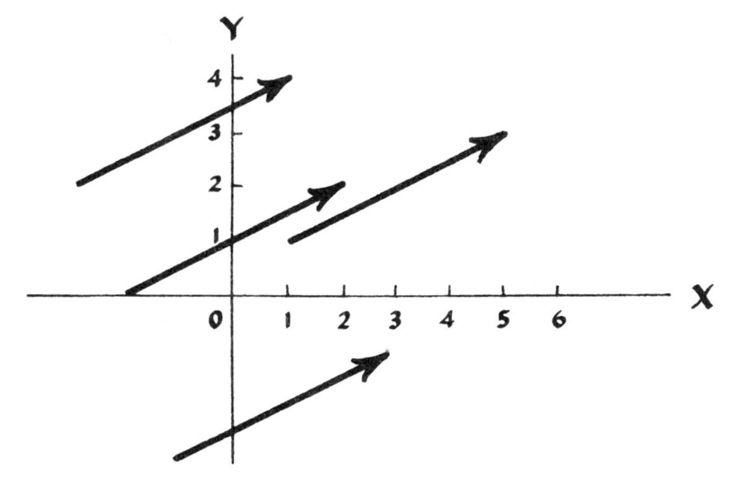

Fig. 3-20 *A set of identical vector arrows. They are identical because all have the same length and the same direction relative to the coordinate frame of reference.*

Now consider the convenience of this. Suppose we wish to add two of these arrows. We merely move the tail of one to the head of the other and the total length from the origin is our answer, as in Figure 3-21. If the two arrows we wish to add do not have the same direction, as F_1 and F_2 in Figure 3-22, the principle still holds, as shown. Here,

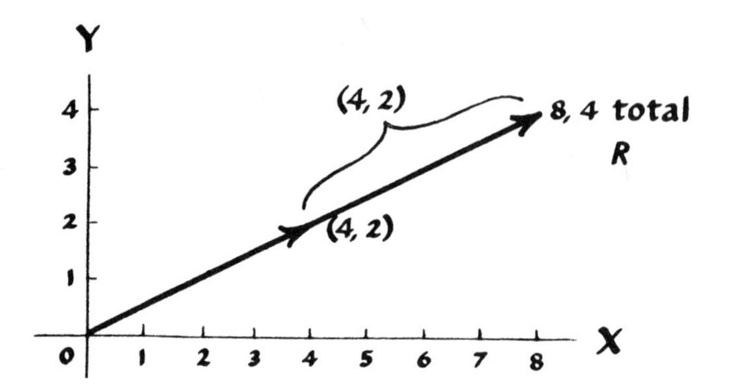

Fig. 3-21 *Addition of vectors that have the same direction.* R *is the resultant, or total, of the addition.*

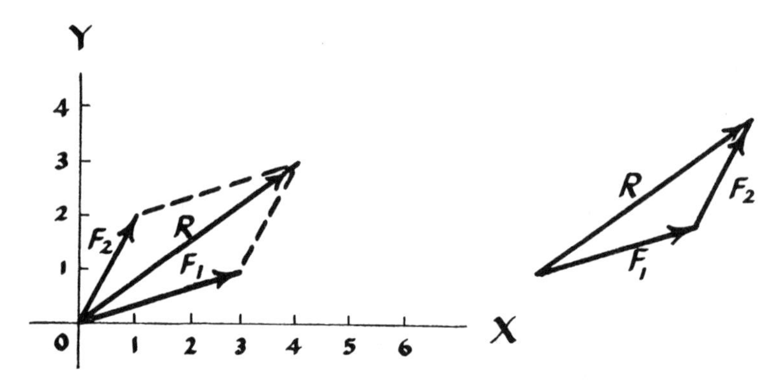

Fig. 3-22 *Addition of vectors which have different directions.* R *is the resultant of the addition.*

however, we must use trigonometry to find R, the resultant force, or we can get R by geometrical construction, drawing F_1 and F_2 at the correct angles and of the correct length, then measuring R. If we wished to subtract two vectors, as in the example of two equal forces acting in opposite directions on the same object U in Figure 3-23, it is evident that the same principle holds. In Appendix 1 we have reminded you of the trigonometry you studied in secondary school.

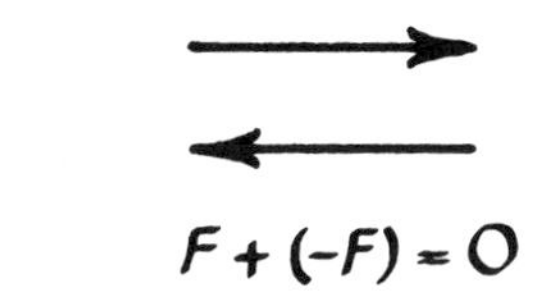

Fig. 3-23 *Subtraction of two vectors of equal magnitude and opposite direction.*

Returning to the example of the charged particles, suppose we have two electrons A *and* B *5 cm apart as in Figure 3-24, and a proton at* P, *4 cm from* A *and 3 cm*

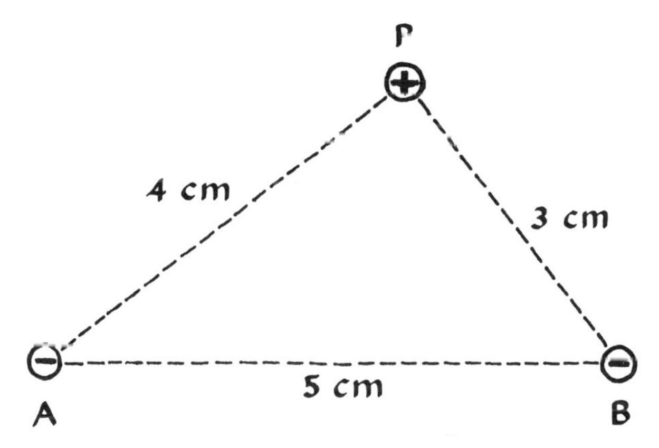

Fig. 3-24 *The geometric diagram of a proton* (p^+) *at point* P *and single electrons* (e^-) *at points* A *and* B.

from B. *The charges on the electrons are* -1.6×10^{-19} C *each, and on the proton* $+1.6 \times 10^{-19}$ C. *We want to calculate the force "felt" by the proton because of the presence of these two electrons each of which attracts the proton. Using the Coulomb equation, and calculating the forces pairwise, we have:*

$$\mathbf{F}_{A,P} - K\, q_A\, q_P/s^2_{A,P} =$$
$$9 \times 10^9 \times -1.6 \times 10^{-19} \times +1.6 \times 10^{-19}/(0.04)^2 =$$
$$-1.44 \times 10^{-25}\ \mathrm{N}$$

$$\mathbf{F}_{B,P} = K\, q_B\, q_P/s^2_{B,P} = -2.56 \times 10^{-25}\ \mathrm{N}.$$

Resultant force on the proton, R, *is* $\sqrt{\mathbf{F}^2_{A,P} + \mathbf{F}^2_{B,P}}$.

Geometrical construction would give us the same value (Figure 3-25).

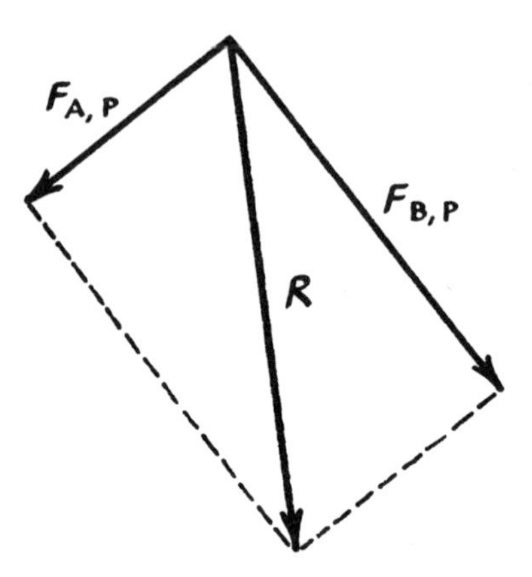

Fig. 3-25 *Vector representation of the forces,* FAP *and* FBP, *on the proton due to the presence of the electrons. The resultant can be calculated by trigonometry (Appendix 1) or may be obtained by geometric construction and measurement.*

3.10 Conceptual approach: field intensity

We said that a field is characterized by a force-giving property at any point in the field. The field has a 'source.' By definition this is the (+) charged particle, source of the lines of force in Figure 3-9. Or, the field has a 'sink' which in the case of charge-fields is the (−) charged particle illustrated in Figure 3-10. We said that the force at a point distant s meters from a charged particle, with charge q_1 depends not only on the magnitude of the charge q_1 but on that of the entity (of charge q_2) which detects the force and thus certifies the field. Since any particles of charge q_2 might be present in the field at a distance s from q_1, and if the q_2's had different magnitudes, then the consequent varieties of **F** would be measured at this point. Conceptual grasp of these possiblities is gained by a mathematical device. The force **F** is divided by whatever the q_2 is. This gives a 'specific' force; a force per unit of charge, and then one can say of q_1 that it has this *specific* field-intensity at the distance s, no matter what magnitude of charge q_2 is in question. This ratio $\mathbf{F}/q_2$ is so useful that it is called 'field intensity,' or 'field strength', and given its own symbol E.

$$E = \mathbf{F}/q_2 = Kq_1/s^2 \tag{3-5}$$

By similar reasoning the field-strength in the neighborhood of particle

2 would be calculated $E_2 = \mathbf{F}/q_1$. E is a vector quantity directed along the lines of force, or tangent to them if they are curved.

In the example in Section 3.7, $q_1 = 2 \times 10^{-7}$C,
$q_2 = 1 \times 10^{-8}$C,
$s = 0.1$ m,
and $\mathbf{F} = 18 \times 10^{-4}$ N.
The field intensity at 0.1 m *from particle* 1 *is:*
$E_1 = 18 \times 10^{-4}/1 \times 10^{-8} = 18 \times 10^{4}$ N C^{-1};
that at 0.1 m *from particle* 2 *is* $E_2 = 18 \times 10^{-4}/2 \times 10^{-7}$
$= 9 \times 10^{3}$ N C^{-1}.

Conceptual approach: potential 3.11

The field is defined, we said, by a force at any point P, and a potential at that point. By 'potential' is meant that a particle of charge q_2 brought to the point, say P_1 at a distance s, from the source of the field (a (+) particle of charge q_1) is capable of doing work: it experiences an energy giving property of the field. Since work and energy are essential constructs in any intelligent person's mental armamentarium we must now discuss them with some care. Later we shall return to the P-plane.

The potential, the energy-giving property of a field, is measured in the following way. Suppose that one has a particle with charge q_1 at some fixed point, and one brings up to it a particle with charge q_2 with the same sign. Suppose that at first particle 2 is so far away from particle 1 (say at P_0) that they do not substantially affect each other (Figure 3-26). But as they come closer together they repel, and so work (W)

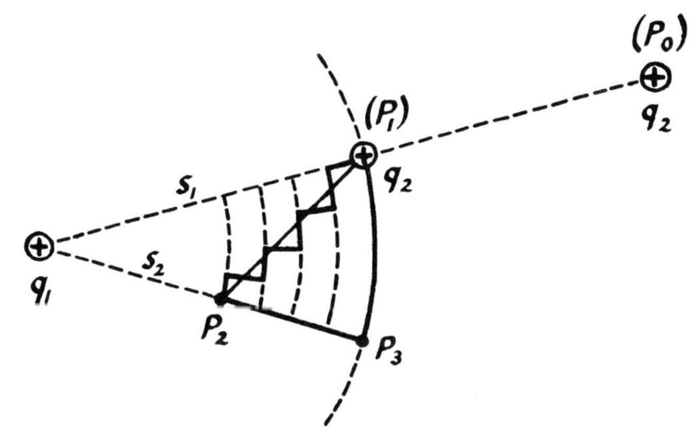

Fig. 3-26 *Conceptual approach to 'potential.'*

has to be done on particle 2 to bring it up to P_1, a distance s from particle 1. Work is defined as force applied multiplied by the distance through which the force acts in the direction of motion:

$$W = (\mathbf{F} \cos \theta)s, \quad \text{(in newton meters, or joules, J.)} \tag{3-6}$$

Here θ is the angle between the force and the direction of the motion produced. If **F** is parallel to the direction of motion, $\cos \theta$ is 1 (see Appendix). If **F** is at right angles to a direction of motion then that force is doing no work with respect to that motion, for $\cos \theta$ is 0 at $\theta = 90°$.

This can be shown by a mechanical model. Suppose you have an object 'Q', Figure 3-27a, which you pull along a platform a distance s,

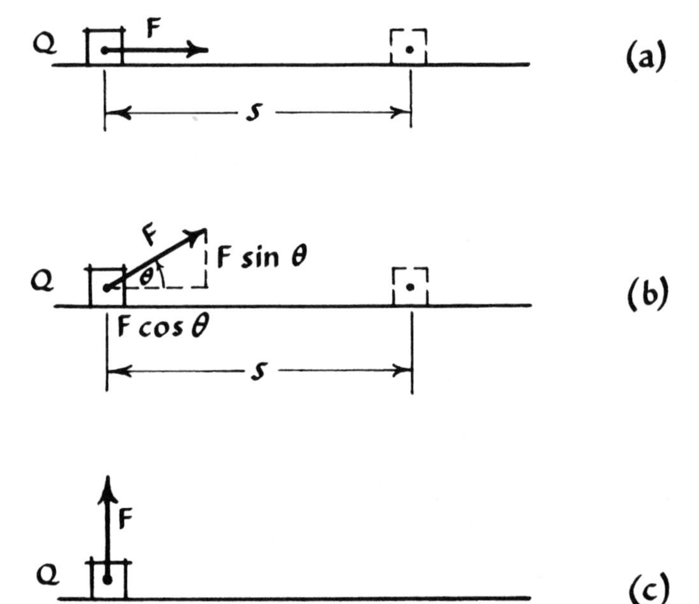

Fig. 3-27 *Illustration of the definition of work.*

applying a force **F** newtons in the direction of the motion. Then $\theta = 0$, and $\cos \theta = 1$, so the work done will be $\mathbf{F}s$. But suppose that you had applied the force in the direction shown in Fig. 3-27b, that is, at an angle θ to the direction of motion—as though you might be pulling a toboggan with a rope. The force **F** that you are applying now has two obvious components: one, $\mathbf{F} \cos \theta$, which is less than **F** in magnitude tends to move the object along the platform; the other, $\mathbf{F} \sin \theta$, also smaller than **F**, tends to lift the object. Only $\mathbf{F} \cos \theta$ moves the object along the platform. If θ were 30°, cos θ would be 0.866, and 0.866 of the applied force would be available to move the object along the platform. Finally, suppose that **F** is applied directly upward. Obviously, this will not move the object along the platform, only tend to lift it. Here $\theta = 90°$, and cos θ is zero. The work in the direction along the platform is clearly zero. (Other work is being done, perhaps, such as lifting work.)

Force between the charged objects changes with their distance apart. We have to take this into account in calculating the work done in moving the charge q_2 from P_0 to P_1. We have to sum the expression for the work over this distance. We note, also, that the work is done *against* the coulomb force, and thus $\mathbf{F} = -Kq_1q_2/s^2$ in this physical situation. Let us represent the distance from q_1 to P_0 by a capital S (we shall soon drop it from the equation), then we can write:

$$\overline{W} = \Delta\mathbf{F}\,\Delta s = (-Kq_1q_2/s^2)s - (-Kq_1q_2/S^2)S$$

Now it is evident that if we put P_0 farther and farther away, so that S becomes larger and larger, approaching infinity, the last term in the equation approaches zero. We can see, then, that the work done in bringing q_2 from a large distance up to position P_1 at s_1, will be

$$W_1 = -Kq_1q_2/s_1 \qquad (3\text{-}7)$$

Similarly, to bring q_2 up to point P_2 at s_2 requires $W_2 = -Kq_1q_2/s_2$. Therefore to move from P_1 to P_2 may require $W = W_2 - W_1$. That the magnitude of the work done in going from P_1 to P_2 is independent of the path taken, and therefore that we can replace 'may require' in the previous sentence with 'requires' follows from considerations such as these (see Figure 3-26). Consider the path P_1 to P_3, which is an arc of a circle of radius s_1. No electrical work is required to move the particle along this path for it is at a right angle to the radius direction s_1 (see above). The actual work, then, is equal to that done in going from P_3 to P_2. This is identical to that done in going from P_1 to P_2 by the direct path. For consider: this path may be broken up into little steps, as shown; a step radially, one tangentially (no work), another radially, another tangentially (no work), and so on until the radial steps add up to the distance equal to that from P_3 to P_2. Make the radial and tangential steps (in your imagination) so small that essentially you traverse the direct line from P_1 to P_2 and you have solved the conceptual problem. W is a scalar quantity and in this convention, work done *on* the system is labeled (−), while work done *by* the system is labeled (+).

What we have done so far in this section is to show that an amount of work $W_1 = -Kq_1q_2/s_1$ is done to bring particle 2 up to the point P_1. What has this to do with the energy-giving property of the field? There are two energy constructs of interest to us at this point: kinetic energy,

KE, and potential energy, *PE*. *KE* is a single symbol. Sometimes it is written E_k. Likewise for *PE*, E_p. These *E*'s are not to be confused with field intensity, so we shall write *KE* and *PE*. Kinetic is energy of motion, and one of its definitions is

$$KE = (1/2)mv^2 \tag{3-8}$$

KE is shorthand for $(1/2)mv^2$; it is a construct measurable in terms of m and v. *PE* is given various definitions depending on the source of potential energy. One commonly experienced source is gravitational. This is an intuitively satisfactory way of introducing *PE* and we use it here for this reason. In a grandfather clock run by a chain from which is suspended a heavy 'weight,' the clock is wound by lifting the weight: work is done on the weight to raise it. It then has potential energy; energy of position. "Potential" because this weight can now be arranged to do work running the clock as it falls. The average amount of work done, by definition, is $\Delta\mathbf{F}\,\Delta s$, or $W = \mathbf{F}s$ in simplified cases. Here it is done in lifting the weight against gravitational force, so we turn to the fundamental gravitation equation (3-9).

$$\mathbf{F} = Gm_1m_2/s^2 = \mathbf{F} = m_2g \tag{3-9}$$

Recall: G *is the universal gravitation constant. In reference to the Earth (with mass constant* m_1 *and average radius* s*) constants are gathered together to give* g, *the earthly gravitation constant,* 9.8 m/sec^2.

Work done on the weight, to raise it to a height h (raising it from some point P_1 to another point P_2, displacing it through a distance s, now called specifically a height: $s = h$) is

$$W = mgh = PE. \tag{3-10}$$

As another example, a pendulum-bob (Figure 3-27) hanging motionless 'has' no *KE*. Raise it to the top of its swing (a distance h above the lowest position). It now 'has' $PE = mgh$ by virtue of the work done on it in raising it against gravity. Since it is motionless at this point it has no *KE*. Release it, and as it swings down, *PE* goes over to *KE*. At the bottom of the swing all the energy is kinetic because here $h = 0$, whence $mgh = 0 = PE$. At the top of the swing, where $v = 0$, $KE = (1/2)mv^2 = 0$. In between, the system displays *PE* and *KE*. When we say 'has' we mean that the system displays this behavior; we do not mean to reify *PE* and *KE*. They are not things, but properties.

Applying these constructs to the interaction of charged particles, and returning to the above illustration at P_1 the particle has potential energy equal to W_1. (The work done by a static charge in raising a charged gold leaf is mgh, where h is an average.) A machine could be devised such that if the particle were released it would do work. If the

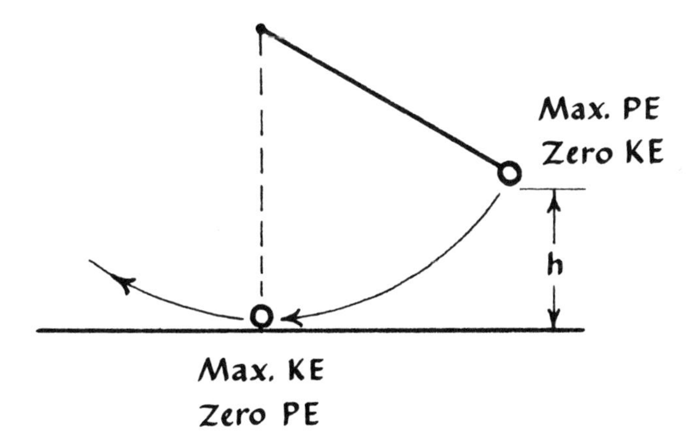

Fig. 3-28 *An idealized pendulum, to show the meaning of potential energy* (PE = mgh) *and kinetic energy* (KE = *1/2*(mv^2).

particles 1 and 2 were of the same sign, repulsion would drive them apart (*PE* might change to *KE*); if the signs were opposite, attraction would bring them together. We now use the device already employed to define E. Since the work done in moving the particle 2 up to 1 is dependent on the quantity of charge q_2 (Equation 3-7) we divide through by q_2. We then obtain the work *per unit charge*, the *PE* per unit charge. This is named the 'potential.' It is conceived as a property of the field at that point. It is a construct, designated by its own symbol 'V', for convenience. The units (joules per coulomb) are named 'volts.'

$$V = W/q_2 = Kq_1/s \tag{3-11}$$

From Equation (3-5) may be derived a useful relation:

$$V = Es \tag{3-12}$$

The potential that drives electrons through a household electrical system is about 110 V.

In all this discussion the charges are conceived to occupy very small regions of space (points, or particles) or centers of charge, and q_2 is taken to have so small a magnitude that it does not produce much distortion in the force-field of particle 1. These idealizations match what is found in practice quite well, which justifies their use. Also, the mathematical manipulations are much simplified by using tiny particles because, for example, s can be measured exactly since we know where the center of the little particle is, and thus $\mathbf{F}$ can be determined without having to find a mean value of $\mathbf{F}$ and s, over a large or irregularly shaped object if such were being considered. Of course, engineers may have to deal with more complicated problems, and have to be more sophisticated than we need to be. The principles remain unchanged.

3.12 Current electricity

Charges in motion are currents. They move from places of higher to places of lower potential. A charge on an electroscope, or insulated particle, has nowhere to go except as it slowly leaks off to the air. A (−) charge on a pole of a dry-cell will pass to the other pole only if a conducting wire connects them (except for small internal leakages). Then electrons flow from the minus to the plus pole. This is *direct current* for it flows one way only. The current that comes into our homes is usually *alternating current* for it changes direction, alternately going one then the other way. This change through the complete cycle usually occurs 60 times per second. In this chapter we have been concerned only with direct currents.

In Figure 3-29 is shown the circuit of a 3-cell flashlight with the the current 'on.' If the switch were open a state of unstable equilibrium

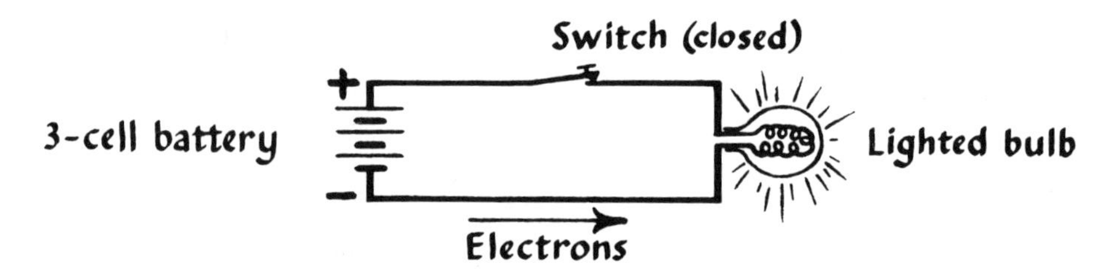

Fig. 3-29 *Diagram of a simple 3-cell flashlight.* (*Each cell contributes 1 1/2 volts of potential.*)

would exist with electrons piled up on one pole (here, the key) and withdrawn from the other under the drive of the battery. No current would flow through the wires. The mobile electrons would be in rapid random motion within the wire. When the switch is closed a potential (4 1/2 volts in the illustration) is impressed across the circuit from pole to pole of the battery, and electrons flow in a directed way. The current does not reach its maximum flow at once, for the inertia of the electrons has to be overcome as well as the resistance of the metal to their motion, but within a very small fraction of a second the current reaches its maximum value. The filament of the bulb becomes incandescent. Steady flow of current continues as long as the battery maintains the same voltage. The amount of electricity that passes is q, and the rate of flow is q/t in amperes. One ampere (A) is one coulomb per second.

To draw all these considerations together, a current of electricity is measured as a flow of q coulombs per sec, or I amps. This occurs under a driving potential difference V. In Section 3.11 we defined mechanical work as $W = \mathbf{F}s$. Electrical work may be defined as qV

(Equation 3-11); then electrical work per unit of time, named 'power,' P, is defined:

$$P = Vq/t = VI \qquad (3\text{-}13)$$

The unit of power is the ampere-volt, which is given its own name, the 'watt.' One watt is equal to one joule per second. A 40-watt light bulb operating in a 110 V circuit uses $40/110 = 0.36$ amps of current. The householder and industrialist buy electric power in kilowatts and watts.

Charge on the electron 3.13

We have had to complete a long excursion through the C-field of physical theory in order to justify our notion of 'electron' and to arrive at the point of directly measuring the amount of charge on this ultimate particle of electricity. ('Ultimate,' that is, at the present time.) We now return to some of the observations that convinced scientists that electrons exist as measurable entities. The architect of that conviction was Robert A. Millikan, a remarkably inventive, creative experimental physicist. He first determined the charge of an electron relative to the coulomb. This was done in what has become famous as "Millikan's oil-drop experiment."

By means of an atomizer A (Figure 3-30) a spray of minute droplets of non-volatile oil is produced above a metal plate B. This contains

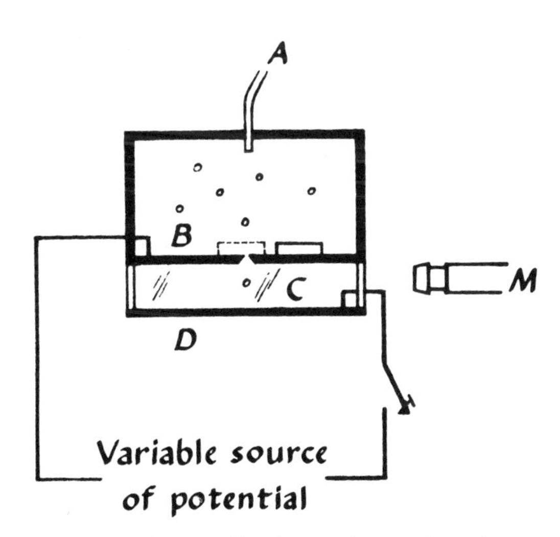

Fig. 3-30 *Millikan's oil-drop apparatus (schematic). An atomizer at A produces a few minute, charged oil-droplets in the chamber. These settle slowly and one of them is allowed to pass through a hole in the metal plate B. The hole is immediately covered with a movable block. The charged drop is manipulated in the space C between the oppositely charged plates B and D, and its motion is observed and measured by means of a microscope optical system M. The plates B and D are connected through a source of high potential that can be varied in magnitude, reversed in sign, or removed from the plates.*

a pinhole over which a small cover can be slid. The plate B is connected to one pole of a source of potential. Below and parallel to B is another plate, D, connected through a switch to the other pole. The source is so designed that the potential can be varied in magnitude, reversed in polarity, or completely removed from the plates. With the pinhole open, a single oil drop is allowed to drift down into the chamber C between the plates. This chamber has windows at each end, which insulate B from D. The cover is then slid over the pinhole. Ingenious illumination of the chamber makes the oil drop visible in a microscope, M, as a bright star. It falls slowly under gravity. Cross hairs in the microscope enable the distance of fall to be measured. The speed at which the drop falls freely is related to its radius and mass and to the viscosity of air in the chamber by a law which enables its size to be measured.

When a drop is produced by the shearing effect of the atomizer on the oil, it is usually charged in the process, as a resinous rod is charged when rubbed with fur. Suppose the drop arriving in C is charged ($-$). It falls under the force of gravity. The upper plate is now charged ($+$). This attracts the drop upward. By adjusting the potential on B a force $\mathbf{F} = qE$ can be exerted on the drop to just balance the gravitational force, $\mathbf{F} = mg$. (In some experiments the drop was slightly accelerated by leaving one of the forces unbalanced.)

Electrostatic force upward = gravitational force downward—

$$qE = mg.$$

The gravitational force remains constant. The electrostatic force depends on q, which may be written $q = ne$, where e is the unit electron charge and n is some whole number. If n decreases, E must be increased proportionately, and *vice versa*, if the droplet is to be held stationary. The droplet frequently loses or gains electrons from the air, or from bursts of X-rays or other energetic radiation that may be shot into the chamber. At each change of charge it moves upward or downward and E must be adjusted in steps of ne. After making an extraordinary number of such adjustments, on a large number of different drops, Millikan took the smallest as $n = 1$. There are no smaller steps and all the others were divisible by this. The electronic charge was calculated as 1.60×10^{-19} C.

3.14 Caveat

At this point, it would seem, any reasonable person would have to agree that electrons 'exist' in some defensible sense of the word. After all, we have a tremendous body of data at the P-plane that is nicely rationalized and ordered by a set of constructs that fit together in a web of theory in the C-field. What more does one want? Further, a huge industrial complex exists fruitfully on the basis of these perceptions and constructs as they are embodied in hardware. However, the conservative person, while accepting light from his electric bulb, and writing on his electric typewriter the thoughts generated in his mind and electrically transmitted to his fingertips, has been here before. He is cultured, and thus has a sense of history; and he knows that the four million books in the library embalm untold theories. So he holds open a conceptual door for . . . what? In any event, he is not satisfied until he knows more about this ubiquitous electron that pervades all matter. The following chapters develop further information about the electron. Inexhaustible Nature even then remains barely touched.

Summary 3.15

Two kinds of charges, (+) and (−), are intimately concerned with the structure of matter. They are ultimately associated with electrons (−) and protons (+). These particles give electricity its particulate nature. Solid materials vary widely in their ability to conduct electricity. A current, a stream of electrons, flows readily through conductors; not at all, or poorly, through insulators. Like charges repel and unlike attract. This behavior is pictured by means of lines of force, and is quantitatively expressed by Coulomb's law of electrostatic attraction and repulsion. The charged particle is conceived to show a field intensity at any point to which another charged particle is brought. To bring this other particle to that point requires positive work if the particle has a like charge, and negative work if the charges are unlike. From this construct arises the concept of potential, and potential energy at a point in the neighborhood of the particle. Under the drive of this potential a particle will move if set free. A stream of particles is a current of electricity, the familiar current in an electric wire being a stream of electrons. It flows under the drive of a potential difference between the ends of the wire. Current electricity may be direct or alternating.

As far as this Chapter goes, electric particles obey Newton's laws. A charge is recognized by its ability to exert a force and cause an acceleration, for example, of the leaves of an electroscope. The gold leaf of an electroscope may be raised by this force and given potential energy.

By applying these ideas in a creative experiment Millikan was able to measure the charge on a single electron.

Exercises

3.1 When we charge a glass rod by rubbing it with silk, and then touch it to a pith-ball, we say that the ball is positively electrified. How do we *really* know this?

3.2 In Figure 3-28 the field intensity at P is $E_p = \mathbf{F}/q_2$. What is the field intensity at q_1?

3.3 A particle, 1, with charge $q_1 = +2 \times 10^{-7}$ coul is fixed 20 cm away from another fixed particle, 2, of $q_2 = +4 \times 10^{-7}$ coul. At what point on the line between them must a particle, 3, of charge $q_3 = -1 \times 10^{-8}$ coul be placed so that it is attracted equally by both 1 and 2?

3.4 A certain electrical appliance draws three amperes of current. a) How much charge passes through per second? b) How many electrons flow past a point in the circuit per second?

3.5 100 meters of #10 copper wire contain about 8×10^{23} mobile electrons, allowing about 2 mobile electrons per copper atom. If one ampere of current flows through the wire, how long, on the average, would it take for electrons entering one end to emerge at the other end, other things being equal? (The current continues to flow during this period.)

3.6 Three strings are tied together at a knot K, and the forces shown in the figure are applied to each string. a) In what direction will the knot start to move? b) Retaining the "downward" force at 500 new, what must the other forces be to keep the knot in equilibrium with these angles (*i.e.*, not moving)?

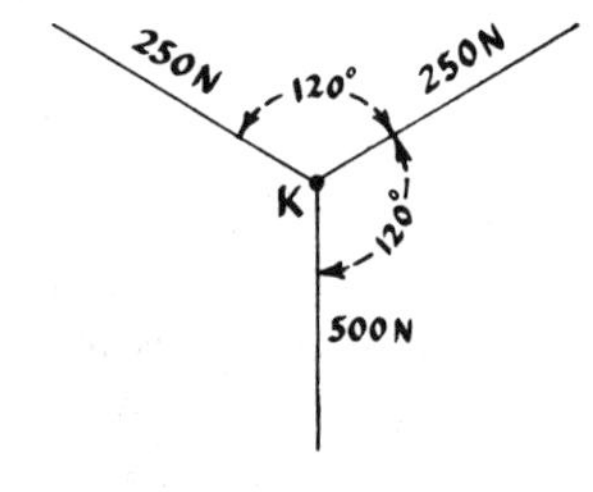

3.7 Suppose the string in problem 3-17 were hung over light fixed pulleys, and weights as shown were attached. What would the angles between the strings be if the resultant force on the knot were zero?

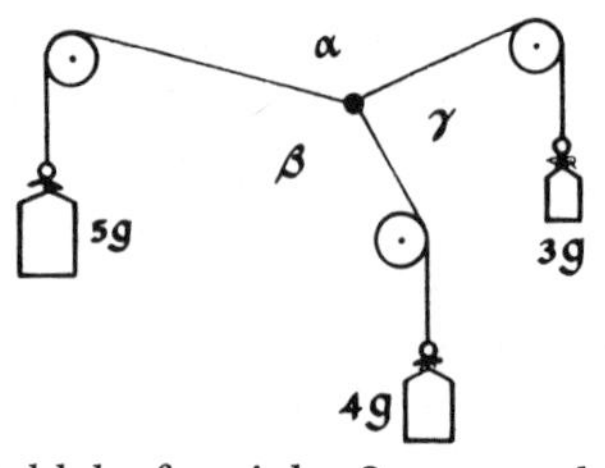

3.8 In an electroscope (Fig. 3-1b) each gold leaf weighs 2 mg, and when charged, each stands out at an angle of 30° from the vertical. Each leaf is 3 cm long from the point of attachment, and its center of gravity is half-way along it. What potential energy is given to each leaf? What assumptions have we made in the calculation?

Answers to Exercises

3.1 We defined it that way, following Franklin.

3.2 It is $E = \mathbf{F}/q_1$.

3.3 $\mathbf{F}_{1,3} = \mathbf{F}_{2,3}$ at equilibrium. Thus, $q_1q_3/(r_{1,3})^2 = q_2q_3/(r_{2,3})^2$; 0,067 m from particle 1.

3.4 a) 3 coul/sec. b) Approximately 19×10^{18} electrons per sec.
3.5 About a day-and-a-half.
3.6 a) In the direction of the 500 N force. b) Make all forces 500 N.
3.7 The resultant must sum to zero. $\alpha = 127°$; $\beta = 143°$; $\gamma = 90°$.
3.8 The potential energy of each leaf is $mgh = 3.9 \times 10^{-8}$ joule. (If you have forgotten how to determine h, see the Appendix on trigonometry.) It is assumed, for one thing, that no energy is expended to bend the leaf.

Bibliography

Vectors are discussed lucidly in Irving Adler, *The New Mathematics*. Signet Science Library, The New American Library, New York, 1960 (John Day Co., 1958).

Robert A. Millikan, *The Electron*, with an introduction by J. W. M. DuMond. University of Chicago Press, Chicago, 1963. This is a facsimile of the 1917 edition. The introduction by DuMond, dated 1963, is superb. The book makes interesting reading.

Some of Coulomb's experiments are described, along with his original data, in M. H. Shamos, *Great Experiments in Physics*, Holt, New York, 1960.

Chapter 4

4.1 What is observed
4.2 Magnetic lines of force
4.3 Moving charges
4.4 Magnetic flux density
4.5 Geometry of uniform circular motion
4.6 Acceleration in uniform circular motion
4.7 The ratio q/m for the electron
4.8 The cyclotron
4.9 The electron volt
4.10 Induced current
4.11 Summary

Magnetism ~ 4

Nearly everyone must have played with a magnet at some time and must have observed that it attracts some things—an iron nail, an opposite magnetic pole—and is unaffected by nearly everything else. These behaviors of magnets have been known for a long time, having been observed with natural magnetic rocks (lodestones) thousands of years ago. That magnetism and electricity are closely related has been known for several hundred years in an empirical way. Electric motors and generators, telegraphy and telephony, were developed out of this empirical knowledge. It has been only within this century, however, since the advent of relativity, that the relationship between magnetism and electricity has been clarified at a deep level. We need to study, or review, some aspects of magnetism to prepare for connecting it with electricity, unifying the two fields. We also need fundamental knowledge of some aspects of magnetism in order to understand atomic structure, and how the great atom-smashing machines work. As we get into this subject we shall see familiar patterns appear. Look for them, for not all will be painted too explicitly. We intentionally leave room for discovery.

What is observed 4.1

Take a straight copper wire and connect the ends through a switch to a strong battery. With the switch open, bring up to the wire a charged electroscope. Bring up also a small compass, and place it below the wire, as in Figure 4-1. The compass points north in the Earth's magnetic field. Neither instrument gives any indication that the presence of the wire has any effect on it. That is, no unbalanced forces due to static charge or magnetic interaction with the wire are present. We

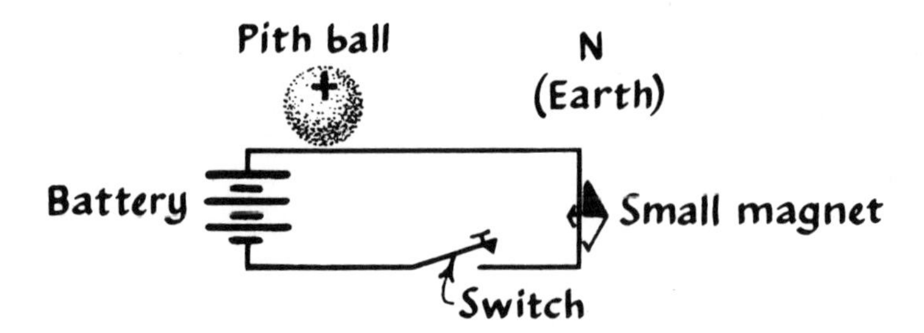

Fig. 4-1 *Diagram of a simple circuit in the open state. Next to the straight wire is suspended a pith-ball, and below the wire lies a compass, the undisturbed needle of which points from South to North in the Earth's magnetic field.*

know that there are atomic charges in the wire, but they are exactly balanced, (+) against (−). The compass needle points north and south in its normal manner.

Now close the switch and let current flow along the wire. A marked change occurs. The electroscope remains unaffected, but the compass needle swings round and tends to line itself up at *right angles* to the direction of the current (Figure 4-2). If we transpose the battery con-

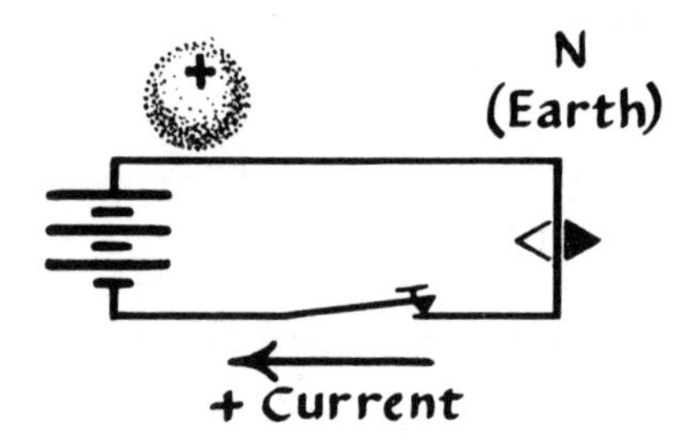

Fig. 4-2 *The same circuit as Figure 4-1, except now closed. (+) Current flows from right to left. The needle of the compass has swung to a point at right angles to the wire. The pith-ball is virtually unaffected.*

nections so that current flows in the opposite direction, the magnet, still tending to line up at right angles to the flow, points in a direction opposite to that previously. The compass needle clearly indicates the presence of a magnetic effect associated with the flowing current. It gives evidence of the appearance of a new force which is able to swing the magnetic needle into a new position in spite of the Earth's field.

The small compass needle is the test instrument that performs for magnetism what the electroscope does for static charges. It consists of a light permanent magnet, straight and pointed on the ends. Its jewelled bearing is balanced on a pointed rod attached to a firm base (Figure 4-3). In the absence of other magnets except the Earth, it points toward the north. The end that points in that direction is marked N and may be painted red.

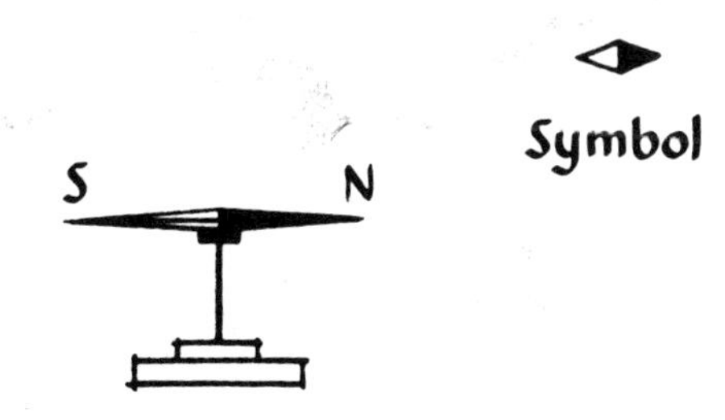

Fig. 4-3 *A simple compass needle on a stand. The pivot is jeweled.*

Magnetic lines of force 4.2

If a small compass is used to explore the neighborhood of a bar magnet, its behavior is that its S end is oriented to point towards the N end of the bar magnet, and the N end of the compass toward the S end of the bar magnet: unlike poles attract (Figure 4-4). If N is brought

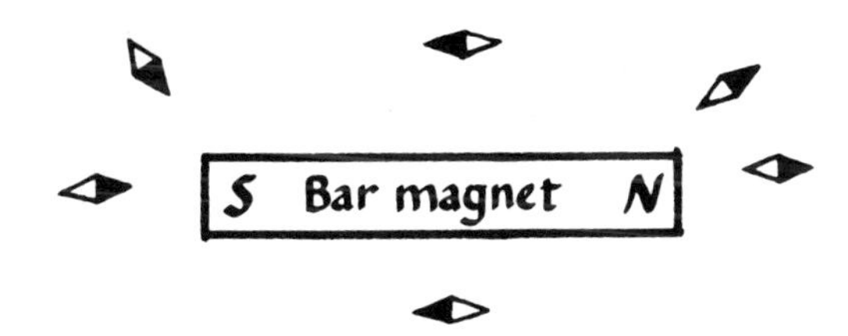

Fig. 4-4 *The orientations taken by the compass needle when it is used to explore the neighborhood of a permanent bar magnet.*

towards N, or S towards S, repulsion behavior is observed. If a piece of glass is laid above the magnet, and sprinkled with magnetized iron filings and then tapped to allow the filings to become oriented, these line up as shown in Figure 4-5, giving a more complete visual image of the pattern suggested by the compass needle. Michael Faraday conceived of lines of force that might be visualized to emerge from the N pole, bend around through space and enter the S pole, and to travel *within* the metal from S to N, forming closed loops. This can be visualized by putting arrowheads on the lines of oriented filings.

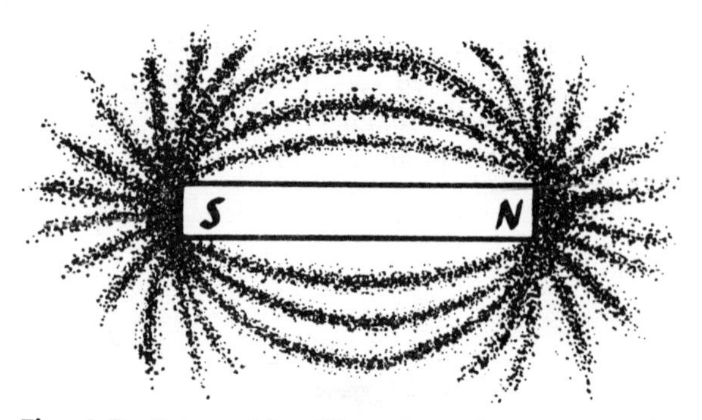

Fig. 4-5 *Pattern of iron filings about a bar magnet, to show pattern of lines of force, defined to emerge from* N *and enter at* S.

A law of force analogous to Coulomb's Law can be written for the magnet, but its use is not without difficulties. One problem is that an isolated pole has never been obtained: there is no N unassociated with an S, as there may be with electrostatically charged particles. For example, the proton and the electron, each with its own charge; or other charged atomic or molecular particles (ions) may be obtained separately, by suitable means. Thus the field in the neighborhood of the pole is not uniformly radial, as that in the neighborhood of a charge (Figure 3-14), but curved because of the always-present opposite pole. We need not go into this very deeply, since all we need, really, for subsequent use is the mathematical relationship itself. What we do need to clarify is the north and south nomenclature. Suppose we take a magnetized compass needle similar to that in Figure 4-3, but unmarked. We notice that in the open air, far from buildings and power-lines, it lines up with one end pointing *Geographic* North—as we can judge because we know this from the position of the North Star. If we push the needle around, then let it go it lines itself up again in the same direction. Obviously, there is a force on it; a force which we recognize as arising from the magnetic field of the Earth. We mark the end of the compass needle that points towards the Geographic North 'N,' and paint it red for easy visualization. If we bring this up to another compass needle which has been marked as a result of the same sequence of operations, we find that the compass needles when close to each other, align in such a way that we have to conclude that N repels N, and attracts S, while S repels S, as we saw above. The nomenclatural die was cast when we marked 'N' on the compass needle that points to the Geographic North. Notice that we have been careful to speak of this pole as an 'N' pole. Since it points toward Geographic North, the magnetic pole there must be 'S.' (It is interesting that the point where the lines of force enter the Earth in the north has moved around a good deal over geologic ages. This can be inferred because when molten rock, containing atomic or molecular-size magnets, cooled and set, the magnets mostly pointed to the Geographic North of that time. Rocks of the same age in different parts of the Earth indicate the same north, but not the same as found at another age. The magnetic pole has wandered over thousands of miles.)

Moving charges 4.3

If we explore the neighborhood of a current-carrying wire with a compass and plot the magnetic lines of force we find that they form concentric cylinders, shown in cross-section in Figure 4-6. The direc-

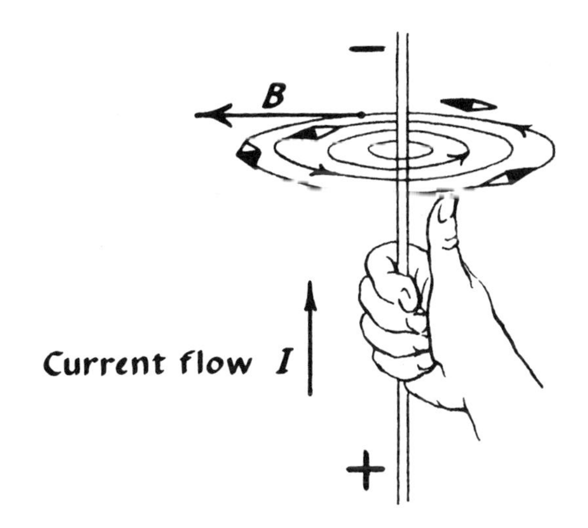

Fig. 4-6 *The pattern of lines of force about a current-carrying wire to show the right hand rule for remembering. The magnetic flux vector* B *is tangent to the lines of force at any point outside of the wire.*

tion taken by the compass needle may be remembered by imagining the wire grasped with the right hand, the thumb pointing in the direction of the flow of *positive* current, and the fingers curled about the wire. This is a convention. Accept it. We know that the current consists of a flow of negative electrons, but when the convention was set up it was thought that the current consisted of the flow of a (+) fluid, like water in a pipe. Conventions die hard. We rationalize the convention by thinking that if electrons flow in one direction, that is from (−) to (+) poles outside the battery, we could *think* equally well of a '(+) current' flowing in the opposite direction. The fingers point in the direction of the lines of force, Figure 4-6. In this figure are shown the directions of the (+) current and of the magnetic field strength. This latter is directed tangential to the line of force at the point, as with electrostatic field strength.

In the permanent magnet the source of the lines of force is a complicated interaction of the lines of force associated with moving electrons and the (+) charged nuclei of the metallic atoms of the magnet. The electrons behave as though they are spherical distributions of charge spinning about an axis. The atoms behave as though there are present negative charges moving around the positively charged nucleus (Figure 4-7). The positively charged nuclei behave as though they have

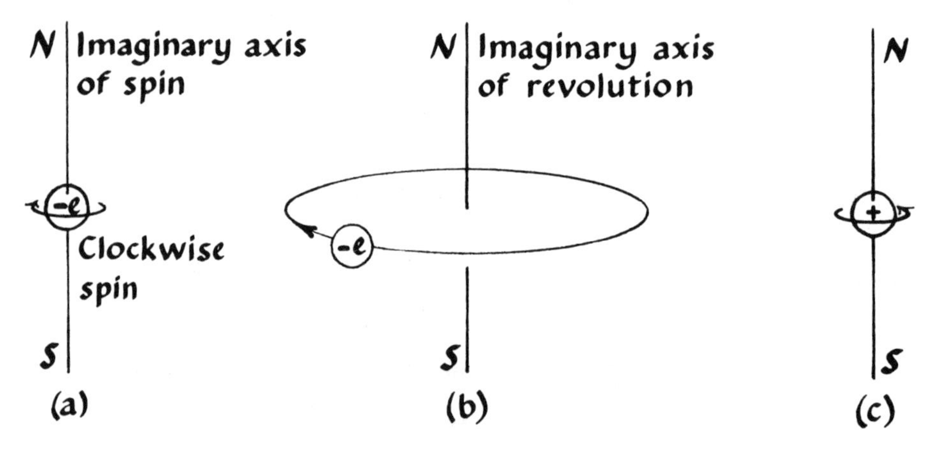

Fig. 4-7 *Some ways of imagining electronic spin* (a); *electronic movement about a nucleus* (b); *and nuclear spin* (c), *with the associated magnetic poles.*

an intrinsic spin. In the process of making a permanent magnet (magnetizing the shaped metal) the net fields of these moving atomic charges are lined up, more or less in parallel and more or fewer in number depending on the process, to give a stronger or weaker magnet. The same sort of thing occurred with the cooling rocks, referred to above.

4.4 Magnetic flux density

The magnitude of the force at a point outside of a current-carrying wire depends on the magnitude of the current in an interesting way. Consider a permanent magnet between the poles of which we place a wire (Figure 4-8) through which a current may be passed. On passing a quantity of positive current (coulombs per second) measured as I

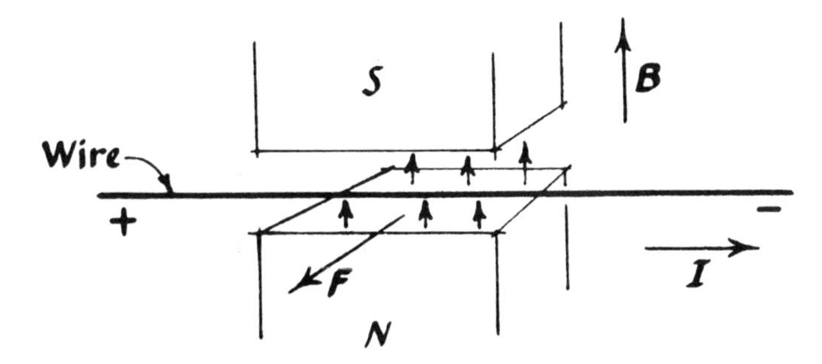

Fig. 4-8 *Diagram to show the direction of the force* F *on a wire carrying a current* I *in a field of magnitude* B.

amperes through the length of the wire in the magnetic field measured as l meters, from left to right in the figure, the wire experiences a force **F** tending to throw it out of the field in a direction "out of the paper." If the current is reversed, the force is directed oppositely. If lI is at right angles to the magnetic lines of force, then the force on the wire is found to be at right angles to *both* directions. In brief, if the thumb of the *right* hand is extended, pointing in the direction of the positive current, and the extended first finger is pointed in the direction of the lines of force, *i.e.*, from N toward S between the poles of the magnet, then the second finger, held at right angles to the plane of the other two, will point in the direction of the force experienced by the wire or other element carrying the current (Figure 4-9).

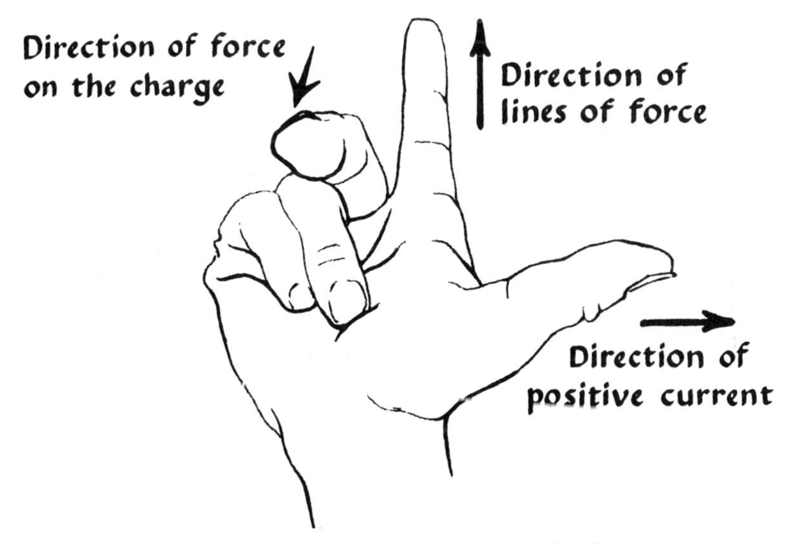

Fig. 4-9 *The right hand rule for remembering the directions in Figure 4-8.*

Here is an interesting situation: a current flows through a wire in the X direction, as in Figure 4-8, in a region of magnetic lines of force directed in the Y direction, and it experiences a force in the Z direction at right angles to the directions of generating current and force field (Figure 4-9). How is this conceptualized? (It has been pointed out that the relation 'three right angles' between current, magnetic lines, and resultant force gives an electromagnetic embodiment to the Euclidean concept of three-dimensional space.)

For reasons of convenience a term B, called the flux density, is defined. It is the 'quantity of magnetism per unit area' between the poles of a magnet, and is measured in webers per square meter. It is used in the following way. Suppose that a wire lies perpendicular to, and in, the lines of force of a magnet of flux density B, and that the length of the wire in the effective magnetic field is l meters. If a current of I amperes flows, the wire experiences a force **F** described in magnitude by

$$\mathbf{F} = |lIB| \tag{4-1}$$

What we are saying here is that the force **F** (which can be measured, of course) is proportional to the current through the wire (I) and the length of wire in the magnetic field (l). The constant of proportionality is B. B is a vector, directed from N toward S. Alternatively, this important equation makes it possible to calculate the magnitude of the force if we know l, I, and B. As it stands it says nothing about direction. But lI has a direction, B has a direction at right angles to it, and **F** has a direction at right angles to both. lI and B are vectors, which must be *multiplied*. In this kind of multiplication it makes a difference which term is multiplied into the other. The rule based on observation, in the first place, says that if lI is multiplied into B, then the direction of **F**, the 'cross product,' is such as a right-handed screw would advance if turned in the sense lI into B (Figure 4-10). In other words,

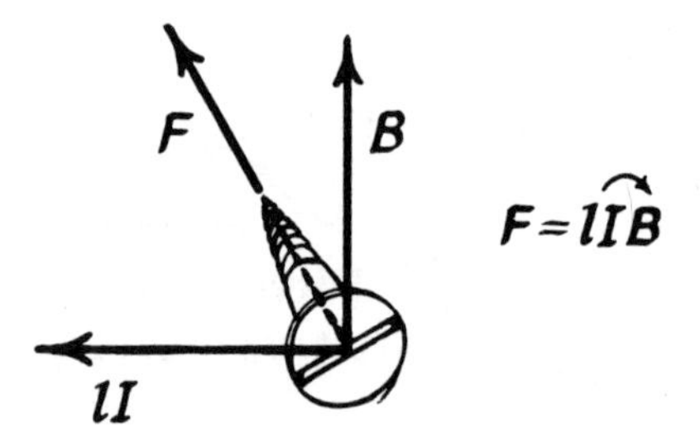

Fig. 4-10 *Definition of the direction of the cross-product* F *obtained by vector multiplication of* lI *into* B.

clockwise turning of the direction of lI into that of B, with the observer facing the clock, produces a cross product directed away from the observer: as though he were driving a right-handed screw. (If the multiplication were B into lI, the force would, of course, be oppositely directed.) To show the direction of the multiplication the force equation is written by convention:

$$\mathbf{F} = \overrightarrow{lIB} \tag{4-2}$$

Sometimes students have difficulty understanding what is being done in an equation such as (4-2). In the first place, why are l, I, and B *multiplied* instead of, say, being added. The answer takes us back to the discussion of a protocol in Section 2.7. It is the observed case (the way Nature is), given the data that a force in this direction is experienced by a current-carrying wire in a magnetic field with the directions shown, that the functional relation between the variables can be simplified into this equation. Put in another way, the observed behaviors (P-plane) can be described in simple mathematical form (C-field) by means of this equation. You might dredge out of your memory of high school algebra the concept of commutation. Recall that in ordinary algebra $(a + b + c) = (b + c + a) = (c + a + b)$, and so forth. The order of adding the terms doesn't change the sum. Similarly, $(a \times b) = (b \times a)$. The terms are said to commute. But in some kinds of operations it makes a difference in what order the manipulations are carried out. Here A $\times$ B does not give the same result as B $\times$ A. As a homely example, consider the operations of going through a closed door: you open it first, then go through; not the other way around. The order of carrying out the operations makes a difference: the terms do not commute. Equation (4-2) says that in order to come out with the correct direction of $\mathbf{F}$ *as observed in Nature*, you must multiply lI vector (an area) into B, as shown in the figures.

Consider the behavior of an electron which is fired between the poles of a magnet, the field strength of which is 6×10^{-3} weber/meters^2. Suppose that the electron is travelling with a velocity of 5×10^4 meters/sec. (We made up these reasonable numbers as illustrations.) What force does it experience? As with all problems of this nature, we sketch and label a diagram, making it somewhat like one we are already familiar with as a kind of internal check, see Figure 4-11. We direct the electron from the right, thus imagining a positive current from the left as in Figure 4-8. The electron is free to move in any direction, and the force on it causes it to accelerate as shown in the figure. It thus describes a curved path. We saw that: $\mathbf{F} = lIB$. Also, $I = q/t$, and $l/t = v$, so we can write into terms of the data given (as a result of algebraic manipulation):

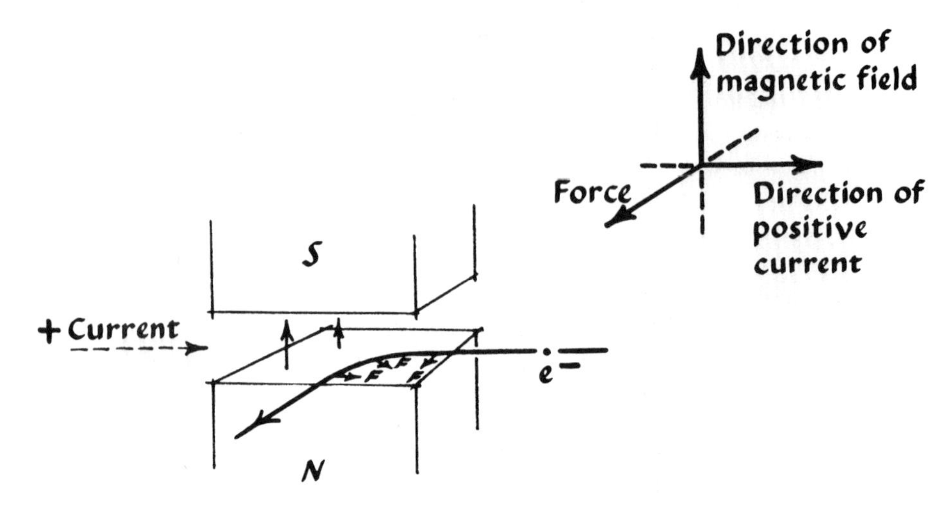

Fig. 4-11 *The trajectory of a negative particle moving through the field of a magnet.* F *is a centripetal force.*

$$\mathbf{F} = l\overset{\frown}{qB}/t = qvB$$

The charge on the electron being 1.6×10^{-19} coul,

$$\mathbf{F} = 1.6 \times 1^{-19} \cdot 5 \times 10^{4} \cdot 6 \times 10^{-3} = 4.8 \times 10^{-17}\ \text{N}$$

It is found that the *direction* of motion of the electron is changed in the magnetic field, though its speed is not affected. The field force F acts always at a *right angle to the direction of motion,* thus the electron is "pulled" around in an arc. But because it acts at a right angle the force does no work in a direction that would change the speed (Equation 3-6). We must recall or learn at this point the ways in which circular motion and its consequences are conceptualized.

Consider a body moving at constant speed, for example a car travelling at 25 mph. If its direction of motion alters, as in going around a curve, this is a change in its velocity even though its speed may remain constant. This change in velocity is an acceleration and must be the result of an unbalanced force acting upon the body. (Recall Newton. In this case the force is between the tires and the road surface.) This type of acceleration, which is associated with *change in direction* of motion is named centripetal acceleration, a_c. Centripetal acceleration may be constant or changing. We shall be concerned only with the former. The analysis of centripetal acceleration must be preceded by a discussion of the geometry of uniform circular motion. These concepts will be used repeatedly in our discussions of atomic structure and the great atom-smashing machines.

Geometry of uniform circular motion 4.5

We think of a circle of radius r (Figure 4-12), circumference C, and s the arc subtended by an angle θ. $C = 2\pi r$; and $s = r\theta$, when θ is measured in *radians*. A radian is the angle subtended at the center of the circle by an arc the length of which is equal to the radius of the circle (Figure 4-12). The circumference of the circle being $2\pi r$, there

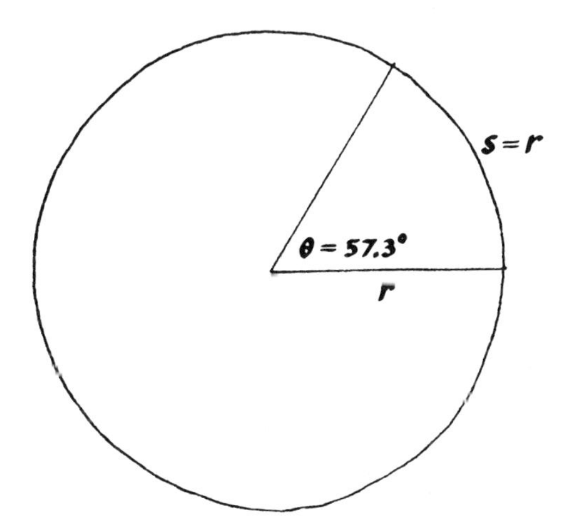

Fig. 4-12 *Definition of the radian:* $\theta = 1$ *radian for* s = r; *the subtended arc* s *is equal to the radius of the circle* r. *For* $\theta = 1$ *radian,* s/r *always equals 1 for a circle of any radius* r.

are 2π radians in 360°. (Thus there are $360/6.28 = 57.3$ degrees in one radian.) The radian is dimensionless—a pure number—for by its definition:

$$\theta = s/r = \text{a length/a length}$$

The name 'radian' is always used to ensure understanding that the angle is measured as a pure number. (π is also a pure number, as seen by the fact that $2\pi r$ is a length, and πr^2 is an area.) Uniform circular motion is the simplest kind of rotational motion in a plane. This is the motion of a particle at constant speed in a circular path about an axis at the center of the circle. To measure the speed of the particle, a pair of Cartesian axes are placed (as a matter of convenience) with their origin at the center of the circle. Certain definitions enable us to conceptualize the behavior of the particle. (Once again we are moving from the particular particle (P-plane) to idealized behavior of *any* particle—C-field.) Since the speed of the particle is constant at $|v|$ m/sec, it moves a distance Δs in Δt, describing an arc which subtends an angle $\Delta\theta$, always measured with reference to the $^+$X axis. The direction of motion is taken as positive if counter clockwise. The time required for the particle to travel the circumference of the circle is the period of rotation, T. In T (sec) the radius r sweeps out 2π radians. The number of periods per second is the frequency of rotation, n, or:

$$n = 1/T \text{ (revolutions per second, or 'per sec')} \qquad (4\text{-}3)$$

4.6 Acceleration in uniform circular motion

If one were to whirl an object at the end of a string and at some instant let go of the string the object would fly off in a direction tangent to the circle in which it had been whirled. This behavior is consonant with (predicted by) Newton's first law of motion: a force, acting on the object, accelerating it as it constantly changes its direction, suddenly ceases to act, and the object continues in the direction of motion with unchanged velocity. The force must have been directed toward the center of the circle of rotation. How is this behavior grasped conceptually and given quantitative expression?

Figure 4-13 shows a particle, supposed in uniform circular motion at a distance r from the center O, of the circle. This particle might be a whirling stone on the end of a string, or an electron whirling around an atomic nucleus at the end of an electrical leash. At any instant, say t, the particle is at P_1. Its velocity may be represented by a vector v_1 of magnitude $|v_1|$ and with a direction tangent to the circular path at P_1, that is perpendicular to the radius from O to P_1.

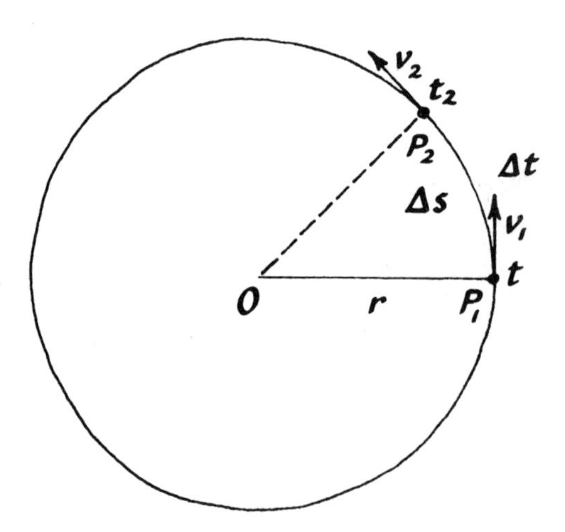

Fig. 4-13 *For the description of circular motion:* $\Delta t = (t_2 - t_1)$. *In* Δt, *the arc* Δs *is swept out.*

During a certain period Δt, the particle moves to P_2, the time at P_2 being t_2. ($\Delta t = (t_2 - t_1)$.) Here the velocity is represented by the vector v_2. Its magnitude is $|v_2| = |v_1|$, but its direction, tangent now at P_2, is clearly changed from its direction at P_1. By definition, then, the particle is undergoing acceleration: the acceleration, as defined for translational motion,

$$\bar{a} = \Delta v/\Delta t = (v_2 - v_1)/(t_2 - t_1)$$

is applicable also to rotational motion, when vectors are subtracted. That is,

$$\bar{a}_c = \Delta v/\Delta t = (v_2 - v_1)/(t_2 - t_1)$$

What is the magnitude and direction of the acceleration?

Consider Figure 4-14. In moving along an arc of length s, the velocity vectors v_1 and v_2, of equal length, have different directions.

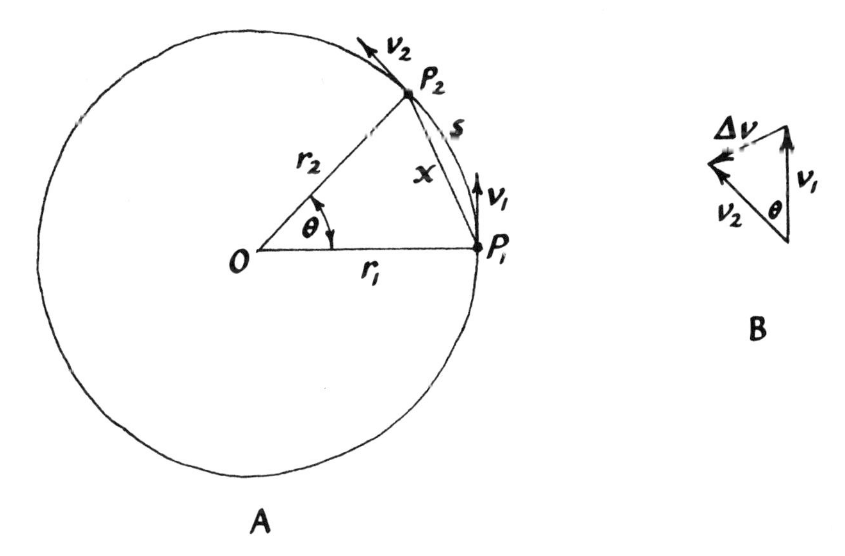

Fig. 4-14 *Derivation of the magnitude and direction of centripetal acceleration in terms of vectors.* (a), *the trajectory of the particle;* (b), *the subtraction of vectors.*

They are perpendicular to the corresponding radius vectors r_1 and r_2, and make the angle θ to each other. There are many ways of demonstrating that the angle between v_1 and v_2 is θ. Here is one way (Figure 4-15). Extend v_1 and v_2 in Figure 4-14 on the circle. They meet at z to form an isosceles triangle with base x. Bisect the angle θ, bisecting x at y, and extend this to the apex of the new triangle at z. The angle $\alpha = 90 - 1/2\,\theta$, therefore the angle $yP_1z = 1/2\,\theta$, since $OP_1z = 90°$. Then $yzP_1 = 90° - 1/2\,\theta$. This requires that P_2zP_1 be $180° - \theta$, whence

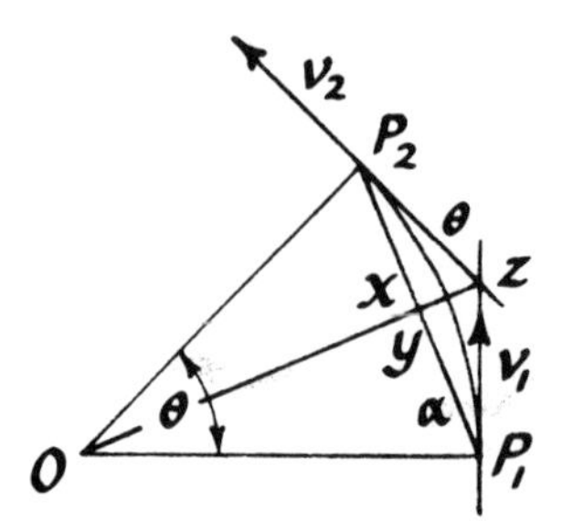

Fig. 4-15 *Demonstration that the angle between* $\mathbf{v}_1$ *and* $\mathbf{v}_2$ *is* θ.

it follows that the angle between v_1 and $v_2 = \theta$, for as we know the sum of the interior angles of any triangle is 180°.

The difference between the two vectors, Δv, is given by vector subtraction of v_1 from v_2. The velocity-vector triangle, with angle θ, and sides $v_1 v_2$, and Δv, is similar to the triangle with angle θ, sides r_1, r_2, and x (the chord from P_1 to P_2). The *ratios* of lengths of corresponding sides of similar triangles are equal; thus we can transfer our attention to the ratios of the magnitudes:

$$\Delta v/v = x/r$$

or, $$\Delta v = xv/r$$

The average acceleration is then (dividing both sides by Δt):

$$\Delta v/\Delta t = xv/r\Delta t$$

This is a re-statement of the equation for $\bar{a}$ given above, and is the result of algebraic manipulation on the way to devising an equation for the centripetal acceleration. The objective is to let the mathematics express the essence of the problem as it is analyzed logically. If you can follow the internal logic of what has been done here, you should have no insuperable trouble with mathematics. If you cannot, then try arguing it out with a friend, or go to your instructor, *but gain control for certain at this point.*

The centripetal acceleration of the particle is a_c at every point on the perimeter of the circle, since the speed is constant and thus the direction is changing in a constant manner. To calculate it the following device is used. We saw that the average acceleration is $\Delta v/\Delta t = xv/r\Delta t$. This we arrived at by algebra and geometry, using the vector formulation in Figure 4-14b. Now we introduce the concept of a 'limit.' If we "make" Δt smaller, which we are at liberty to imagine without doing violence to the physical situation, we can see that in a shorter period the particle has not moved as far, thus P_2 is closer to P_1, θ is now a smaller angle, x is shorter, and Δv is smaller in magnitude.

So let us make Δt still smaller, and smaller yet, letting Δt get as close as we wish to zero. This is symbolized by $\Delta t \to 0$, which is read, "Δt approaches zero." We ask ourselves, what is the limit of the expression $\Delta v/\Delta t$ as Δt approaches 0? We look at the Figure 4-14a. As x becomes smaller because P_2 is getting closer to P_1 in our imaginary measurement, it becomes closer and closer in length to the arc s, and when Δt is small enough, x is indistinguishable from s. This discovery solves half our problem for now we can replace x by s in the limit. The expression for 'the limit as Δt approaches zero of' is written '$\lim\limits_{\Delta t \to 0}$'.

$$a_c = \lim_{\Delta t \to 0} \frac{v}{r} \cdot \frac{x}{t}$$

$$= \frac{v}{r} \cdot \frac{s}{t}$$

$$= \frac{v}{r} \cdot v$$

$$= \frac{v^2}{r} \qquad (4\text{-}4)$$

The instantaneous acceleration at the point P_1 is then v^2/r in magnitude. Some students boggle at this: how can you have an *instantaneous* acceleration when by definition $a = \Delta v/\Delta t$! At least two satisfying replies may be given, and the student may have his choice. Notice that $\Delta v/\Delta t$ is a *ratio*, which means that as Δt is taken smaller, Δv must be commensurately smaller, while *a does not change at all* (see Figure 2-9, where v is plotted against t and a is the slope of the line). Then, make Δt as small as you like, even approaching zero, and a remains a finite number without change. So it is quite all right to say that in the limit, as Δt approaches zero, a "is" thus and such. Another way of looking at the matter is this: you say "that zero bothers me" (indeed, it should). Well, then, just make Δt so small that it is as close to zero as makes no difference (because your stop-watch can't even measure it); then, say that that is the value of the 'instantaneous' acceleration a_c. Since the acceleration is constant, either view gives the same numerical result. In any case, *you* have the decision as to how small you demand Δt to be: just "make" it small enough.

The other half of the problem concerns the direction of the acceleration. Look again at Figure 4-14b. Δv is the base of an isosceles triangle. As θ becomes smaller, approaching a limit, the base angles come closer and closer to right angles, and in the limit are indistinguishable from right angles: right angles to the v vectors. In the limit, Δv is perpendicular to the vector v. The direction of a_c is the same as that of Δv, because Δt has no spatial direction. At P, then, where a_c is measured, the direction of the acceleration vector is along the radius toward the center of the circle because it is at 90° to the tangent to the circle. We recognize intuitively that this is correct, having whirled things on the ends of strings and noticed the pull on our hands in the direction of the string. Now you have an example of how the observation (P-plane) may be conceptualized.

4.7 The ratio q/m for the electron

The cathode-ray tube was described in Section 3.4. J. J. Thomson designed one shown in Figure 4-16. It was slightly different from the conventional type described before because the anode A consisted of a

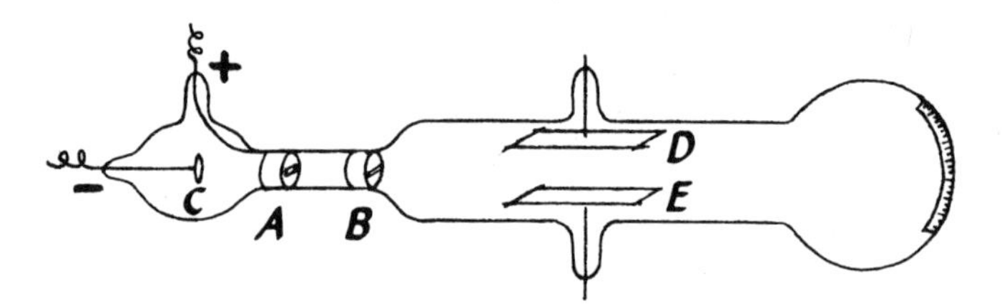

Fig. 4-16 *Thomson's cathode-ray tube.* C, *cathode;* A, *anode;* B, *collimating slot;* D, E, *parallel plates.* [*After J. J. Thomson, "Cathode rays."* Philosophical Magazine V, 44, *293* (*1897*), *Fig. 2, p. 296.*]

metal cylinder with a thin slot through it. Beyond this was another, similar slot, the two being lined up along the axis of an evacuated glass tube and in line with a metal cathode C. When a high potential difference is applied between the anode and cathode, a stream of electrons pours out of the cathode and accelerates to the anode, reaching high velocity. Most of the electrons hit the metal anode but those moving in a straight line along the axis of the tube shoot through the slot. Some are deflected a little by the (+) charged walls of the slot, but they lose little velocity in passing through. Those that are still moving in a straight line pass on through the second slot and emerge as a ribbon of high-speed electrons. Some of them cause the residual gas in the tube to glow (as in a 'neon' tube) by a mechanism we shall shortly explain. The "ribbon" passes between two plates, D and E, parallel to the stream, and strikes the end of the tube where a patch of light appears on the glass. A piece of graph paper pasted on the glass allows the position of the patch of light to be read off. Thomson first showed that the "ribbon" could be deflected by charging D and E. If D was (+) and E (−), the ribbon was deflected upward, and if (−) and (+) then downward; thus the ribbon clearly bore a negative charge. He then did a most ingenious thing: he put a tube such as in Figure 4-16 between the poles of a strong magnet, as in Figure 4-17.

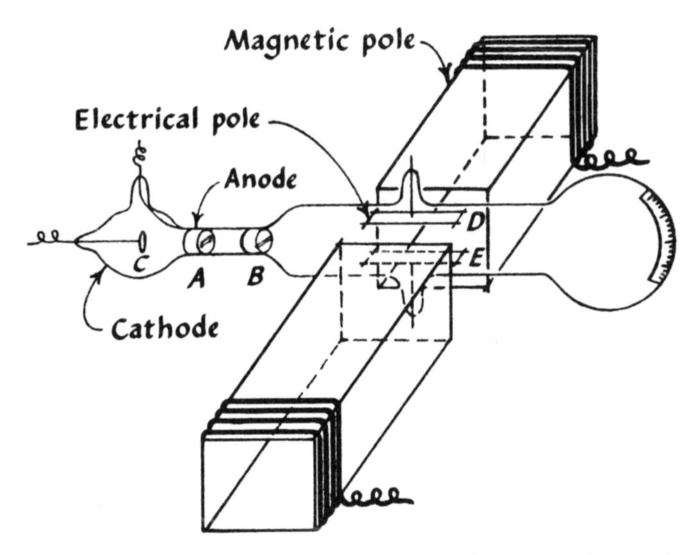

Fig. 4-17 *Thomson's tube for measuring* s/m *(diagrammatic). Opposed electrostatic and electromagnetic fields are used to control the electron beam.*

He used an electromagnet. An electromagnet makes use of the magnetic effect produced by a coil of wire with a current flowing through it. The lines of force about a loop of current-carrying wire are shown in Figure 4-18. (Refer to Figure 4-6.) When many loops are wound to produce a coil, Figure 4-19, and a rod of soft iron is placed inside the coil, insulated from the wires so as not to short-circuit them, the lines of force, following the metal, convert it into a magnet. Soft iron magnetizes when the current is turned on, demagnitizes when it is

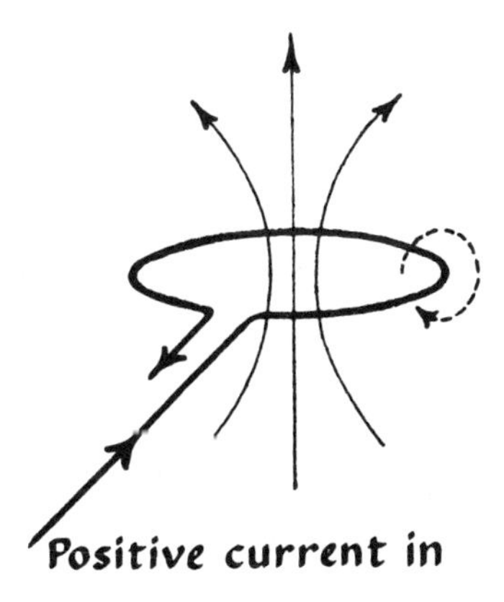

Fig. 4-18 *The lines of force about a wire loop.*

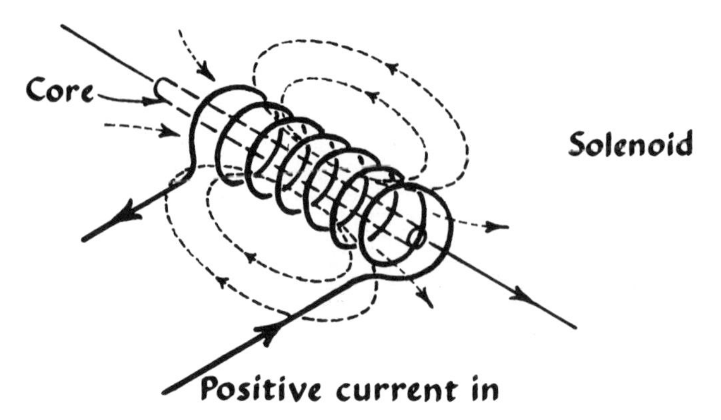

Fig. 4-19 *The lines of force about a wire coil (a solenoid). If a soft wire rod (a core) is introduced through the coil, an electromagnet is produced.*

turned off, and so yields an electromagnet the magnetic flux of which can be controlled sensitively by the amount of current flowing through the coil.

Now suppose that with the *magnet alone* turned on to a certain flux density B, the beam of electrons is deflected vertically *upward* in the tube. This behavior must be produced by a Force $\mathbf{F}_B$ acting on the electrons. (Refer to Figure 4-11.) As we saw in Section 4.4 this force may be written:

$$\mathbf{F}_B = q\overset{\curvearrowright}{vB}$$

Letting $q = e$, where e is the electron charge, we can write this equation as:

$$\mathbf{F}_B = e\overset{\curvearrowright}{vB}$$

Were the electrostatic field alone turned on, with the upper plate negative, the beam would be deflected vertically *downward*. The electrostatic force here would be written (Equation 3-5, rearranged):

$$\mathbf{F}_E = qE = eE$$

where E is the electric field intensity.

Now suppose that with the magnet turned on, the electric field intensity is adjusted to bring the beam of electrons *exactly back to its undeflected position*. Under these conditions,

$$\mathbf{F}_E = \mathbf{F}_B, \text{ that is } \mathbf{F}_E - \mathbf{F}_B = 0;$$

thus $eE = e\overset{\curvearrowright}{vB}$,

and $v = E/B$

In Thomson's apparatus all the electrons that form the cathode beam have essentially the same velocity, which makes such balancing possible, depending as it does on v. Were there a range of velocities, no sharp return of the beam to its undeflected position would be possible by adjusting E and B.

With the electrostatic field turned off, the deflection of the beam is due to the magnetic field alone. Here a centripetal force provided by the magnetic field takes the electrons in a curved path (Figure 4-11). One may write, recalling that $F = ma$, and that here acceleration is a_c:

$$\mathbf{F}_B = evB = mv^2/r \tag{4-5}$$

We now have an expression for e/m in terms of B and r, both of which can be measured; and in terms of v:

$$e/m = v/Br$$

But we also have an expression for v in terms of E and B, namely $v = E/B$. Eliminating v between the two equations, there remains:

$$e/m = E/B^2r \tag{4-6}$$

By measuring E, B, and r, Thomson was able to determine e/m. He showed that the same value was obtained whatever the nature of the cathode (whether of aluminum, iron, or platinum) and whatever the gas in the tube (whether air, hydrogen, or carbon dioxide), and he drew the conclusion that he was dealing with electrons, the existence of which had been a subject of some speculation. He concluded that electrons are present in all metals and other substances, and that the mass of the particles was probably very small, though he did not know what was the value of their mass.

The cyclotron 4.8

By the nineteen-twenties it was pretty generally agreed that if anything could be learned about atomic nuclei, it would have to be through bombardment of them with controlled high-speed particles. We shall see the reasons for this conclusion later. Actually, the pioneer in this area, Rutherford, had used high-speed particles given off by radium to bombard atoms, but what was needed were high-speed particles the energy of which was known and controllable. The first success came in Rutherford's laboratory when Cockcroft and Walton built a linear accelerator (to be described later). The first cyclic accelerator was described in 1932 by E. O. Lawrence and M. Stanley Livingston. The heart of the instrument is shown in Figure 4-20. It consists of two flat, hollow, D-shaped metal cans ('dees') A and B. Each is made of an upper and a lower half-circular plate held apart but close together by a semi-circular strip around the circumference. The two cans are open along their diameters and are fastened, insulated from each other, with the open faces opposite and close together. Thus there is provided a flat, cylindrical chamber. The two

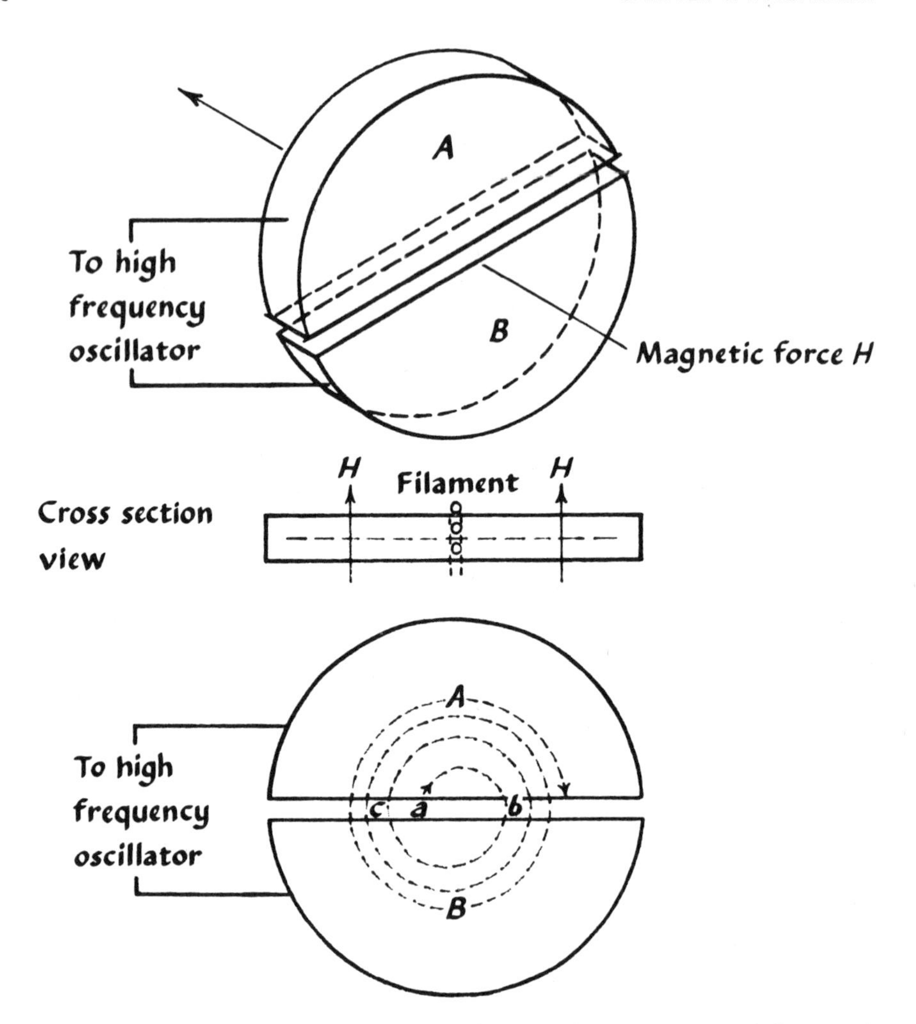

Fig. 4-20 *Diagram of the heart of a cyclotron. An ion, produced at* a, *circles about in the field of a magnet (not shown; poles above and below paper) and is accelerated by alternating charges on the dees* A *and* B *as it passes across the gaps* c *and* d. *A cross-sectional view is shown above the circular pair of dees. (From E. O. Lawrence and Stanley Livingston, "The production of high speed light ions without the use of high voltage."* Physical Review Series 2, 40, *19 (1932)*.]

halves can be given opposite electric charges. The pair of dees is placed between the poles of a large electromagnet. The dees are connected to a circuit which changes their polarity relative to each other alternately at a rate that can be adjusted. When a dee is charged, say (−), the hollow space inside the dee is field-free: no lines of force cross it since it is everywhere (−). A charged particle entering this space experiences no *electrostatic* force; it is shielded by the metal dee.

Suppose that the ion to be accelerated is a proton. Hydrogen (H_2) is admitted to the chamber. A hot filament at *a*, giving off energy, dissociates a molecule of H_2 and strips off an electron from a hydrogen atom, leaving a proton. The dee *A* may at this instant be negative relative to *B*, and the proton accelerates into its hollow space. Here it is in an electrical field-free region, but it is moving in a magnetic field. A magnetic force at right angles to the direction of motion and to the field H (in the diagram) takes it in the curved path to *b*. With proper timing of the oscillation of polarity, the proton emerges from dee *A* just as dee *B* becomes negative (and *A* positive). As a consequence of the potential difference, then, it accelerates across the gap into the electrical field-free region inside of *B*. Continuing to be bent by the superimposed magnetic field the particle swings around to *C* where, just as it arrives, *B* becomes positive and *A* becomes negative. The proton again accelerates and so, at step-wise increasing velocity it performs wider and wider half-circles and finally, when it approaches the edge of a dee, it is acted on by an electrostatic potential (not shown) that deflects it out of the dee and to the desired target.

Any plus-charged particle will be accelerated across a potential drop from (+) to (−) electrodes. That the cyclotron works depends on two key factors. The shape of the magnetic field imposed on the dees, keeps the protons within the median plane of the dees so that they do not become lost to the upper and lower walls. This is achieved by so shaping the field that a proton straying upward, say, finds the magnetic force upon it pointed slightly downward; thus it swings towards the middle. Overshooting, it finds, again, a force directed towards the median plane. Thus, as Lawrence and Livingston say, "As the ion spirals around, it will migrate back and forth across the median plane and will not be lost. . . ." In fact, the beam of high-speed protons was less than 1 mm in cross-section and about 1 cm wide.

The second key factor has to do with the timing of the oscillation of polarity of the dees so that the oscillation is in resonance, or is tuned, to the passage of *every* proton across the gap. That this is achievable—and indeed with comparative ease—can be shown by calling upon our previous discussions. Let the charge on the particle be q, and let it move with a velocity v in a magnetic field of magnitude B. (The field is marked H in Figure 4-20.) Then as we saw (Section 4.4) the particle will experience a force $\mathbf{F} = qvB$. This is a centripetal force. The particle moves in a circular path, and thus we can write:

$$\mathbf{F} = q\overrightarrow{vB} = mv^2/r. \text{ Or, } r/v = m/Bq \tag{4-7}$$

As usual, m is the mass of the particle, r, the radius of the path in a dee, and B, the magnetic flux density. Since the particle changes its velocity only at the gap, and not within the dees, where the only force on it is the centripetal force at right angles to its motion, its speed is constant in the dees. While in the dee, it travels a distance substantially (πr), for the separation of the gap is small compared to the length of the path in the dee. Thus the time that it is in a dee, $t_{1/2}$, making the semicircular transit, is:

$$t_{1/2} = \pi r/v$$

Combining this equation with that for r/v, to eliminate this ratio,

$$t_{1/2} = \pi m/Bq \qquad (4\text{-}8)$$

Thus *the period of time between each crossing of the potential gap does not depend upon the velocity of the particle or the radius of its path:* as the velocity increases, the radius increases by the same factor. You can see that this is so because Equation (4-8) does not contain either r or v!

The kinetic energy of the particle can be calculated from the relation for r/v:

$$KE = (1/2)mv^2 = B^2q^2r^2/m$$

In practice, the tuning of the oscillating electric field so that a particle with a given m/e ratio is caused to travel in phase with it can be done by adjusting the magnetic field. Lawrence and Livingston reported that in experiments in which the electric field reached a peak of 4000 V, and the protons spiraled around 150 times, receiving 300 increments of energy, they achieved "a speed corresponding to 1,200,000 V." This means the speed the particle would reach were it to accelerate through a potential difference of that many volts.

Since the timing of the alternation of electric polarity between the dees depends on the mass of the particle being accelerated (Equation 4-8), the cyclotron just described requires mass constancy for its proper functioning. This means that the speed of the particle must not be so high that relativistic mass increase becomes an important factor. Actually, the cyclotron will operate up to about 15 million *electron* volts with protons. At higher energies the particles, reaching these energies near the periphery of the dees, and now "more massive," will require a longer period to traverse the half circles. This can be allowed for by designing the magnetic field to increase toward the periphery proportionately to the increase in mass, or by arranging to decrease the frequency of the oscillation toward the periphery, or both. Instruments of this kind are called 'synchrocyclotrons.' Mass increase is discussed in Chapter 6.

4.9 The electron volt

The electron volt is a quantity of *energy*. The instruments discussed in the previous section bring us to atomic-level phenomena, where we shall have to discuss energies of single atoms, or of single electrons or protons. When an electron falls through a potential of one volt, it acquires one electron volt of energy (eV). In Equation (3-11) $V = W/q$. Here $V = 1$ and $q = e = 1.6 \times 10^{-19}$ C, whence one electron volt equals 1.6×10^{-19} joules. To determine the speed acquired by the electron, note that $E = (1/2)mv^2$. The mass of the electron is 9×10^{-31} kg, so that v works out to be 6×10^5 m/sec. When protons are accelerated by one electron volt of energy they reach substantially less speed because their mass is nearly two thousand times greater than that of the electron.

These quantities of energy are small compared with the energy of a two kg mass moving 1 meter/sec, but relative to the scale of the particle they can be large. Cyclotrons, as noted, may accelerate particles to energies of 15 million electron volts (MeV); other accelerators raise the limit into the BeV, or billion electron volt range.

From time to time we describe in detail some particular piece of apparatus which played a determinative role in suggesting and supporting theory. This is to re-emphasize the connections, both ways, between the plane of perceptions and the field of constructs. We re-emphasize at the same time the close link between theoretical science and technology. The two are closely linked in bonds of mutual survival. And we would also like to suggest that neither modern science nor the modern philosophy of science could have developed without these instruments: and they could not have been built until people learned to blow glass in unusual shapes; till they learned to seal electrical wires (leads for leading-in electricity) through the walls of the glass vessels without developing a leak. This was no mean accomplishment. The expansion on heating and contraction on cooling of both wire and glass have to be closely matched, and the glass must wet the metal, not pull away from it on cooling. Also the vacuum pump had to be invented. Hundreds of techniques for working glass and metals and plastics have to be mastered to make modern instruments. Without this technology, no instruments; without instruments no good theory because no test of theory. The relation of technology to pure science is as close as that.

Induced current 4.10

In summary, current flow in a metallic conductor is due to the passage of electrons. In an ordinary piece of wire, say a copper wire, there are some electrons associated with each metal atom which are relatively free and which exchange between atoms somewhat freely, in a random way, some moving in one, some in another, direction. When a current flows in the wire, electrons enter at one end, and other electrons exit from the other end. There is, clearly, a force exerted upon the electrons which accelerates them, superimposing on the essentially random, rapid motion a drift in the direction of the drop in potential along the wire. This is not instantaneous in its effect but, starting at zero, builds up to a steady state. During this period, the charges are being accelerated. At the same time the electrical field about the wire is building up, and the magnetic field due to the *moving* charge is building up. Similarly, when the current is broken there is a finite period of decay in which the reverse of the process occurs. There is an additional effect, due to the acceleration of charges, which is the subject of this section.

If a current is made to flow through a coil of wire, separated from another coil, then it may be observed that a current will flow in the other coil during the period in which the first current is starting to flow. Thus in Figure 4-21 a current may be made to flow in the lower coil,

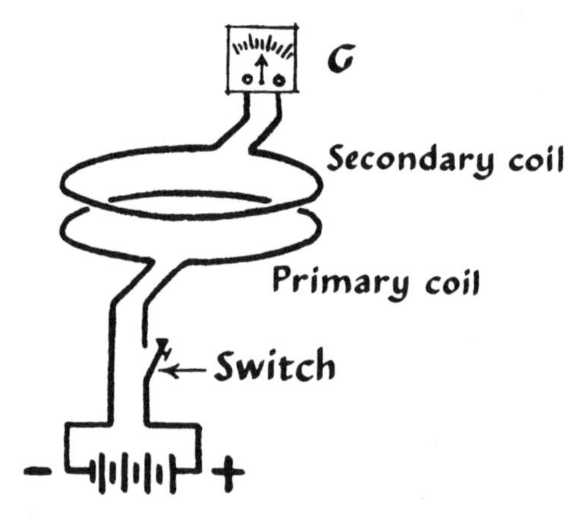

Fig. 4-21 *Demonstration set-up for induction. The needle of the galvanometer* G *moves as the switch in the lower circuit is closed or opened.*

which is connected to a battery, by the closing of the switch. As the switch is closed, or opened, a momentary current flows in the upper coil. This is shown by the deflection of the sensitive current-measuring instrument (a galvanometer) in the circuit of the upper coil. This phenomenon is known as 'induction.' A current is induced in the upper coil, associated with the *change* in current flow in the lower. When the current in the lower coil becomes steady, no current is induced in the upper.

Induction occurs when a magnetic field *changes* in the neighborhood of a conductor (Figure 4-22). It is important to notice that when induction occurs *energy* is transported across space. For example, when in Figure 4-21 the switch is closed, the galvanometer needle moves. Work is done on it. This energy (work) could only have come from the

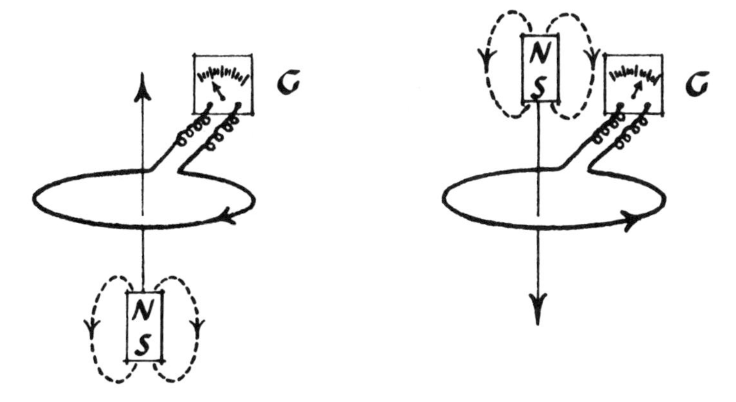

Fig. 4-22 *The effect on the galvanometer obtained by moving a permanent magnet through the wire loop, up and down.*

energy put out by the battery in driving electrons through the wire of the primary coil. The phenomenon supported the concept, prevalent toward the end of the last century, of an aether, a subtle material capable of carrying energy and pervading all space. This aether, present between the two coils, was supposed to transmit energy from the one to the other.

4.11 Summary

Electricity and magnetism are intimately connected in that magnetic force effects appear in the presence of moving charges. The magnetic force can be recognized by its ability to do work on a compass needle, turning it out of its normal north-south orientation (unless, of course, the magnetic field happens to be lined up with the Earth's field). It may be measured in terms of the mechanical force on a wire produced by passing a current through it while it lies in and perpendicular to the magnetic lines of force. By the opposition of electrostatic and magnetic forces the ratio of charge to mass of the electron, that is e/m, may be measured. If an alternating electrical field is applied to charged particles in a magnetic field it becomes possible to accelerate them to high velocity. In this way an available potential difference such as 1000 V may be repeatedly applied to accelerate an electron to a velocity such as it might have in falling through a potential difference of 15 million V. Instruments required for testing and suggesting scientific theory, such as cathode-ray tubes and accelerators, require a high development of the technological arts.

When a charge is accelerating it induces motion in neighboring charges not in mechanical contact with it.

Exercises

4.1 Is the north geographic pole also a north magnetic pole?

4.2 Considering the source of the magnetic field of a permanent magnet, discuss the observation that such a magnet may be demagnetized by mechanical jarring, or by heating.

4.3 In Figure 4-8 we have a wire at right angles to a magnetic field. Suppose this wire were 10 cm long in the field and when a current of 20 amperes flowed through it it experienced a force of 6 newtons. What is the magnetic flux density of the field?

4.4 A positively charged particle moving in a horizontal plane, from right to left, enters a magnetic field and is deflected upward, out of the horizontal plane. What is the direction of the magnetic field?

4.5 During a certain small interval of time an electron, passing horizontally from east to west with a velocity of 2×10^4 m/sec through an instrument, is 20 cm from and has been moving parallel to, a long straight wire in which a positive current of 5 amps is flowing from east to west. a) What is the direction of the force on the electron? b) Suppose the electron passed in a direction at right angles to the direction of the current in the wire, what force and direction of force are experienced?

4.6 How can we tell when a suitable test object in a field experiences a force?

4.7 An undisturbed test-compass needle points N-S. a) A wire is placed just above and parallel with the needle. When a (+) current is sent from S to N through the wire, in what direction is the needle deflected? b) When the wire is placed below the compass needle and (+) current is sent through in the S to N direction, what is the needle's deflection? c) In what direction must the wire be placed and in what direction the current flow to leave the needle pointing undisturbed N and S? d) The wire is placed vertically next to and parallel with the support of the compass needle, and (+) current is sent down the wire. In what direction is the compass needle deflected?

4.8 Suppose two loose vertical wires are side by side and parallel. A strong current is sent up one wire and down the other. How, if at all, will they interact?

4.9 What is the velocity of a proton ($m = 1.67 \times 10^{-27}$ kg) with one eV of energy?

Answers to Exercises

4.1 No. The north geographic pole is at a different point on the Earth from the north magnetic pole, which is S, since the test magnet's N pole points to the north magnetic pole.

4.2 Both treatments shake up the atoms and cause their fields to become unaligned, neutralizing each other.

4.3 $\mathbf{F} = \overset{\frown}{lIB}$. $B = 4$ webers/square meter.

4.4 

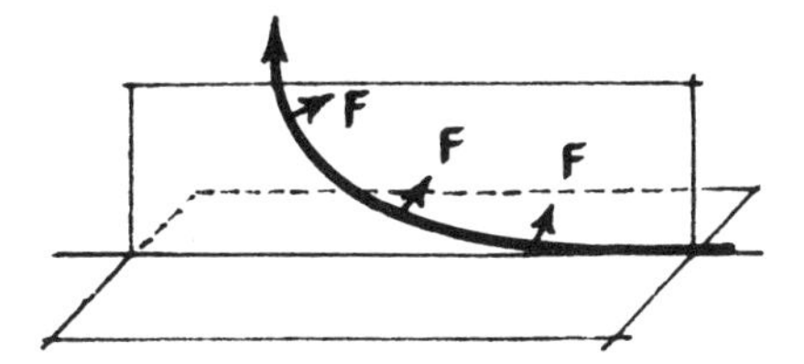

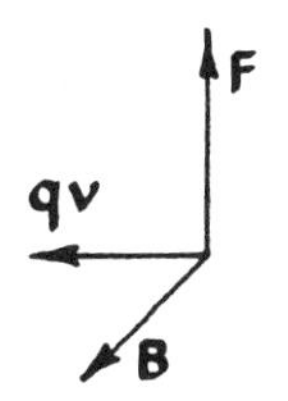

B is directed toward the observer. The planes are introduced to help visualize in three dimensions.

4.5 a) Upward, away from the wire. b) No force. No magnetic lines of force are being cut.

4.6 It accelerates; or it acquires potential energy.

4.7 a) The N end of the compass points west. b) The N end of the compass points east. c) The wire must run from west to east, and if it is placed above the needle, current must run from west to east. d) It depends on which side of the needle the wire is placed. See Figure 4-6.

4.8 The wire carrying current upward will have B as shown in Figure 4-6. Now let another wire with current going down lie side by side with it. B will be in the same direction and the wires will repel each other.

4.9 $E = (1/2)mv^2$. One eV $= 1.6 \times 10^{-19}$ J. $(1/2)mv^2 = 1.6 \times 10^{-19}$ J; $v^2 = 3.2 \times 10^{-19}/1.67 \times 10^{-27} = 1.93 \times 10^8$. Velocity $= \sqrt{1.93 \times 10^8} = \sim 1.39 \times 10^4$ m/sec. ($\sim$ means 'approximately'.)

Bibliography

The historical background of the subjects in Chapters 3 and 4 may be traced in Sir Edmund Whittaker's *A History of the Theories of Aether and Electricity. Vol. I, Classical Theories, Vol. II, Modern Theories, 1900–1926.* Thomas Nelson & Son, London, 1951–1953; Philosophical Library, New York,

1954. These two volumes make fascinating reading, especially the first volume after our Chapters 3 and 4, and the second after Chapters 5 and 6.

For a report of J. J. Thomson's experiments, see J. J. Thomson, "Cathode Rays," *Philosophical Magazine, Series V, 44,* 293 (1897). See also M. H. Shamos, *Great Experiments in Physics,* Holt, New York, 1960.

For the early work on the cyclotron see Ernest O. Lawrence and M. Stanley Livingston, "The Production of High Speed Light Ions without the Use of High Voltage," *Physical Review, Series 2, 40,* 19 (1932).

Chapter 5

5.1 Light
5.2 The seeds of discontent
5.3 The geometry of light beams
5.4 Waves
5.5 Interference
5.6 The Michelson interferometer
5.7 The constructs of absolute space and time
5.8 The Michelson-Morley experiment
5.9 The sequel
5.10 Summary

Light ~ 5

Light 5.1

"Light," said Goethe, "is an 'Einziges'." It is something "given," an "elemental entity," an "inscrutable attribute of creation," which should not be profaned by experimental examination. Goethe drew the conclusion that Newton blundered in attempting to analyze light. Sir Charles Sherrington has written perceptively on Goethe. He says: "In following Goethe's 'science' we are helped by his having laid down principles which in his view should govern scientific observation. One of them is that the conditions for observation be kept as simple as possible, and for that reason should eschew apparatus. Prominent in his objection to the prism experiment [in which Newton decomposed white light to colors of the rainbow] was that the prism introduced heaven-knows-what complications. Essential for scientific observations was *Anschaulichkeit*, 'obviousness' or 'naked clarity.' This clarity could dispense with mathematics. Goethe was not himself equipped in mathematics, and he regarded the role of mathematics in science with distrust. Mathematics led to the introduction of propositions which were not truly contained in the original proposition. They had brought calamity to optics. He did not see that a use of apparatus is to simplify conditions. Nor again, that mathematics can be a main means toward obtaining *Anschaulichkeit*."

Light is, indeed, an inscrutable attribute of creation. As far as we can tell, without sunlight and other radiations that fall on the Earth, there could be no life as we know it. If a scientist were to think about it at all in these terms—and many do—he would surely think his investigations of Nature the very opposite of profanation.

The light that bathes the Earth when she turns toward the face of the Sun has held special fascination for men from the earliest times.

Our absolute dependence upon the Sun for the maintenance of life on Earth must have been realized long ago—after all, green plants cannot grow in the dark. The fact is that Sun-worship is of ancient origin, and that the Sun was—and probably still is—held sacred in some cultures. Mircea Eliade points out that solar symbolism dominated the "diurnal domain of the mind" with its chains of reasoning, while the symbolism associated with the Moon appealed to "the nocturnal domain of the mind," "a layer of man's consciousness that the most corrosive rationalism cannot touch." With the development of the medieval World-Allegory, displayed for example in the *Divine Comedy* of Dante, a consistent view of the universe was developed that comprehended all of known Nature, from the imperfection of earthly things to the ethereal perfection of the heavens, and that was consonant with the religious hierarchy of the Ladder of Nature. In this world-picture light played an important symbolic role.

From time immemorial the flame of the lamp that held away the dark shadows of evening has been an analogue of the living creature. The flame, though consuming its oil or wax and its wick, flaring up and subsiding, and blown aside by drafts, maintains its integrity. No wonder that the poet, as well as the scientist, has been preoccupied with lights and light, and that the "eternal flame" can be such a noble symbol. And little wonder, too, that Newton's decomposition of white light to its colors and his other studies with light might seem sacrilegious to Goethe and others. Yet Newton of all people could hardly be accused of sacrilege: he surely belongs to a long and illustrious line of scientists whose search for truth about Nature is devotional.

In a sense, light is the theme of this textbook. For we are committed to modern physical science, and it was in a very central sense discontent with the state of knowledge about light, with respect particularly to strange properties that classical views could not comprehend, which ushered in the modern era in Physics and Chemistry.

5.2 The seed of discontent

To hear and to pay attention to the small voice of creative discontent is one of the most difficult things to learn—yet this may be the distinguishing characteristic of an innovative pioneer in all areas of knowledge and experience. Big bangs, startling paradoxes, intellectual fireworks may come later. First there is a still small voice in someone's mind; so still and small and subtle that it may not be recognized as the potential source of great things. Yet a thought attendant on the fall of an apple pre-figured a change in cosmology; a patent-clerk's lucubrations changed much more than a cosmogony.[1]

---

[1] The patent-clerk was Albert Einstein.

Toward the end of the 19th century it was clear that things were amiss in the physical sciences. It was not that the classical Newtonian theories of Physics and the automistic theories of Chemistry were wrong. Their successful practical fruit in the burgeoning industrial revolution testified to their pragmatic truth. But areas of misfit between experiment and theory, referred to below, forced scientists to re-think fundamentals. For example, there seems always to have been tension between two views of the nature of matter, between the continuum and the atomistic views. Was matter infinitely divisible, or was there some point beyond which no further division to finer or more fundamental entities might be made? With respect to light, is it particulate, a stream of fine particles travelling in a straight line and casting sharp shadows, or is it wave-like and continuous in character?

Newton himself had favored a corpuscular model for light—the model of a stream of particles—but his researches led him at the same time to attribute wave properties to these particles. Newton insisted upon the importance of following the lead of experiments and, says N. R. Hanson, would have readily admitted either wave or particle properties, or both together, to light if the phenomena so demanded. But over the succeeding years a bias had developed among most scientists who studied the properties of light. In Newton's time, F. M. Grimaldi had observed experimentally that light would bend into the shadow behind a hair, much as water-waves bend around a piling. This suggested wave character. And when Thomas Young, in 1807, demonstrated interference (see below), and particularly when Jean Foucault in 1850 demonstrated that the speed of light in water is less than that in air, a wave model for light seemed obligatory. In Foucault's demonstration (which we shall not pursue beyond this point) it was shown that light slows down on entering water. The particle theory predicted that as particles approach a dense medium they should be attracted, and so speed up as they pass from less dense air to more dense water. Foucault's experimental findings were interpreted to demolish the corpuscular theory of light.

However, further experiments yielded results that were not compatible with a simple wave model for light. These and their interpretation are the burden of the following three chapters. They led eventually to the concept of light quanta—and to a present uneasy resolution of the continuum-atomistic question.

We shall first consider certain properties of light which led scientists to attribute wave character to it. The way will thus be prepared to discuss relativity (Chapter 6). Then we shall be ready to tackle light as an electromagnetic phenomenon, in Chapter 7. Our approach having been broadened in this way, we go to quantum phenomena and their interpretation. The consequences of some of these ideas are then worked out in subsequent chapters.

5.3 The geometry of light beams

A light beam is a 'pencil' or 'ray' or 'ribbon of light' and may be defined by an obstruction with an aperture through which light may pass. The facts that unobstructed light travels in a straight line; and that its intensity falls off as the square of the distance from its source; that white light may be dispersed into colors that may be recombined to white light (Newton's striking demonstration); that light may be diffracted into the region behind an opaque object; and the laws of reflection and refraction, are matters of common knowledge, or are readily reviewed. When one of these properties is invoked will be the time and place to discuss it.

Recall: *When a ray of light, incident on a surface, is reflected, the angle of incidence, α, is equal to the angle of reflection, α', measured from a line normal to the surface (Figure 5-1). If the surface is rough, or if reflection occurs from fine particles such as motes, fog, or smog, the reflection is diffuse, the light is scattered (Figure 5-2). When light strikes a translucent surface,*

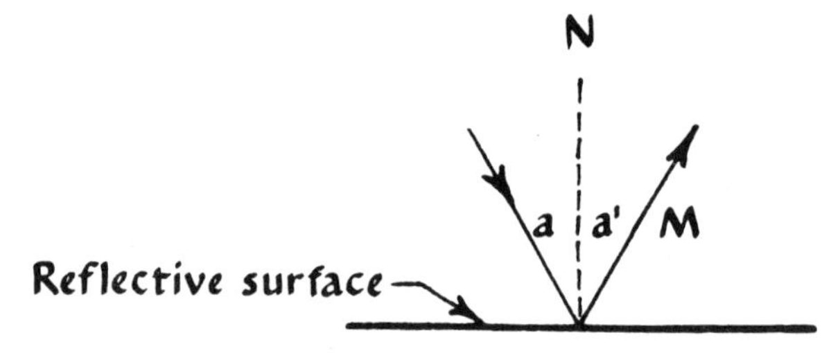

Fig. 5-1 *Reflection of light at a plane surface.*

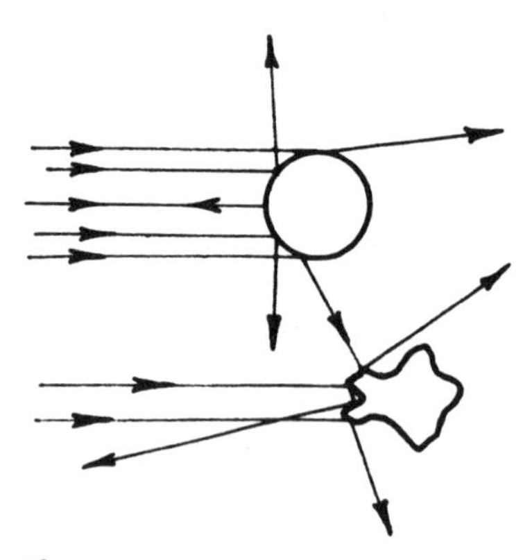

Fig. 5-2 *Scattering of light by small particles.*

say between air and a glass plate, part may penetrate and be refracted out of its original path (Figures 5-3, 5-4). The law of refraction is that sin $\alpha_1 = \mu_{1,2}$ *sin* α_2, *where* $\mu_{1,2}$ *is the index of refraction for the two media that make the surface. The index of refraction* $\mu_{1,2}$ *for two media, say air (1) and water (2), is the ratio of the speeds of light in the two media, that is* $\mu_{1,2} = v_1/v_2$. *It is different for different wavelengths (for reasons we shall come to) and thus permits dispersion of white light into a spectrum. When white light falls on a prism it may be decomposed to a colored spectrum ranging from red at the least refracted to violet at the most refracted edge (Figure 5-5). By means of a monochromator (Figure 5-6) it is possible to select out a narrow range of color from the spectrum.*

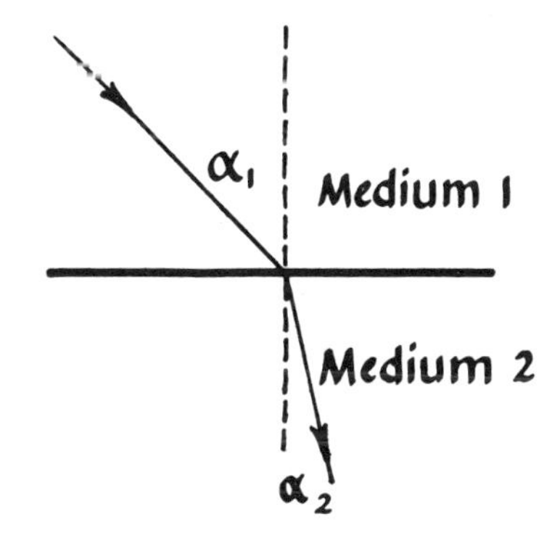

Fig. 5-3 *Refraction of light at the surface of two media. Medium 2 is denser than medium 1.*

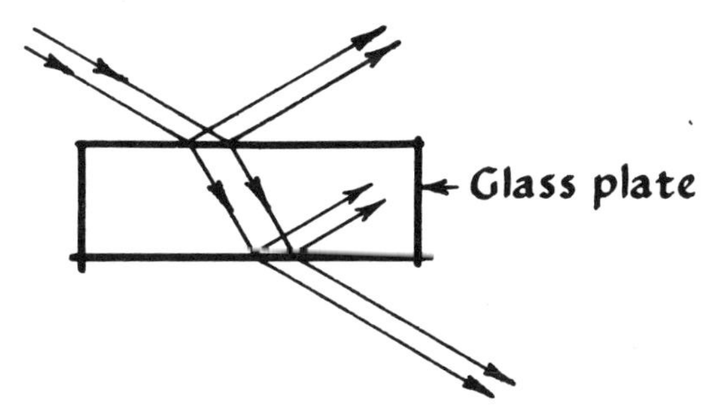

Fig. 5-4 *Light passing through a block of glass with parallel faces. [From P.S.S.C.* Physics, *D. C. Heath, Boston, 1960. Figure 13-7, p. 216.]*

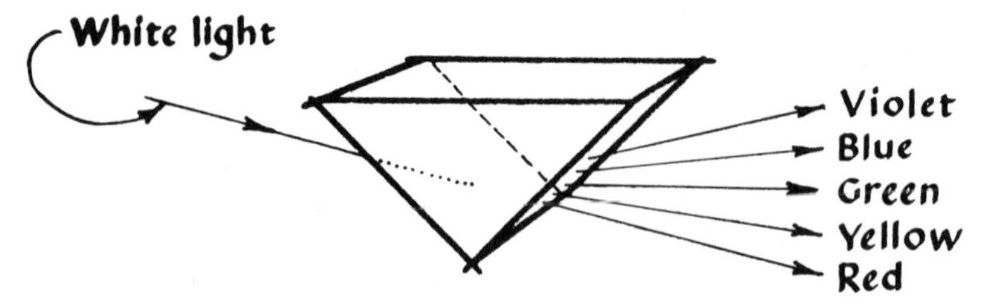

Fig. 5-5 *Dispersion of white light by a prism.*

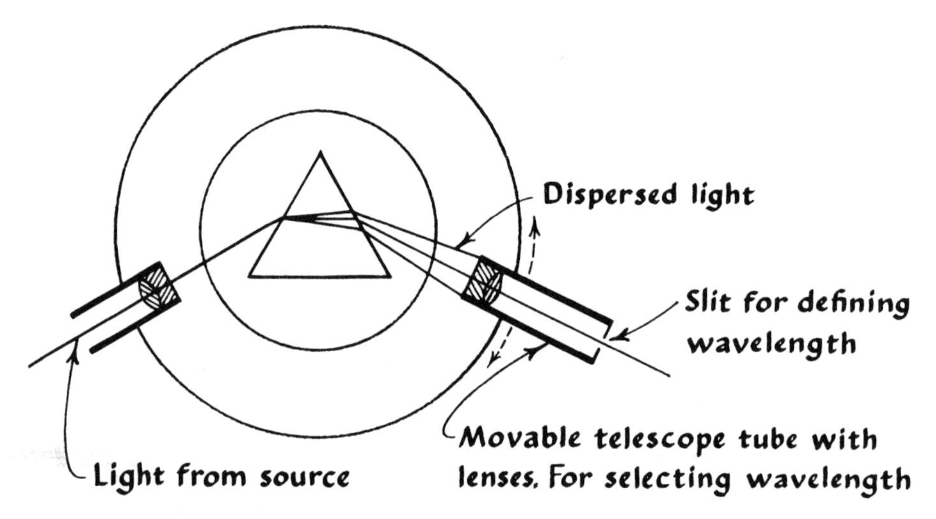

Fig. 5-6 *A monochromator by means of which a beam of light, dispersed by a prism into a spectrum, may be scanned by a telescope with a slit to pick out narrow bands of light of essentially the same color.*

5.4 Waves

Suppose that we have a long rope, attached firmly at one end to a support. The other end we hold, stretching the rope taut. We suddenly jerk the end that we hold up, down, and back (say, a distance of about a foot (Figure 5-7a). A wave pulse travels along the rope to the end where it is fastened, is reflected there, and travels back, giving a small jerk to our hand as the wave reaches it. Evidently energy was transmitted along the rope, and obviously a wave-form travelled along the rope. How can we grasp this behavior conceptually? It turns out that a single pulse is harder to deal with than a continuous train of waves, but it does serve well in describing and visualizing how a wave is formed.

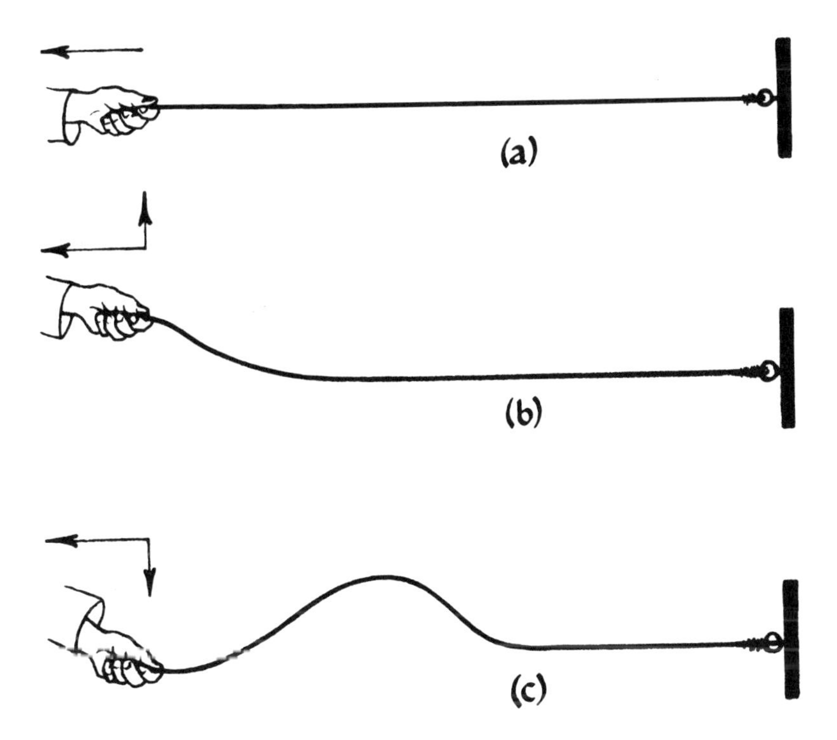

Fig. 5-7 *Initiation of a wave in a stretched rope.*

When the end of the rope is jerked upward, the rope close to the point of disturbance also moves upward because rope is, after all, a solid material: forces between the constituent molecules of the material tend to preserve its form. But it is somewhat elastic and can stretch a little; also it is heavy, and displays inertia; so that, as the end is jerked up, successive portions of the rope respond (Figure 5-7b). We have, in effect, an elastic medium to which we have applied an unbalanced force, accelerating a portion of the medium, and causing adjacent portions to respond to the disturbing force. As we jerk the end down we see the adjacent parts of the rope respond in sequence, at each successive point a little later, so that some parts are still rising while others are moving downward. The result is a wave-form (Figure 5-7c) which moves along the rope. [It is described as a 'mechanical transverse' wave. Transverse because the motion of any small section of the rope is perpendicular to the direction of propagation of the wave, and mechanical because it is the result of the oscillation of the portions of the rope about their equilibrium points under elastic restoring forces.]

Recall: *Newton's first law is that any material object will continue in its state of rest, or of motion in a straight line, unless acted upon by an unbalanced force. His second law tells that if the object has a mass* (m), *and is acted upon by an unbalanced force of magnitude represented by* (**F**), *acceleration (a) in the direction of the force will ensue according to the relation* **F** = m **a**.

The elastic restoring forces may be visualized by means of vectors (Section 4.6). The rope is under tension, horizontally. Any point on the rope, say *P*, is initially stationary, and thus by Newton's first law it must have no *unbalanced* force on it. It is pulled equally toward the hand and the wall (Figure 5-8a). When a wave crest passes this point

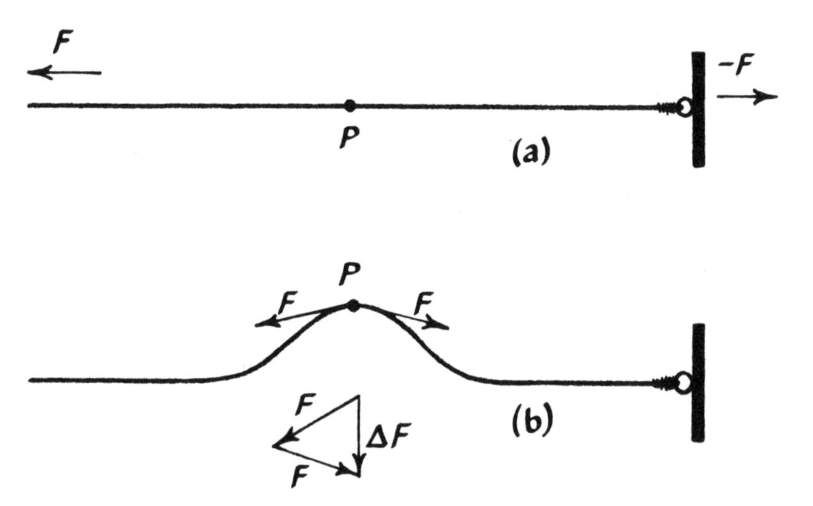

Fig. 5-8 *Diagram of restoring forces in the development of a wave form along a stretched rope, or string.*

(Figure 5-8b) the rope is stretched. A downward restoring force over a short length of rope is the resultant of the tensile forces pulling at angles downward. If we measure the acceleration of this small section and if we know the mass of the rope section, we can determine the force on it.

Imagine what happens if the tension on the rope is increased. A trial or two would show that the wave pulse travels faster with increase in tension. But as the tension is increased the effect is to make the rope more rigid. As a limit to this, we might imagine a rigid steel rod, hinged at the support. As we lift the free end, the rod responds as a whole, and we lever it up and down. Conventional wave-behavior has disappeared. Going in the other direction, we may imagine relaxing the tension. We might relax this to the point where the rope sags onto the floor. Now when we attempt to generate a wave in it the energy we put in damps out in friction with the floor.

We can't very well relax the mechanical tension on the rope any further. Suppose, then, we move in imagination to a liquid such as water, where there is no cohesion that maintains its shape, as with the rope. But there is elastic cohesion between the molecules of the water. If we excite surface waves by plunging an object into the water, or pulling one out, we see the familiar wave-form rippling away from the disturbance, and reflecting from the walls of the container. (The weight of the water, operating against gravity and with it, plays a role also.)

Close examination of wave behavior, for example by watching the behavior of a light-weight object—a cork fragment—moving with the liquid, as it bobs up and down in response to the passage of a wave, shows that the object moves in an elliptical path. On the crest of a water-wave, for instance, the particle moves in the direction of propagation of the wave; in the trough, in an opposite direction. The axis of this ellipse is perpendicular to (transverse to) the path of the wave.

Continuing still further with our imaginative relaxation of 'tension,' we arrive at a gas, which fills any container into which it is placed and is quite incapable of displaying mechanical transverse waves because cohesion between adjacent molecules is minimal.

Returning to our original demonstration, an up-and-down motion of an elastic medium in response to an unbalanced force may generate a wave which propagates perpendicularly to the axis of the motion. The properties of the material have converted the transverse motion into wave motion. To grasp this behavior conceptually, we must move away from the P-plane (Figure 1-4) into the field of constructs. We first simplify and generalize the problem in our thoughts. Assume that a train of waves passes by a point of observation. The crests and troughs march by at a steady pace. The observation point is so far away from the origin of the waves as well as from the point where they are absorbed, that the disturbing extra oscillations, stretchings, and other results of the imperfections of brute matter have damped out. These waves are travelling along an idealized rope somewhere out in Space, perhaps, where gravity does not affect it. The rope has mass (a measure of its inertia) and we imagine the physical properties of the rope to be such that if we stretch it a little we do not change its physical properties appreciably.

With all these provisos one might think that we have gone very far from reality. Yet it turns out that the conclusions we reach, using them and navigating about the field of constructs, apply remarkably well when we return to actual ropes, vibrating strings, and water waves: we make "hits" on the plane of perception that are quite close to the theoretical predictions. This, of course, justifies us, and makes us more confident. Besides, some of the problems that we approach are so complicated that, if we fix our attention spastically upon the complexities, we would never get very far forward. For example, we pluck a guitar string: waves, the fundamental, and harmonics flow along the string, losing to the air energy, which we hear as sound, losing energy at the ends of the string, where the nut and the bridge transmit to the body of the instrument vibrations, decreasing in amplitude, finally damping out, as the energy of the original impulse is dissipated, but affected by changes in tension as the string stretches, and affected by macroscopic and microscopic imperfections in the string. . . . The complications could be endless. Thus we learn the value of the general construct that conceives of ideal conditions for an experiment with nature; later we can consider the special conditions, the complicating factors, the unique cases.

Figure 5-9 represents five sequential snapshots of this idealized travelling wave. The observation post takes in the waves between the two vertical lines, which are three wavelengths apart. The wave is travelling to the right with a velocity u. That it is moving to the right can be seen by examining the relative motions of particles. For example, particle 5 at time zero is stationary. It has just come up to this point because by time "plus $1/4\ T$" it will be on the way down. This means that it is at the crest at time zero. Concomitantly, particle 6 was moving upward at time zero, and at $1/4\ T$ it is at a crest. Hence the crest has moved from position 5 to position 6 in $1/4\ T$, as the wave is travelling to the right. Particle 5, for example, moves up and down in the Y direction. (This is of course idealized. The particle will in general as we said move in a flat ellipse—it might under some circumstances undergo nearly circular motion. The projection of the motion of the particle onto the Y axis is shown here, as it was analyzed and described above.) The amplitude A of the wave is the same as that of the individual particle. It is defined as half of the total distance up and down.

It is at this step of symbolic transformation that many beginning students experience difficulty. We fix our attention on particle 5 (for example) out of the thirteen illustrated in Figure 5-9. The behavior of this 'particle' is idealized from that of a bit of cork, say, riding on a water wave. The particle is a construct: it is imagined so small that there is no ambiguity in saying where it is. That is, we don't have to worry about irregular shape, or where we measure from, on a diameter, as we would with an actual flake of cork. We then define some terms

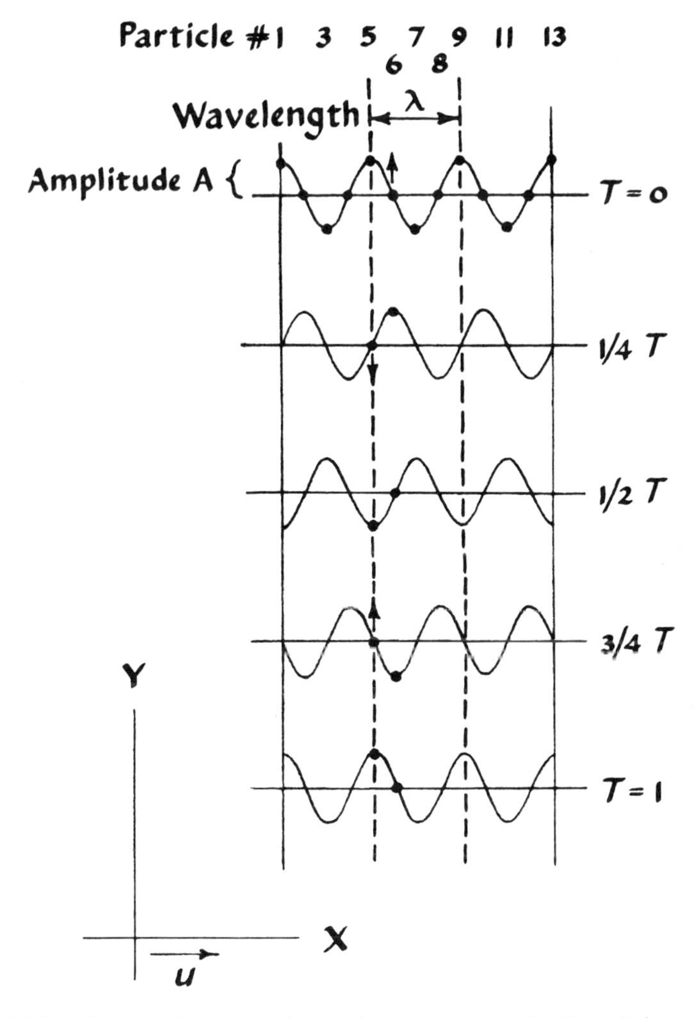

Fig. 5-9 *Motion of a particle in a moving continuous wave as a function of time. "Snapshots" at 1/4 of a period.*

so that we can talk about, or write down, what is happening. Amplitude of the wave is measured from the unperturbed level, above and below it. In a *particular* case of an observed wave we correlate the observation with the construct by finding a numerical value for A, say 1 meter in a rough sea. We have to measure this in a particular case, or imagine it. We also notice that there is an invariant-relation that simplifies the discussion: particle 5 comes back to the same top position after a period of time, repetitively. This constant length of time is designated the 'period' of oscillation and represented by T.

It also is a construct, applicable to any wave and capable of being given a numerical value in a specific, observed instance. We also, looking at a train of waves moving by, notice that the distance between crests is constant. Here again is an invariant-relation, and we name it the 'wavelength,' represented by the Greek letter lambda, λ. The time that it takes for particle 5 to make one complete oscillation is exactly that required for the wave to move one wavelength forward. Therefore we can define the speed of the wave, represented by u, as the distance travelled (λ) divided by the time (T) required to travel this distance: $u = \lambda/T$. It turns out, also, that a very useful construct is the reciprocal of the period, namely $1/T$. We therefore give it a special name, for convenience: the 'frequency,' represented by the Greek letter nu, ν. Then by algebra we can write the expression $u = \lambda/T = \lambda\nu$.

This is not yet a full conceptual description of a wave. It is only a set of relations between defined measurable quantities. We leave a fuller description for a Note at the end of this chapter. It must be emphasized at this point that each of the above constructs must be accompanied by a statement of the terms in which it is measured. Thus amplitude is measured in meters. The wavelength is also a length and is measured in meters. The time is measured in seconds. Therefore the speed u will be given in meters-per-second, and the frequency in 1/seconds, or sec^{-1}, or 'per second,' or 'cycles per second.'

We shall be concerned in this book with five fundamental constructs—length $[l]$, time $[t]$, mass $[m]$, quantity of electricity $[q]$ and temperature $[T]$. All constructs can be specified in terms of these. Thus amplitude has the fundamental property of a length; velocity is a length divided by a time, *i.e.*, $[l]/[t]$. We use these constructs to check equations, as we shall show repeatedly. Our unit of length is the meter, unit of mass the kilogram, and unit of time the second. This group of units belongs to the mks (meter, kilogram, second) system of measurement, and we shall adhere consistently to it. There are other systems in use: the centimeter-gram-second, cgs, and the foot-pound-second, fps, as examples. All of these are $[l][t][m]$ systems.

5.5 Interference

If two waves run along together, that is, if they are superimposed, the resultant wave is the algebraic sum of the two. If both have the same amplitudes A the resulting wave would show twice the amplitude of a single wave. When one wave *increases* the amplitude of another, the phenomenon is called 'constructive interference.' Suppose, however, that the two waves are superimposed so that the crest of one falls on the trough of the other. The result is cancellation, called 'destructive interference.' There are various degrees of interference from 100% to zero. In Figure 5-10 two waves, as dashed and dotted lines, with dif-

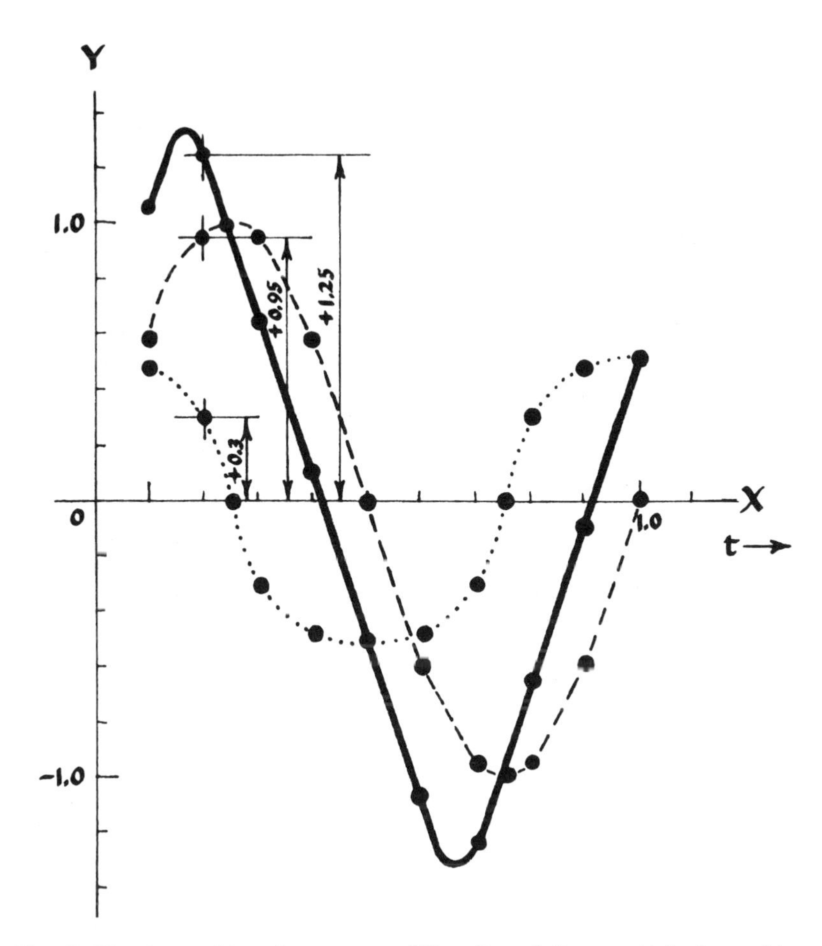

Fig. 5-10 *Superposition of two waves. When* ○ *and* ○ *are added, the resulting wave is* ○.

ferent amplitudes and periods are superimposed. The resultant, shown as a solid line, is the sum of the two. It shows the effects of constructive and destructive interference. This phenomenon was discovered as a property of light by Thomas Young and, as we mentioned, it convinced physicists that light must be a wave manifestation. It was difficult to think of particles being superimposed in such a way that they cancel.

Suppose a source of light. It radiates waves in all directions except those blocked by the socket or any opaque parts of the source. Now, up to the present we have been considering waves in two dimensions, propagating along the X axis, say, with amplitude extending in the Y direction. Here we have waves spreading out from a source in three dimensions. The light travels a straight path as shown in Figure 5-11. An opaque object casts a shadow because it prevents the rays imping-

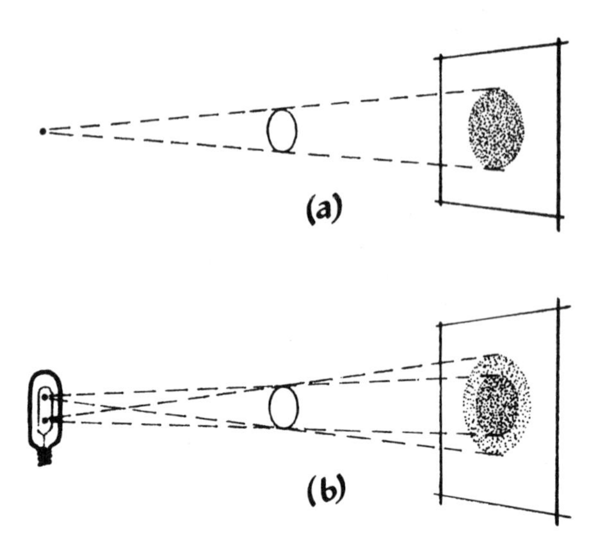

Fig. 5-11 (a) *Sharp shadow thrown by a point source.* (b) *The central umbra and surrounding penumbra cast by an extended source.*

ing on it from passing. (It was evidence of this kind that led Newton to describe light as a stream of particles moving in a straight line and either weightless or moving so fast that any effect of gravity is unnoticeable. This paradox or contradiction, as you will think of it, was one of the anomalies in classical physics that led to dissatisfaction or dissent, as we will shortly see.)

The light wave spreads out, however, in a spherical way, and so either all the rays that reach a spherical surface must have been present at the original source, or new sources must be available along the wave surface, or front. The front may be thought of as an envelope of crests all equidistant from the source. Christiaan Huygens suggested that at a wave front every point acts as a new source, so that from any given point on a wave front one could draw rays spreading out. This concept permits drawing a wave front by a geometrical method which is used as shown in Figure 5-12.

In front of a light source is placed an opaque shield S with a sharp, parallel-edged slit in it. The slit is very narrow—of the order of 0.01 to 0.05 mm—and is perpendicular to the section through the shield in the drawing. In front of this, and parallel to it, is another shield with two parallel slits, S_1 and S_2, parallel to S and equidistant from S. These are also narrow, with sharp edges, and quite close together, say 1 mm apart. Behind this is placed a screen on which an image will appear. The diagram in Figure 5-12 is exaggerated, so as to aid labeling and show what is imagined to be occurring.

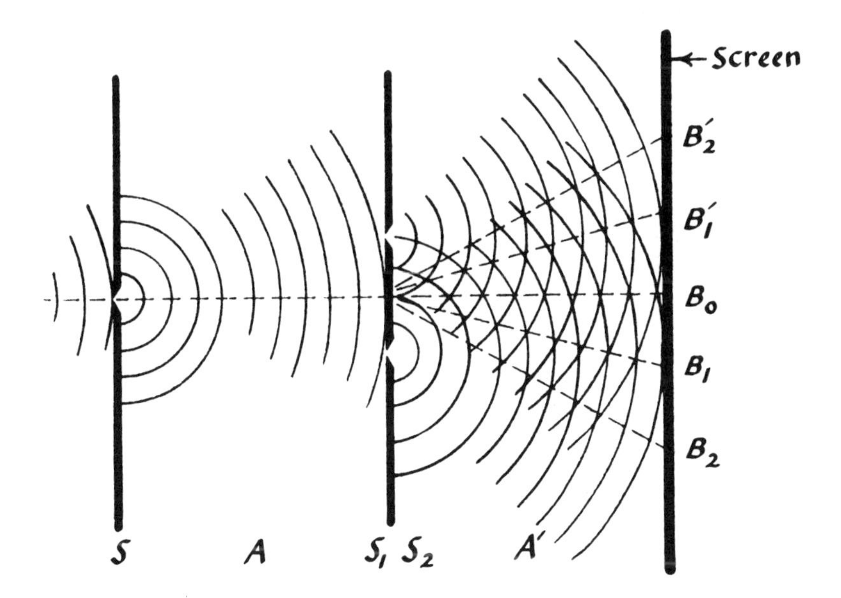

Fig. 5-12 *Huygens' construction of a wave front, showing interference pattern produced by two slits. The distance apart of the slits is greatly exaggerated.*

According to Huygens' principle, any wave front moving out from the source, and impinging on the first slit has, isolated from it and allowed to pass, a small wave front as wide as the slit. This becomes a source of waves, and produces its own curved fronts as shown at A. Each curved line is traced along a crest. So we have a sequence of crests of the marching waves. These fronts impinge on the double slit system, and parts pass the two slits. S_1 and S_2 now act as sources for waves which will be in phase and have the same wavelength since they came from the same source S. Moreover the waves from S_1 and S_2 will have the same amplitude. Because they spread out as they do in hemispherical fronts from S_1 and S_2 it will happen, because of the geometry of the system, that at certain points on the front of the superposed waves from S_1 and S_2 constructive interference occurs, and at other points destructive interference. This is shown in part A' of the figure. On the screen will appear alternate bright and dark lines where the two sets of waves constructively and destructively interact at this distance.

The geometry of the image on the screen (which can be observed directly) is shown in Figure 5-13. Let the distance apart of slits S_1 and S_2, measured from their centers, be d, and the distance from slit to screen be l. The angle made between the line from 0 to B and from 0 to B_1 is θ. Suppose that B_1 is the first bright line next to the central bright line.

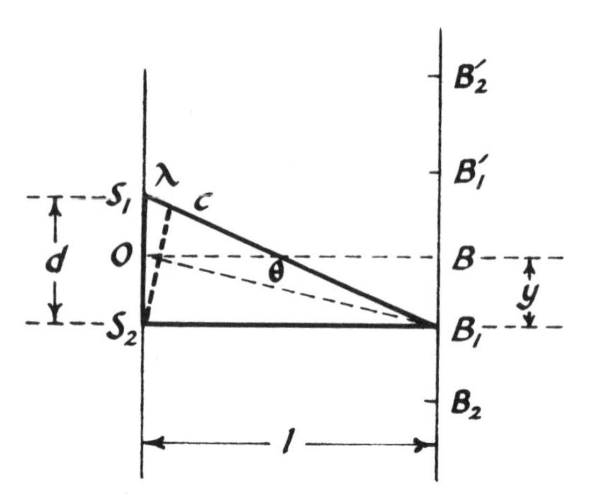

Fig. 5-13 *Geometry of interference. The figure is far out of proportion.*

This must result from constructive interference between the rays from S_1 and S_2. Since it is the first such line outside of the central line, the path length from S_1 to B_1 must be just one full wavelength longer than that from S_2 to B_1. This path difference is shown as λ. The pattern of light and dark, shown by an actual photograph in Figure 5-14, gives striking evidence for ascribing wave character to light, since it shows the alternate constructive and destructive interference of the light.

From data obtained with an instrument built on the principle of Figure 5-12, it is possible to determine the wavelength of any wave impinging on the double slit, provided only that it is a simple wave

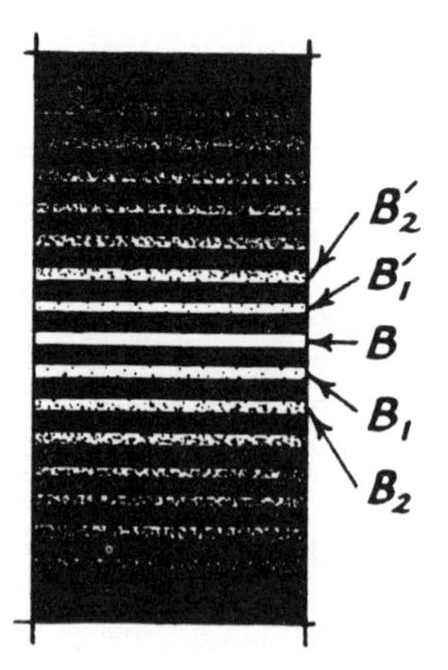

Fig. 5-14 *Interference fringes. From a photograph.*

and not a mixture. Since the wavelength of ordinary, visible light is of the order of 4 to 7 $\times 10^{-7}$ meters, one can see that the double slit instrument must produce quite large magnification if this wavelength is to become measurable on the macroscopic scale of the screen.

It was shown to be eminently reasonable, by the geometry in Figure 5-13 and the associated physical phenomena in Figure 5-14, that the path S_1B_1 is one wavelength longer than S_2B_1. This is because (1) the two wavesources S_1 and S_2 are in phase at any instant, having derived from the same source S, which is equidistant from them; (2) the point of maximum intensity of the bright line at B_1, a result of constructive interference, confirms that the two wave trains are in phase at that point; and (3) there is no bright line between B_1 and B_0. B_0 is the point of maximum intensity of the central line. Here the paths from S_1 and S_2 are of equal length. At B_1, then, the first point of superimposition of the wave trains from S_1 and S_2 occurs with constructive interference. Thus the distance S_1B_1 is equal to $S_2B_1 + \lambda$.

The angle θ in Figure 5-13, between OB_0 and OB_1, is quite small. For example, the distance l might be one or more meters and the distance B_0B_1, which is designated 'y,' might be 0.002 m. When the angle is very small the values of sine and tangent are almost identical (see Appendix 1). Thus if $l = 1$ m and $y = 0.002$ m, $\theta = 0^\circ 7'$; $\tan \theta = 0.0020$, and $\sin \theta = 0.0020$. Because l is large compared with S_2C, this line, S_2C, struck off as an arc with center at B_1, is essentially straight and perpendicular to S_1B_1, making the triangle S_1S_2C a right triangle similar to OB_1B_0. With these limitations, λ is related to y through the ratios:

$$S_1C = d \sin \theta$$
$$B_0B_1 = l \sin \theta$$
$$\lambda = (y/l)\, d$$

If a bright line is designated B_n,

$$B_0B_n = nl \sin \theta$$

Whence:

$$\lambda = (y/nl)d \tag{5-1}$$

In a particular apparatus with slits 0.5 mm apart, the center of the first bright line on a screen 1 m away from the slits was found to be 1 mm from the center of the ('zero-order') central bright line. The wavelength of the light could then be calculated to be:

$$\lambda = yd/nl = 0.0005 \times 0.001/1.00 = 5 \times 10^{-7} \text{ m}$$

It follows from this discussion that the interference phenomenon provides a most sensitive tool for measuring a wavelength, given accurate measurements of y, l, and d. Very ingeniously, A. A. Michelson turned the concept around and developed an instrument, an 'interferometer' for using light to measure small differences in length between two light paths. This difference in length might be mechanical or optical: the two paths might differ in length as measured by a ruler, or they might differ in "length" because some material interposed in one path slows down the light, and thus in a given period of time prevents a light wave from travelling as far as it would without this retarding factor. This enabled the measurement of small differences in the speed of two light beams.

5.6 The Michelson interferometer

The interferometer may be described as follows. A source of light (S) (not a slit or pinhole, but an extended source such as a strong lamp) is placed behind a ground-glass diffusing screen (Figures 5-15, 5-16). The diffused light falls on a half-reflecting mirror (P_1). This is a piece of highly polished plateglass upon one side of which has been deposited a uniform thin film of silver or aluminum. The film of the 'half-silvered mirror' is just thick enough so that half of the impinging light is reflected and half is transmitted. The light from (S) passes into (P_1), is refracted, strikes the half-silvered surface on the side away from (S),

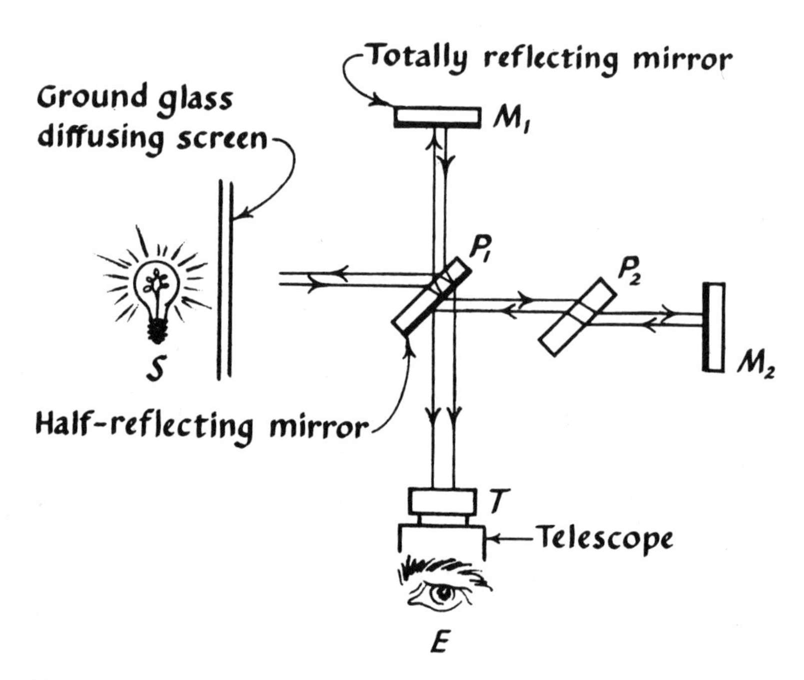

Fig. 5-15 *An interferometer. (Schematic. Some of the light paths are left out.)*

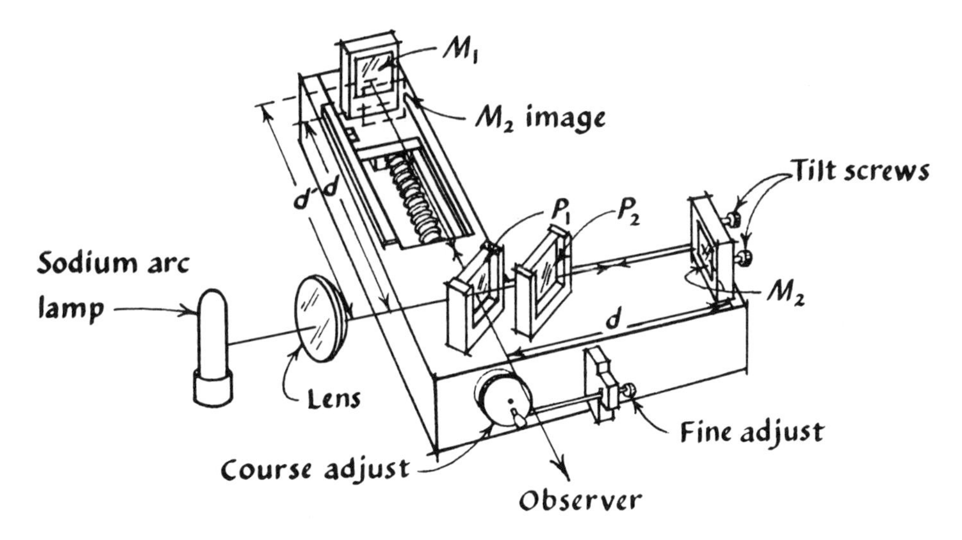

Fig. 5-16 *A sketch of the interferometer. [From C. Harvey Palmer,* Optics, Experiments, and Demonstrations. *Johns Hopkins Press, Baltimore, 1962, Figure B4-1, p. 148.]*

half is reflected, and half is transmitted. The reflected light goes to the totally reflecting mirror (M_1) and back through the plate (P_1) with one-half being reflected back towards (S) and the rest passing on to a telescope (T) through which the observer looks with his eye (E). The telescope is not actually necessary except for precise measurements. (M_1) is mounted on a screw so that it may be smoothly moved toward or away from (P_1) while being kept exactly vertical and at right angles to the light path. The distance from the reflecting surface of (P_1) to that of (M_1) can be read off from a vernier head attached to the fine screw that moves (M_1) in or out.

The transmitted half of the light from (S) passes from (P_1) to (M_2), another totally reflecting mirror, and back to the telescope and eye. In the path of this light is placed a highly polished unsilvered plate (P_2), of the same glass and the same thickness as (P_1) to compensate for the fact that the beam to and from (M_1) went three times through the glass of (P_1) while that to (M_2) passed only once. Now the optical paths in glass are the same. The mirror (M_2) has adjustment screws which allow it to be turned sideways or tilted, so that it may be adjusted relative to (M_1). What has been done is that the light from the source (S) has been divided into two parts and sent out along paths at right angles to each other, then brought back and recombined. Because of the types of reflection occurring along the paths of the beams, they return exactly out of phase ((c) in Figure 5-17).

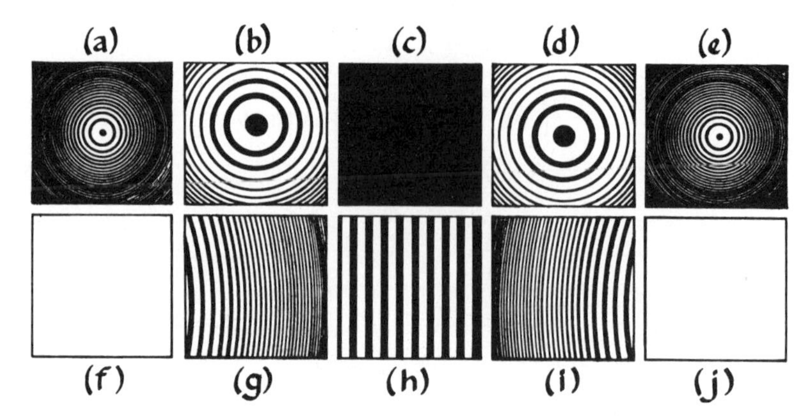

Fig. 5-17 *The appearance of the various types of fringes observed in the Michelson interferometer. If the length of the two paths is nearly the same, that to* d′ *being a little longer, and the two mirrors are exactly at right angles, the type of pattern seen is* (a). *As the mirror* d′ *is moved toward the source, to bring the path lengths more nearly the same, the rings contract, vanishing at the center. At the same time, they become more widely spaced* (b) *until at exactly equal path lengths, the central dark fringe covers the whole field. Moving beyond this, with the path to* d′ *becoming steadily less than that to* d, *fringes, alternately light and dark, grow out from the center:* (d) *and* (e). *If the mirrors are not quite parallel, the fringes are not circular, but eccentric, and can be made to appear as in* (g), (h), (i). *As the mirror* d′ *comes closer to exact equality of path length with* d, *the fringes move to the left*—i.e., *inward as before, becoming straighter. At exact equality of path length,* (h) *is observed, and if the mirror is brought inward still further, the fringes curve in the opposite direction. "The blank fields* (f) *and* (j) *indicate that this type of fringe cannot be observed for large path differences."* [*From Francis A. Jenkins and Harvey E. White,* Fundamentals of Optics, *Ed. 2, McGraw-Hill, New York, 1950.*]

When one looks through (P_1) with or without a telescope, at (M_1), one sees superimposed on it a *virtual image* of (M_2) (Figure 5-16). A virtual image makes an object appear to be at a certain place, as a reflection in a mirror does, when it is not actually there. Imagining, however, that (M_2) is at the position of its virtual image, one can then bring (M_1) up to this image, by means of the vernier screw, and superimpose them. The superimposition can be made exact by the use of the tilt and turn screws on (M_2) if needed.

Suppose that (M_1) and (M_2) are exactly at right angles to each other, but the path involving (M_1), which we shall call the (M_1)-path, is a little longer than the (M_2)-path. The difference in distance read off on the vernier is d; then the path-length difference is $2d$ because of reflection. If d is not too great—a few wavelengths for white light; many centimeters for monochromatic light—fringes appear, as in Figure 5-17*a*. As (M_1) is brought closer to the virtual image (M_2), the rings of fringes contract, successively vanishing at the center as the (M_1)-path-length approaches that of (M_2) while at the same time the fringes become more widely spaced (*b*). The effect is very striking. At exact equality of path-lengths, the central dark fringe occupies all of the field (*c*). This point can be reached only with the most delicate adjustment. Going beyond this point, away from path-length equality, fringes emerge from the center, growing outward and retracing (*b*) and (*a*) as in (*d*) and (*e*).

As one could compute from the visual image of the distance between (M_1) and the virtual image of (M_2), a movement of (M_1) by $1/4\ \lambda$ lengthens the optical path by $1/2\ \lambda$, and if started in phase brings the wave train to (M_1) just out of phase with that from (M_2). Further movement of (M_1) to a distance of $1/2\ \lambda$, brings the two wave trains back into phase.

If the mirrors are not exactly at right angles, so that the "space" separating (M_1) and the virtual image of (M_2) is wedge-shaped, then the fringes are no longer circular but eccentric, becoming nearly straight as d becomes smaller (Figure 5-17g). In some cases the fringes at exact equality of path-length (where the virtual image of (M_2) intersects the image of (M_1)) are straight (*h*). They march across the field as (M_1) is moved toward coincidence with the virtual image of (M_2). If a pointer is placed in the telescope so that its image is projected on the field, one can count the fringes as they march past the pointer when d is made larger or smaller.

The sensitivity of the instrument is clearly very great, for the distance between fringes corresponds to a path difference of one wavelength of the light being used, and it is possible to detect a displacement of 0.1 to 0.05 of a fringe. The wavelength of the yellow sodium D line in air is 5.9×10^{-7} m.

The constructs of absolute space and time 5.7

In Aristotelian theory objects are conceived to fall because all matter strives toward the center of the Earth. This, so to speak, gives the center of the Earth privileged status in the spatial Universe: an absolute status. Here was an anchor-point; a place to make measurements from. Copernicus initiated the breakdown of this theory, with his conceptual removal of the Earth from the center of the Universe. There was also a fixed point from which to measure the passage of time: the moment of creation of the Universe. But this concept, too, began to be questioned, and was destined to be abandoned in its original form. By the time of Galileo, and more fully in Newton's theories, it was recognized that as a base for *theoretically* valid measurement, the Earth frame of reference would not suffice. *Practically* speaking, there is no serious problem, of course, and everyday life goes on as usual. But the theoretical problem is roughly this.

By Newton's first law of motion, an object not acted upon by an unbalanced force will continue in its state of rest or uniform rectilinear motion indefinitely, or until some applied force changes its state of motion. By the second law, the acceleration that ensues will be directly related to the applied force in the direction of motion, and inversly related to the mass of the object: $\mathbf{F} = m\,a = m\Delta v/\Delta t$. But in what frame of reference do these laws hold? In what frame a rectilinear path? A straight path taken in our laboratory would seem far from straight to an observer on another planet, for the moving body would partake also of the Earth's rotation and orbital motion. Practically speaking the Earth's surface is taken as a suitable platform to which to attach a set of coordinates from which to make measurements. We know, too, that any point on the surface of the Earth, because it is changing its direction of motion all the time, is undergoing acceleration. For this reason no object on the face of the Earth is free from centripetal force. Therefore the Earth could not, theoretically, provide an inertial frame—one free from acceleration, wherein Newton's first two laws of motion would hold *exactly*. It was suggested that an inertial frame, with the coordinates located relative to the fixed stars, would suffice, and it does. This reference frame enables the orbits of the planets to be calculated exactly from Newton's laws. However, why should this frame be especially privileged? Newton conceived of space that exists independently of anything in it: absolute space. He conceived also of even-flowing time, independent of events that happened in time: absolute time. No relation between the two was proposed.

Neither of these constructs seemed amenable to test, a situation that always disturbs physicists. It was suggested, for example, that the center of gravity of all matter in the Universe might be considered to be absolutely at rest, and a plane through this point chosen as a fixed direction; how would one determine that center? Other expedients were advanced. One attractive proposal said that since light has a wave-nature, there must be something that waves. This 'substance,' a 'luminiferous' or light-bearing aether, must permeate all space since light from the farthest stars reaches the Earth. The proposal, then, was that the aether serve as reference body. It was thought that the Earth, in her annual journey around the Sun, must at some time, and perhaps at all times, move relative to the aether, especially as it was known that the Solar System is moving relative to the fixed stars. One was asked to accept the image of the Solar System rushing through aether-filled space. Compounded to this motion were the orbital and spinning motions of the Earth. With this model in mind, many experimenters tried, through a wide variety of approaches, to detect this motion by finding an aether stream, or aether "wind" rushing past the Earth. For if it could be detected, and if a natural direction of motion could be found, an absolute system of reference could be defined, assuming the aether stationary. It was to this problem that A. A. Michelson, and later with him his colleague E. W. Morley, was attracted.

The Michelson-Morley experiment 5.8

One can make a model of this reasoning in terms of the behavior of boats on a river. Given a river flowing with velocity v, and two boats that go at exactly the same speed c; one goes upstream and back, the other directly across the stream the same linear distance, and back. Which will arrive first? (Exercise 5-23)

Suppose that light passes along the arm bearing mirror (M_2) (Figure 5-15 or 5-16) which is turned parallel to the direction in which the Earth moves through the aether (and thus the aether 'drifts' past the Earth along the direction (S) to (M_2)). If the velocity of the aether drift is v, and that of the light is c, then the time t_1 required to go from (P_1) to mirror (M_2) is $L/(c - v)$, where L is the length of the path (P_1M_2). The aether drift increases the speed of the return journey from (M_2) to (P_1); now the time $t_2 = L/(c + v)$. Going at right angles to the drift (Figure 5-18), with velocity c and drift velocity v, the net velocity is $\sqrt{c^2 - v^2}$. The time of transit each way is $L'/\sqrt{c^2 - v^2}$, where L′ is the distance (P_1M_1). The total time parallel to the drift is then $T_{\parallel} = 2\,Lc/(c^2 - v^2)$ and the total time perpendicular to the drift

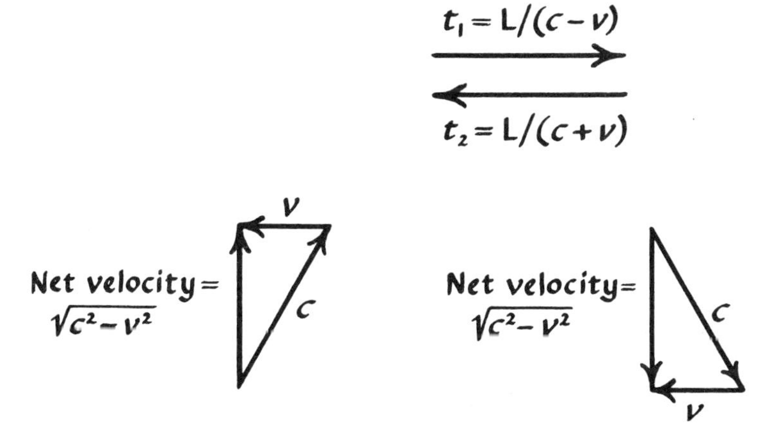

Fig. 5-18 *Vector diagrams of reasoning in ether-drift experiment.*

is $T_\perp = 2L/\sqrt{c^2 - v^2}$. That the former is a larger number can be shown by dividing the two:

$$T_{||}/T_\perp = 1/\sqrt{1 - (v^2/c^2)}$$

It should be noticed that this ratio is larger than 1 for $c > v > 0$. If v were zero, the ratio would be 1. This is what would be expected if Earth and aether were stationary with respect to each other (other things being equal). If v were greater than c, the equation would be indeterminate. This is also meaningful in the sense that with $v > c$, the light would never be able to reach the mirrors but would be swept downstream.

Michelson planned to utilize the orbital velocity around the Sun of the Earth to provide v. This is about 3×10^4 meters per second and is four orders of magnitude (that is, 10^4) smaller than the velocity of light, which was known at the time to be close to 3×10^8 meters per second. At first he set up his interferometer in the Physical Institute of Berlin, but the vibrations of the building even at night defeated him. He finally solved his instrumental problems by setting up the interferometer on a massive block of stone which was floated on mercury. This apparatus was built and the work done at the Case Institute of Technology in Cleveland, where Michelson and his colleague E. W. Morley carried out definitive experiments for many years.

The apparatus is shown in the three figures reproduced from the original illustrations (Figures 5-19, 20, 21). A stone block, about 1.5 meter square, and 0.3 meter thick, resting on a ring-shaped piece of wood floated on mercury in a circular cast-iron trough. There was a clearance of about 1 cm around the float, and the whole was kept centered with a pin which, however, did not bear any weight. The interferometer mirrors were mounted on the stone, as shown. When the entire apparatus was lined up, with a set of fringes appearing through the telescope, it was ready for observation.

The procedure was to give the apparatus a slight push so that the stone would rotate, floating on the mercury, making one full turn in about six minutes. After getting it smoothly moving, the experimenter would walk round with it, keeping it gently moving and taking readings at exact intervals, at sixteen equidistant points around the circle. Six revolutions would be made, and the six readings at each point averaged. The readings were made near noon, and near six o'clock in the evening. At the noon readings the stone was rotated counterclockwise; in the evening, clockwise. Results are shown in Figure 5-22, in graphical form. The upper curve reports the noon, the lower the evening observations. "The dotted curves represent *one-eighth* of the theoretical displacements." Said Michelson and Morley, "It seems fair to conclude from the figure that if there is any displacement due to the relative motion of the earth and the luminiferous aether, this

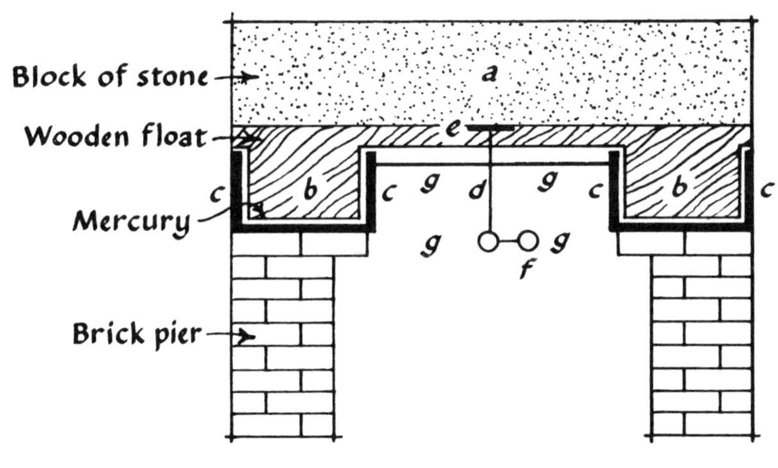

Fig. 5-19 *Michelson's interferometer for the ether-drift experiments. A block of stone,* a, *"is about 1.5 meter square and 0.3 meter thick. It rests on an annular wooden float* bb, *1.5 meter outside diameter, 0.7 meter inside diameter, and 0.25 meter thick. The float rests on mercury contained in the cast-iron trough* cc, *1.5 centimeter thick, and of such dimensions as to leave a clearance of about one centimeter around the float. A pin* d, *guided by arms* gggg, *fits into a socket* e *attached to the float. The pin may be pushed into the socket or be withdrawn by a lever pivoted at* f. *This pin keeps the float concentric with the trough, but does not bear any part of the weight of the stone. The annular iron trough rests on a bed of cement on a low brick pier built in the form of a hollow octagon."* [*From A. A. Michelson and E. W. Morley,* Am. J. Science *III,* 34, *333* (*1887*), *p. 339, Fig. 5.*]

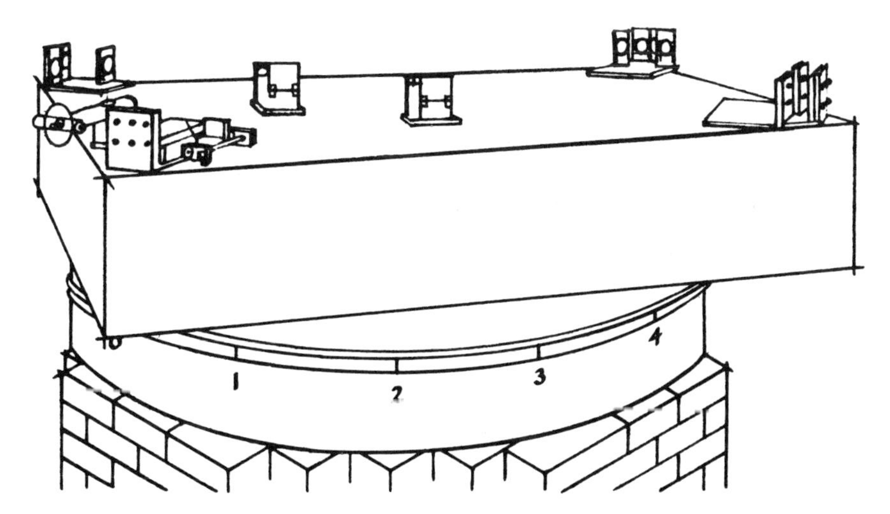

Fig. 5-20 *Appearance of the interferometer. The telescope and adjustable mirror are at the front left.* [*From Michelson and Morley,* ibid., *p. 337, Fig. 3.*]

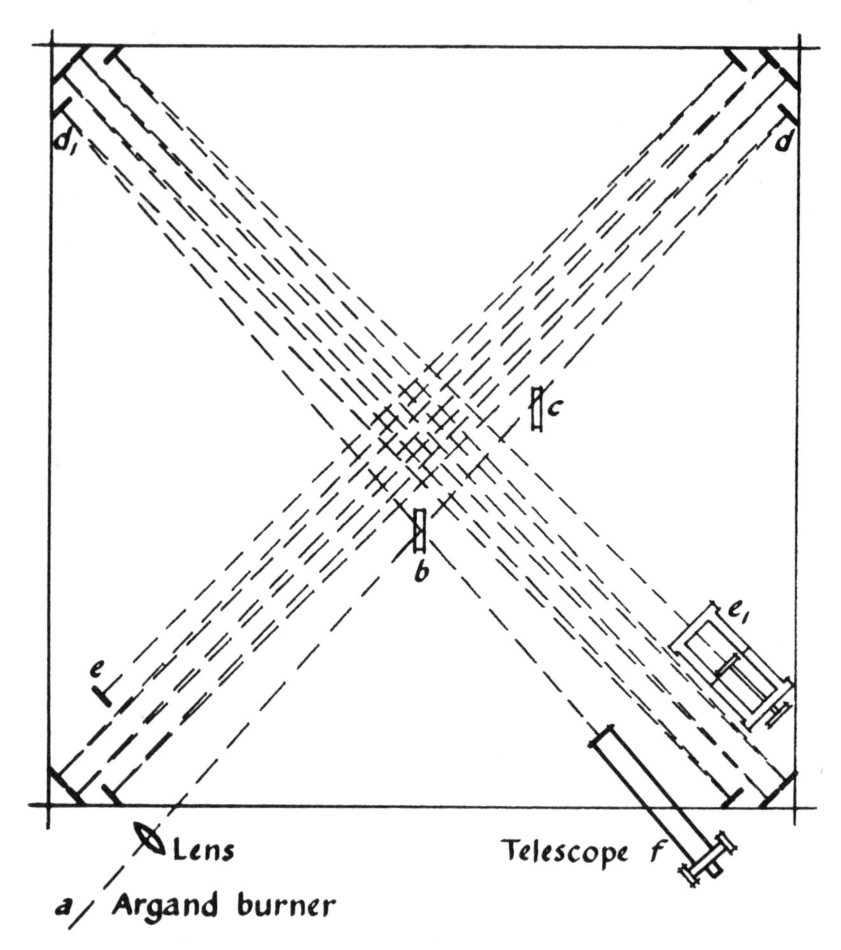

Fig. 5-21 *The light-path. "At each corner of the stone were placed four mirrors* dd, ee, ... *Near the center of the stone was a plane-parallel glass* b. *These were so disposed that light from an Argand burner* a, *passing through a lens, fell on* b *so as to be in part reflected to* d: *the two pencils followed the paths indicated in the figure,* bdedbf *and* $bd_1e_1d_1bf$ *respectively, and were observed in the telescope* f. *Both* f *and* a *revolved with the stone. The mirrors were of speculum metal carefully worked to optically plane surfaces five centimeters in diameter, and the glasses* b *and* c *were plane-parallel and of the same thickness, 1.25 centimeter; their surfaces measured 5.0 and 7.5 centimeters. The second of these was placed in the path of one of the pencils to compensate for the passage of the other through the same thickness of glass. The whole of the optical portion of the apparatus was kept covered with a wooden cover to prevent air currents and rapid changes of temperature."* [*From Michelson and Morley,* ibid, *pp. 337–338, Fig. 4.*]

cannot be much greater than 0.01 of the distance between the fringes." In short, any relative motion of the Earth and the luminiferous aether did not exceed about one-quarter of the Earth's orbital velocity. Later

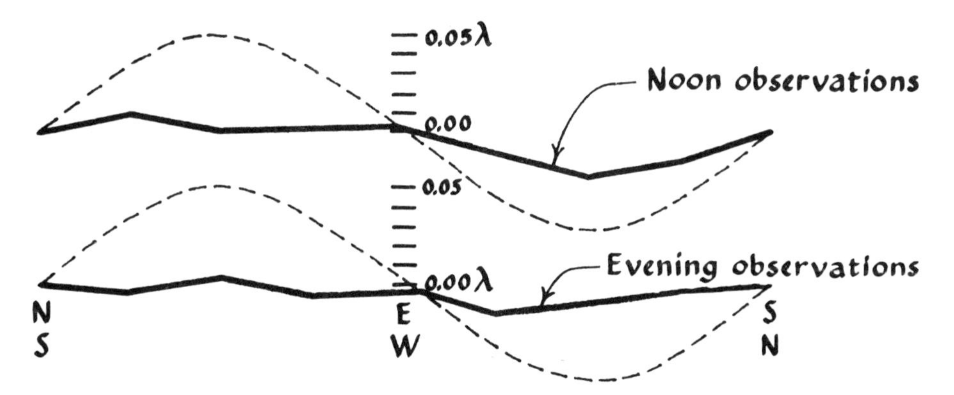

Fig. 5-22 *The heavy lines represent the results of observations, the upper for observations made at noon, the lower for those made in the evening. "The dotted curves represent* one-eighth *of the theoretical displacements."* [*From Michelson and Morley*, ibid, *p. 340, Fig. 6.*]

experiments using other methods have confirmed the absence of aether-drift

The sequel 5.9

Michelson was disappointed in the negative results of his attempt to demonstrate an aether-drift, and it was only years later, after Einstein had created the special theory of relativity, that the importance of the experiment came to be realized. It is sometimes suggested that Einstein's theory was developed because of the Michelson-Morley results. Michael Polanyi queried Einstein on this point and was assured that the theory was not founded to explain these results, nor did the Michelson-Morley experiment play any role in the development of the theory. This is not to derogate the experiment, of course, but to be sure that the record is straight. Einstein had developed the theory as a result of dissatisfaction with the lack of symmetry of certain physical relationships; his driving motive seems to have had a large esthetic component.

Michelson and Morley had carried out their experiments at different times of the day, and at different times over a year, so that if there were a stationary aether the effect of the drift would be doubled in comparing the results. When, say, the Earth was moving in one direction relative to the aether, and six months later in the opposite direction as it circled around the Sun, the effect of an aether wind would be observable in one direction or the other, would it not?

That no appreciable effect was found in the Michelson-Morley experiment might have been explained in several ways. It was suggested, for example, that the Earth carries its contiguous aether along with it; but this was not in agreement with experiments which had shown the slowing down of light in transparent materials, and thus the unlikelihood that the aether was carried along with the same velocity as the material object. H. A. Lorentz and G. F. FitzGerald suggested, quite independently, that the lengths of the arms of the interferometer might be affected by their velocity relative to the aether. Lorentz calculated from theoretical considerations what the shortening should be, assuming that the forces that hold atoms and molecules together are related to stresses in the aether. Thus a solid material moving into the aether should be contracted in the direction of motion. (The analyses and calculations are more complicated than we can present here.) A ruler, for example, would be shortened. Also an observer in the laboratory would experience the same ratio of shortening or flattening in the direction of motion. Moreover, the analysis showed that clocks would be slowed, and presumably metabolic processes in the observer would be slowed, so that one would not be aware of any change: everything would change in commensurate ratios. When these calculations were applied to the Michelson-Morley experiment it was shown that (if this hypothesis of contraction were correct) there should be no shift of fringes. We shall return to the Lorentz calculations in the following Chapter.

5.10 Summary

Light is hardly noticed until our attention is called to it. This chapter on light begins a discussion of those aspects of science that are relevant to our whole purpose. Especially it sets the context for the modern era by suggesting the state of affairs at the end of the nineteenth century, and by reporting in some detail the experiment of Michelson and Morley in which it was not possible to find an aether-drift; a 'wind' of aether 'blowing' past the Earth in its passage.

When the state of a discipline becomes complex because of an accumulation of discrepancies which the prevailing pattern of thought cannot handle, the stage is set for the appearance of a synthesis that resolves those discrepancies in the old theory by answering them in a new pattern of theory. Either this or an alternative suggested by Lancelot Hogben: that when science cannot provide ways of discussing things that interest them, people always fall back on the language of magic. As he remarks, "Magical views of the world have declined not because science disproves them, but because science provides better ways of discussing the same issues." What happened was a grand synthesis, shaped by Planck, Einstein, Rutherford, and others, going under the names of relativity and quantum theory. We now turn to these, beginning with relativity.

Note: Description of a wave. What we wish to show is the way of thinking about an observed wave that enables it to be conceptualized in mathematical terms. This recalls Section 4.5, and some trigonometry. Consider the continuous wave-form in Figure 5-9, in which the waves move at a steady pace, with amplitude A, period T, frequency ν, wavelength λ, velocity u in the X direction, and the vibrating particle (*e.g.*, No. 5) with the same amplitude, r, and period, moving in the Y direction with velocity v and angular velocity ω. (The angular velocity is defined analogously to linear velocity, i.e., $\omega = \theta/t$: the angle swept out in the time t, in radians per second. The linear speed around the circle in Section 4.5 is $s/t = r\theta/t$. From the two equations, $s/t = v = r\theta/t$. Then, given $a_c = v^2/r$, one can derive by algebra $a_c = \omega^2 r$. This equation is used below.)

The motion of a wave of this kind is doubly periodic: it is periodic in time and in distance. Thus to describe such motion SHM (simple harmonic motion) must be combined with linear motion at constant velocity perpendicular to the SHM.

Simple harmonic motion is an oscillation such as the up-and-down motion of an object suspended from a good spring: at the top of the stroke the object stops, reverses direction, falls to the bottom of the stroke, stops, reverses and starts up again: up and down. This is the motion that would be observed if a particle were moving in a circle at constant velocity and its position were projected onto the Y axis, as in Figure 5-23. At some

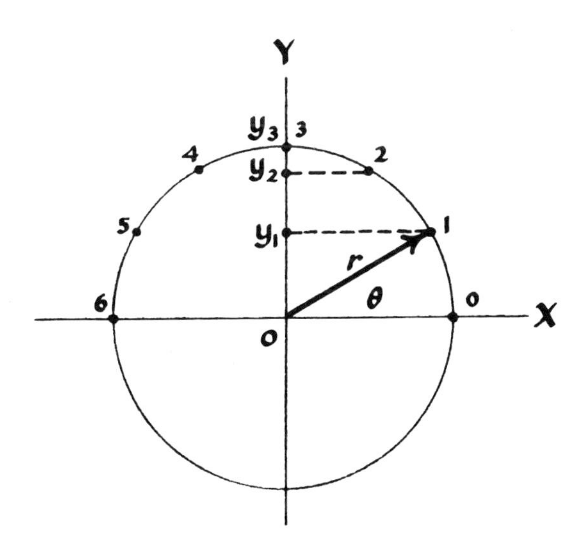

Fig. 5-23 *Projection of a point in uniform circular motion onto a Y axis. The motion of the point on the Y axis is SHM.*

instant, say when the particle is at '1' on the circumference of the reference circle of radius r, the projected position is y_1. As the particle moves along, the angle θ, the 'phase angle' changes, but always the corresponding positions on circumference and Y axis are related by the phase angle θ, and r. At $\theta = 90°$ (position 3 in the Figure) $y_3 = r$. This is the 'top' of the stroke, and the projected point now starts downward as θ increases. At $\theta = 180°$, or π radians, position 6, $y = 0$; at 270°, $y = -r$. Thus the projected point oscillates, with amplitude r. The angular velocity $\omega = \theta/t$, thus the 'phase' of the point on the Y axis, oscillating with SHM, is $\theta = \omega t$. Since one period is the time taken for θ to go through 2π radians, the frequency of oscillation is $\nu = \omega/2\pi$.

The 'displacement' of the particle 5 (say) can in this way be related to position on a reference circle with amplitude r. We now look at a particle on the Y axis and refer to a point on the reference circle. This displacement, by the geometry of the Figure 5-23 is $y = r \sin \theta$, and, introducing the angular velocity into the above relation:

$$\text{displacement} = y = r \sin \omega t \qquad (5\text{-}2)$$

Speed around the circle is constant (Figure 5-24). The Y component of the speed, projected from vectors on the circle, clearly changes. It is equal to the speed at $\theta = 0$; decreases to 0 at $\theta = 90°$; reverses sign; and at $\theta = 180°$ is again equal to speed, but in the opposite direction, becoming 0 at 270° and reversing direction again. By definition the reference point has the speed ωr around the reference circle, so the velocity of the particle on the Y axis, projected from the constant speed on the circle, is:

$$\text{velocity } v = \omega r \cos \theta = \omega r \cos \omega t \qquad (5\text{-}3)$$

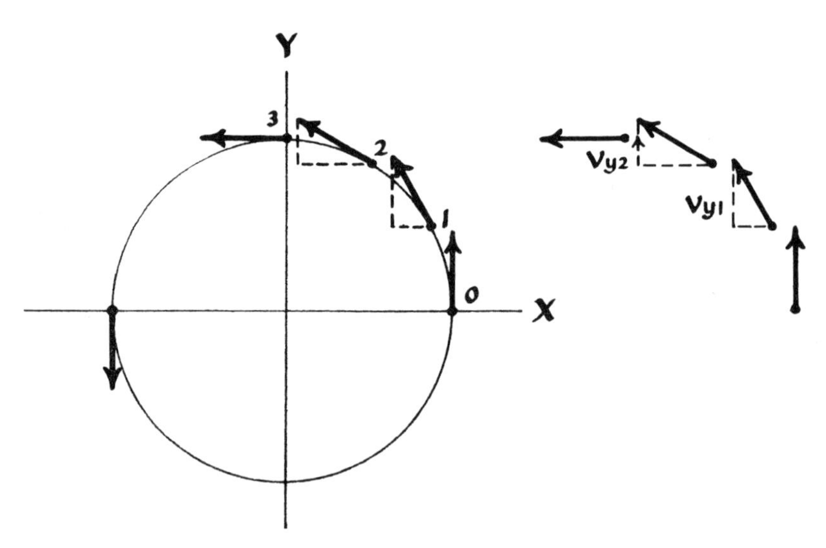

Fig. 5-24 *Projection of velocity vector of a point in uniform circular motion onto the Y axis to show velocity in SHM.*

Acceleration of the point on the circle, we saw is $a_c = \omega^2 r$. Its direction may be represented by a vector pointing towards the center of the circle (Figure 5-25). Its projection on the Y axis then varies as sin θ, that is sin ωt. The relation between the acceleration of the oscillating particle and the phase angle thus is:

$$\text{acceleration} = a = -\omega^2 r \sin \omega t = -\omega^2 y \quad (5\text{-}4)$$

The minus sign indicates that the direction of the acceleration is opposite to that of the displacement. From the figure one can see that a is greatest at the top of the stroke, becomes zero and changes direction at $\theta = 0$ and $\theta = 180°$, and reaches a maximum again at $\theta = 270°$.

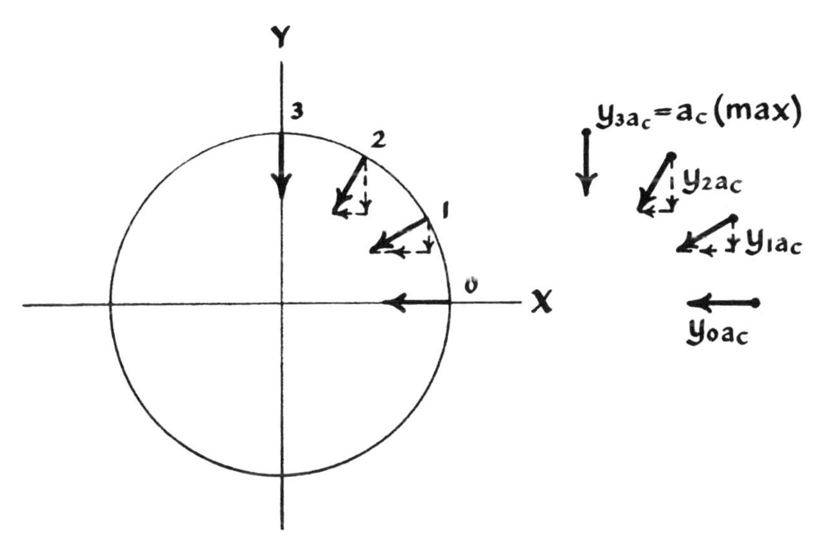

Fig. 5-25 *Projection of acceleration vector of a point in uniform circular motion onto the Y axis to show acceleration in SHM.*

With this introduction to a device for speaking of a linear oscillation, where s, v, and a are all changing in terms of circular motion with constant $|v|$ and $|a|$, we are ready to return to considering the wave form (Figure 5-9).

Look, for convenience, at particle 5. Its displacement in the y direction is:

$$y = r \sin \theta = r \sin \omega t$$

Consider this the reference particle at a time zero, and phase angle zero. Look along the wave to particle 9 which is in the same phase. Its phase angle is 2π less than that of particle 5. Thus the displacement of a particle in the direction of the wave motion is:

$$y = r \sin (\omega t - \theta)$$

Here we can write θ instead of $\Delta\theta$ because by taking $\theta_0 = 0$, $\Delta\theta = \theta - \theta_0 = \theta$. The negative sign shows that the particle ahead of the reference always shows a lower phase angle. This equation holds as time goes on. The phase angle θ of reference particle continually increases, and so does that of the particle down-wave, but always behind by the quantity θ of the reference particlc.

Now by the model, and by actual observation, θ is proportional to the distance of this particle down-wave from the origin; that is, to the value of x. Thus:

$$\theta = Kx$$

Here K is a constant. Then:

$$y = r \sin (\omega t - Kx)$$

This is the equation for a sine wave travelling to the right along the X axis. It gives the profile of the wave at any instant t, by relating the displacement in the y direction of a particle to its position along the x axis. It can be related to a particular wave of amplitude A, period T, and wavelength λ, by noticing that $r = A$; that at $x = \lambda$, $\theta = 2\pi$, whence $K = 2\pi/\lambda$; and that $\omega = 2\pi/T$. Thus:

$$y = A \sin 2\pi[(t/T) - (x/\lambda)] \qquad (5\text{-}5)$$

If the wave had been travelling in the opposite direction, namely toward the left, the phase of the reference particle at the origin would be behind that of a particle to the right of the origin, so that for its equation:

$$y = A \sin 2\pi[(t/T) + (x/\lambda)]$$

To plot this equation in a particular case certain data are needed. The amplitude A, or r, is in a sense an arbitrary constant: it could be larger or smaller, depending on the amount of displacement that originates the wave. But ω, or T, depends on the properties of the thing that undergoes the wave motion. It is not under our control—given a particular rope, say. It is related to the physical properties of the material and the way that the material is fabricated. So we must return to perceptions if we wish to describe some particular wave. But if we wished we could imagine values of ω or T for some hypothetical material to calculate what would be the result.

Consider the wave equation (5-5) in a most general way. Observe that y is a quantity that specifies displacement. The essential feature is that the value of y specifies the displacement of a reference particle, or of a particle in a wave. A is an arbitrary constant which has the physical meaning of amplitude. It could be stated as the radius of the reference circle, r, or as the maximum value of y, for a given wave form. The term x/λ is, physically speaking, the phase of the motion at time zero, and is arbitrary in the sense that it may be desirable to set $t = 0$ at $\theta = 0$, but it could be set at any value. What is not arbitrary is the restoring force. This depends on the nature of the thing that is undergoing wave motion. It is reflected in ω, T, λ.

Energy of the wave. If a wave-pulse is sent down a rope (Section 5.4), and reflects back to our hand, we receive a jerk, an acceleration produced by an unbalanced force. We must conclude that energy travelled down the rope and back. There was no bulk movement of the whole rope, just the oscillatory motion of the parts, yet energy was transmitted along the rope from our hand, the source or origin of the wave, back to our hand which is now the receiver. The path from source to receiver was, of course, occupied by the medium (the rope) that transmitted the wave. We might even find it difficult to think of wave motion in the absence of a medium ("something that waves"). This was, in fact, made a source of crisis in connection with the behavior called 'light,' as we shall see. At this point, however, it is clear that waves may be transmitted along strings, ropes, on water, and so on, and that they furnish a mechanism for the transmission of energy from one point to another.

To make this concept quantitative, return to the idealized wave-train and enquire about the total energy that it carries. For SHM: for example of an object oscillating at the end of a spring:

$$\boldsymbol{E}\text{ (total)} = \text{KE} + \text{PE} = (1/2)mv^2 + (1/2)ky^2 \qquad (5\text{-}6)$$

where k is the force constant of the spring, and m the mass of the oscillating particle. The wave equation combined SHM with linear velocity (Equation 5-5) to yield an expression for y. This requires that in the equation for energy the expression for y must be introduced. This involves algebraic manipulation and leads ultimately to an expression for the energy of the wave:

$$\boldsymbol{E}\text{ (total)} = (1/2)m\omega^2A^2 \qquad (5\text{-}7)$$

A derivation of this equation rests on the observation (Figure 5-24) that when y is maximum, v is minimum. Thus, returning to circular motion and Equation 5-3, v is a cosine function of y—that is, at $\theta = 90°$, v of the oscillating particle is zero, and y is maximum and equal to r. For convenience in doing the algebra, let G be the code designation of $2\pi[(t/T) + (x/\lambda)]$; then $y = A \sin G$ (from Equation 5-5). In the energy equation v occurs. By the use of the reference circle, where $v = r\omega$, $v = \omega A \cos G$, remembering that y (maximum) equals r.

Then, substituting in the equation for the total energy (Equation 5-6):

$$\begin{aligned}\boldsymbol{E}\text{ (total)} &= (1/2)mv^2 + (1/2)ky^2 \\ &= (1/2)m\omega^2A^2(\cos G)^2 + (1/2)m\omega^2A^2(\sin G)^2.\end{aligned}$$

Trigonometry helps to simplify this equation. In a right triangle, by the Pythagorean theorem, $x^2 + y^2 = r^2$. By the definitions of sine and cosine, (by convention (sine $\alpha)^2$ is written $\sin^2 \alpha$)

$$x^2/r^2 = \sin^2 \alpha \qquad y^2/r^2 = \cos^2 \alpha$$

and thus, by algebra, $\sin^2 \alpha + \cos^2 \alpha = 1$. Also $k/m = \omega^2$, whence $k = m\omega^2$, thus we can write

$$\boldsymbol{E}\text{ (total)} = (1/2)m\omega^2A^2$$

(It is interesting to mention that studies by army ordnance physicists of the energy carried by sound waves show that within 20 or so feet of many jazz combos that use electronic amplification the energy is sufficient to cause permanent impairment of hearing on short exposure.)

The value of this equation is this: it says that *the energy that is carried by the wave is proportional to the square of the amplitude of the wave*. This will turn out to have an interesting interpretation in quantum theory and atomic structure. At the same time we must wonder what m refers to. The quantity must be related to the mass of the rope (or other vibrating object) or its density (mass per unit of volume). We shall not need to entertain this question further here, *e.g.*, how long a piece of rope is a 'particle,' since when we actually need to answer it we shall be dealing with oscillating electrons, which do have a mass; or we shall deal with it in some other way.

Exercises

5.1 Using the principles discussed in this chapter, can you find where the image would appear if a source of light at O (or any small object at O) were reflected in a curved mirror with radius of curvature r?

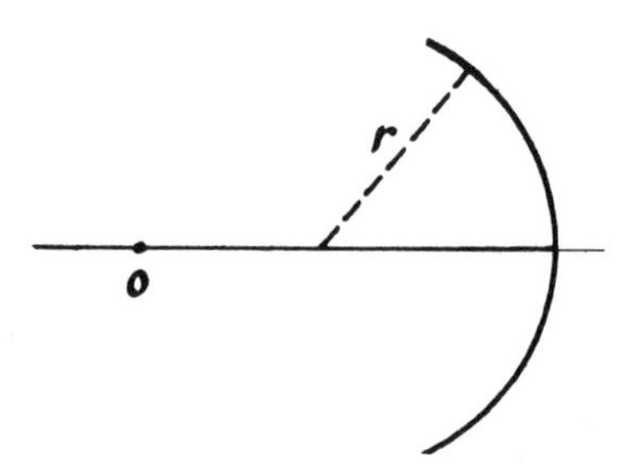

5.2 The mean distance of the Earth from the Sun is 1.49×10^{11} meters (one astronomical unit). How long does it take for a pulse of light, originating at the Sun, to reach the Earth?

5.3 Why is it necessary to point out that the line S_2C in Figure 5-13 is virtually perpendicular to S_1B_1—that is, why isn't it *exactly* perpendicular?

5.4 If white light, consisting of a mixture of wavelengths, is used in the double slit experiment it is observed that the central line (the zero'th order line, or "fringe") only, is white. The others show spectral colors. (Red is about $\lambda = 6.5 \times 10^{-7}$ m; violet is about 4.3×10^{-7} m.) a) Why is the central line white? b) In what order, from the center outward, do the colors appear in the fringes? Explain.

5.5 What is the mathematical device by means of which a microscopic measurement, *e.g.*, wavelength of light, is brought to macroscopic measurement, *e.g.*, distance between two fringes? (This question refers to Young's double slit instrument.)

5.6 In Figure 5-15 or 5-16 we say that the light paths are at right angles. Show exactly why this is a correct statement; that is, trace the paths in detail.

5.7 Compare the principle of the Michelson interferometer with that of Young's double-slit demonstration, Figure 5-12. Show similarities and differences.

5.8 Go through the algebra in connection with Figure 5-18, including the ratio of $T_{\parallel}/T_{\perp}$.

5.9 Given a river flowing at 3 miles per hour and two identical boats going at 5 m.p.h. Boat 1 goes across the river one mile and back; the other, boat 2, goes up-stream one mile and back. Which will arrive sooner, and in what ratio of times, assuming that they take essentially no time to turn around? They start at the same place and return to it.

Answers to Exercises

5.1 Consider the geometrical construction shown in the accompanying figure and proceed from there.

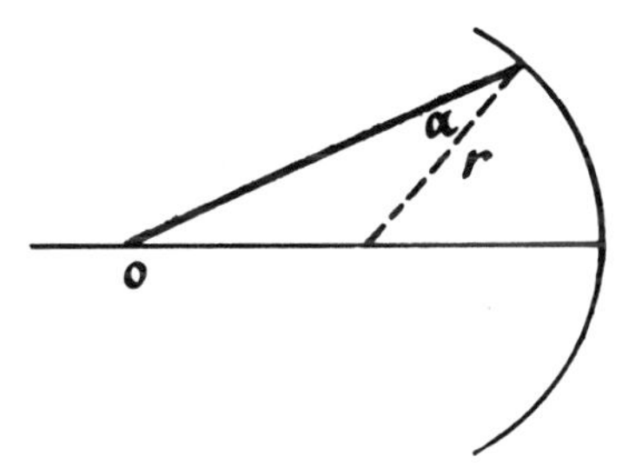

5.2 500 seconds.

5.3 The tangent to B, C would be perpendicular. This line is a chord, from C to S_2.

5.5 Ratios are used.

5.6 Trace the refraction and reflection angles and draw geometric conclusions.

5.9 #1. $T_{\parallel} = 1/(4 - 3) + 1/(4 + 3) = 1 + 1/7 = 1.14$

#2. $T_{\perp} = 2/\sqrt{7} = 2/2.65 = 0.754$.

Thus boat 2 takes less time than boat 1, and the ratio of times is close to 1.5.

Bibliography

Sir Charles Sherrington, *Goethe on Nature and on Science*. Cambridge University Press, 1942; 2nd ed. 1949.

Mircea Eliade, *Patterns in Comparative Religion*, Meridian Books, World Publishing Co., 1963. Sheed & Ward, 1958.

L. L. Whyte, *Essay on Atomism: from Democritus to 1960*. Harper & Row, New York, 1963. Wesleyan University Press, 1961. This is a good, concise survey, with a useful "Annotated Selected Biography."

Willard C. Humphreys, *Anomalies and Scientific Theories*. Freeman, Cooper & Co., San Francisco, 1968.

A. A. Michelson, *Light Waves and their Uses*. University of Chicago Press, 1903 (The Lowell Lectures of 1899).

Lancelot Hogben, *The Nature of Living Matter*, Kegan Paul, Trench, Trubner & Company, London, 1930, p. 250.

Part III

Chapter 6. Relativity

Chapter 7. Electromagnetic Radiation

Chapter 8. Quanta and Photoelectric Theory

The Fruit of Discontent ~ PART III

Toward the end of the last Century it was fairly evident to many scientists that the foundations of Science needed repair. It was not that Classical Science was wrong, so much as that it was limited, inadequate to explain some of the strange discoveries that were being made. For example, in Physics, it had been found that when light is shone on certain metals it becomes easier to pull electrons out of their surfaces; and that other substances—called radioactive—might give off penetrating "rays," with no explanation for this behavior. In chemistry, the whole subject had been classified in a comprehensive scheme called the Periodic Classification of the Elements, without real, basic knowledge of why this was possible. Also strange new molecules were being made by research chemists. These things seem prosaic and unexciting as we write them down coldly, yet they were the "still, small voices" that whispered of mysteries to those with ears to hear; the "clouds no larger than a man's hand" that presaged the storm that swept through the World in the next several generations. These and other discoveries generated and nourished in many physicists a growing discontent with older theories, and led eventually to the modern era which we begin to explore in this Part.

Scientists tend, in general, to be conservative, holding on to that which is good. Some of them, at least, are likely to be discontented conservatives. This Part is given over to a discussion of the fruit of their discontent. It may be added that during the period covered in this Part, namely from about 1900 to the present, the ferment in science was part of correlative ferments in all aspects of our culture. Moreover the roots of discontent may be traced in all fields, deep into the preceding generations of thoughtful people. Thus Einstein's theory of relativity fell on fertile ground which had been prepared long before. Even though some of its conclusions were quite extraordinary nevertheless they were subject to test, and it did not take many years for the power and fruitfulness of the theory to manifest itself. We concern ourselves in Chapter 6 with what is known as the Special Theory; the simpler of his two theories of relativity. This one shows us how from simple, yet previously not obvious, propositions and rigorous reasoning the most far-reaching conclusions can follow. Starting with an enquiry about how two events can be certified to be simultaneous, the argument leads eventually to the famous mass-energy relations of the nuclear age. Typical of this kind of scientific advance, it does not destroy or contravene the classical science. What happens is that classical physics turns out to be a special case of a much wider, more comprehensive physics. What was good is conserved—as part of a greater whole.

In Chapter 7 two ideas are developed. First we show with the simplest argument how elegantly relativity ties together electricity and magnetism. This is one of the most convincing demonstrations of the power of the relativistic view. And at the same time it shows how the invariant-relation that, indeed, was the crux of Einsteinian relativity, the "Lorentz transformation," can act as the absolute that gives meaning to relativity. The second idea is that of the wave nature of electromagnetic radiation. We need this development—introduced through the elegant experiments of Heinrich Hertz—to emphasize the break with the past that begins with Chapter 8.

In Chapter 8 we introduce modern physics and chemistry with the fashioning of one of its cornerstones: quantum theory. On the basis of this theory, and the associated photoelectric effect, we find large areas of physics and chemistry making a great deal of sense, and all but falling into our laps. By the end of this chapter we are ready to deal with the structure of matter in modern terms.

Chapter 6

6.1 Justification
6.2 The Galileo-Newton principle of relativity
6.3 Stationary systems
6.4 Systems in uniform motion
6.5 Motion in moving systems
6.6 The special theory of relativity
6.7 The concept of simultaneity
6.8 The Lorentz transformation
6.9 Interim remark
6.10 Relativity of time
6.11 Experimental test
6.12 Relativity of distance
6.13 Some implications
6.14 Relativity of mass
6.15 Relativity of energy
6.16 Concluding comments
6.17 Summary

Relativity ∾ 6

Justification 6.1

The theoretical work of Lorentz, referred to in the previous chapter, soon raised fundamental questions about the meaning of measurements of length and of time. The theory argued that if there were an aether, experiments of the Michelson-Morley type could not demonstrate its presence. For if there were an aether 'wind,' blowing past and through an instrument, the atomic structure of the material would be affected by the 'stress' upon it; the material would shorten in the direction of the 'wind' in ratio to the relative velocity; one result, clocks would slow down. This would occur to just the extent that would make the effect unnoticeable relative to other parts of the instrument at right angles to the motion. This was the explanation for the null result in the Michelson-Morley experiments. In spite of numerous attempts, no way was found around this difficulty.

The implication was, in addition, that since the shortening of rulers and slowing down of clocks could only be calculated if one knew the velocity of passage through the aether—and one could not measure this with contemporary instruments or through experimental devices that were conceived to that time—a certain ambiguity resided in the measurements that were made.

From a practical point of view the matter was not serious. For example, the Lorentz equation for the shortening of a measuring rod of length l_0 when at rest relative to the aether, is $l = l_0\sqrt{1 - v^2/c^2}$. Here v is the velocity in the direction of its length, c is the velocity of light, and l the new shortened length. The value of c is 3×10^8 m/sec in vacuum. This is an enormous figure compared with ordinary velocities. For ordinary velocities, then, (v^2/c^2) is so small relative to (1) that the quantity $(1 - v^2/c^2)$ is essentially (1), and the root is essentially (1). Thus $l = l_0$ for all practical purposes. For very high velocities or great distances the problem becomes exigent. For example, 'now' to us has a very definite meaning: we can compare two events happening now at two places within the range of our vision. We say that those events, coming to perception at the same instant, are simultaneous. But if we look at a moon of Jupiter in a certain phase what we are seeing 'now' happened many minutes ago. The light signals that tell us of the 'now' appearance of a star may have been in transit thousands of millions of years. Moreover, not being able to measure the velocity of the Earth relative to the aether, we cannot make the corrections needed to find out how far away these events were when they occurred. So we remain able only to be certain that 'now' means different things in different contexts.

Einstein approached the problems of relativity with the basic assumption that the velocity of light has the same value in all inertial systems, independent of the speed at which the systems may be moving relative to each other, just so the motion is uniform. It had been demonstrated experimentally that the speed of light is unaffected by the speed of the source of the light. Out of his theoretical investigations came the 'Special Theory' of relativity. Later he developed the 'General Theory' to take account of non-inertial frames of reference. It is the Special Theory that we are concerned with in this chapter.

The special theory of relativity has had such an extraordinary effect in science, philosophy, and technology that it should be thoroughly understood by all educated people. The best exposition of the theory remains Einstein's short book, first published in 1916. The elements of this theory can be grasped with no more mathematical equipment than high school algebra and geometry, plus a certain amount of faith.[1]

Our purpose in this chapter is to present the special theory in such a way that its implications for all thought are brought out. We shall adhere closely to Einstein's method of exposition. What we wish to show are the facts and the theory, since these will be repeatedly referred to in later chapters. Also, we would like the power inherent in the interplay of fact and theory to become evident: a theme to which we shall also repeatedly revert. We make use of thought-experiments (which could be performed in the laboratory) that are presented formally. The student should work all the algebra through by hand, for conviction.

---

[1] Albert Einstein, *Relativity, The Special and General Theory*. Trans. Robert W. Lawson, Crown Publishers, New York, Sixteenth Edition, 1961. The original paper on the Special Theory is Albert Einstein, *Annalen der Physik*, *17*, 1891 (1905).

The Galileo-Newton principle of relativity 6.2

So far in this book we have treated space and time as they are treated in classical mechanics, assuming familiarity with these notions. In Chapter 2 Newton's first law was discussed in connection with a Cartesian frame of reference implicitly set up by the person doing the experiment. In Chapter 5 there were noted some of the suggestions for providing an absolute spatial frame of reference for mundane measurements. The theory of relativity, combined with empirical tests, has made it highly likely that no such absolute frame of reference exists, and in any event has made it unnecessary to require such a frame.

Consider a train running with constant velocity along an embankment (Einstein's model). A person in the train, using the walls and floor of his coach as his coordinate system, looks out the window and sees the embankment moving with respect to him. Reason tells him that embankments don't move, and that really the train is moving relative to the embankment. Indeed, to a person on the embankment, watching the train go by, this is clearly the case: the train is moving relative to him. However, if there were just these two persons making their observations it would be quite legitimate for each to consider the other moving relative to him. For example, two spaceships might pass: each moves relative to the other and neither (not looking at the Earth, or stars) has a fixed frame except his own. Moreover, in each case the velocity of these motions is measured with reference to the fixed, rigid measuring-rod which gives distance, and a clock which gives time, in the particular system from which the measurement is made; or by observation of a measuring-rod and clock or clocks in the other system. Either train or embankment, or either space-ship, may equally well be chosen as the site of the reference system for describing what is going on. Moreover, the laws of nature, for example Newton's laws of motion, have exactly the same form in each system. When the matter is put to experiment it turns out that there is no privileged frame that can serve as absolute standard for all the measurements of the universe. We must live with the fact of relativity: that there is no privileged frame of reference. The theory of relativity provides an invariant-relation which connects all frames and makes the fact of relativity tolerable.

6.3 Stationary systems

Consider, first, two stationary coordinate systems K and K' which differ only in that their origins, O and O' are a distance d meters apart along an X axis. These two systems might be set up on the laboratory table, or they might be drawn on a blackboard. The X and X′ axes coincide; the Y and Y′ are in the same direction, and the Z and Z′ are in the same direction (Figure 6-1). For convenience we omit the Z axes

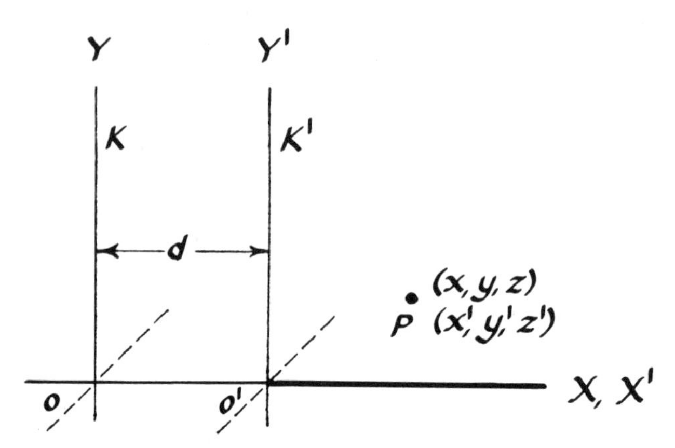

Fig. 6-1 *Fixed Galilean frames of reference* K *and* K′. *The frames are fixed a distance* d *apart, along the X axis. The point* P *has coordinates* x, y, z *in* K *and* x′, y′, z′ *in* K′.

in later diagrams. The coordinates of a certain stationary point P can be given in each system as, respectively, (x, y, z) and (x', y', z'). The data in one system can be translated to that in the other system simply by adding or subtracting d from the value of the X coordinate.

Thus:
$$x - d = x'$$
$$y = y'$$
$$z = z'$$
and:
$$x = x' + d$$
$$y = y'$$
$$z = z'$$

6.4 Systems in uniform motion

Suppose, now, that the frame of reference K is fixed to the embankment and that K' is fixed to a railway car which moves with a constant velocity v relative to the embankment and parallel to it, and that P is fixed in the K' system—say a point marked up forward in the car (Figure 6-2). Relative to the K' system the coordinates of this point do not change with time, but relative to the K system they do, in that the train is travelling along X with a uniform velocity $v = \Delta d/\Delta t$. Since the velocity has no Y or Z component, y and z do not change. For simplicity we assume that there are identical clocks in K and K', and that they read 0 seconds at the instant that O and O' "coincided." We can let the

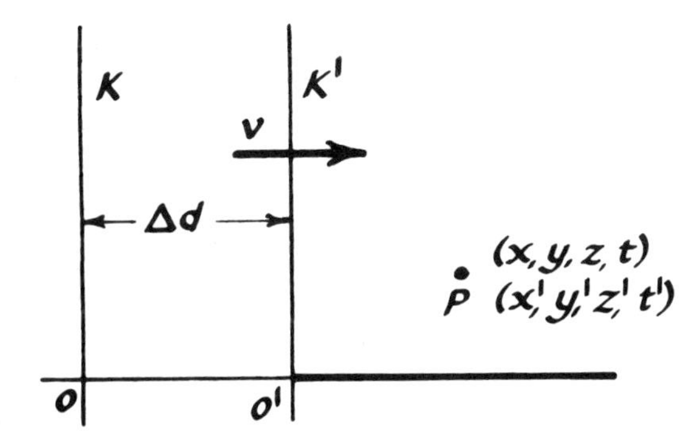

Fig. 6-2 *Moving Galilean frames of reference.* K *and* K′ *are moving with a velocity* v *with respect to each other. If* K *is considered fixed* K′ *moves with* +v *with respect to* K. *If* K′ *is fixed,* K *moves with* −v *with respect to* K′. *The point* P *is fixed in* K′.

two frames move as close to each other as we wish, and thus we can make coincidence of O and O' so close that the two identical clocks start instantly and at the same instant. Then at any time t on the K clock, the clock t' will have the same reading (we may suppose). At such a time t (after an elapsed period $\Delta t = t - t_0$) the origin of K' will have moved a distance $\Delta d = d - d_0$. Since t_0 and d_0 both have the value zero, we can write $v = d/t$, and at any given time t the origin of K' will have moved a distance $d = vt$. The data in one system may be translated to that in the other by the relations:

$$x - vt = x'$$
$$y = y'$$
$$z = z'$$
$$t = t'$$

and:

$$x = x' + vt'$$
$$y = y'$$
$$z = z'$$
$$t = t'$$

This translation scheme is based on the classical mechanical principle of the additivity of distances, a principle to which no exceptions are found in classical mechanics. (The change in sign of vt between x' and x in these formulae is recognition that the directions of the relative velocities are different.)

Motion in moving systems 6.5

Suppose, now, that instead of being at a fixed point P in the railway carriage, a person representing the point P walks with a constant velocity u' in the direction of motion (X′) of the car which is moving with uniform velocity v relative to the embankment. That is to say, K' moves with constant velocity v relative to K, and P moves with constant velocity u' relative to K' (Figure 6-3). The translation scheme between

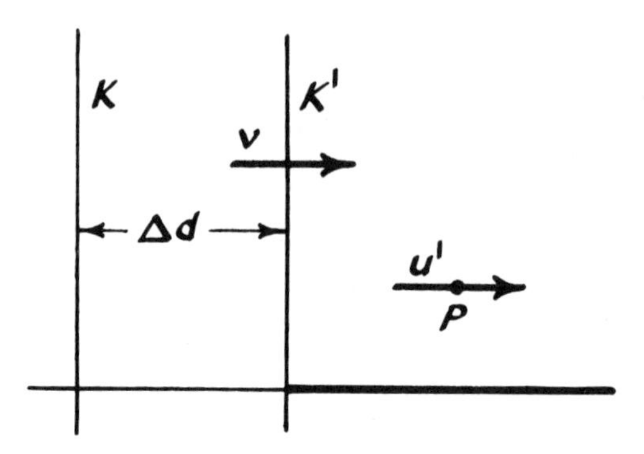

Fig. 6-3 *Moving Galilean frames with an object* P *moving with a velocity* u′ *with respect to* K′.

K and K' is derived as follows. Again, for simplicity, the identical clocks start instantly in K and K' at $t = 0$ when O and O' coincide. P is moving in the frame K' along the X′ axis, so his velocity can be found by measuring the distance he has travelled $(x'_2 - x'_1)$ in a certain period of time $(t'_2 - t'_1)$. The velocity of P in K' is by definition $(x'_2 - x'_1)/(t'_2 - t'_1) = \Delta x'/\Delta t' = u'$. Note that at the same time the railway carriage, the frame K', is moving with a velocity v in the X direction, and we saw from the previous translation scheme, Section 6.4, that $x' = x - vt$. By algebraic manipulation,

$$\begin{aligned} x'_2 - x'_1 &= (x_2 - vt_2) - (x_1 - vt_1) \\ &= x_2 - vt_2 - x_1 + vt_1 \\ &= (x_2 - x_1) - (vt_2 - vt_1) \end{aligned}$$

Also: $(t'_2 - t'_1) = (t_2 - t_1)$

Thus: $$\frac{x'_2 - x'_1}{t'_2 - t'_1} = \frac{(x_2 - x_1)}{(t_2 - t_1)} - \frac{v(t_2 - t_1)}{(t_2 - t_1)} = u - v$$

The translation scheme, then, is:

$$\begin{aligned} u' &= u - v \\ u &= u' + v \end{aligned}$$

Here the scheme is based on the classical principle of the additivity of velocities: that one velocity may be added to or subtracted from another.

This makes sense. If the person in the railroad car were to stand still, he would be carried forward with a velocity v relative to K. If then he were also moving forward with a velocity u' relative to the car, his total velocity relative to K would be $v + u'$. No exceptions are found to this principle in classical mechanics. It is used to calculate relative astronomical speeds. All of these transformations may be related to a frame of reference based on the "fixed" stars, as "absolutely at rest." For the practical purposes of classical mechanics any motion of the fixed stars could well be taken as zero, since relative to this reference system the laws of Newton are obeyed. An object set in motion in the absence of unbalanced force (all nearly enough achievable) does move in a straight line without acceleration of any kind relative to such a system. Such a reference system is called a Galilean frame of reference, or a Newtonian inertial frame.

The special theory of relativity 6.6

An analysis of what is meant by the constructs 'time,' 'simultaneity,' 'distance,' together with the acceptance of an empirical observation that the velocity of light does not depend on the velocity of motion of the body emitting it, as well as other considerations bearing on symmetry in physics, led Einstein to enunciate the Special Theory of relativity.

This theory says in effect that no frame of reference of the kinds discussed above, namely those moving relative to each other uniformly, or not moving relative to each other, has privileged status for the physical description of any natural process: any frame is as good as any other. As a second premise it holds that the velocity of light in a vacuum is constant and, as we noted, independent of the speed of its source, if that source is moving uniformly. This premise raises the difficulty that if a pulse of light started out in the railroad car (Section 6.5) with a velocity c' in the direction of its motion, the velocity measured in K would be $c = c' + v$ by the classical rules. Or, if a pulse started out along the embankment, its velocity as measured in the car would be $c' = c - v$. These arguments violate the experimental finding that $c = c'$. Einstein showed that by retaining the law of the constant velocity of light *and* the principle of relativity (Sections 2 through 5) a theory of relativity could be arrived at consonant with both.

The concept of simultaneity 6.7

Suppose (Figure 6-4) a person situated on the embankment (reference system K) at a point M, measured to be equidistant between points A and B. He has a device for observing A and B, and at some instant he sees both a flash of light from A and one from B. Both flashes arrived at M at the same time and each travelled over the same distance.

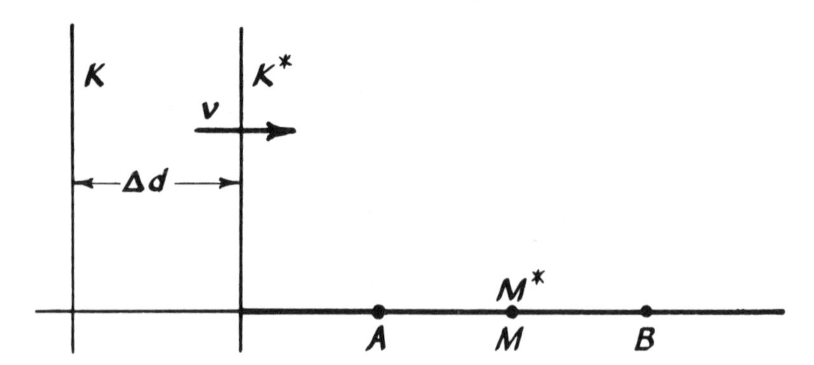

Fig. 6-4 *Relativity of simultaneity. In frame* K, *simultaneous flashes at* A *and* B *arrive at the same instant at* M, *exactly half-way between them by measurement in* K. *In the frame* K*, *moving with velocity* v *with respect to* K, M* *comes opposite to* M *just as the flashes at* A *and* B *are generated.*

Since each travels, but in opposite directions, with the same speed c, the observer at M is justified in saying that the flash-events at A and B were simultaneous in K, for he knows that the distances are the same.

Consider, now, an observer on the train. His reference system is K^* (the asterisk warning that the Galilean, primed, approach is not being used). Suppose that he is at M^*, coincident with M when the events, the flashes, occur at A and B. If he were not moving (K^* stationary with respect to K) the two flashes would meet at M^* as they do at M. But he is moving toward B and away from A and in his reference frame he will conclude that B occurred before A, for the light having less far to go as he moved toward it would arrive at M^* sooner than the light from A. Thus events that are certified to be simultaneous in K need not be in K^*; and are not if K^* is moving relative to K. That the two flashes in K were certified simultaneous when their light traversed equal distances required that the velocity of each flash be the same, otherwise the time of transit would not be the same. We saw, however, that this condition is fulfilled. The problem is to discover how the principle of relativity may be made compatible with the constancy of the propagation of light.

6.8 The Lorentz transformation

The velocity of light in K is $(x/t) = c$, where x is the distance travelled and t the time of travel. Similarly, in K^*, we remember that c is constant and has the same value for any reference system, $(x^*/t^*) = c$. It is, of course, necessary that the same length and time scales are chosen in the two frames, and that the same types of instruments are used: that is, the two frames are brought into coincidence and clocks and meter sticks are shown to be identical. The law of the propagation of light *in vacuo* is:

$$(x/t) = c = (x^*/t^*) \qquad (6\text{-}1)$$

This must be made compatible with the principle of relativity (Section 6.4), namely, to begin with, that

$$x' = x - vt$$

and $x = x' + vt$

and with the principle that neither K nor K^* has privileged status with respect to the other. In other words, some function k must be found, independent of the frame of reference (*i.e.*, an invariant-relation) which connects events in one frame with those in another so that an observer in one can calculate what is observed by an observer in the other frame of reference.

Let $kx^* = x - vt$

and $kx = x^* + vt^*$

Since $x = ct$

and $x^* = ct^*$

then $kx^* = ct - vt$, and $(kx^*)/t = (c - v)$ (6-2)

also $kx = ct^* + vt^*$ and $(kx)/t^* = (c + v)$ (6-3)

Multiplying Equations (6-2) and (6-3):

$$(k^2xx^*)/tt^* = c^2 - v^2$$

or, from Equation (6-1):

$$k^2c^2 = c^2 - v^2$$

Then $k^2 = (c^2 - v^2)/c^2 = (c^2/c^2) - (v^2/c^2)$

and $k = \sqrt{1 - v^2/c^2)}$

This series of manipulations shows what the value of k is (k can be shown to be unique) and enables a set of rules to be written:

$$kx^* = x - vt, \text{ or } x^* = (x - vt)/\sqrt{1 - (v^2/c^2)}$$

and $kx = x^* + vt^*$, or $x = (x^* + vt^*)/\sqrt{1 - (v^2/c^2)}$

These two equations are part of the set of transformations for x, y, z, t to x^*, y^*, z^*, t^* known as the Lorentz transformations. For ease of writing, the expression $1/k$, that is $1/\sqrt{1 - (v^2/c^2)}$, is named γ.

Then: $x^* = \gamma(x - vt)$ (6-4)

$x = \gamma(x^* + vt^*)$ (6-5)

It is to be observed that if v is small with respect to c, which is always the case with the rectilinear motions with which classical mechanics concerns itself, then v^2/c^2 becomes a very small number, and $\sqrt{1 - (v^2/c^2)}$ becomes practically indistinguishable from 1. Thus the Lorentz transformation reduces to the Galilean translation rule for systems moving at ordinary velocities, as we saw in another context in Section 6.1.

By algebraic or geometric reasoning, the transformations for the time dimensions are found to be:

$$t^* = \gamma(t - vx/c^2) \tag{6-6}$$

and

$$t = \gamma(t^* + vx^*/c^2) \tag{6-7}$$

In these relations vx/c^2 has the dimensions of time, that is $(l^2/t)/(l^2/t^2)$, as it must have if it is to be subtracted from or added to t. When v is small with respect to c, as we saw, γ reduces to 1; if x is small, the expression $\gamma(t - vx/c^2)$ reduces to t, similarly for t^*, and this yields the Galilean relations.

The Lorentz transformations, the coding devices or rules that enable measurements in K to be related to those in K^*, may be summarized for the system illustrated in Figure 6-4, where the y and z coordinates being perpendicular to the direction of motion are not affected by the motion:

$K^* \quad K$

$$x^* = \gamma(x - vt) \tag{6-4}$$

$$y^* = y$$

$$z^* = z$$

$$t^* = \gamma(t - vx/c^2) \tag{6-6}$$

$K \quad K^*$

$$x = \gamma(x^* + vt^*) \tag{6-5}$$

$$y = y^*$$

$$z = z^*$$

$$t = \gamma(t^* + vx^*/c^2) \tag{6-7}$$

6.9 Interim remark

The student coming upon this subject for the first time may well be offended by this assault on his common sense. He may feel that not only is his Newtonian common sense offended, but also the algebraic manipulation is no more than that, and rather tedious to boot. If he has these reactions he has had plenty of company in them. When Newton's laws were new, many people felt offended in their traditional Aristotelian sensibilities and common sense. Also many looked upon the laws, which were clearly very useful and powerful, as mere numerical devices without explanatory power.

To such an understandable reaction two responses may be made. The first is that common sense is what one grows up with: his mental map of his relationship to his habitat which he constructs from childhood onward and which provides at least part of the network of ideas through which he learns to see the world. He is continually, usually in small ways, occasionally in large ways, modifying this map as he changes through childhood to maturity and beyond, and as his habitat changes. He and his habitat affect and change each other retroactively. Moreover there is no guarantee that his map is able to fit regions he has not encountered before. That he be alert, interested, and flexible are requirements for growth. There are some broad principles on which he may safely rely in foreign territory: for example, the metaphysical principles that out of the flux of experience (at the P-plane) there may be derived orderly relations between constructs which have stability and permanence—Newton's laws are not abrogated by Einstein's; that these constructs gain validity and meaning through the multiplicity of their interconnections; that it is out of Nature's book that he must read physical science; and that what he reads will have the qualities of simplicity and elegance if he reads truly.

The second response is that when relativity was new it was regarded by many laymen as a kind of higher witchcraft. They parroted the statement that only very few people in the whole world could understand Einstein (it is by no means a guarantee of profundity not to be understood: look around!), and implied by this an excuse for not trying. But time has passed, and relativity is better understood by more people and has shown its power and elegance. So we continue. We now shall imagine clocks, mirrors, sources of light, and so on, attached to the laboratory equipment. We follow expository methods worked out by Feynman, Sherwin, and others (references at the end of the chapter). Visualize what is going on.

Relativity of time 6.10

Consider reference frames K and K^*, as before. K^* moves to the right with a constant velocity v relative to K (Figure 6-5). A^* is a mirror on the end of a rigid rod fixed in the Y direction. The mirror is parallel to the X axis. The length of the rod from the origin O^* to the mirror surface is y^*. The two systems, with their frames affixed, move close to each other. As O^* comes into coincidence with O, a light pulse of very short duration is emitted from O, and at the same instant, triggered by this signal, a clock riding on the moving system at O^* is started. The two origins are supposed so close together that the signal starting the clock passes between them over an infinitesimal distance: it could as well arise at O^* as at O. The event occurs in both systems at definite points in time and space in coincidence, the one system with the other.

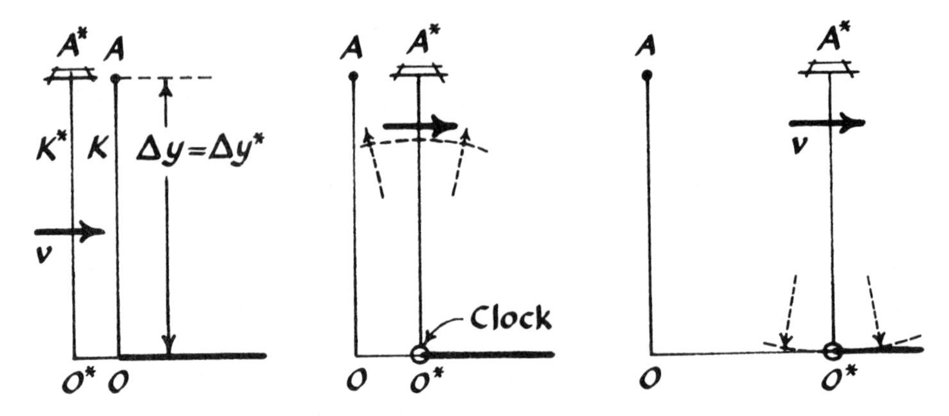

Fig. 6-5 *Measurement of a proper time interval in the moving system* K*. *The origin* O* *of* K*, *moving with a velocity* v *approaches the origin* O *of the* K *system along the X axis.* [*Figures 5 through 9 are adapted from C. W. Sherwin,* Basic Concepts of Physics, *Holt, Rinehart ∩ Winston, New York, 1961.*]

Fig. 6-6 *Measurement of a proper time interval. A flash of light set off at the coincidence of* O* *with* O *starts a clock travelling at the origin* O* *in* K*. *The light flash is shown moving toward the mirror A*.*

Fig. 6-7 *Measurement of a proper time interval. The light flash, reflected from* A*, *has returned to* O* *where it stops the clock instantly.*

The pulse moves out as a spreading wavefront at 3×10^8 meters/sec (Figure 6-6), reaches A^*, is reflected back instantaneously (that is, any delay is so much smaller than the effects being measured as to be negligible) and arrives back at the clock, at O^*, stopping it instantly (Figure 6-7). This is an inertial frame, for O^* to A^* are at rest relative to each other. The measurement of the time period from O^* to A^* and back took place *at the clock* which remained *in the same place in its K^* inertial system:* the two events that bounded the interval occurred at that same place. If the distance y^* were 1 meter, then the time interval $\Delta t^* = 2y^*/c = 2/3 \times 10^{-8}$ seconds.

The phenomenon is now examined with the aid of instruments in the reference system K. The question is to measure the time interval between the same events: the flash of light at O, and the arrival back at O^*, by means lying in the K system. For this purpose, many small clocks, all as close as possible in properties to each other, are mounted at definite positions x_1, x_2, x_3, . . . along the X axis of the system K. These clocks are started one after another by the light pulse as it reaches them along the X axis (Figure 6-8). The light spreading out as described is reflected from A^*, and returns to the X* axis where it stops the clock at O^* (Figure 6-9). At this instant, too, it stops the nearest clock on the

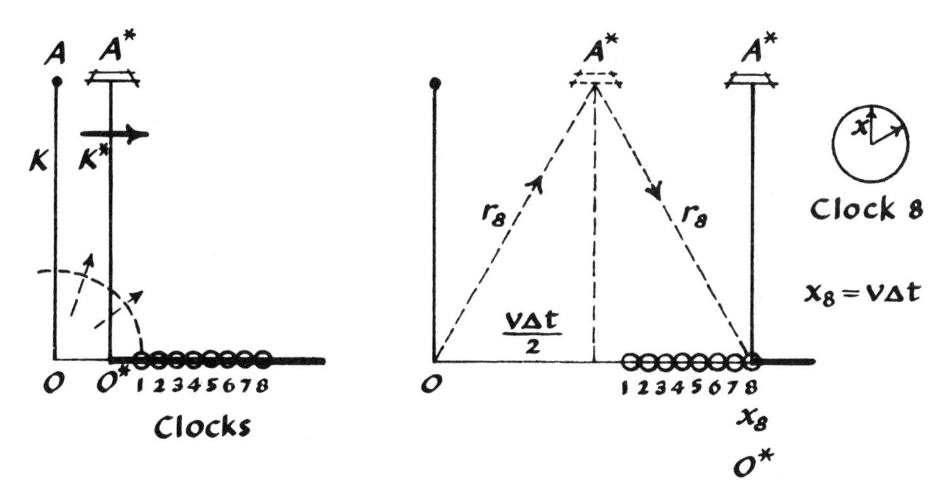

Fig. 6-8 *Relativity of time. The light flash, originating at* O, *moves out in* K *and starts clocks along the X axis as it comes to them.*

Fig. 6-9 *Relativity of time. The light flash reflected from the mirror* A* *in* K* *returns to the origin* O* *and stops clock No. 8 on the X axis in* K. *To the observer in* K *the light has taken the path shown by* r_8.

X axis, and marks it with an x for identification. In the example, this is clock number 8. The two events that bound the time interval, Δt,[1] to be measured, are the initial light flash at the coincidence of O and O^* and the arrival of the reflected flash at x_8, where O^* is when the flash arrives. Thus both spatially and temporally the events are common to both systems. They occur at the same points and the same times within negligible differences. The path of the light in K^*, as seen from K, is the dotted path marked r_8.

Clock #8 reads the time elapsed from the moment it was started by the flash moving along X to the moment it was stopped by the flash reflected from A^*. The period up to its start (which it did not measure) was x_8/c. It stopped when the flash had traversed the path $2r_8$, namely at $2r_8/c$. The *clock* measured an interval $\Delta t_8 = (2r_8 - x_8)/c$. The total time interval $\Delta t = \Delta t_8 + x_8/c = 2r_8/c$.

From the geometry of the figure,

$$r_8 = \sqrt{(y_0^*)^2 + (v\Delta t/2)^2}$$

Here $v\Delta t/2$ is half the distance from O to clock 8. The distance from O to clock 8 is that traversed by O^* at the velocity v between O and clock 8, during the time Δt, and thus it is $v\Delta t$.

---

[1] These sections closely follow Sherwin's method of presentation.

Eliminating r_8 between the above equations for Δt and r_8,

$$\Delta t = (2/c)\sqrt{(y_0^*)^2 + (v\Delta t/2)^2}$$

And solving for Δt (first square both sides, solve for Δt^2, then for Δt):

$$\Delta t = (2(y_0^*)/c)[1/\sqrt{1 - v^2/c^2}]$$

which may be rewritten:

$$\Delta t = \Delta t^* \cdot 1/\sqrt{1 - v^2/c^2} = \gamma \Delta t^*$$

This illustration shows two time intervals Δt^* and Δt that are defined by the *same* pair of space-time events occurring at the same points. These events are located exactly with respect to the reference frames K and K^*. The interval Δt^* is measured by an instrument at rest in K^*; the interval Δt is measured by instruments at rest in K. But two different kinds of time are being measured. The interval Δt^* is measured at the same place by a clock in the K^* system: the events that bounded it occurred at the clock. This is named a measurement of *proper time*. The interval Δt is named *non-proper* time because it is computed from a recorded time, *and* a distance divided by a velocity: *it involves a distance measurement.*

In this illustration a time interval was defined by the same events occurring at the same places. It was measured in two different frames of reference moving rectilinearly at constant velocity relative to each other. The time intervals were found not to be the same. The data for each measurement was 'there' to be examined by anyone. The value of Δt^*, read on the face of the clock at O^* could have been photographed at the same instant that the face of clock 8 could have been photographed. Since $y = y^*$, each distance, measured in its frame would have the same value. Being perpendicular to the direction of motion, the length would not be affected by the velocity. The velocity itself, measured from either system, has the same magnitude, and upon squaring, any difference in sign disappears. Yet the interval measured in K, namely Δt, is larger than Δt^*, measured in K^*.

For example, let $v = 0.75c$ and $y = y^* = 1$ meter.

Then $\Delta t^* = 2y^*/c = 0.67 \times 10^{-8}$ sec.

$$\gamma = 1/\sqrt{1 - v^2/c^2} = 1/\sqrt{1 - 0.55} = 1.5$$

$$\Delta t = \gamma t^* = 1.5\Delta t^* = 1.0 \times 10^{-8} \text{ sec}$$

The non-proper time interval inferred in K is 1.5 times as long as the proper time interval measured in K^*. This is what is meant by "time dilatation."

Suppose, now, that K^* is taken as the system at rest. Then the frame K is moving to the left. Let a light flash at O occur when O and O^* are in coincidence, and let it start a clock at O in K. The array of clocks now must be in K^* since one of them is to receive the flash reflected from A, which is moving with K. Proper time is measured in K; non-proper in K^*, and it is found that:

$$\Delta t^* = \gamma \Delta t$$

What this means is that for any pair of frames:

$$\text{(non-proper time interval)} = \gamma \text{ (proper time interval)} \qquad (6\text{-}8)$$

Thus it makes no difference whether K or K^* is taken as the rest frame, the other moving with respect to it. In either case the non-proper and proper time measurements bear the relation shown in Equation (6-8): the "clock in motion" runs slow to an extent dependent on its *velocity* relative to the "clock at rest."

Experimental test 6.11

There have been a number of tests of the relationship in Equation 6-8.[1] For example, there is a type of meson, a nuclear particle, that can be produced at will by bombardment of suitable atomic nuclei. These particles decay with a certain probability; that is, with a certain overall rate. They are, in effect, small clocks. If one has a thousand of these, at rest, then at the end of 2.6×10^{-8} sec there will be 368 left (632 having decayed to two other particles), on the average, and after 2.6×10^{-8} seconds more, 135 will be left, the number falling off in a regular way that depends on the number present. Decay of a particle is analogous to stopping a clock: a meson detector shows no meson when it has decayed: that is, it is not sensitive to the products of meson decay. The internal clock of the meson measures proper time, and although it is not possible to tell when a clock will stop, it is possible to say very accurately that if one has 1000 of them at rest at 0 sec, there will be 368, give or take a few, left 2.6×10^{-8} sec later.

If mesons are produced from a suitable source they can be obtained as a beam, travelling with a velocity 0.75 c. When a meson detector is used to examine this beam at various distances from its origin it is found that the mesons have decayed much less rapidly than the same detector would find if they were at rest: the measurements at various distances are non-proper measurements since here Δt = (distance from origin)/0.75 c. The mesons are the K^* system; the detectors are at rest in the laboratory in the K system. The phenomenon is that described in detail above. What is found experimentally is that if a detector is placed at, say, 22.5 meters from the origin O, it takes $22.5/0.75\,c = 10 \times 10^{-8}$ sec for the transit to the detector. This is non-proper time: Δt. During this interval, 1000 mesons at rest would have decayed to only

[1] R. P. Durbin, H. H. Loar, and W. W. Havens, Jr., "The Lifetimes of the π^+ and π^- Mesons." *Physical Review*, Series 2, *88*, 179 (1962).

about 20 left. The detector at 22.5 meters reports that there are about 100 still present. By Equation (6-8) Δt (non-proper) $= \gamma\Delta t^*$ (proper). Here $\gamma = 1.5$. Therefore in the meson frame of reference the transit time is 6.67×10^{-8} sec instead of 10×10^{-8} sec. In 6.67×10^{-8} sec the rate of decay (because the mesons are at rest in that frame) leaves some 100 still present, in agreement with experiment. The experiment clearly supports relativity theory.

6.12 Relativity of distance

Suppose that in the K^* frame, moving with velocity v, there is a rigid rod lying parallel to the direction of motion. The people in K^* can measure this rod with a meter-stick. They find it to have a (proper) length Δx^*. Applying Equation (6-4):

$$\Delta x^* = \gamma(x_2 - vt) - \gamma(x_1 - vt) = \gamma\Delta x$$

(Note that at the points x_1 and x_2, vt has the same value because the rod is rigid.)

The measurement made from the K system is non-proper length. It would involve the use of clocks, that is of time, to measure length. It develops that the lengths measured in the two systems would therefore not be the same. Two different kinds of operation are being used to measure the lengths. If the proper length Δx^* were one meter, and v were $0.9\,c$, so that $\gamma = 1/\sqrt{1 - (0.9)^2} = 2.3$, the length measured in K would be:

$$\Delta x = \Delta x^*/\gamma = 1/2.3 = 0.435 \text{ m}$$

Therefore one finds that the (non-proper) length of the rod measured in K is less than the proper length. To the instruments in the rest system the moving rod appears shortened:

$$(\text{proper length}) = \gamma \text{ (non-proper length)} \qquad (6\text{-}9)$$

If the analysis is made of systems in which K^* is considered at rest and K in motion, so that the proper length measurement is made in K and the non-proper in K^*, then it is found, in accord with Equation (6-9) that $\Delta x = \gamma\Delta x^*$.

6.13 Some implications

One implication of these discoveries is that space and time are not independent dimensions but that, when phenomena in two systems moving with respect to each other are being measured, both space and time are inextricably united. Another implication is that neither relative direction of the motions of the two frames of reference is privileged. The instruments in the two frames disagree. Suppose two frames are at rest and contiguous, so that two rods, one in each frame may be compared and certified to be of the same length. (One might imagine one on a stationary railroad car, the other on the embankment next to it—to continue Einstein's illustration.) Now the frames are put in motion relative to each other parallel to the direction of the rods. Whichever is taken as the stationary reference frame—the embankment to the man on the embankment; the car to the man in the car—those instruments will measure the other rod, the moving rod, shortened. They make a non-proper measurement. Moreover, because of this fact, it is impossible to detect a state of absolute motion or to determine the absolute direction of relative motion if it is uniform.

It may be added that the same arguments as used with the X coordinate apply to the Y and Z coordinates. Further, because of the nature of γ, if v is small compared with c, the Lorentz equations for distance reduce to the classical equations (Sections 6.3 to 6.5).

Relativity of mass 6.14

Everyone today has heard of the equation $E = mc^2$. How is it derived? Basically we are concerned with time dilatation. Suppose two systems K and K^*. Again, these might be embankment and railroad car. Persons S and S^* stand one in each, and each with a ball in his hand. Ahead of time, when K and K^* were side-by-side and stationary, the balls were compared and found to have exactly the same mass. They were found to be perfectly elastic: on collision each bounced back with the same velocity as before collision except that the direction was exactly reversed. And the two persons agreed to throw the balls in a Y direction exactly perpendicular to the direction that K and K^* would later move, and with such skill that the two balls would collide half-way between the two persons at the exact moment that they would come opposite to each other. K^* was then set in motion relative to K.

The person in K^* threw his ball with a velocity u^*; the person in K threw his ball with a velocity u; the balls collided, bounced back, and each person caught the ball he had thrown. The person in K^* threw his ball straight along Y with a velocity u^*. This was clear to him. But it appeared to him that the other ball, the one in K, followed the dotted path shown in Figure 6-10a, for the ball left that person's hand before they both came opposite, then it bounced back into the hand of the person in K after they had passed. The person in K threw his ball straight out, and saw the ball in K^* follow the dotted path in Figure 6-10b.

Suppose that K is taken as the rest frame. The problem is to measure u^* in K^* and then to measure it with instruments in K. In K^* the ball will have moved a distance Δy^* in the period Δt^*, and therefore

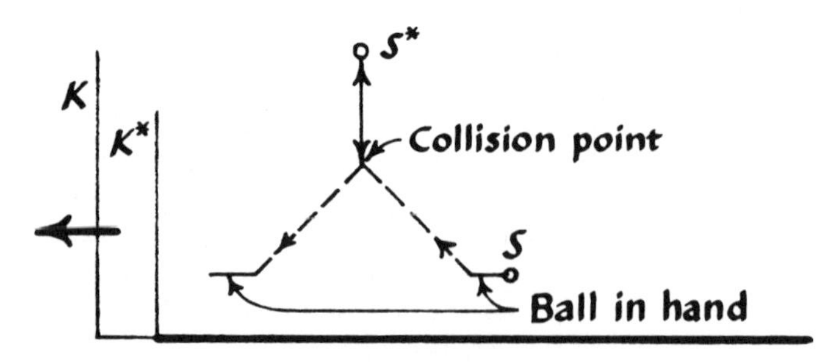

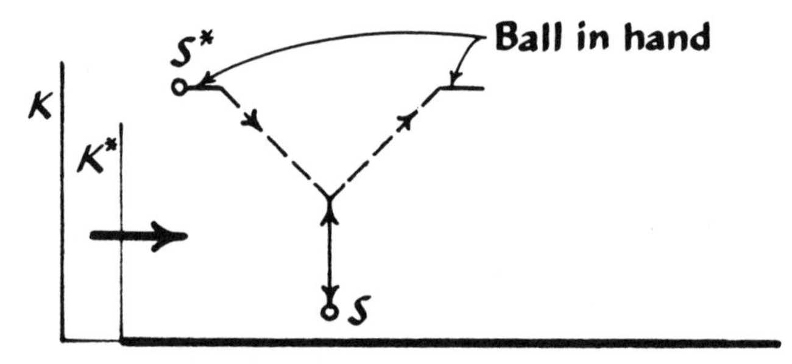

Fig. 6-10 *Relativity of mass.* Upper figure, *to the observer in* K* *the ball in* K *follows the course shown, colliding with the ball thrown directly along the Y* axis.* Lower figure, *from the point of view of an observer in* K *the ball in* K* *follows the peth shown, colliding with the ball thrown directly along the Y axis.*

$u^* = \Delta y^*/\Delta t^*$. This involves a measurement of proper time since the distance Δy^* could have been from hand to collision and back to hand, so that Δt^* was measured at the same point in space, and was a pure time measurement. Not so, however, when measured from K, for here the quantity measured in K^*, which we designate w, for clarity, is $w = \Delta r/\Delta t$. The time interval was measured over two different positions in space (Figure 6-10b). By Equation (6-8) $\Delta t = \gamma \Delta t^*$, and we saw that $\Delta y = \Delta y^*$; thus $w = \Delta y^*/\gamma \Delta t^*$. But $u^* = \Delta y^*/\Delta t^*$. Therefore $w = u^*/\gamma$, or $u^* = \gamma w$.

The person in K now compares the measurements he has made of the total change in velocity in the two systems: the velocity u out to collision, and $-u$ back, $\Delta u = u - (-u) = 2\,u$. In his own system $\Delta u = 2\,u$. Measured for K^*, $\Delta w = 2\,u^*/\gamma$. However it was known that $|u| = |u^*|$, so he can write $\Delta w = 2\,u/\gamma$.

Now the two balls were known to have the same mass when measured in the rest system (or when measured at rest in respect to the coordinates of the K system). This is called the rest mass, m_0. When they are thrown and made to collide, they are given momenta, and one way of determining inertial mass is to bring about the collision of an object of known mass m_A, traveling with velocity v_A with the object of unknown mass m_B, traveling with velocity v_B, and to measure the change in velocity of both. By the principle of the conservation of momentum in an isolated system (one not subject to external forces):

$$m_A \Delta v_A = m_B \Delta v_B$$

Thus, $m_B = m_A \Delta v_A / \Delta v_B$. In the case of the balls in K and K^*, the person in K knows m_0, Δu, and Δw (which is the number given to Δu^* by his instruments). He enquires what the value of m^* is. (It was m_0 when at rest.) Now, however,

$$m^* = m_0 \cdot 2u/(2\,u/\gamma)$$
$$m^* = \gamma m_0 \qquad (6\text{-}10)$$

He may therefore conclude that the inertial mass of the moving body is because of its motion larger than the mass of the body at rest.

An important conceptual point must be emphasized here. Either $m_0 = m^*$, in which case the law of the conservation of momentum must be abandoned, or inertial mass becomes dependent on velocity (better yet, observed *momentum* becomes dependent on velocity), in which case, $m^* = \gamma m_0$. It was assumed that the law of the conservation of momentum would hold at the high velocities where γ becomes significant, as well as at low classical velocities for which the law was derived. Thus *mass had to be modified accordingly*. Nowadays it is customary to define a relativistic momentum $p = m^* v$, as:

$$p = \gamma m_0 v$$

This is because what is customarily met with experimentally is the momentum of the object, *e.g.*, of an atomic particle. Thus, to save the law of the conservation of momentum the constancy of mass was sacrificed. Whatever the conceptual decision, the need for *a* decision was there.

Relativity of energy 6.15

To recall what was discussed in Section 3-17, energy and work are commonly expressed in the same units. Very generally, $\Delta W = (\mathbf{F} \cos \theta) \Delta s$ where ΔW is work, in newton-meters, or joules; $\mathbf{F}$ is force in newtons; and Δs is displacement, in meters, through which the force acts. The magnitude of the angle between the direction of the displacement and the direction of the force is θ. If the direction of $\mathbf{F}$ coincides with that of s, then $\cos\theta = 1$ (Figure 6-11). The force produces a velocity v, and work is done upon the body: it gains energy. If it started from rest, $v_0 = 0$, and accelerated to v_1 then, classically, the kinetic energy gained by the body would be:

$$KE = 1/2\, m(v_1^2 - v_0^2) = 1/2\, mv^2$$

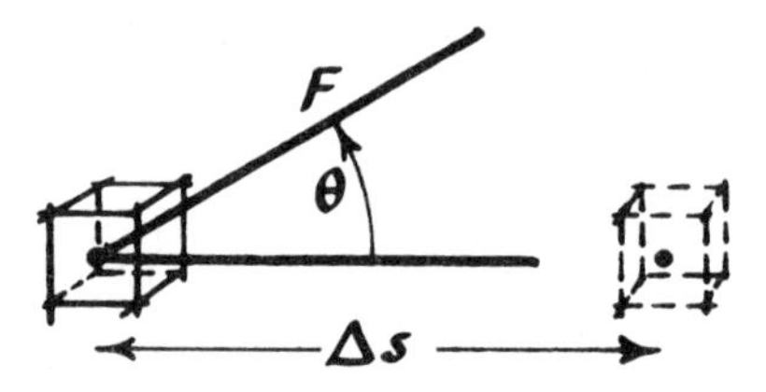

Fig. 6-11 *The definition of work as* w = (**F** *cos* θ) Δs.

Since this expression relates mass, time, and distance, it must be subject to the relativistic transformations. Remember, however, that we are manipulating definitions here. We have agreed to sacrifice the constancy of inertial mass so as to retain the law of the conservation of momentum at the high velocities where the relativity of space and time force us to abandon the Galilean principles. At ordinary velocities the decision is not exigent from a *practical* point of view. From a *conceptual* point of view it is both exigent and far-reaching in its effects since it changes our conceptual framework. The C-field (Figure 1-4) is vastly enriched, and richly reinterpreted under the special theory of relativity.

Return to the equation:

$$m^* = m_0/\sqrt{1 - (v^2/c^2)} \qquad (6\text{-}10)$$

This equation can be subjected to algebraic manipulation known as series expansion to yield the expression

$$m^* = m_0\left(1 + \frac{v^2}{2c^2} + \text{other terms with higher powers of } v \text{ and } c\right).$$

The expression can be written, with little error, if v is much smaller than c:

$$m^* = m_0 + \frac{mv^2}{2c^2}$$

Multiplying both sides by c^2,

$$m^*c^2 = m_0c^2 + 1/2\, mv^2$$

Or, recalling that $KE = 1/2\, mv^2$, the equation may be written:

$$m^*c^2 = m_0c^2 + KE$$

The equation says that the total energy of the system (m^*c^2) is equal to the rest mass-energy (m_0c^2) plus any kinetic energy (KE) that the system might have. If it should not be moving, then to that frame of reference its total energy is m_0c^2. This equation holds for all frames of reference, and so the restriction to small v can be removed. It is evident that by giving up the conservation of mass so as to save the conservation of momentum a tremendous gain has been achieved: two laws, that of the conservation of mass and that of the conservation of energy, have been combined into one sweeping law, the conservation of mass-energy. This law holds for all kinds of energy—mechanical (potential and kinetic), chemical, electrical, or nuclear. We shall not pursue the experimental verification of the law since this is widely known in the large, and since it will be thoroughly explored in a later chapter with better background preparation (Chapter 13).

Concluding comments 6.16

Consider the effect that relativity has had upon ideas of time and space, and thus upon what we can know. The classical view, implying the classical idea of simultaneity, conceives of us here-now as at a point, say O in Figure 6-12, located on an interface between what is past and what will be future. This interface moves forward, as time moves, or in time. What we can know now, here, lies on this line, the time axis, and extends indefinitely, spatially, in the X, Z plane if we think of time in the Y direction, for the moment, for the sake of visualization.

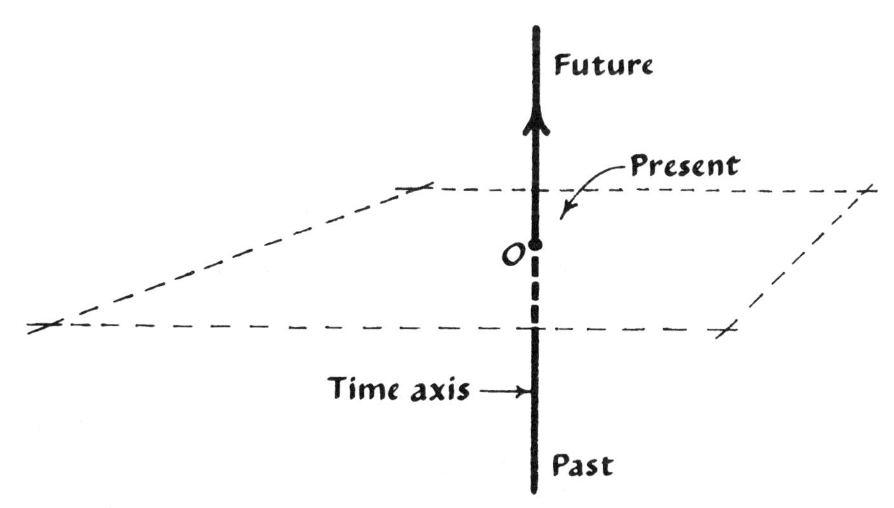

Fig. 6-12 *The classical view of present, future, and past to an observer at O who has instant communication with all present events.*

Minkowski names a-point-in-space-at-a-point-in-time a 'world-point,' with coordinates x, y, z, t, and he names the sum of all such points 'the world.' One such world-point is O in the Figure 6-13.

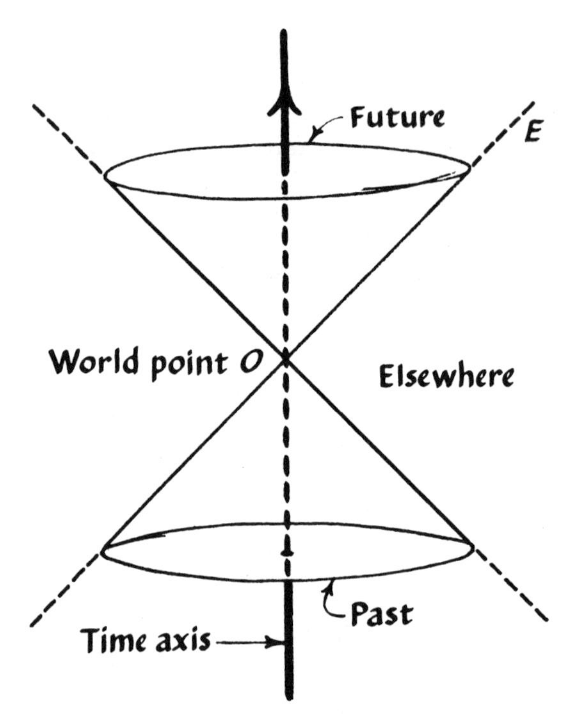

Fig. 6-13 *The Minkowski relativistic view. The observer at* O, *linked with events by light is restricted in his knowledge to events within the cones of past and future travelling with him along the time axis. The name 'elsewhere' was given by Eddington.*

Suppose that the coordinates of this point change by a small amount, Δx, Δy, Δz, Δt. This describes an element, or short section of a world-line. Each point traces out such a world-line; the universe is made up of world-lines. The "distance" from O to some point in this four-dimensional manifold is found by a kind of hyperbolic geometry invented for the purpose:

$$s^2 = x^2 + y^2 + z^2 - c^2t^2$$

Here again we see the speed of light turning up, the limitation on our means of communication. Out of the application of this geometry there arises the relativistic view diagrammed in Figure 6-13. The time axis is vertical, passing through O, and three spatial dimensions are 'represented' on the X, Z plane. Now, however, there is none of the Newtonian simultaneity suggested in Figure 6-12 as the plane of known events. Instead, a ray of light moving out from the origin O, fixed in its frame but moving along the time axis, spreads out as a light 'cone.' The axes are laid out in units of $\tau = ct$, which accounts for the 45° angle to the cone. A light ray leaving O traces out a line such as OE. The conical surface of all such possible rays encloses the world-lines of events with lesser velocities and includes all that can be known about events with respect to the future behavior of the four coordinates of O. Correspondingly a similar but inverted conical surface comprises all light signals that could have come to O out of the past and arrived at this point, here-now. O, the here-now, the present of a world-point, thus lies at the intersection of the X, Y, and Z axes with the T axis. This section so far has been a rather sketchy presentation of the Minkowski ideas.

The theory of relativity clearly places certain limitations to what one can know at a given time. It would take us too far from our set task to trace in any detail opinions about these limitations (with their associated responsibilities and freedoms). Excellent discussions are to be found in Bohm and Fraser; and in the latter, suggestive references to the literature. The main purpose here is to intimate that there are differences in interpretation of the meaning of the generally accepted facts.

A number of philosophers, some of whom are also physicists, interpret the Minkowski diagram as representing an essentially static space-time. This may happen when it is forgotten that the field of constructs is not the plane of perceptions. Actually, a person is part of the universe, and cannot 'stand aside' except in imagination. The observer at O has a past, conventionally extending to his day of birth. But he cannot see it for as this diagram implies the past no longer exists at any given moment of the present except as a trace in our memories, or as artifacts of various kinds, such as newspapers and other documents, photographs, and so forth. Or, on a larger scale, as the archeological and geological deposits that allow the past to be reconstructed.

Of course, if time can be represented on the OT axis, then perhaps, by an induction, the past might be said to exist in the way that we know our home exists (barring catastrophe) in some x, y, z region when we leave it (in the 'behind' of our spatial progress) and that the office or classroom exists in the 'before' to which we will come if we pursue our course. 'Now', so to speak, is immediate awareness of what is and was there 'before' and 'after' according to this view.

Other physicists and philosophers interpret the Minkowski diagram as representing a process of becoming. The observer at O stands at the prow of his past light-cone, navigating his world-line, knowing nothing 'at this time' of events in the absolute elsewhere, and able to predict events in the absolute future only by means of knowledge based on past experience and subject to imperfections (because of imperfect knowledge) and to contingencies which may arise out of the unknown and, at the moment in question, unknowable elsewhere.

We sense in these points of view evidences of the perennial questions about being and becoming, state and process, form and function. There are physicists and philosophers who insist that the theory of relativity is by no means satisfactorily established—this is not the same as saying that it will be subsumed in some wider synthesis, as classical mass and energy are united in Einsteinian mass-energy. It is more like the position of those who held the Earth to be flat after Galileo, Kepler, and Newton had lived.

The theories of relativity began to be discussed in popular works near the beginning of this century. These powerful, and indeed world-shaking, ideas gripped the imagination, and people tried to express their thoughts in the idiom. One feels certain that the "fourth dimension" in discussions of art at that time did not correspond with the constructs of Einstein's theories. Yet there must have been certain points of contact. It seems that one could sense what might be called "correlative movements in the evolution of the Western mind." Some authors have traced this movement to the advent of non-Euclidean geometries, when it became evident that the postulates of Euclid were not fixed and immutable, but man-made, and thus subject to investigation and change by men. The suspicion that these postulates were not absolute has a long history, but actual evidence in the case was presented just before the middle of the nineteenth century by Gauss, Lobaschevski, Bolyai, and Riemann. More and more intellectuals became aware of this loosening up of postulates and assumptions, and together with the fantastic advances in science and technology this created a climate that affected even the most non-scientific persons.

It certainly was the case that at about, and shortly after, the turn of the century there was a spate of innovation in the arts. This can only be suggested, here. This was the period of the Cubists, and a painter could well exclaim that classical perspective was dreadfully Euclidean: what was needed was to represent all sides of the object—and perhaps its interior as well—in a single work. A long step had then been taken in the direction of still more subtle ventures into the realms of mystery and psychology. Such basic changes in postulates in the hands of Cezanne and other masters like Picasso and Braque led to fresh and moving art. It was as though shackles had been struck from men's imagination. Naturally it could not possibly be claimed that Einstein's relativity theory led to this flowering of art; the roots go far back in history—as did the roots of the Reformation, and of the Renaissance. But one is not prevented from seeing relationships and possibly correlative movements which help to confirm that neither science nor art exists in a vacuum.

One might, for example, cite the plays of J. M. Barrie which so far departed from the tenets of Aristotle, and even the Elizabethans, that time could stand still, regress, or proceed with accelerated speed. J. B. Priestly treats time almost as a character in the play. In music Ravel, Bartok, Schoenberg, Stravinsky, and von Webern reflect this new freedom. Bartok spoke, for example, of the need for an utter break with the musical postulates of the nineteenth century. So it was, also, in all of the other arts.

These movements provide fascinating areas for further investigation, and the subjects of many interesting books.

We have ventured into this addendum to physical relativity in order to make it clear that physical science does not exist and evolve in an intellectual and cultural vacuum.

Summary 6.17

In Einstein's theory of relativity an invariant-relation is developed that connects all possible (relative) frames of reference. This must be clearly understood: it does not leave the observer floating freely without any conceptual anchors. Such license is invoked sometimes in the name of relativity. "Anything goes because everything is relative," one might hear as a result of such misunderstanding. On the contrary, the theory has demonstrated a constraint, an invariant-relation, in the relative frames that Nature allows us to choose from. This invariant-relation connects space and time with the absolute velocity of light. It is presented in the form of the Lorentz transformations.

Exercises

6.1 In the Figure 6-14 some unimaginative persons have calibrated frame K in feet, and K' in centimeters. Given that the (x_1', y_1') coordinates are (8.5, 91.5) cm, and the (x_2', y_2') coordinates are (100, 91.5) cm, a) what are the coordinates of the points in the K system, in feet? Assume that the origin of K' is at 3 ft on the X axis of K, and that the X and X' axes coincide.

6.2 In Exercise 1, communication between people in K and K' is ensured (within reason) by the agreed coding that one inch equals 2.54 cm. How else might this be ensured? Ponder (but no need to report unless you feel the urge to do so) whether if both use the same units there has been a loss in "enrichment" of the "lives" or "experiences" of the people in K and K'.

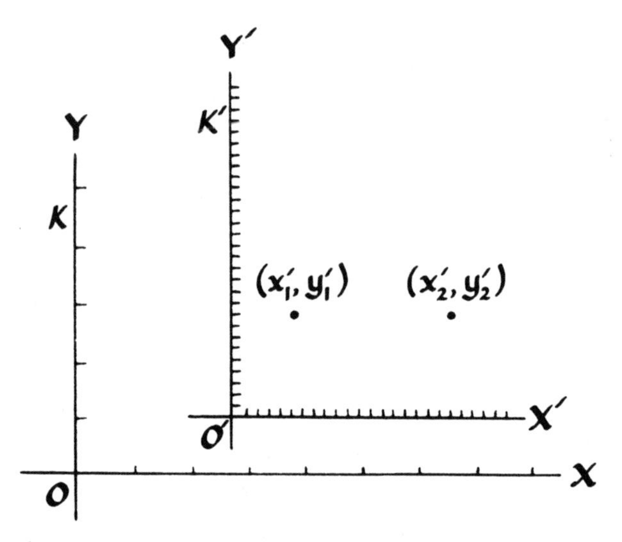

Fig. 6-14 *Two frames of reference calibrated in different units.*

6.3 To ride the theme of Exercise 1 a little further, a particle moves from x_1', y_1' to x_2', y_2' along the X direction with uniform velocity, in 12 sec. An observer in K times this movement in feet per hour. The frames K and K' are stationary with respect to each other. What velocity does each measure for the particle?

6.4 We assume, now, that both observers (Exercise 1) agree to standardize on centimeters and seconds. a) Set up the coding transformation group, in the form below, that allows them to translate from one system to the other:

System K	to	System K'
$x - d$	$=$	x'
y	$=$	y'
System K'	to	System K
		x
		y

b) Suppose, now, that the origin O' of K' were at (x, y) (4,3) cm in the K system. What would the transformation groups look like? c) Where is the origin of the K' system, relative to K, if the transformation group of K' to K is:

$$x' + 2 = x, \quad y' - 3 = y$$

6.5 Michelson said, in one of his papers, that "the relative velocity of the Earth and the ether is probably less than one-sixth the Earth's orbital velocity, and certainly less than one-fourth." This was the result of a set of measurements. He continued, "In what precedes, only the orbital motion of the Earth is considered. If this is combined with the motion of the solar system, concerning which but little is known with certainty, the result would have to be modified; and it is just possible that the resultant velocity at the time of the observations was small though the chances are much against it. The experiment will therefore be repeated at intervals of three months, and thus all uncertainty will be avoided . . ." Explain, with a diagram, what he means.

6.6 Given that the relativistic mass m of a body with rest mass m_0, moving with a velocity v is calculated by:

$$m = m_0/\sqrt{1 - v^2/c^2}$$

make a table showing the relativistic mass increase of a high-speed particle at the velocities shown below. That is, complete the Table. This gives a convenient table of values for γ for use in other problems. (Recall that v^2/c^2 is equivalent to $(v/c)^2$.

Velocity of particle, m/sec	% of velocity of light	γ	% increase in mass
3×10^6 . .	1		
3×10^7 . .	10		
9×10^7 . .	30		
1.5×10^8 . .	50		
2.1×10^8 . .	70		
2.7×10^8 . .	90		
2.97×10^8 . .	99		

6.7 Let P_{cl} (classical momentum) $= mv$, where m is the mass of an object and v is its velocity; and P_r (relativistic momentum) $= \gamma m_0 v$. Calculate P_{cl} and P_r for the conditions in the Table of Exercise 6-6, letting m and m_0 equal 1 kg for convenience. Plot these values versus v (v along the X-axis). To what limiting value of v must P_r become asymptotic?

6.8 A jet plane, moving at 400 m/sec at a very high altitude, fires a shell straight ahead. The muzzle velocity of the shell is 500 m/sec. a) What is the velocity of the shell relative to an observer on the ground? b) If the shell were fired straight to the rear, what would be its velocity to the observer on the ground? c) Are relativistic considerations pertinent? Explain.

6.9 A particle in an accelerator is moving with a velocity (v) of 1×10^7 m/sec, relative to the accelerator, which is fixed to the Earth. It ejects an electron directly ahead, in its line of motion, with a velocity (u^*) of 2×10^8 m/sec relative to the particle. What is the velocity of the electron that would be expected to be measured by a laboratory observer on the a) Galilean, and b) relativistic transformations? c) Suppose that instead of ejecting an electron the particle ejected a pulse of light. Show that the velocity of the pulse is c. *Hint:* First determine K and K^* frames of reference. Write appropriate equations. Write Lorentz transforms. Obtain u in terms of v and u^*. Then substitute numerical values.

6.10 Two fixed accelerators, pointed directly at one another, produce electrons, each with velocity 2.5×10^8 m/sec relative to the accelerator that produced it. Calculate the velocity of an electron from one accelerator relative to an electron from the other using a) Galilean, and b) relativistic methods. (See hint in Exercise 9.)

Answers to Exercises

6.1 a) (3.28, 3) (6.28, 3)

6.2 a) By putting both systems into the same units of whatever kind. b) What was lost was the knowledge of a measuring system that was given up.

6.3 In K, $s = 3$ ft; $v = 900$ ft/hr
In K′, $s = 91.5$ cm; v = 274.5 m/hr

6.4 a)

System K	to	System K'
$x - 91.5$ cm	$=$	x'
y	$=$	y'
System K'	to	System K
$x' + 91.5$ cm	$=$	x
y'	$=$	y

b)

K	to	K'
$x - 4$	$=$	x'
$y - 3$	$=$	y'
K'	to	K
$x' + 4$	$=$	x
$y' + 3$	$=$	y

c) $(2, -3)$.

6-5 See Figure 6-15. Motion of the Earth (E) at (1) would annul part of the resultant flow; at (3) it would increase it, and at (2) and (4) would change its direction. Drawn much out of scale, of course.

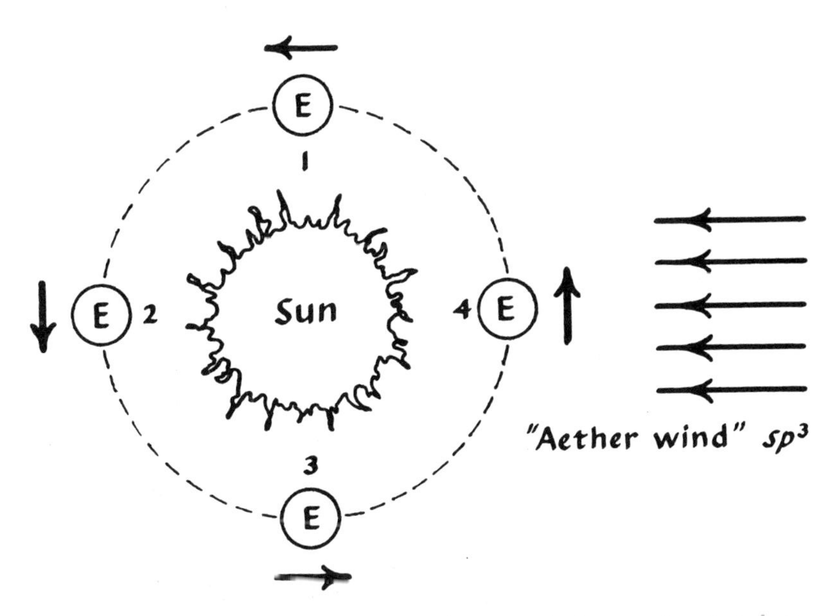

Fig. 6-15 *Relation of the Earth in its orbit around the Sun to a hypothetical ether wind. At 1 the Earth is going with the wind; at 3 against; and at 2 and 4 across.*

6.6

Velocity of particles	% of velocity of light	$\sim\gamma$	% increase in mass
3×10^6	1	$\sim 1^+$	$\sim 0^+$
3×10^7	10	~ 1.01	~ 1
9×10^7	30	~ 1.05	~ 5
1.5×10^8	50	~ 1.16	~ 16
2.1×10^8	70	~ 1.41	~ 40
2.7×10^8	90	~ 2.3	~ 130
2.97×10^8	99	~ 7.1	~ 610

6.7

P_{cl} kg m/sec	P_{rel} kg m/sec
3×10^6	3×10^6
3×10^7	3.03×10^7
9×10^7	9.45×10^7
1.5×10^8	1.74×10^8
2.1×10^8	2.96×10^8
2.7×10^8	6.2×10^8
2.97×10^8	2.1×10^9

P must become asympototic as v approaches c.

6.8 a) 900 m/sec. b) −100 m/sec, *i.e.*, in opposite direction. c) No; γ for this velocity is so close to 1 that the effect is negligible.

6.9 a) 2.1×10^8 m/sec.

b) $u^* = x^*/t^*$; or, $x^* = u^*t^*$

Translating to the K system by means of Equations 6-4 and 6-6:

$$(x - vt) = u^*[(t - vx/c^2)]$$

simplifying, $(x - vt) = u^*(t - vx/c^2)$

rearranging,

$$x - vt = u^*t - x(u^*v/c^2)$$
$$x + x(u^*v/c^2) = u^*t + vt$$

$x(1 + u^*v/c^2) = t(u^* + v)$ from which is obtained

$$x/t = u = (u^* + v)/(1 + u^*v/c^2),$$

the correct form of u for the K system. Numbers may now be inserted.

$$u = (2 \times 10^8 + 0.1 \times 10^8 \text{ m/sec}) \Big/ \frac{1 + 2 \times 10^8 \times 1 \times 10^7 \text{ m/sec}}{9 \times 10^{16} \text{ m}^2/\text{sec}^2}$$
$$= 2.1 \times 10^8 \text{ m/sec}/1 + 2.2 \times 10^{-2}$$
$$= 2.1 \times 10^8/1.022$$
$$u = 2.05 \times 10^8 \text{ m/sec}$$

6.10 a) 5×10^8 m/sec

b) $u = (u^* + v)/1 + (u^*v/c^2)$

$$= \frac{2.5 \times 10^8 \text{ m/sec} + 2.5 \times 10^8 \text{ m/sec}}{[1 + (2.5 \times 10^8 \times 2.5 \times 10^8)]/9 \times 10^{16} \text{ m}^2/\text{sec}^2}$$
$$= \frac{5 \times 10^8}{1 + 6.95 \times 10^{-1}}$$
$$= 2.95 \times 10^8 \text{ m/sec}$$

Bibliography

There exist many books on relativity. Here is a selection, more or less in order of preference.

1. Books in whole or in part on relativity.

Albert Einstein, *Relativity, The Special and General Theory*. Trans. Robert W. Lawson, Crown Publishers, New York, sixteenth edition, 1961. Translations of the original papers found in H. A. Lorentz, H. Minkowski, and others are to be found in H. A. Lorentz, A. Einstein, H. Minkowski, and H. Weyl, *The Principle of Relativity. A Collection of Original Memoirs on the*

Special and General Theory of Relativity, with notes by A. Sommerfeld. Tr. W. Perrett & G. B. Jeffery, Methuen and Co., London, 1923.

David Bohm, *The Special Theory of Relativity*, Benjamin, New York, 1965. This is an excellently written book, highly to be recommended.

C. W. Sherwin, *Basic Concepts of Physics*, Holt, Rinehart & Winston, New York, 1961, pp. 105 ff.

David Bohm, *Causality and Chance in Modern Physics*, Harper Torchbooks, Harper & Brothers, 1961. (Van Nostrand, 1957)

J. T. Fraser, editor, *The Voices of Time*. A cooperative survey of man's views of time as expressed by the sciences and the humanities. George Braziller, New York, 1966.

Clement V. Durell, *Readable Relativity*, Harper & Row, New York, 1960. (G. Bell & Son, London, 1926)

P. W. Bridgman, *A Sophisticate's Primer of Relativity*, Harper Torchbooks, Harper & Row, New York, 1965. (Wesleyan University Press, Middletown, Conn., 1962)

Robert Katz, *An Introduction to the Special Theory of Relativity*. (Commission on College Physics), Van Nostrand, Princeton, New Jersey, 1964.

Edwin F. Taylor and John A. Wheeler, *Introductory Relativity*. (Fifth draft) W. H. Freeman & Company, San Francisco, 1963.

2. Books and articles that bear on Section 6.16, and the effects of relativity theory on contemporary thought.

Gillaume Apollinaire, *The Cubist Painters, Aesthetic Meditations*, 1913. Trans. Lionel Abel, Wittenborn, Schultz, New York, 1949.

H. G. Cassidy, *The Sciences and The Arts. A New Alliance*. Harper & Brothers, New York, 1962, p. 138.

H. G. and L. R. Lieber, *Non-Euclidean Geometry: or Three Moons in Mathesis*, Academy Press, 1931, and *The Einstein Theory of Relativity*, Farrar & Rinehart, 1945.

James R. Newman, ed., *The World of Mathematics*, Simon & Schuster, New York, 1956, essays by W. K. Clifford, H. von Helmholtz.

George Gamow, *Matter, Earth, and Sky*. Prentice-Hall, 1958.

Henry Margenau, *The Nature of Physical Reality*, McGraw-Hill, New York, 1950.

H. Margenau, *Open Vistas. Philosophical Perspectives of Modern Science*, Yale University Press, New Haven, 1961.

R. G. H. Siu, *The Tao of Science. An Essay on Western Knowledge and Eastern Wisdom*. Technology Press, 1957. These are just a few of the many interesting treatments of this subject.

H. G. Cassidy, "The Muse and The Axiom," *Am. Scientist*, *51*, 315 (1963).

Paul M. Laporte, "Cubism and Science," *The Journal of Aesthetics and Art Criticism*, 7, 243 (1948–1949).

J. M. Barrie, "Peter Pan" and "Dear Brutus."

"Dangerous Corner," in *The Plays of J. B. Priestly*, Vol. I. William Heinemann, London, 1948. There are many other examples. Michael Mattil, in a term paper at Yale, 1964, cited Max Frisch, "The Chinese Wall"; and the moving picture "Last Year at Marienbad," as using time in new ways.

Christopher Gray, *Cubist Aesthetic Theories*. Johns Hopkins Press, Baltimore, 1956.

E. H. Gombrich, *The Story of Art*, Phaidon, New York, 1956.

Roland Penrose, *Picasso: His Life and Work*, Harper, New York, 1958.

Adrian Stokes, *Cezanne*, Faber & Faber, London, 1947.

Gyorgy Kepes, see for example, *Education of Vision*, Braziller, New York, 1965. *Language of Vision*, Theobald, Chicago, 1951. *The Nature and Art of Motion*, Braziller, New York, 1965. *Structure in Art and in Science*, Braziller, New York, 1965.

Chapter 7

7.1 Introduction
7.2 Summary of progress
7.3 A preliminary demonstration
7.4 A somewhat simpler model
7.5 Relativistic interpretation, qualitative
7.6 A closer examination
7.7 Electromagnetic waves
7.8 Test of theory (Hertz's experiments)
7.9 Electromagnetic spectrum
7.10 Summary

Electromagnetic Radiation ~ 7

Introduction 7.1

In Chapter 4 it was shown that there is an intimate connection between electricity and magnetism. How intimate that connection is becomes evident when the contributions of relativity are taken into account. Einstein's first paper on relativity was entitled "On the Electrodynamics of moving bodies." We are now at a point where electricity, magnetism, and relativity can be drawn together.

Electromagnetic theory is the organized knowledge of electromagnetic radiation. Electromagnetic radiation is a disturbance, periodic in nature, which travels with the speed of light c, 2.997925×10^8 m/sec. This is the speed *in vacuo*, and it is taken as the maximum speed at which any signal can travel. Electromagnetic radiation is not heat or light. What we call heat is a sensation produced by a certain range of radiation in the infrared region when it interacts with our skin, or other appropriate sense organs, or when it raises the temperature of material objects. Light is the effect produced in our brain when the range of electromagnetic radiations in the "visible region" impinge on and are received by our eyes.

The distinction made here is important. We can receive the sensation of light if struck on the head: we "see stars." The fact is that all our sense transducers transform inputs from our environment into nerve impulses. The transducer, broadly speaking, accepts energy in one form, e.g., electromagnetic or sound or chemical, as in odor and taste, and transforms it into the electrochemical energy of the nerve impulse. Apparently our brain interprets the impulse into "light," "sound," "odor," etc., according to its source transducer.

Electromagnetic radiation comprises a range, or spectrum, of phenomena from the shortest wavelength in cosmic radiation to the longest radio waves, and beyond on both ends of the range. At first the electromagnetic spectrum was understood chiefly in its visible and tangible effects. But it became evident that besides light and heat, long waves we now call radio waves and the short X-rays belong to that spectrum, and gradually, as suitable instruments for detecting the radiation have become available, the range has been extended. It has also been discovered that the Earth is continuously bathed in electromagnetic radiation of many sorts.

Electromagnetic theory, which was largely developed a generation before Einstein, is the only classical theory which stands unchanged in the face of relativity. This is because relativistic effects were present, observed, and taken into account at that time, though not then recognized as such. We take up some aspects of electromagnetic radiation at this point in our work because it is naturally connected to relativity (Chapter 6). We were not ready, in Chapters 3 and 4, to go this far, nor yet in Chapter 5. We needed to prepare the basis for discussion by considering the laws of motion applicable to particles and the conceptualization of harmonic motion and of the oscillation of particles. We needed to review the properties of charged particles and moving charged particles. All this because *electromagnetic radiation appears when charged particles accelerate.*

The electromagnetic theory was developed by James Clerk Maxwell. He brought together and unified all the then known experimental facts about electricity and magnetism. Out of this work came the suggestion that electromagnetic waves carrying energy would propagate out from an oscillating electric charge, and that light is an electromagnetic phenomenon. One aspect of Maxwell's predictions was tested successfully some fifteen years later by Henry Rowland and verified experimentally. It was not until another ten years had elapsed that another striking verification, which caught the imagination, was made. Heinrich Rudolf Hertz demonstrated unequivocally that electromagnetic waves did indeed propagate from an oscillating current exactly as predicted by Maxwell.

Our objectives in this chapter are to show how a relativistic approach united electrical and magnetic phenomena, and to describe how Hertz demonstrated experimentally what Maxwell had predicted, namely that non-uniformly accelerating (oscillating) charge would produce radiation. Maxwell's theory is beyond our mathematical competence, and we cannot develop it here. However, with the preparation given in this chapter we can embark on the study of the phenomena that led to quantum constructs. First we recall important relationships developed (for these purposes) in earlier chapters.

7.2 Summary of progress

The fundamental law of electrostatic interaction is:

$$\mathbf{F} = Kq_1q_2/r^2 \tag{3-4}$$

The corresponding law of magnetic interaction is[1]:

$$\mathbf{F} = \phi_1\phi_2/kr^2$$

From these were derived the specific strengths of force fields as experienced by a particle with a charge q_2 at a point P distant r from the charged source q_1 at the point O, namely:

$$\mathbf{E} = \mathbf{F}/q_2 \text{ or } \mathbf{F} = Eq_2 \tag{3-5}$$

Similarly the force on the particle of charge q_2, moving with a velocity v at the point P in a magnetic field of field strength $\mathbf{B}$, is

$$\mathbf{F} = q_2\overset{\frown}{vB} \text{ or } \mathbf{F} = \overset{\frown}{lIB} \tag{4-4}$$

Recall that the quantities F, E, v, and B are vectors, with magnitude and direction. E and B are, respectively, measures of the specific field strengths about a charged particle and a magnetic pole: the electric field strength and the magnetic field strength. These may be added to show the force on a particle of charge q_2:

$$\mathbf{F} = q_2E + q_2\overset{\frown}{vB}$$

If the charged particle is at rest with respect to another charged particle (the source of the field with which the charge q interacts) then $v = 0$, and there is no magnetic field: $q_2vB = 0$. If the charges are in steady motion parallel to each other a magnetic field appears. Further, if the particle of charge q_2 is considered the source then the same form of equation holds, and the force on the other particle (with charge q_1) will be:

$$\mathbf{F} = q_1E_2 + q_1\overset{\frown}{vB_2}$$

where now the values of E and B depend on the charge q_2. This is to say that neither particle provides a privileged frame of reference.

Unlike mass, energy, length, and time, charge is relativistically invariant. The total charge of a particle, or of several interacting particles, does not depend at all on their motion. This has been tested to high velocities. Also the magnitudes of the unit (+) (protonic) and (−) (electronic) charges are to a high degree of certainty exactly equal. In addition, charge is conserved. In a system of charged particles if no particles enter or leave the system the total charge remains constant (unless annihilation occurs—Section 12.10); then *two* opposite charges simultaneously disappear).

A preliminary demonstration 7.3

In an easily made demonstration two long flexible copper wires are hung close together and parallel. If no current flows, or if current

[1] This equation is given for the first time here. Accept it. ϕ represents a quantity of magnetism at a magnetic pole.

flows in one wire only, the wires remain unmoved. But if current is made to flow in the same direction in both wires, the two are observed to move toward each other (Figure 7-1). This behavior is diagnostic of

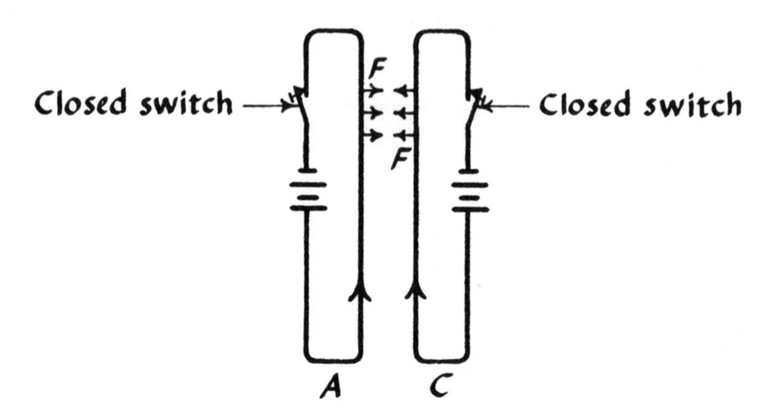

Fig. 7-1 *Between two wires* A *and* C, *parallel and close together, with current flowing through both in the same direction, there is a (magnetic) force of attraction.*

unbalanced force acting on the wires. This force is magnetic. If the current flows in opposite directions in the two wires they are repelled from each other. To visualize the effect, apply the right-hand rule to each wire, with currents flowing in the same direction, as in Figure 7-2. The field directions, predicted by the rule and visualized with

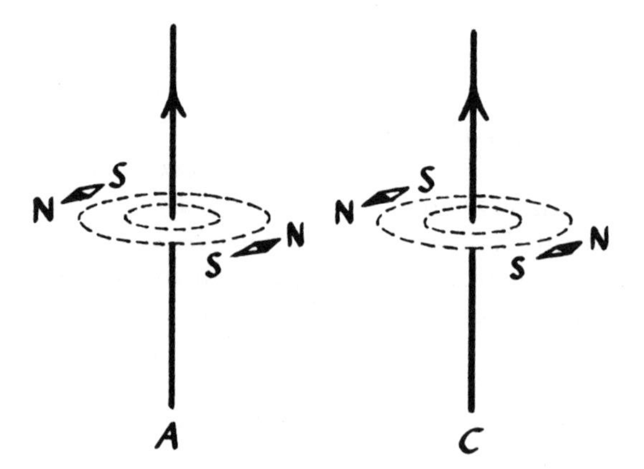

Fig. 7-2 *The magnetic force, Figure 7-1, is visualized as lines of force, the directions of which are shown by compass needles. (The needles lie in the* X, Z *plane when the current is along the* Y *axis.)*

compass needles in the figure, show that between the two wires the N-S directions of the two fields are lined up in an attraction orientation. Of course the inner N-S orientation of wire A repels the outer N-S field of wire C, but this is farther away, and since by the fundamental equation the force falls off as the square of the distance, we can see that the attraction interaction of the fields between the wires will be larger than the repulsion interaction between A and the outer field of C.

A somewhat simpler model 7.4

Suppose that a special kind of "wire" replaces A of Figure 7-1. This "wire" (Figure 7-3) is a linear array of (+) charges next to and

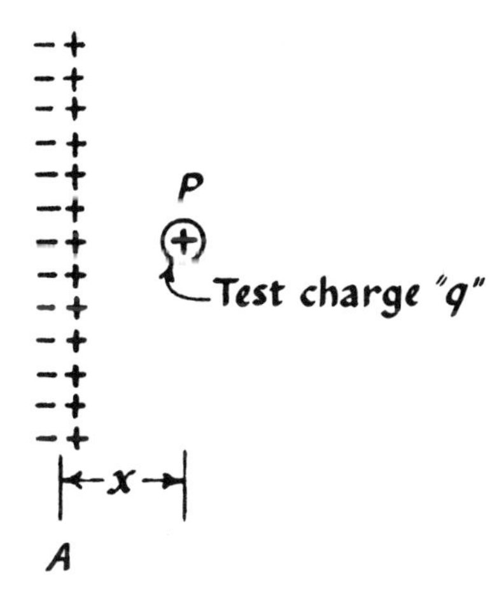

Fig. 7-3 *A neutral wire with no current in it has no effect on a test charge.*

parallel with a linear array of (−) charges. Both arrays have the same linear charge density d. That is, along any length of the "wire" the number of (+) charges is exactly equal to the number of (−) charges. At a point P distance x from this array is a stationary (static) (+) test charge, q. The test charge experiences no unbalanced force. A "current" of the following kind is now imagined to flow through this "wire." The (+) charges move in a +Y direction and the (−) charges in a −Y direction each with the same speed $|w|$ relative to P. This artificial model is devised for ultimate simplicity. As we are making a model we arrange it to suit our needs. Since the two kinds of charges in the wire are always in balance, the wire remains neutral. The test charge under this new state of flowing charges shows no evidence of unbalanced force. It shows no net effect of the flow of these charges (Figure 7-4).

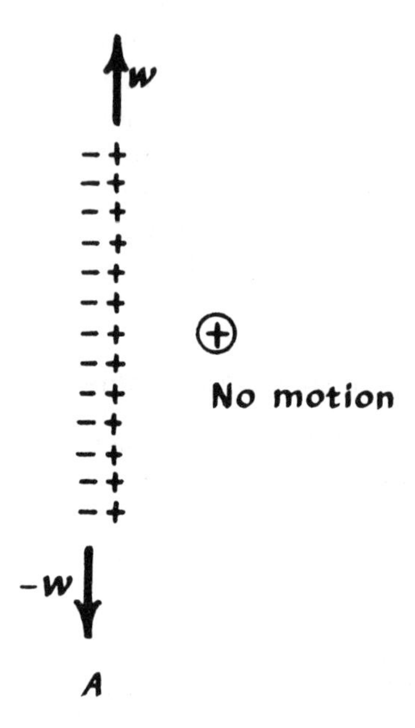

Fig. 7-4 *A neutral wire with current flowing in it,* (+) *charges in the* +Y *direction and* (−) *in the* −Y *direction, with the velocities* w *with respect to the wire frame of reference, has no effect on a test charge.*

We imagine that we can cause a current to flow in the "wire" A at a constant rate. Lest this seem too artificial it should be remarked that in a Crookes tube one can have a current of electrons in one direction and of (+) ions in the opposite direction; in solution, currents of (+) and (−) ions may flow in opposite directions (Section 3.4). We imagine that the test charge is instrumented so that if it experiences a force not only the magnitude of this but its direction will be available. Also, we imagine that the test charge may be held stationary (a static charge) or that it may be given a velocity in a specified direction relative to the "wire."

After the "current" has been flowing we cause the test charge to move with a velocity v parallel to, and in the direction of, the (+) current. *Immediately* it experiences a force $\mathbf{F}_v$ in the direction of the "wire," and at right angles to the direction of v (Figure 7-5). When the test charge was stationary there was no force on it. Now we must conclude on the basis of the discussion in Chapter 4 that it is experiencing a magnetic force, and we draw a field intensity vector arrow B in a −Z direction (into and behind the plane of the figure). This, of course, agrees with the requirement of the right-hand rule. We re-emphasize: the reason for this conclusion is that there are only two kinds of force under this type of situation, electrical and magnetic. Static charges do

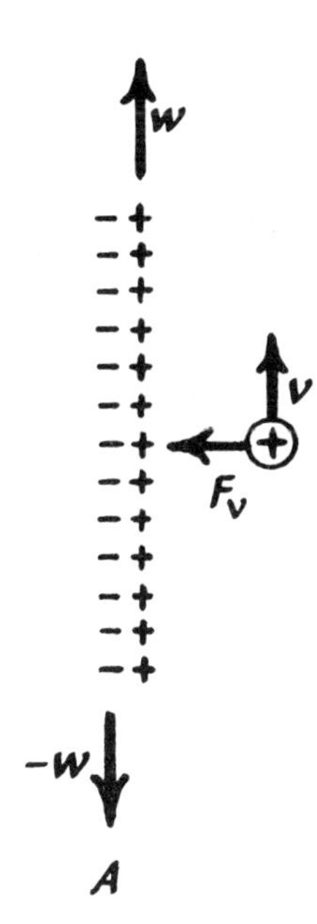

Fig. 7-5 *In the arrangement of Figure 7-4 the test charge, given a velocity* v, *experiences a force* F *toward the wire.*

not respond to magnetic forces (Section 4.1). Only if charge is in motion may it respond in a magnetic field.

We say "after the current has been flowing" because investigation of the behaviors just described shows that when the current is turned on in the wire it takes a finite time, namely x/c, for the field to establish itself at the point P, distant x from O. c Is the velocity of light. However, the moment that the test charge moves with velocity v its experience of $\mathbf{F}_v$ is instantaneous, provided only that there has elapsed x/c seconds since turning on A. Thus the field to which the test charge responds must be a property of that point (P) in space and it must be a consequence of the steady current in the wire A. Further investigation shows that it is the movement of the test charge that brings out this behavior: the force is proportional to v, as well as to the current in A, and inversely to a power of the distance x. That the field is a property of the location P whether or not a test charge is present suggests that in the absence of motion of the static charge (Figure 7-4) whatever field is there, there is no net E intensity.

Relativistic interpretation, qualitative 7.5

Consider the "wire" (through which (+) and (−) charges are moving) as the K reference frame. In this frame, as we saw, the test charge, moving with velocity v, exhibits behavior which must be interpreted as evidence for the presence of a magnetic field.

Now consider the phenomenon from the frame K^* of the test charge. In this frame the test charge is stationary: we imagine a coordinate system attached to it. Yet it experiences a force toward the wire. This can only be interpreted as response to an E field!

It is a remarkable consequence of the relativistic view that whether a field of force is interpreted as electrical or magnetic depends on the frame of reference and the state of the test instruments. In the K frame, a magnetic field is experienced; in the K^* frame, an electrical.

It may help understanding if one places himself in imagination first close to the "wire"—holding on to it, perhaps, and looking over at the (+)-charged test pith-ball. First, when the pith-ball is stationary nothing happens to it. But the moment it starts to move parallel to the wire it also moves toward the wire. It is like the illustration in Figure 7-1, only here instead of a (+) current in wire C we have a charged pith-ball, the motion of which is a (+) current. The resulting interaction must be magnetic. Now change your point of view. Ride, in imagination, the charged pith-ball. It is stationary in this frame, and stationary charges do not generate a magnetic field. They do, however, display an electrostatic field (Section 3.6). Therefore when the wire A in Figure 7-5 moves toward the pith-ball, which is the interpretation made from the pith-ball frame of reference, the force must be electrical. Both interpretations are correct.

7.6 A closer examination

In the K frame of the "wire," with the pith-ball stationary, it is reasonable to conclude that it experiences no net charge on the "wire" because (+) and (−) charges are moving with equal speed in opposite directions, so that there is always a (−) close to every (+) to yield net neutral. But in the K^* frame of the pith-ball in the case in Figure 7-5, where the "wire" as a whole is moving in a −Y direction in that frame, the two streams of charges in the wire are moving with different speeds. The (+) charges move past the pith-ball in the +Y direction and the (−) charges move faster in the −Y direction. The charges in the wire are moving with a speed v, and the wire as a whole is moving with a speed w: we have to take relativistic account of two speeds:

$$w^*_+ = \frac{w - v}{1 - wv/c^2} \text{ and } w^*_- = \frac{w + v}{1 + wv/c^2}$$ [1]

---

[1] Referring to equations in Chapter 6:

$$\Delta x^* = \gamma(\Delta x - v\Delta t) \qquad \text{(From 6-4)}$$

$$\Delta t^* = \gamma(\Delta t - v\Delta x/c^2) \qquad \text{(From 6-6)}$$

This is what explains the force on the particle with charge (+) q, for in its frame K^* the particle experiences a stream of charges which is net negative. The interaction is electrical in this frame. The moving stream of charges sets up a field E^*_w at P; the test charge sets up a field E^* at the "wire."

We have gone through this examination of linked electrical and magnetic interpretations of an experienced force in order to show the unifying power of the relativistic approach. It obviously remains convenient to talk about electrical fields and magnetic fields. The E field is a vector in the direction of the force; the B field is a vector at right angles to the force. This one example is a hint of the extraordinary way in which relativity penetrates and unifies all of electrodynamics, and also it shows, once more, what a tremendous difference a change in frame of reference may make.

Electromagnetic waves 7.7

A wire with a current flowing in it has associated with it a magnetic field which may be drawn as lines of force concentric to the wire. This field appears at a point P outside the wire with a delay x/c when current is made to flow in the wire. That there is an E field was implied in the discussion in Section 7.5. This field can be demonstrated by means we shall not pursue. Outside of the wire it is directed parallel to the wire and opposite to the flow of the (+) current. Suppose that in this wire current is made to alternate (oscillate). That is, electrons flow for a time in one direction, slow down and stop, start up in the opposite direction, reach maximum (net) speed, slow down, stop, and reverse. Alternating current is of this kind (Section 4.10). When this occurs, the two fields, at right angles to each other, build up and decay, change sign and repeat (Figure 7-6). Maxwell postulated that under conditions of this kind an electromagnetic wave would propagate out with the velocity c from the oscillating (non-uniformly accelerating) charge. He also predicted that the electromagnetic wave would carry energy away

---

$$w^*_+ = \lim_{\Delta t \to 0} \frac{\Delta x}{\Delta t}$$

$$= \lim_{\Delta t \to 0} \frac{\Delta x - v\Delta t}{\Delta t - v\Delta x/c^2}$$

$$= \lim_{\Delta t \to 0} \frac{\frac{\Delta x}{\Delta t} - v}{1 - \frac{\Delta x}{\Delta t}(v/c^2)}$$

$$w^*_+ = (w - v)/(1 - wv/c^2)$$

Similarly for w^*_-.

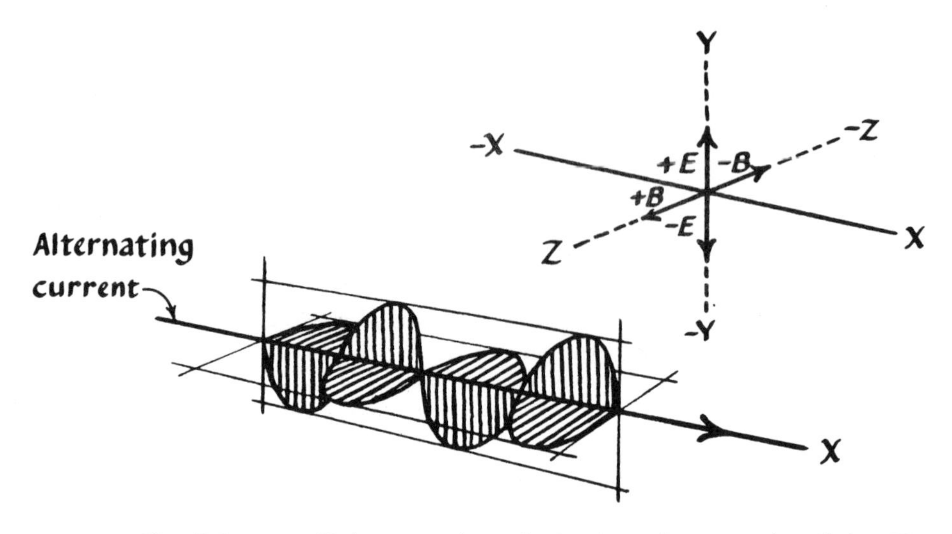

Fig. 7-6 *An oscillating current in a wire broadcasts electromagnetic radiation. The corresponding* B *and* E *vectors are shown.*

from the wire. This has been fully confirmed, as we shall see in later chapters. Today we might think of the wire with electrons surging up and down it as an antenna, broadcasting radio waves.

7.8 Test of theory (Hertz's experiments)

If a spark is made to jump between the terminals of an open circuit as at B in Figure 7-7, and if a wire loop, a, b, c, d, forming a circuit open at M is close by and properly oriented toward B, a spark is observed to jump between the terminals at M. It was this discovery that led Hertz to demonstrate, first that energy is propagated, and then that radiation with a wave character is involved.

Consider the key elements of this discovery. Referring to Figure 7-7, A is a spark-coil: a common means for building a potential difference between two terminals. C and C' are metal objects to accumulate charge. When the spark jumps, breaking down the insulating air in the gap at B, electrons surge across from the negative to the positive side. Because of inertia, and inductance effects, excess electrons flow, so that the side that was positive becomes negative. Electrons surge back, again overshooting, but less. These surges are rapid and changing accelerations of current, quickly damping out, as shown in Figure 7-8. They can be violent, rapid, and energetic: just what Maxwell's theory required. In the loop with the small gap at M, Hertz had a detection device—a transducer which, interacting with the electromagnetic ra-

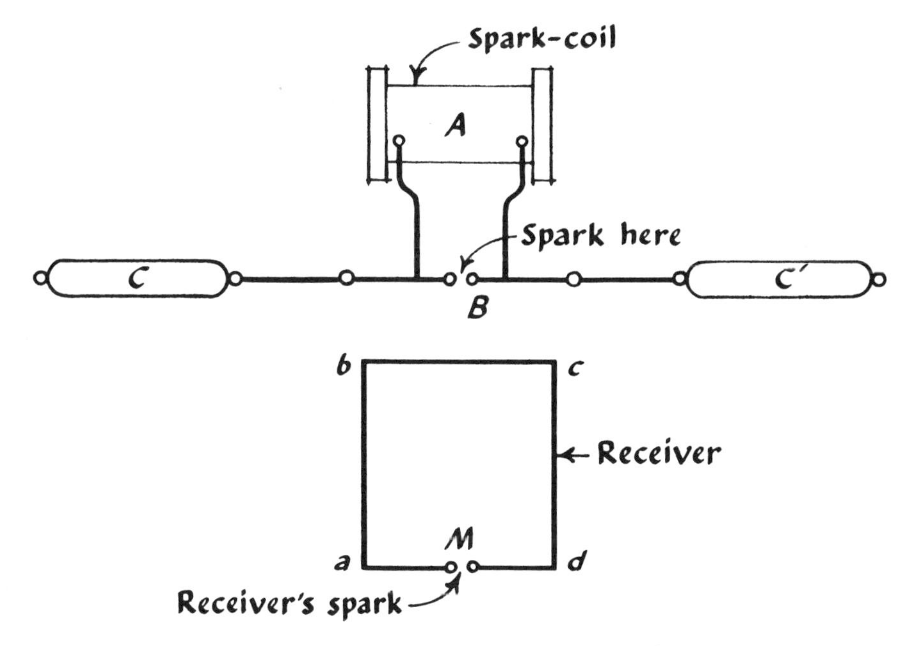

Fig. 7-7 *The spark-coil* A *produces sparks at the electrodes* B. *Two large insulated conductors* C *and* C′, *three meters apart, connected by 2 mm thick copper wires to the electrodes, helped to give good sparks at* B. *The secondary circuit, 120 cm long and 80 cm wide, at a distance of 50 cm from the primary, yielded a stream of sparks 2 mm long.* [*From H. Hertz,* Electric Waves, *Trans. by D. E. Jones. Macmillan, London, 1893, p. 40, Figure 9.*]

diation from the spark-gap, converted its energy to electrical energy and this, via the small spark that jumped across M, to visible and audible evidence of the invisible disturbance emanating from the originating spark-gap. In more modern terms he had discovered a broadcast from B and had invented a receiving antenna, the loop with the gap M.

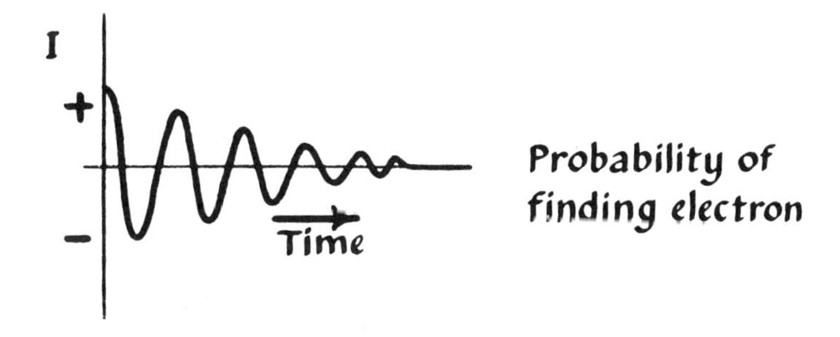

Fig. 7-8 *Damped oscillations of a current* I, *as might occur at a spark gap.*

Moreover, he could carry out quantitative experiments by measuring the maximum width of the gap M that would just let a spark cross: the greater the distance the spark would jump, the greater the energy collected by the loop.

He proceeded to vary the length of wire in the loop, not moving the position or orientation of the loop while keeping the intensity of the sparks at B fixed. This showed him that at a critical point in the length of wire he obtained the longest sparks (Figure 7-9). Lengthening or

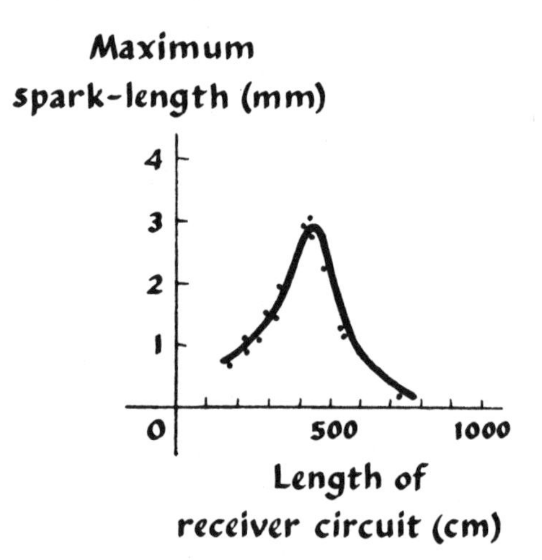

Fig. 7-9 *Evidence of resonance as the length of the secondary circuit wire, from electrode to electrode of the spark-gap, is lengthened. The length of the receiver circuit in cm is plotted against the maximum spark-length that can be obtained, in mm. A maximum occurs when the secondary circuit is tuned to the primary.* [*From H. Hertz,* Electric Waves, *Macmillan, London, 1893, p. 45, Figures 10a and 10b.*]

shortening the wire in the loop decreased the maximum length of spark obtainable at M. He recognized that he was 'tuning' the loop to the spark source; that he was bringing the two circuits into resonance.

Having refined his detection device Hertz was now able to investigate the properties of the radiation which carried energy from the spark-gap outward. It must be remembered that Hertz was working in the dark in more ways than one. Light was not known definitely to be an electromagnetic radiation. Maxwell's hypothesis was available, but not any substantiating experiments except Rowland's, which were not widely known. The radiations studied by Hertz belong to the radio band, and unlike the light of the spark were invisible to him. Clearly there was *something* there at a distance from the spark-gap, but he had only his test loop and his mind to guide him. In the idiom of Chapter 1, he was proceeding back and forth between the plane of perceptions and the field of constructs, weaving a pattern of experimental-theoretical science.

Hertz was able to show that the radiation propagated in a straight line. He found that a sheet of metal interposed between spark-gap and coil sharply cut off the reception of energy when the edge extended beyond the line between gap and coil. This was like an opaque object cutting off light and casting a shadow. He found also that insulators such as wood and paper did not appreciably attenuate the radiation. Further analogies to light were found. He showed that the invisible radiation could be reflected from a metal sheet. Carrying this further, he built two parabolic metal sheets, one around but insulated from the source B, and the other around but insulated from M (Figure 7-10).

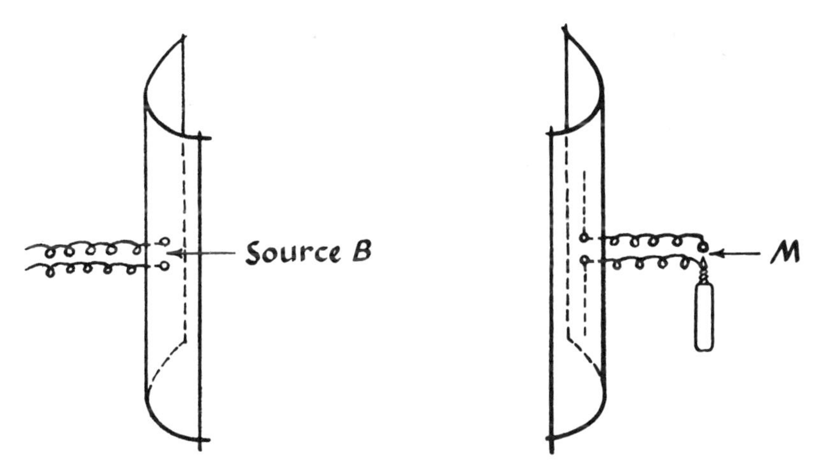

Fig. 7-10 *The parabolic mirrors such as those used by Hertz. The mirror is a sheet of zinc 2 m long, 2 m wide, and 0.5 mm thick, bent over and supported on a wooden frame (not shown) to give a depth of 0.7 m, and an aperture of 1.2 m. The primary oscillator was fixed at the middle of the focal line, the insulated wires being led out behind the mirror; the same for the secondary circuit, the receiver. [Redrawn from H. Hertz,* Electric Waves, *Macmillan, London, 1893, pp. 183, 184, Figures 35 and 36.]*

Mirrors of this type are known to focus light (Figure 7-11); he found that the apparatus greatly enhanced the effectiveness of the receiver, as judged by the length of sparks obtainable at M. Moreover, on rotating the receiver about the line between M and B the sparks at M became more and more feeble until when the two were at right angles no sparks were obtained even if the two were moved close to each other. This indicated a directional quality to the transmission, and enabled Hertz to show not only that the radiation could be reflected from a metal surface but, by using the directional parabolic detector, that the angle of incidence was equal to the angle of reflection (Section 5.3). He

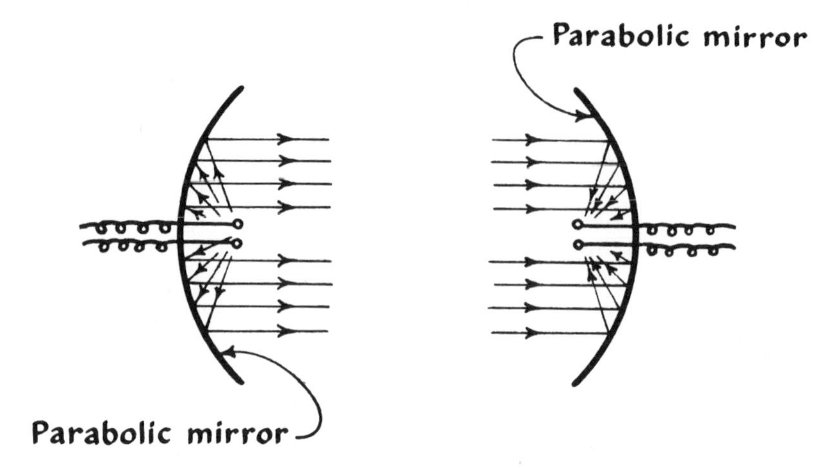

Fig. 7-11 *Detail of reflection at parabolic mirrors.*

built a large prism of hard asphalt, and was able to show that the radiation was refracted in analogy to the way that a glass prism refracts light (Section 5.3). He was by this time quite convinced of a relationship between light, his radiation, and electromagnetic radiation.

During some of his work he had picked up sparks in such a way that he was led to think of wave-action, particularly interference (Section 5.5). The sparks would be stronger as he approached a wall, then would disappear close to the wall. He was able to demonstrate the correctness of this inference. He put a metal sheet against the wall of a large lecture room. Across from it he placed the source A (Figure 7-12). Loop in hand he then walked along, holding the loop in the line perpendicular from wall to A (the line AX) and measuring the intensity of the

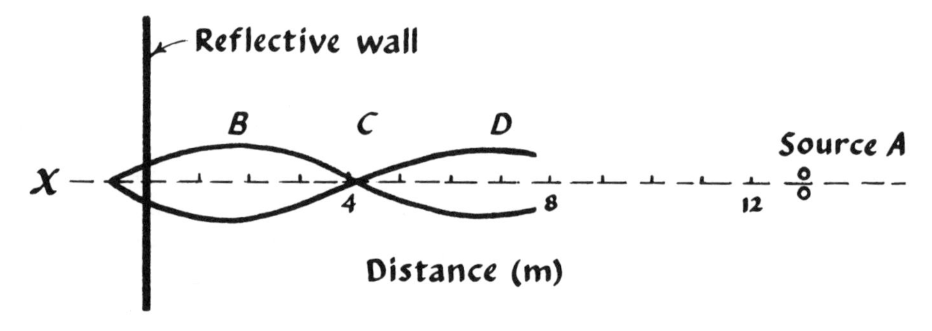

Fig. 7-12 *Schematic diagram of an electromagnetic standing wave reflected from a wall. [Redrawn and simplified from H. Hertz,* Electric Waves, *Macmillan, London, 1893, p. 128, Figure 26.]*

spark. Figure 7-12 shows what he found. The spark was strong at B, feeble at C, again strong at D, and so on. Reflection seemed to come from a small distance into the wall. Assuming the phenomenon to be interference, B to D was a half-wavelength. By careful measurements Hertz found the half-wavelength of some of his radiations to be 4.8 meters. Today we recognize these as short radio waves.

Electromagnetic spectrum 7.9

Electromagnetic radiation is the name of a set of phenomena, vast in range, defined by the properties that all of the entities of the set travel with the same velocity, that of light; all originate (as far as is known) in non-uniform acceleration of charge; and all transport energy through what appears to be otherwise empty space. These radiations are classified in subsets related to the methods of detecting them. Hertz's short radio waves, the first radiations unequivocally certified to be electromagnetic, are detected by means of an antenna; the shorter wavelengths are recognized by a host of specialized instruments such as the thermocouple for heat; the photographic plate for visible and shorter waves; photoelectric cells and radiation detectors (which we shall study later) for visible and shorter waves; and others. Most thoroughly studied is the narrow range directly accessible to us through our eyes. A conventional classification of electromagnetic radiations is given in Table 7-1. The wavelengths range from over several kilometers

Table 7-1

The Electromagnetic Spectrum

Wavelength, λ in meters	Designation
.	
.	
More than 10^6 to about 10^4	Long electric waves
About 10^4 to about 10^{-2}	Radio waves
About 10^{-2} to about 10^{-4}	Short electric waves
About 10^{-4} to about 7×10^{-7}	Heat waves or infrared
About 7×10^{-7} to about 4×10^{-7}	Light waves; visible radiation
About 4×10^{-7} to about 10^{-8}	Ultraviolet radiation
About 10^{-8} to about 10^{-10}	X-radiation
About 10^{-10} to beyond 10^{-14}	Gamma radiation
.	
.	
.	

Note: λ represents wavelength, ν represents frequency, $\tilde{\nu}$ represents wave number. Then $\tilde{\nu} = 1/\lambda = \nu/c$. Data are in mks system.

to shortness below the grasp of present instruments (one must predict). We shall be concerned chiefly with the mechanisms attending the appearance of the shorter wave-length radiations.

7.10 Summary

In respect of moving charged particles, behavior that in one frame of reference is interpreted as magnetic may from another frame be interpreted as electrical. The two constructs are intimately related. When charged particles are non-uniformly accelerated, electromagnetic radiation, characterized as a disturbance periodic in nature that travels with the speed of light, makes its appearance. In the last hundred years there has been a vast extension of the electromagnetic spectrum on both ends of the narrow range that is accessible to our eyes: on the one side, the infrared, short and long radio waves; on the other, ultraviolet, X-rays, and gamma rays. Necessarily the application of these discoveries in radio, television, radar, X-ray, maser, and laser technologies has been even more recent.

Exercises

7.1 In some carnivals there is a large wheel, shaped somewhat like a baseball pitcher's mound, made with a highly polished hardwood surface. You sit on this wheel, and the operator starts it turning. As the speed increases the occupants slide off, one after another—or in batches. The person who sits right at the center on top of the wheel merely becomes dizzy. As we stand outside and watch the people going round, we see them bracing themselves. They push themselves inward in order to keep from sliding off the wheel. a) Discuss the physical situation (in terms of forces) from the viewpoint of an observer standing in the room and watching the turning wheel. b) Discuss the physical situation from the frame of reference of the person sitting firmly attached to the wheel. What is the explanation, for example (from each point of view), of the fact that if the person on the wheel puts a ball down on a flat surface on the wheel (so we can neglect gravity) it tends to roll toward the periphery, and he has to exert force on it to hold it at a given position.

7.2 What is meant by saying that the speed of light is an absolute?

7.3 If sunlight falling on the Earth on a bright day delivers 0.7 kilowatts per square meter where it falls perpendicularly, estimate how much energy falls upon the Earth in one day.

7.4 If you were working with Hertz's parabolic mirrors, how would you go about showing that the radio waves could be refracted, and that the angle of incidence was equal to the angle of reflection?

7.5 Sketch the conditions for proving that the rays could be refracted.

7.6 In Figure 7-5 discuss the magnetic lines of force. In what frame(s) would you expect to draw them?

Answers to Exercises

7.1 a) To the observer the ball is behaving like an object whirled on the end of a string. In order to hold it in position in which it is undergoing acceleration (because of the changing direction of motion) the person on the wheel has to apply a centripetal force to it. b) To the person on the wheel the ball when put down next to him is behaving peculiarly. He puts it down beside him and it rolls away from him. He invents a centrifugal force so that opposing it with an equal force he can hold the ball still in his frame of reference. His frame is not an inertial frame; Newton's law of motion is not expected to hold. (Even if he could not see the walls of the room, the person on the wheel would know from muscular and other senses that he is being accelerated—as one knows it in an elevator, for example.)

7.2 It is meant that it is an upper limit of the velocity of material particles.

7.3 Don't forget that there are clouds over some of the Earth, and that not all rays are perpendicular.

7.4 Use some such arrangement as having the source behind a massive wall, with its conducting axis vertical and the ray directed through a door to a mirror—such as a sheet of metal. This is arranged to reflect the ray to another parabolic mirror, with axis vertical, and set so that its normal is at right angles to that of the source. When the mirror is at exactly 45° to both paths, or when the angles of incidence and reflection are equal, maximum sparking is obtained. (This, in fact, was Hertz's method.)

Bibliography

Heinrich Hertz, *Electric Waves. Being Researches on the Propagation of Electric Action with Finite Velocity through Space*. Trans. D. E. Jones, Macmillan, London, 1893. This work makes interesting reading if one is concerned about the human side of science, the frustrations, the wrong turns, and false starts, and discouragements that stand in the way of success. Radio technology took its start with Hertz's work.

Chapter 8

8.1 Introduction
8.2 The problem
8.3 Planck's contributions
8.4 Einstein's contributions
8.5 Millikan's contributions
8.6 Some implications
8.7 Compton's contributions
8.8 Summary

Quanta and Photoelectric Theory ~ 8

Introduction 8.1

This chapter has to do with the discovery of the electromagnetic quantum and quantum of energy, where the word quantum means a unit, or packet: implying something with particulate properties. The old problem of whether matter is ultimately continuous or particulate has not been solved. Indeed, it would seem that the problem has entered a new phase, a phase in which particle and continuum—or as we shall designate it later, particle and field—are not placed in an either/or relationship. Instead, Janus looks out of Nature, and the face we see depends on our stance—our instruments and their state. In other words a kind of counter-revolution against Galileo and Newton has occurred, and the distinction between primary and secondary qualities has been blurred, if not removed.[1] Not only is modern science a science of process, as we have suggested and will explore further in Chapter 17, but the role of the observer has been emphasized in his interaction with Nature's processes.

[1] The doctrine of primary and secondary properties of things: The primary are conceived as intrinsic to the things: shape, having a boundary, being in a particular place and state of motion or rest, being one, or many, in or not in contact with other bodies, and so on. The secondary are dependent on the observer, and not in the thing, such as color, sound, taste, warmth. This categorization was introduced by Galileo. Now, however, it has been concluded by certain physicists that the intervention of mind on matter must be taken account of at the atomic level (J. von Neumann, M. R. von Smoluchowski, Leo Szilard), and "that it is not possible to formulate the laws of quantum mechanics in a complete and consistent way without a reference to human consciousness." Max Jammer, *The Conceptual Development of Quantum Mechanics*, McGraw-Hill, New York, 1966, cf. p. 373.

The resulting implications for scientific thought provide a basis for understanding and judging patterns of thought, the impacts of which are now beginning to be felt in our culture, particularly the emphasis on probability and on process. These new developments in physics and chemistry are important because they have an effect on thought—in some respects an unwonted effect. As one textbook says, "Physics is the science through which man views the most fundamental aspects of the universe." Enough people believe this that the impact of modern physical hypotheses is very great—even in spite of warnings that these principles such as indeterminacy are applicable in any practical sense only to atomic-level phenomena. So we issue this *caveat*: it may well be that there is no 'the most fundamental aspect of the universe' unless it is the cognitive process of man himself.

8.2 The problem

The discontent that led eventually to quantum theory resulted from an inability to explain certain experimental data by fundamental classical laws (such as $F = ma$, and the like). When a piece of metal or high-melting substance such as carbon, or a ceramic, is heated it goes through a sequence of color changes as it gets hotter: cherry red, orange, yellow. Eventually it becomes white-hot. The radiation that it gives off is responsible for these colors. First there is infrared, the long wavelengths of which are not visible; then as the shorter wavelengths appear and increase in intensity light begins to be given off and the color changes as described. This change is correlated with temperature, and although the general pattern is that described, yet each substance contributes its own characteristics to the color that it gives off (as we shall see later). These characteristics make it possible to examine the light given off by the Sun, or a star, and to recognize the chemical elements present in the emitting body.

But these characteristics also complicate matters because of their diversity of detail. Early investigators were overwhelmed by an embarrassment of riches. What was needed was some source of radiation, or some absorber of radiation, which would give results independent of its chemical nature—what it was made of—and dependent only on its physical state. It was known at the time that substances would emit or absorb given wavelengths of radiation in fixed ratio, so a perfect absorber would also be a perfect emitter. Gustav R. Kirchhoff had proven this, and he went on further to introduce the idea of a perfect absorber; a perfectly black body. This is illustrated in Figure 8-1. It is a hollow sphere lined with carbon black (soot), or some other finely divided particles, and with a small hole on one side. Radiation entering this hole is trapped inside the 'black body' and absorbed, even though it may be reflected or scattered inside.

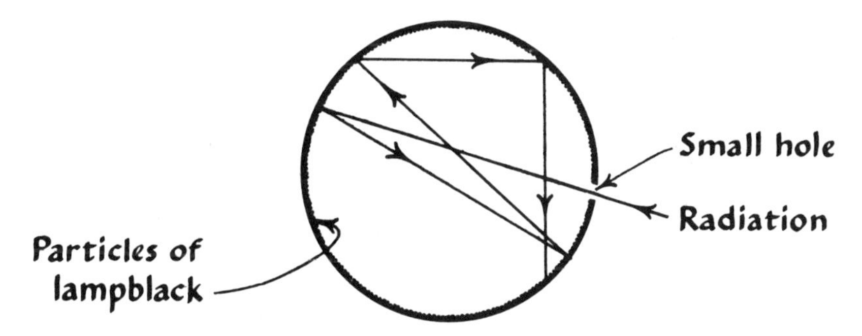

Fig. 8-1 *The 'black body.' A nearly perfect absorber of radiation.*

At this time no-one understood the mechanism of absorption or emission of radiation. But the black body was significant because it behaved in the same way no matter what substance it was made of. Kirchhoff had said that whenever he came upon some relation, such as that of emission power to absorption ability, that was independent of the chemical nature of the substance it turned out to be a simple relation. These considerations might seem esoteric to the non-physicist, but there are practical implications also. It had been discovered, for example, that the total radiant energy given off by a hot body is directly proportional to the fourth power of the temperature (provided this is measured in absolute degrees, degrees Kelvin, which are centigrade +273). This discovery not only made it possible to estimate the temperature of the Sun and other stars (and no-one at that time except perhaps Jules Verne would have thought that it might soon be possible to send a probe to find out), but it also led to means by which, peering into a blast-furnace, one could measure its temperature.

The ratio in question is the Stefan-Boltzmann law, $W = \sigma T^4$*, where the total radiant energy emitted per unit of area per second by a glowing body is represented by* W *in watts per square meter,* T *is in degrees Kelvin, and the proportionality constant* σ *is close to* 5.7×10^{-8} *watt/m*^{2}*. The accuracy of this relation over a range of temperatures is illustrated in Figure* 8-2, *where the fourth root of* W *is plotted against* T. *In the diagram* E *is equivalent to* W.

When a black-body sphere is placed in an oven and heated, the radiation that comes out through the hole—that which is emitted—has the fundamental characteristic that it is independent of the composition of the black body. Whether the body is made of iron or ceramic

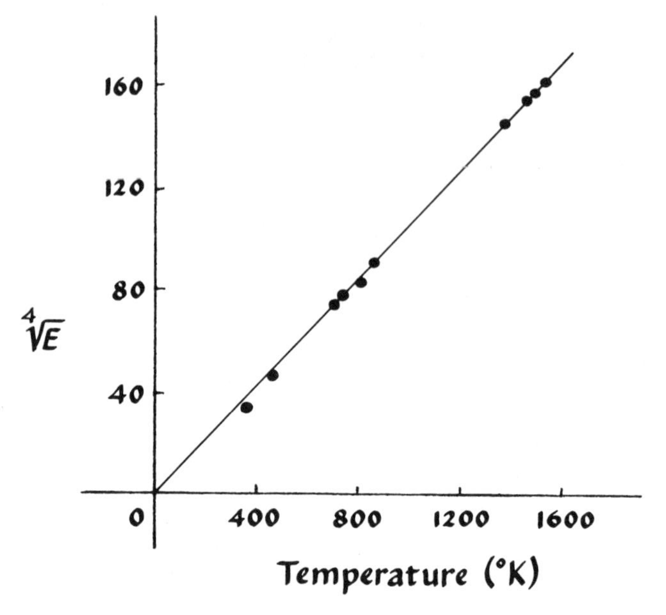

Fig. 8-2 *Verification of the Stefan-Boltzmann total radiation law. [From F. K. Richtmyer,* Introduction to Modern Physics. *McGraw-Hill Book Company, New York, 1928, p. 201 (Figure 52).]*

it gives the same spectrum at a particular temperature. The radiation produced has a characteristic distribution of wavelengths. Figure 8-3 illustrates the amount of energy emitted at a range of wavelengths in the red and infrared. The wavelengths are given in Angstrom units (Å). One Å $= 1.0 \times 10^{-10}$ meters. The curves are plotted for different temperatures of the emitting body. These temperatures are measured independently, to test the law. To obtain the total emissive power one measures the area under the curve, estimating the shapes at the ends where extrapolation is necessary, and one calculates from this value.[1]

The curves, as Figure 8-3 shows, all have a maximum and tail off at the extremes: sharply in the short-wavelength region, slowly in the long-wavelength region. The heights of the curves increase with temperature, as the Stefan-Boltzmann law implies, and also, the position

[1] We may use any of a number of expedients to measure area under a curve. If one has an equation for the curve, then one may use mathematical means. One might plot the curve on graph paper and count up all the little squares under the line, estimating parts. One might use a planimeter and measure it that way, or submit the problem to a computer. Or one might cut the curve out along the line and weigh the piece with the desired area; then calculate from the weight of a piece of the same sheet that has a known area.

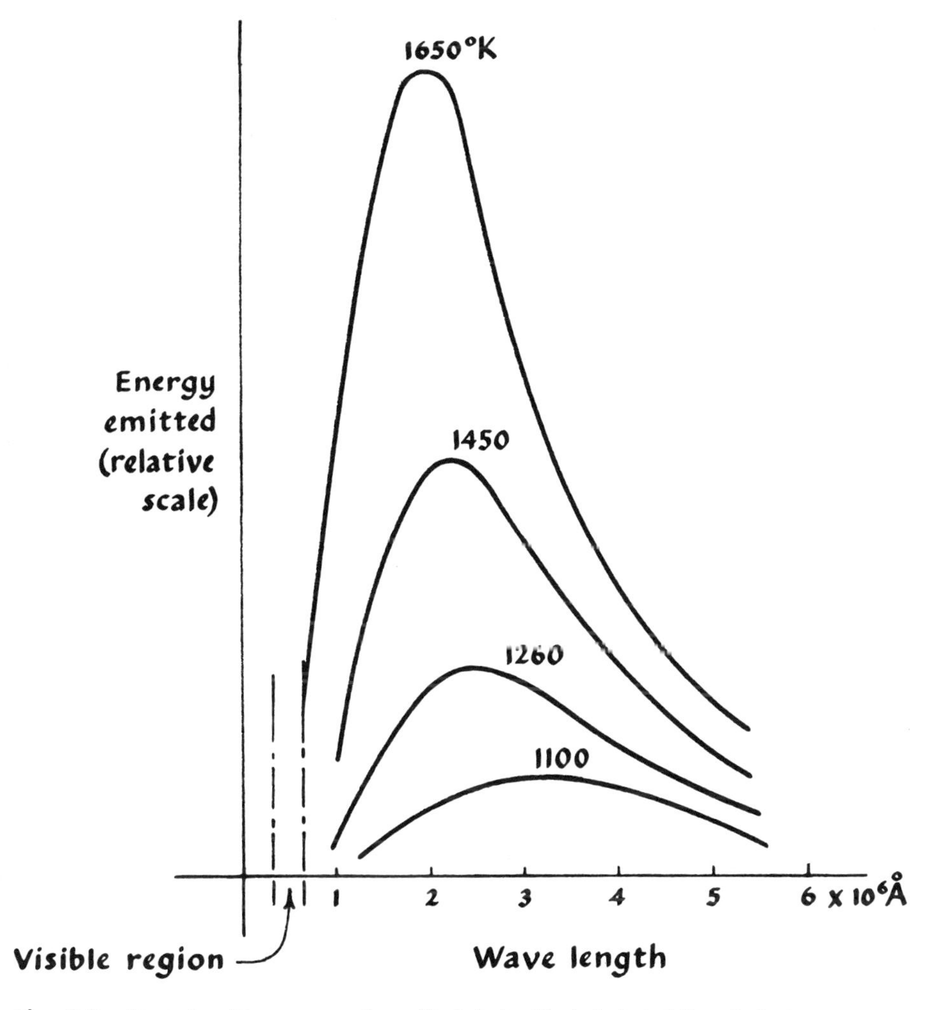

Fig. 8-3 *Spectral emittance curves for a black body. Vertical dashed lines indicate limits of visible spectrum. Figure 48.9, p. 722, from H. Margenau, W. W. Watson, and C. G. Montgomery,* Physics: Principles and Applications. *McGraw-Hill Book Company, New York, 1953.*]

of the maximum changes in a regular way with temperature. This latter discovery, that the wavelength of maximum absorption (λ_{max}) is displaced toward shorter wavelength with increase in temperature, is known as Wien's displacement law. It is stated:

$$\lambda_{max}T = \text{constant}$$

The constant is 2.898×10^7Å°K. This relation also makes it possible to determine the temperature of an emitting body such as a star or a furnace. Notice, though, that in making these estimates one assumes the emitter to behave as a perfect black body.

In Figure 8-3 we have data that summarize the results of countless *experiments*. Ideally a physicist would like to see these same curves capable of *calculation* from first principles. We have already suggested that this is the method by which theories in the field of constructs are validated. By calculation from first principles is meant this. Having learned from Hertz that electromagnetic radiation is emitted by oscillating charges (as predicted by Maxwell, who supplied a theoretical basis for this behavior) one takes this as a mechanism. Now, in the hot body there are atoms, with their (+) and (−) charges, jittering about, oscillating. At a given temperature, a certain amount of kinetic and potential energy is associated with these oscillating atoms. Moreover, the number of atoms in a given black-body is astronomical—one gram of iron contains 1.1×10^{22} atoms—so that it is safe to think of their behavior in probability terms. Some atoms will have little energy, having recently collided with others and given up energy; others will have a large amount, having been recently struck by many atoms; the majority will have intermediate amounts. The theoretical distribution of energies had been worked out by Ludwig Boltzmann and J. Willard Gibbs from probability considerations. If, then, in the hot body there is this vast company of atoms, ranging in energies from a few with very high through a majority with intermediate to many with very low energies, and if the wavelength of the radiation that is emitted depends on the energy of the oscillating atom, then at that temperature, with that probability distribution of energies of the oscillating atoms, it should be possible to *calculate* a distribution of emitted wavelengths that should fit the *experimental* observations quite exactly.

8.3 Planck's contributions

Reasoning along these lines Lord Rayleigh derived an equation, to which James Jeans added a correction, which fitted an emission curve at very long wavelengths. Wien did somewhat better (Figure 8-4). Both of these attempts started from classical considerations of the kind sketched, involving Boltzmann's probability distribution of energies. Both failed (the calculated curve rose indefinitely). Both implied *unlimited increase* of energy at the shorter wavelengths, while experiment, the data actually found, showed a maximum in the curve and a sharp *fall* in intensity at the shorter wavelengths (Figures 8-3, 8-4). (The situation described by the Rayleigh-Jeans equation was named the "ultraviolet catastrophe.")

Max Planck, too, started from classical considerations and was led by logic into the same difficulties as Rayleigh and Wien. By a mathe-

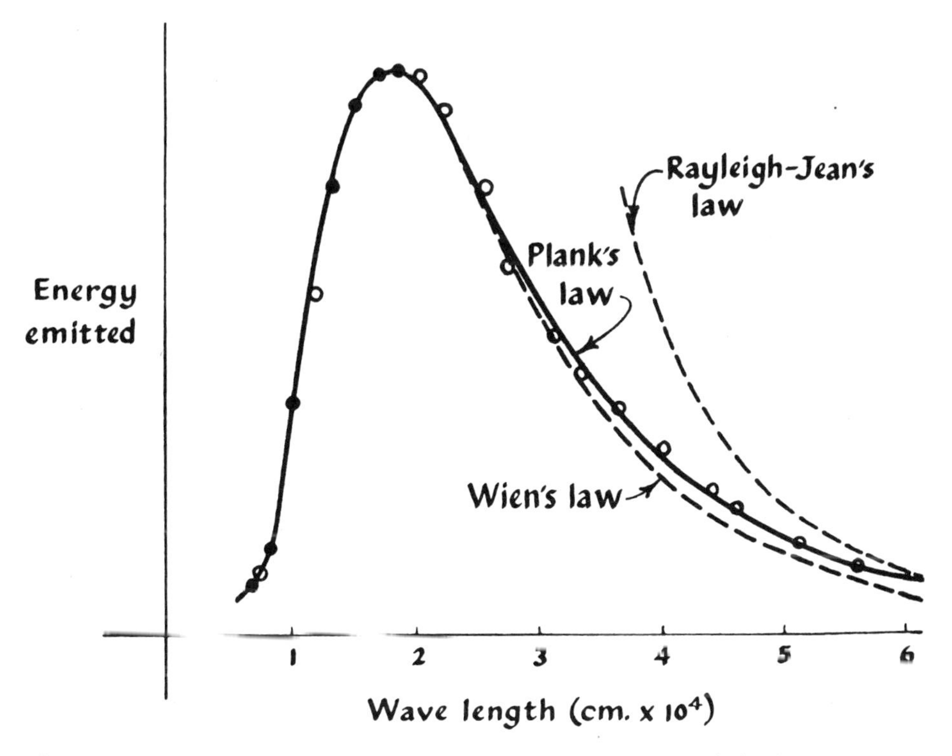

Fig. 8-4 *Comparison of three radiation laws with experiment, for the radiation from a black body at 1600°K. The experimental values are the open circles. From F. K. Richtmyer,* loc. cit., *Figure 65, p. 246.*

matical device he produced an equation which exactly fitted the data. When the equation was plotted for a given temperature the calculated curve that was obtained rose along the line of experimental points to a maximum at the right place and fell off in the manner observed in the experiments (Figure 8-4). The equation agreed completely with experimental results of the most carefully done kind. Moreover, Planck was able to derive from it several fundamental physical and chemical constants including the Stefan-Boltzmann law and Wien's law, and he showed the inner meaning of the constants in both these laws in terms of his equation. But it was not classically based. In order to fit experimental data, in other words, an equation had to be devised which, in turn, required that the conceptual bases for explaining what was happening had to be changed. The new concepts can be introduced in the following way.

A pendulum may be started swinging. There is no reason that can be given on the basis of Newton's laws why it should not have any one of an infinite number of degrees of kinetic energy between zero (when it is not swinging) and some maximum value that depends on the length of the suspension. Recall that the bob may be given potential energy by drawing it aside, lifting it to some height h above its lowest position (Section 3.11). The bob now has potential energy mgh. Nothing in classical experience limits the number of subdivisions that may be made in h. Similarly, because of friction at the fulcrum and with the air, the swinging pendulum gradually loses its energy. Since m and g are constant, h decreases. The pendulum ultimately stops. Nothing in classical theory suggests that the energy is lost in steps. The whole accent is on continuity. We saw the conceptual connections between circular motion, pendulum oscillation, and SHM: all continuous. It would therefore be a natural assumption that an atomic oscillator, rapidly changing rate and direction of acceleration of charge, and emitting in the process electromagnetic radiation, could behave in a continuous manner. That is to say, it could emit an infinite number of different amounts of energy, depending on the infinite gradations of amplitude of possible oscillation. (cf. note to Chapter 5.)

But Planck found himself required by his equation to discard this assumption. Energy of atomic oscillators does not change down or up a ramp but down or up a staircase; each step a single quantum jump. The atom can absorb or emit energy only in "packets," or quanta. These quanta bear an exact proportionality to the frequency of the radiation (and thus to the wavelength) according to the relation

$$\mathbf{E} = h\nu \qquad (8\text{-}1)$$

Here E is the energy carried by the radiation, ν is, as usual, the frequency (Section 5.4), and h is a constant of proportionality which is now known as Planck's constant. Planck's constant has the dimensions energy $\times$ time, and for this reason is called the elementary quantum of action. The name *photon* was given to the quantum of radiation by G. N. Lewis, and is in wide use. Planck's constant is one of the fundamental constants of Nature. Its value is 6.625×10^{-34}. The quantity $h\nu$ describes a given quantum of radiation.

8.4 Einstein's contributions

Planck's first major paper on quanta was published late in the year 1900. This is considered by many scientists to be a convenient marker for the beginning of the modern era in physical science. Planck's equation was subjected to the most careful testing by experimental physicists and found to fit the data for black-body radiation with great exactitude; attempts were made over and over—some by Planck himself—to connect these indubitably correct discoveries with classical theories. It just wouldn't work. Then, in 1905 Einstein proposed the construct of light quanta. Planck's quanta were associated with the interactions of radiation with matter (the black-body absorption and emission); Einstein proposed that radiation is *transmitted* between emitter and absorber as quanta. This was an even more radical idea, because it went against all the firmly based knowledge of the wave nature of light and, in general, of electromagnetic radiation. Einstein said in effect that light is atomic in character; particulate in nature; and that the units in which light is transmitted are Planck's quanta, $h\nu$. It took many years before the two theories, and particularly that of Einstein, were fully accepted. Some of the experimental findings that forced acceptance are described below.

Einstein's contribution was made in connection with what is known as the photoelectric effect. It was known (Hertz had discovered it, and many investigations had been made of it, but there was no explanation available) that if a metal plate is charged negative and ultraviolet light is allowed to fall on it, the charge is rapidly lost, while if it is charged positive the light has no effect. This was called the photoelectric effect. Experiments showed that none of the metal of the plate is being torn off and driven across space by the light. "Corpuscles" are, however, being expelled from the metal. These turned out to be electrons, and P. Lenard showed that the energy with which they are ejected from the metal surface by the light has nothing to do with the intensity of the light but depends on the wavelength of the light. This was a most unexpected observation. One would naturally think that if light is expelling electrons from a metal surface the velocity with which the electrons come out would depend on the intensity of the beam. But this is clearly not so.

Another unexpected observation was that not just any light shining on any metal surface would produce a photoelectric effect, that is, expel electrons; but for a given metal only light with a frequency higher than a lower limit or 'threshold value' ν_0 could expel electrons. Moreover, each metal has its own threshold value. This seemed strange because, according to the concept of continuous waves, light falling on a surface should be able to cumulate energy in the surface until enough was there to expel electrons. But this was not borne out by experiment. Imagine a light source giving out 100 watts of power. The amount of energy that falls on a spherical surface at a radius of 10 meters from this source is $(100/4\pi r^2) = 0.08$ joules per sec m^2. A metal plate contains atoms the radii of which are of the order of 10^{-10} m. Each atom fully in the surface of the plate thus presents to the light a disk with the area $\pi r^2 = 3.14 \times 10^{-20}$ m^2, and in a square meter of surface there will be some $1/3.14 \times 10^{-20} = 3.2 \times 10^{19}$ of these atoms. Dividing the available energy between them (and this is where the classical continuity comes in) we find that each will receive on the average about 2.5×10^{-21} joules/sec. Suppose, to take an average value, that the electrons in the metal need to receive 4.8×10^{-19} joules to be torn loose from the surface, and 0.2×10^{-19} more to give them kinetic energy. Calculation shows that according to classical theory it would take $5 \times 10^{-19}/2.5 \times 10^{-21} =$ about 200 seconds after the light was turned on for enough energy to accumulate at an atom to expel these electrons. Experiment shows that the instant the light is turned on charge begins to leak off the plate. (Of course "the instant" is the time required for light to leave the bulb and reach the plate, namely $10/3 \times 10^{-8}$ seconds.)

Einstein encoded all of the known experimental data in his important photoelectric equation:

$$(1/2)mv^2 = h\nu - p \tag{8-2}$$

This equation says that the kinetic energy of an expelled electron, namely $(1/2)mv^2$, is equal to Planck's constant h times the frequency of the light, ν, less the amount of energy, p, needed to separate it from the surface. That is to say, if an electron is held to a surface with a certain energy p, then the quantum of light which expels it ($h\nu$) must have at least this much energy. If it carries more energy than this the excess will show up as kinetic energy of the electron. In examining this behavior through the experiments of Millikan we shall use the following conventional terms. The quantum of light will be named a photon. The electron which is expelled is indistinguishable from any other electron in the atom, but it will be called a photoelectron to emphasize its antecedents. Its kinetic energy, $(1/2)mv^2$, is measured, sometimes, in electron volts (Section 4.9) through the relation:

$$(1/2)mv^2 = Ve$$

where m is the mass of the electron, V a potential difference, and e the electronic charge. (This equation is another way of putting Equation 3-11.)

Imagine the physical situation. Here is a metal surface many atoms deep. Photons with average energy $h\nu$ strike the surface. Some strike atoms just at the surface; others penetrate below the surface, passing between atoms. Electrons in the atoms just at the surface are least tightly restrained because they are only partially surrounded by other atoms; those below the surface are surrounded on all sides. Therefore it is not surprising that photoelectrons will emerge from the surface with a range of kinetic energies. There will be those with most energy, from just at the surface, where the value of p is that of the metal atoms themselves. Then there will be the photoelectrons that arise from deeper below the surface and have to be extruded along a path lined with atomic obstacles. These will appear with less energy, some having been given up in interactions along the way to the surface. Thus p may be thought of as having two parts: an energy $h\nu_0 = p_0$, constant and characteristic of the metal, and different for each metal; and an energy $p - h\nu_0$ which is variable. The former depends on the chemical structure of the atoms and is the work required to set the electron free from that particular kind of atom right at the surface; the latter is the additional work required to free deeper-lying electrons. According to this analysis, the electrons just at the surface would be expelled according to the equation:

$$(1/2)mv^2 = h\nu - h\nu_0 \tag{8-3}$$

The simplest experimental arrangement to investigate these behaviors is shown in Figure 8-5. A clean metal surface M is connected electrically through a galvanometer to a 'collector' A. The surface and collector are inside a glass bulb with a flat window of quartz, Q, to admit light. The tube is evacuated so that the electrons expelled from M can get to A with a minimum of collisions. This is not a perfect arrangement, because the light may knock electrons out of A, and thus disturb any quantitative measurements of its effect on M, but it shows

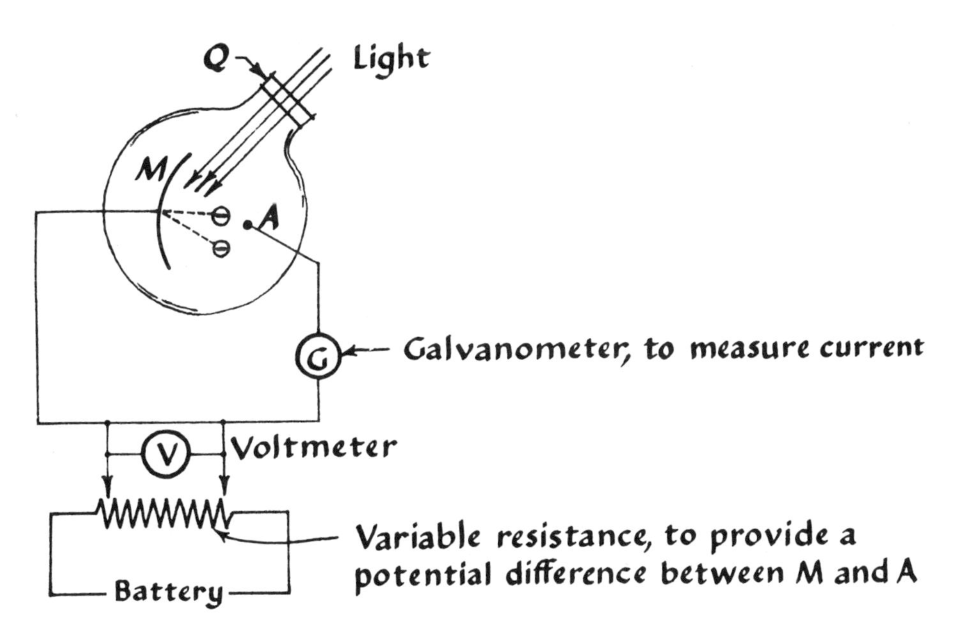

Fig. 8-5 *Schematic diagram of a photocel. Light to be investigated shines through the quartz window* Q *onto the metal surface* M *from which photoelectrons are expelled to the collector* A. *The current so produced passes through the circuit back to* M *via a galvanometer or electrometer* G. *A battery and variable resistance enable a voltage, read off on the voltmeter* V, *to be impressed between* M *and* A *with the desired direction of polarity.*

the principle of the earliest apparatus used. With such an apparatus, in the absence of light, no current flows because the circuit is electrically open: there is no electrical connection between M and A. If photoelectrically active light falls on M, a current flows and reveals its presence and its magnitude by the deflection of the galvanometer G. This deflection, when it occurs, is interpreted to mean that there are electrons flowing between M and A, closing the circuit. The galvanometer tells how much current, (I, in amps.) is flowing. This is called a photocurrent because it is a current produced by light. The direction of the current is consistent with a flow of electrons from M to A. Thus in the absence of an externally applied potential, some at least of the photoelectrons leave M with sufficient energy and in the proper direction to arrive at A.

Now, by means of the battery and variable resistance, a potential difference is impressed between M and A. With M (−), and A (+), electrons are repelled from the metal surface and attracted to the collector, and the current that flows is larger than before. The potential difference, measured with the voltmeter V, may be increased, in which case the current becomes correspondingly larger *up to a point*, at which further increase of potential makes no additional change in the photocurrent. This is interpreted to mean that all the electrons released by the light are being collected: all the photocurrent that can flow is flowing in the circuit, and no amount of added energy can produce more. If, now, the intensity of the light is increased, the photocurrent increases. This means that the *number* of electrons released depends on the intensity of the light. *That* there is a current, as we saw, depends on the frequency being at or above the threshold ν_0. This is interpreted to mean that frequency is related to energy of the photon, as the equation states. Intensity is a measure of the number of photons in a unit cross-sectional area of the beam, and thus the number arriving at the metal surface per second.

Now, if the polarity of the impressed potential difference is reversed, so that A is made (−) with respect to M, the photocurrent is found to decrease. This is readily explained: if M is (+), it is more difficult for the (−) photoelectrons to leave the metal surface; at the same time A is (−), and only the most energetic ones can overcome the repulsion of A and complete the circuit. Increasing the potential difference between M and A under these conditions decreases the photocurrent until at some potential difference V_t no current at all flows. Even the most energetic electrons are suppressed, or completely retarded, and fail to reach the collector A. If the intensity of the light is decreased, less current flows, over the entire range of voltage, but the threshold V_t remains the same. Thus it is independent of intensity—and this has been tested over a wide range of intensities. In terms of the photoelectric equation, at V_t even the most energetic electrons are just opposed. Thus one can write, in terms of energies:

$$V_t e = (1/2)mv^2 = h\nu - h\nu_0$$

Knowing ν, and V_t, and e and h, ν_0 can be calculated. Alternatively, as discussed below, by plotting V_t from a series of experiments at different frequencies ν, one obtains a line the slope of which is h/e, and knowing e, h can be determined. The relation between photocurrent (I) and potential difference between the electrodes M and A of Figure 8-5 leads to curves of the general shape shown in Figure 8-6.

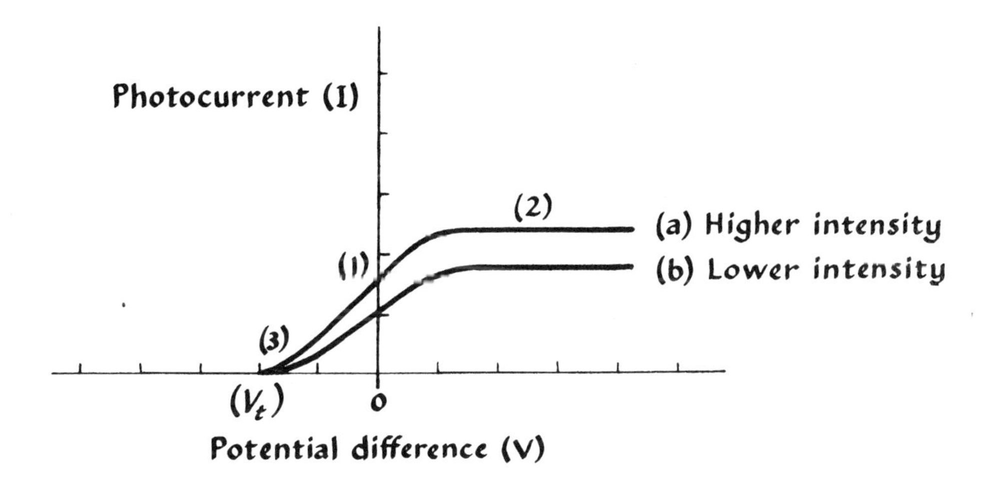

Fig. 8-6 *The type of curve obtained in a photoelectric experiment with two different intensities of light (curves* a *and* b*) but both at the same frequency. This shows the presence of a photocurrent (I) with no impressed voltage between* M *and* A *(Figure 8-5), at (1); at (2) is shown the saturation current, and at (3) the threshold voltage* V_t *that just suppresses photocurrent.*

These curves, Figure 8-6, summarize what has just been described. At (1) a photocurrent is found to flow in the absence of net potential between M and A. At (2) the photocurrent has reached its maximum ('saturation') level, which is higher for the more intense illumination, curve (a), than for the less intense curve (b). At (3) both curves yield the same threshold voltage V_t, with the polarity such that the metal surface is (+) with respect to the collector.

8.5 Millikan's contributions

We turn now to Millikan's work, certain experimental details of which are presented in order to make it possible to appreciate the art of experimentation. A fairly complicated apparatus was needed, made of metals and glass. Its creation, apart from the physics of the design, required a high degree of skilled craftsmanship. This is the analogue of the craftsmanship of a Rodin or a Brancusi, whose artistry includes a connoisseurship of marbles and a sensitiveness to the use of the sculptor's tools, to the touch of a maul on a chisel. Here the artist, Mr. Julius Pearson, to whom Millikan pays tribute, had to know the properties of metals and glasses, and to be able to build a complicated instrument inside of a glass bulb: an accomplishment that required a high degree of inventiveness, or creativity, as well as the skill of a superb craftsman who knows his materials.

The experimental problems that Millikan conquered, and which had raised questions about the accuracy of previous work, were roughly these. The behavior of a metal surface is *extremely* sensitive to the presence of surface films. Therefore the greatest care has to be exercised to produce and maintain clean surfaces. At the same time, some of the best metals to use—because they exhibit a photoelectric effect even with visible light—are also (and for a connected reason) highly reactive chemically. These are the metals lithium, sodium, potassium, and cesium. They are instantly coated with an oxide film in air, and also react in hydrogen gas and in nitrogen gas. Thus they have to be handled in the highest vacuum feasible, other factors considered. (Millikan was able to obtain a vacuum of 10^{-7} mm, or less, of mercury, as needed. That is, the pressure inside his apparatus would support a column of mercury this high.) The wavelength of the light used must be known. This problem is readily solved, as we have shown in principle (Section 5.3). Moreover, with a good spectroscope and a suitable lamp, it is possible to isolate a very narrow band of electromagnetic radiation. Millikan had available the narrow bands centering on $\lambda = 5461$ (green); $\lambda = 4339.4$ (blue); $\lambda = 4046.8$ (violet); $\lambda = 3650.2$; $\lambda = 3125.5$; $\lambda = 2534.7$; and $\lambda = 2399$ (in Angstrom units), from a mercury arc lamp with a quartz bulb. These could be brought at will to the slit of a spectroscope and directed into his photoelectric apparatus.

He had made an apparatus which is shown very schematically in Figure 8-7. It consisted of a glass bulb that could be evacuated, having a quartz window at one side. On the periphery of the wheel W were fastened cast cylinders of the metals to be studied—*i.e.*, sodium, potassium, and lithium. This wheel could be rotated on an axle, the bearings of which were fastened to the bulb, by means of an electromagnet (not shown), until one of the cylinders was opposite the rotary knife, K. The

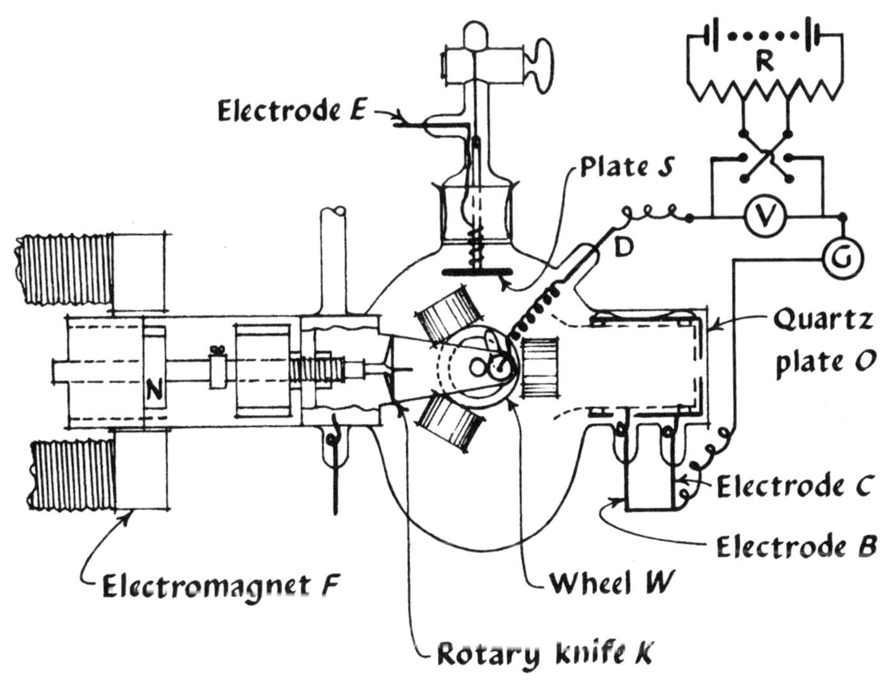

Fig. 8-7 *The "machine shop* in vacuo" *of R. A. Millikan. The drawing is simplified from his drawing, Figure 2,* "A direct photoelectric determination of Plank's 'h'." *Physical Review Series II,* 7, 355 (1916). *See the text for its description.*

motion of the knife was controlled by the electromagnet F, working upon a magnet N inside the tube. This enabled the knife to be pushed up against the metal cylinder (a little pin fitting in a hole at the center of the cylinder helped to center and steady it) and to be rotated so as to shave off the surface of the cylinder of metal. These metals are quite soft. The shavings, falling to the bottom of the bulb in which this mechanism resides, helped to scavenge out any residual reactive gases. The fresh surface so obtained could be renewed at will by shaving it. The knife was then drawn back and the wheel W rotated to bring the fresh surface accurately opposite to O, and normal to a beam of radiation entering at O, or opposite and parallel to a plate S. It was important to have a minimum of stopcocks because, as Edward Morley once remarked, "A stopcock is merely a located leak." Therefore the work that had to be done inside the tube was controlled by electromagnets from outside.

The light, entering at O through a quartz plate sealed onto the glass bulb, passed through a hole in a copper-sheet cylinder attached to electrode C, through a hole in a copper-gauze cylinder attached to electrode B, and impinged upon the cut surface M of the metal block.

The copper cylinders were carefully covered (by chemical treatment) with a black oxide film, to help prevent the reflection of light back to the cut surface, and the gauze cylinder was extended inward as shown so as to catch all electrons expelled from the metal surface. The two formed an electron trap, *A*, so that electrons, flowing across from the metal surface and caught by the gauze and sheet, could flow along the circuit through the current-measuring electrometer *G* and back to the metal via the electrode *D*. The metal surface is represented by *M* in Figures 8-5 and 8-7, the electron trap by *A*.

One problem that was passed over in describing the simple arrangement of Figure 8-5 is that when two different metal surfaces are close to each other, a potential is present between them. This potential, originating in their different electrical properties, is called a contact potential. It was always present in Millikan's apparatus between the metal surface *M* and the copper oxide-coated electron trap *A*. In order to measure it, the freshly cut surface could be brought under the disk *S*, which was of the same material and treated the same way as the copper-sheet cylinder. A potentiometer, or sensitive voltmeter (not shown), connected to *D* and *E* served to measure this voltage. When a fresh lithium surface was tested in this way (no light of any kind present, of course) a potential of 1.51 volts was found to exist, with the lithium positive with respect to the copper oxide plate *S*. This had to be corrected for, since it would act as a retarding potential. Thus to achieve the state of no net potential difference between *M* and *A* (the point (1) on the curve of Figure 8-6) an equal potential with opposed polarity was impressed on *M* and *A* by means of the battery and variable resistance shown schematically at *R*. The contact potential had to be measured in the same apparatus as the photocurrent so as to ensure as far as possible that the same metal surface was being measured. This is because the contact potential also is extremely sensitive to surface impurities, such as films of oxides or other compounds.

The whole apparatus was carefully shielded to prevent stray light from getting to the metal surface. Also, it was essential to avoid having light fall on *A*. Even so, when the beam from *O* fell upon *M* some of it was unavoidably reflected to *A*, where it might have a photoelectric effect. This action was eliminated wherever possible by using light of frequency below the threshold for the copper oxide surface (about $\nu_0 = 1.1 \times 10^{15}$ sec^{-1}), but still photo-active to the metal. If it could not be avoided, then a correction had to be made in the data.

The importance of being able to obtain a freshly cut surface at *M* is shown by the fact that, as Millikan reported, "the photocurrents from a freshly shaved sodium surface are, in the best attainable vacuum, hundreds of times larger than those from surfaces a few hours old." Many refinements not shown in the simplified diagram, Figure 8-7, were present. Also for measurements of radiant energy a light-measuring probe was introduced into the path of the light emerging from the spectroscope. Everything then known to the art was done to ensure the accuracy and reproducibility of the experiments.

In Figure 8-8 are shown a set of curves for a freshly cut lithium surface, not corrected for contact potential, but as reported by Millikan, and traced from the original paper. These curves are just the region (3)

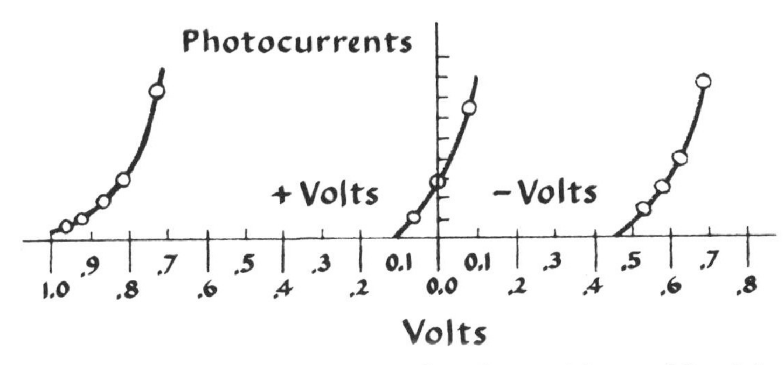

Fig. 8-8 *The curves for photocurrent versus voltage impressed between* M *and* A, *Figure 8-7, to allow extrapolation to* V_t *at zero photocurrent. The dots are experimental observations made with a lithium surface. The data are not corrected for contact potential. Traced from the curves of Figure 7, Millikan,* loc. cit., *p. 376.*

in Figure 8-6, and show the intercepts V_t for a set of frequencies. The uncertainty in location of the intercepts is less than about 0.02 V, and the curves were plotted so as to have about the same maximum value at the saturation region (2) in Figure 8-6. Plotted on the same scale as in Figure 8-6, the entire curve, says Millikan, "would be from three to twenty meters high," so that it is evident that in this figure only the region very close to V_t is shown.

If, now, a plot is made of V_t versus frequency the values of which, corrected for contact potential, are listed in Table 8-1, the straight line

Table 8-1

Values of λ and ν versus V_t for Lithium *

λ Å	ν per sec	V_t volts
4339	6.91×10^{14}	+0.49
4047	7.41×10^{14}	+0.67
3650	8.21×10^{14}	+1.03
3126	9.59×10^{14}	+1.61
2535	11.8×10^{14}	+2.51

* Calculated from curves and data in Millikan's paper, *loc. cit.* These data are plotted in Figure 8-9.

in Figure 8-9 is obtained. According to the photoelectric equation,

$$V_t e = h\nu - h\nu_0$$

or $$V_t = [h/e] \times [\nu - \text{constant}]$$

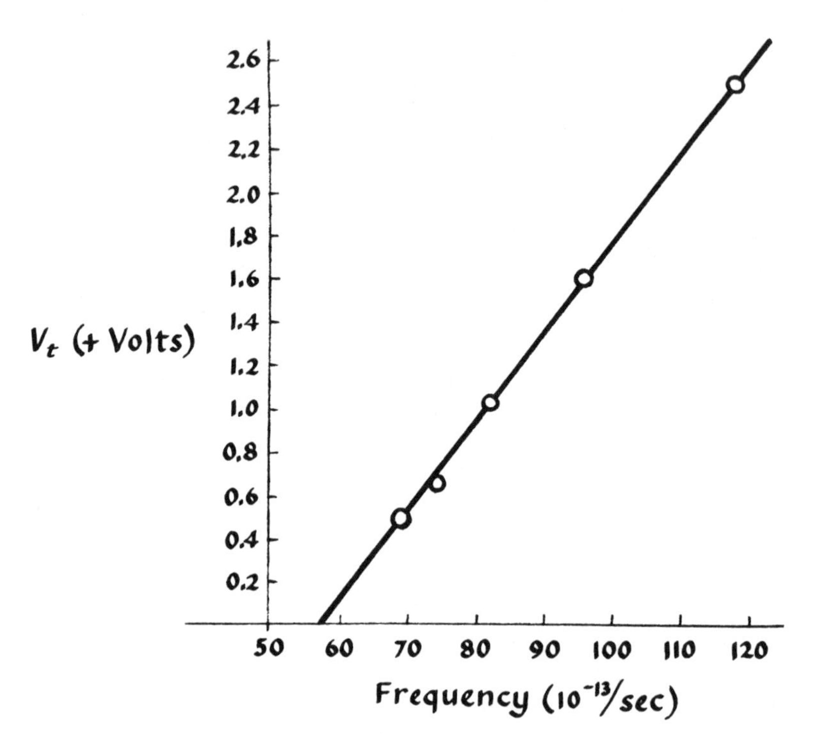

Fig. 8-9 *The data for* V_t *read off from Millikan's curves, Figure* 7, loc. cit., *p. 376, and corrected for contact potential are plotted against frequency, to determine* ν_0 *for lithium. This corresponds to one of the curves in Figure 8,* loc. cit., *p. 377.*

This is the equation of this straight line. The slope is h/e, and the intercept with the X axis is ν_0. Millikan found the best value of h to be 6.57×10^{-34} J. sec. (in mks units), and ν_0 for this lithium surface to be about 57×10^{13} per sec., or λ_0 = about 5,260 Å. The accepted value for h, today, is 6.625×10^{-34} J. sec.

This work, together with other investigations, confirmed the predictions and implications of Einstein's equation, that:

1. There is a threshold frequency ν_0 for a metal which must be exceeded for electrons to be expelled. This means that light energy is not stored up until some threshold is reached, when an electron is expelled. Either the photon carries enough energy, $h\nu_0$ or greater, or no photoelectric effect occurs, no matter what the duration of illumination.

2. For each frequency greater than ν_0 there is a maximum velocity of emission of photoelectrons from a surface. That is, if v_{max} is this maximum velocity of emission, then

$$(1/2)mv_{max} = Ve_t = h\nu - h\nu_0$$

3. For a given surface the constancy of ν_0, h, and e requires a linear relation between V_t and ν.

4. The slope of this line is equal to h/e.

It was further observed that:

5. There is no delay in emission, that is, no "storage" of energy (see 1). If the frequency is ν_0 or larger, electrons appear immediately when the light impinges on the surface.

6. Each surface (for example of each pure metal) has its characteristic threshold frequency ν_0.

7. The number of photoelectrons expelled from a surface is a function of the intensity of the irradiation.

8. The photoelectric effect, reflected in ν_0 and v_{max}, for example, is unaffected by temperature. This implies that the photoelectrons do not partake appreciably of the kinetic motion of the atoms or molecules out of which they come.

Some implications 8.6

By quantum theory we mean the theoretical and experimental studies such as those described in this chapter that make use of Planck's constant. Quantum theory made it possible, in the hands of Niels Bohr and others, to understand atomic structure and the spectroscopic behavior of gases (as we shall see). However, it raised many very difficult problems for those who wished to obtain a picture of the nature of light that could be reconciled with implications of interference phenomena, that is, implications that electromagnetic radiation has wave character (Section 7.7). The quantum energy $h\nu$ is clearly related to wave notions, because frequency, ν, is after all a wave construct. At the same time, however, the entity $h\nu$ behaves in a particulate way.

It was difficult to reconcile the concept of propagation in the form of photons with the classical picture of energy spread out over a wavefront. The photon is clearly not a revival of Newton's light corpuscle because it has no rest mass (Section 6.14); it travels with the speed of light. When it is brought to rest, as occurs, one must suppose, when it is absorbed in matter, it changes to something else: an energetically equivalent amount of kinetic, potential, or chemical energy.

In an ordinary beam of light there are so many photons that perhaps one might think of a uniform distribution of energy over the front of the beam. But nevertheless photons are present, as can be shown by attenuating the beam until the intensity is so low that separate photons can be recognized when they arrive at an instrument. Moreover, photons, travelling with the speed of light and endowed with energy, behave in ways that classically imply a momentum: the momentum expected of a particle.

A gas exerts pressure on the walls of a container because its atoms, in energetic motion, collide with the atoms of the walls and rebound, imparting momentum to the wall. The energy of the particles of a gas is proportional to the absolute temperature. But the energy of radiation, as the Stefan-Boltzmann law states, is proportional to the fourth power of the absolute temperature. Thus one is prepared to find that the construct 'momentum,' which is related to energy (see below), may have special implications when applied to a photon. That is, the photon is a special kind of "particle." That a photon may indeed display momentum was shown explicitly by Arthur H. Compton.

8.7 Compton's contributions

When an electrode of some heavy metal is bombarded with a stream of sufficiently energetic electrons it gives off X-rays in a reverse photoelectric effect. By sufficiently energetic is meant that the electrons bombarding the metal must have kinetic energy sufficient to provide $h\nu$ of the emitted X-ray photons. From Equation 8-2 rewritten:

$$(1/2)mv^2 = Ve = h\nu$$

The wavelengths of X-ray photons are in the range of 0.1 to 1.0 Å. The principal wavelength given off by molybdenum is 0.707 Å; the frequency is $\nu = c/\lambda = 3 \times 10^8/7.07 \times 10^{-11} = \sim 4.25 \times 10^{18}$ sec^{-1}. Therefore the kinetic energy needed to produce these photons is $6.63 \times 10^{-34} \times 4.25 \times 10^{18} = 2.82 \times 10^{-15}$ J. per electron. This energy is obtained by accelerating electrons, as in a cathode-ray tube, against the molybdenum metal target by means of a potential difference (Figure 8-10). In this example the potential difference is $V = 2.82 \times 10^{-15}/1.6 \times 10^{-19} = \sim 18{,}000$ volts. Thus X-rays are high energy photons, with energies of the order of tens of thousands of electron volts.

Suppose, said Compton, that an X-ray photon of frequency ν_0 collides with an electron. If the energy of the photon is $h\nu_0$, the relativistic mass equivalent will be $h\nu_0/c^2$ (Section 6.14). Since the velocity of the photon is c, then by the definition of momentum as mass $\times$ velocity, the *momentum* p of the incident photon will be $h\nu_0 c/c^2 = h\nu_0/c = h/\lambda_0$. In the collision suppose that the photon "bounces off" from the electron at an angle θ from its initial path. It has changed direction, and also changed momentum to a new value $p_\theta = h\nu_\theta/c = h\lambda_\theta$. (These subscripts are just labels to tell us that we are referring to properties of the recoil-

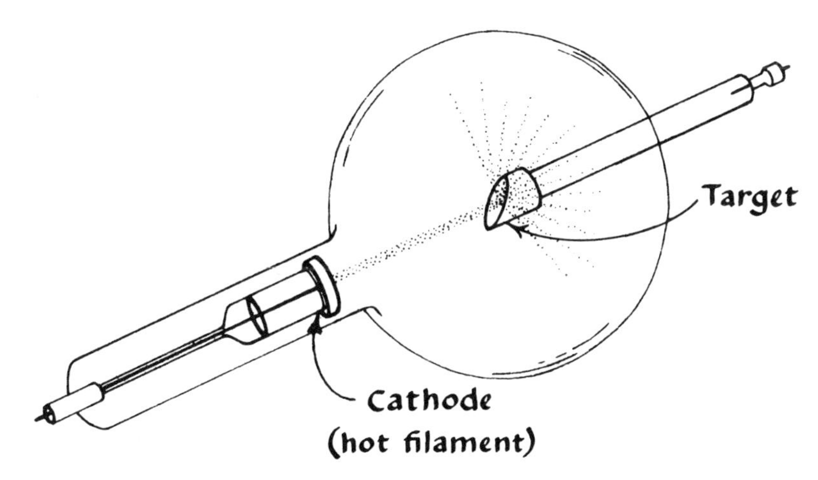

Fig. 8-10 *The essence of an X-ray tube. Electrons, from a hot filament are accelerated to the metal target and generate X-rays. The filament and target are in an evacuated tube.*

ing photon.) The electron recoils from the collision with its own momentum $p_e = mv$, where m is the mass of the electron and v its acquired change in velocity. All three of these quantities, initial momentum of photon and recoil momenta of photon and of electron, are vectors.

In an actual experiment with X-rays of wavelength 0.708 Å, the wavelength of the ray scattered at $\theta = 90°$ was measured as 0.730 Å; the difference should calculate to the momentum p_e of the recoil electron. But these electrons were not observed because there were too few of them in the presence of many photoelectrons. (The major portion of the X-ray beam acted photoelectrically.) Therefore the value of the difference between incident and scattered ray had to be approached in a different way. The problem of measuring mv for a recoiling electron under the conditions of this experiment being too difficult, recourse was had to mathematics. The idea was to use available relations between constructs such as momentum, relativistic mass, frequency, and so on to find an equation entirely in terms of what could be measured, namely, the wavelengths of incident and scattered X-rays, and of what was known about the electron, namely, m, and other necessary data such as the values of h and c. In effect, to 'step' conceptually from one group of constructs to another more experimentally accessible group with the aid of algebra and invariant-relations. When this was done (see *Note* at the end of this chapter), there was found to be excellent agreement between theory and experiment.

This experiment marked a turning point in the development of quantum theory because it convinced even the most skeptical scientists of the particulate nature of high-energy quanta.

8.8 Summary

At the turn of the twentieth century an area of misfit had appeared between experimental findings for the absorption and emission of radiation and the classical laws which were invoked to explain experiment and which failed to. Planck showed that experimental findings could be exactly fitted, and predictions of new experiments could be made, if instead of thinking of the interaction of matter with radiation taking place with an infinitely divisible range of energies, the interaction was actually limited to finite, stepwise changes; if energy interacted in packets, or quanta. The most careful experiments possible exactly supported this idea. A radical extension to Planck's theory was made when Einstein showed that the photoelectric effect (and many other phenomena) could be exactly explained if one assumed that light was transmitted through space as quanta. This blow to the wave-theory was not accepted until exhaustively careful experiments such as those of Millikan and Compton made disbelief impossible. There were other areas of misfit between theory and experiment (Section 2.8). That described here led to the quantum theory: the theory of physical science which invokes Planck's constant. Other such areas will appear as we pursue modern ideas about the structure of matter.

Note: Taking the simplest case first, assume that the collision is head-on, and the photon bounces directly back along its original path, the electron recoiling directly ahead (Figure 8-11). The momentum of the incident photon is h/λ_0; that

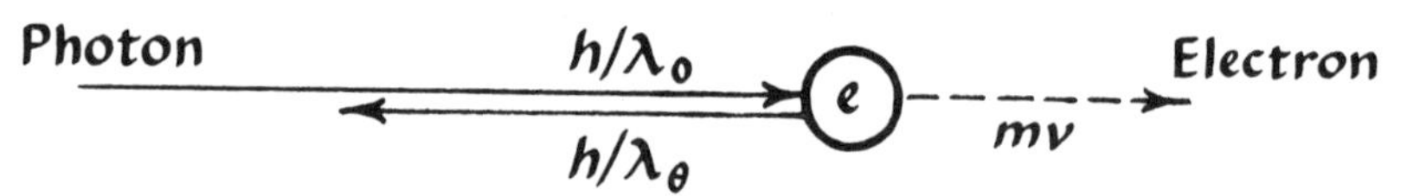

Fig. 8-11 *Collision of a photon with momentum* h/λ_0 *with an electron* e, *head-on. The photon scatters with momentum* h/λ_θ *and the electron recoils with momentum* mv.

of the rebounding photon is $-h/\lambda_\theta$, for it has changed its direction through 180°. The momentum of the electron is mv. By the principle that momentum is conserved:

$$\frac{h}{\lambda_0} = \frac{-h}{\lambda_\theta} + mv$$

The wavelength change is observed to be small. For this reason we may write:

$$mv = 2h/\lambda \text{ (approximately but quite closely)} \qquad \text{(a)}$$

By the principle that energy is conserved,

$$h\nu_0 \text{ (photon)} = h\nu_\theta \text{ (recoil photon)} + (1/2)mv^2 \text{ (recoil electron)} \qquad \text{(b)}$$

By the device of squaring the momentum equation (a), to yield

$$m^2v^2 = 4h^2/\lambda^2 \qquad \text{(c)}$$

v^2 can be eliminated in the following way. Rearrange equation (b)

$$(1/2)mv^2 = h\nu_0 - h\nu_\theta$$

multiply both sides by 2, and gather terms

$$mv^2 = 2h(\nu_0 - \nu_\theta)$$

Now multiply both sides by m, and equate to equation (c):

$$m^2v^2 = 2mh(\nu_0 - \nu_\theta) = 4h^2/\lambda^2$$
$$\nu_0 - \nu_\theta = 2h/m\lambda^2$$

To convert this entirely into terms of the wavelength, (Section 5.4), we divide through by c:

$$\frac{\nu_0}{c} - \frac{\nu_\theta}{c} = \frac{2h}{mc\lambda^2}$$

Now, $c = \lambda\nu$, and therefore $\nu/c = 1/\lambda$, whence we can write

$$\frac{1}{\lambda_0} - \frac{1}{\lambda_\theta} = \frac{2h}{mc\lambda^2}$$

Since the wavelength difference, as we saw, is small, we may again take the close approximation that $\lambda_0\lambda = \lambda^2$.

Then: $$\frac{1}{\lambda_0} - \frac{1}{\lambda_\theta} = \frac{\lambda_\theta - \lambda_0}{\lambda^2}$$

Multiplying both sides by λ^2 eliminates this quantity, and yields the simplified result:

$$\lambda_\theta - \lambda_0 = 2h/mc \tag{8-4}$$

The algebra that leads to equation (8-4) is given in detail to show the reasoning. (The chief use of mathematics is to keep from getting confused, as one could easily do if one used purely verbal statements. The mathematical expression can always be converted to a verbal statement if necessary.) Equation (8-4) says that when the experiment is carried out in this way, the difference in wavelength—that is, the change in wavelength of the photon in its collision with the electron—will be equal to twice Planck's constant divided by the mass of the electron and the speed of light.

In the most general case of a glancing blow, where the photon bounces off at an angle θ, and the recoil electron at an angle to the original path of the photon (Figure 8-12), conservation of momentum has to be calculated in the X and Y

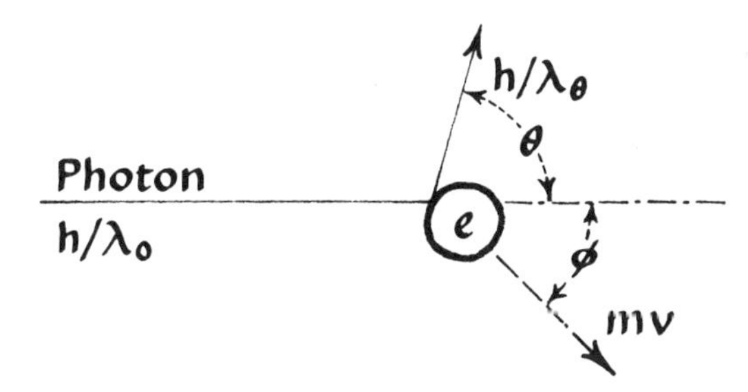

Fig. 8-12 *Diagram for the collision of a photon, with momentum* $h\omega_0/c = h/\lambda_0$ *with an electron. The scattered photon moves off at an angle* θ *to the original path, and a momentum* $h\omega_\theta/c = h/\lambda_\theta$, *while the electron recoils with momentum* mv.

directions (Figure 8-13). (Energy is not directional, not a vector quantity; momentum is.) The more general equation that replaces (8-4) is:

$$\lambda' - \lambda = h(1 - \cos\theta)/mc \tag{8-5}$$

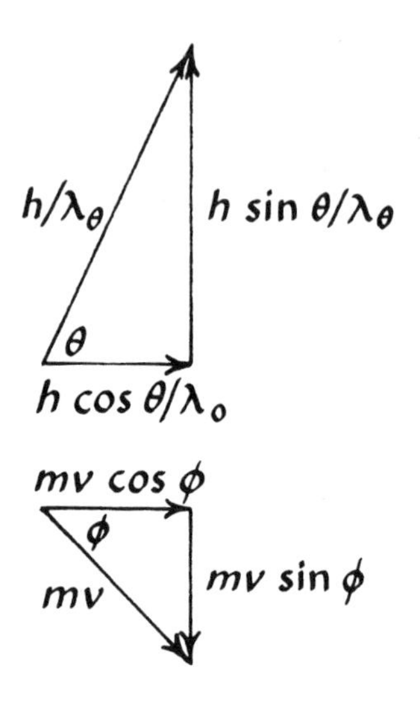

Fig. 8-13 *Vector diagram of conservation of momentum for the behavior shown in Figure 8-12. X and Y directions are taken account of.*

Cosine θ for 180° is -1, so that equation (8-5) becomes identical to (8-4) for the case of head-on collision described above.

This calculation is made as follows. It depends as before on the laws of the conservation of momentum and energy (for simplicity the relativistic mass change for the electron is ignored in this example). Let the incident photon have the energy $h\nu_0$, the scattered photon $h\nu_\theta$, and the recoil electron $(1/2)mv^2$. Then with conservation of energy,

$$h\nu_0 = h\nu_\theta + (1/2)mv^2$$

For conservation of momentum in the X direction (see Figure 8-13):

$$mv\cos\phi = \frac{h}{\lambda_0} - \frac{h\cos\theta}{\lambda_\theta}$$

In the Y direction:

$$mv\sin\phi = \frac{h\sin\theta}{\lambda_\theta}$$

There is a useful relation in trigonometry which says that the square of the sine of an angle plus the square of the cosine of the angle has the value 1. That is,

$$\sin^2\theta + \cos^2\theta = 1 \tag{8-6}$$

This is made use of to simplify these equations. Each is first squared:

$$m^2v^2\cos^2\phi = h^2/\lambda_0^2 - 2h^2\cos\theta/\lambda_0\lambda_\theta + h^2\cos^2\theta/\lambda_\theta^2$$
$$m^2v^2\sin^2\phi = h^2\sin^2\theta/\lambda_\theta^2$$

Adding these equations and making use of equation (8-6) yields an equation for m^2v^2. We saw above that according to the conservation of energy we could write $m^2v^2 = 2mh(\nu_0 - \nu_0)$, thus:

$$m^2v^2 = h^2/\lambda_0^2 - 2h^2 \cos\theta/\lambda_0\lambda_\theta + h^2/\lambda_\theta^2 = 2mh(\nu_0 - \nu_\theta)$$

Again we invoke the approximation that λ_0^2, and $\lambda_0\lambda_\theta$, and λ_θ^2 are almost equal, and we also replace ν by c/λ. This leads directly to the equation (8-5).

When equation 8-5, in which the unknown electron velocity has been eliminated, and the measurable quantities are wavelengths, was applied to the experimental data it was calculated that the scattered beam should be 0.024 Å longer than the incident beam. The value found was 0.730 Å − 0.708 Å = 0.022 Å. This was considered "very satisfactory agreement."

Exercises

8.1 We said that the total emissive power of an ideal black body is independent of the material of which it is constructed. Is this really so, as baldly stated? Suppose the box were made of wood? Or, of the metal gallium, melting point about 29.75°C? Qualify the statement suitably.

8.2 For the star Vega, a surface temperature of about 14,000°K is calculated. What is λ_{max} for this star?

8.3 Calculate the surface temperatures for the stars whose λ_{max} values are given: Rigel, about 2630 Å; Capella, about 4458 Å; Antares, 9660 Å.

8.4 Translate the following idiomatically but accurately, and sketch the apparatus described.

Die Versuche, welche ich über den Einfluss des electrischen Lichtes auf electrostatisch geladene Körper angestellt habe, waren meist auf folgende Weise angeordnet. Eine blank geputzte, kreisförmige Zn-Platte von etwa 8 cm Durchmesser hing an einem isolirenden Stativ and war durch einen Draht mit einem Goldblattelectroskop in Verbindung gesetzt. Vor der Zinkplatte stand parallel mit ihr ein grosser Schirm aus Zinkblech von etwa 70 cm Breite und 60 cm Höhe. In der Mitte desselben befand sich ein Marienglasfenster, durch welches die Strahlen einer jenseits aufgestellten Siemens'schen Bogenlampe auf die Platte fallen konnten. Das System aus Platte und Goldblättern isolirte gut: während eines Tages nahm die Ablenkung um etwa 1/4 ab, während der Dauer eines Versuches um keinen merkbaren Betrag. Auch wenn die Bogenlampe im Gang, das Marienglasfenster aber durch geeignete Substanzen bedeckt war, blieb die Isolation völlig erhalten.

Ladet man die Platte sammt Electroskop, welch letzteres von den Strahlen nicht getroffen werden kann, negativ electrisch, so beginnen, sobald die Lichstrahlen auf die Platte auftreffen, die Goldblättchen lebhaft zusammenzufallen; bei positiver Ladung tritt ein Zusammenfallen auf den ersten Blick gar nicht, bei geneauerer Untersuchung erst nach längerer Zeit in merklichem Betrag ein.—From W. Hallwachs, *Annalen der Physik* *33*, 301 (1888) pp. 302–303.

8.5 Translate the following and sketch the set-up.

Versuche über das "Zerstäuben der Körper durch ultraviolettes Licht" haben die Möglichkeit gezeigt, dass wägbare materielle Teile des bestrahlten Körpers jene Träger seien. Dass indessen solche unmittelbare Auffassung der damaligen Versuchsergebnisse doch schwerlich zutreffend sei, habn mir besondere Versuche gezeigt, von welchen einer hier berührt sei.

Eine von Wasserstoffgas umgebene blanke Oberfläche von Natriumamalgam wurde so lange bestrahlt, dass $2{,}9.10^{-6}$ Coulombs negativer Elektricität von derselben entwichen, und das elektrische Feld über dem Amalgam war so gestaltet, dass diese gesamte Elektricitätsmenge an einen reinen Platindraht getragen werden musste. Wären hier Natriumatome die Träger gewesen, so würde, dem als einzigen vorhandenen Anhalte hier maassgebenden elektrochemischen Aequivalente des Natriums entsprechend, deren Gesamtmenge $0{,}7.10^{-6}$ mg betragen haben; wären es grössere Teile von Natrium gewesen, noch mehr. Nun sind nach Bunsen weniger als $0{,}3.10^{-6}$ mg Natriumsalz in der Flamme noch mit der grössten Deutlichkeit erkennbar. Wurde aber der Platindraht, welcher die Träger empfangen hatte, in der Flamme geprüft, so ergab er keine Spur von Natriumreaction.—From P. Lenard, *Annalen der Physik*, Series IV, *2*, 359 (1900).

8.6 V_t for lithium irradiated with light of wavelength 4339 A is +0.49 V. What is the maximum velocity of the ejected photoelectrons?

8.7 Given V_t for lithium irradiated with light of wavelength 4339 A as +0.49 V, and knowing e and h, calculate ν_0.

8.8 The following data are read off from Figure 5 in Millikan's paper, *loc. cit.* (b). V_t for sodium for a set of wavelengths is as follows, where the wavelengths are given in Angstrom units and V_t in volts:

	V_t		V_t
2535	+0.52	4047	−1.295
3125	−0.38	4339	−1.49
3650	−0.91	5461	−2.04

These values were not corrected for the contact potential, which was 2.51 V. a) Can h be determined without correcting for the contact potential? (Explain) b) Determine the value of h. c) Would all metals give curves with the same slope? (Explain) d) Determine ν_0 for the sodium surface.

8.9 In Millikan's apparatus (Figure 8-7) the photoelectric threshold of the copper oxide surface of one particular trap was found to be $\nu_0 = 1.1 \times 10^{15}$ per second. Below what wavelength would he have cause to correct for scattered light?

8.10 Of course, the Compton effect had not been discovered, or probably even suspected, at the time of Millikan's experiment. It was certainly present. Explain why, had it been known, it would not have been a source of difficulty for Millikan.

8.11 Photons of wavelength 0.05 A, on collision with electrons, are scattered, and the radiation scattered at a right angle has a wavelength 0.072 A. a) What is the kinetic energy of recoil of the scattering electron, assuming it to be essentially motionless before the collision?

8.12 A good average wavelength of the visible spectrum is $\lambda = 5000$ A. From the relation $m = h\nu/c^2$, calculate: a) the mass of the quantum of this wavelength. b) Suppose that a hydrogen atom of mass $1.66 - 10^{-24}$ g emits a quantum of radiation of $\lambda =$ 5000 A. What percentage of its mass has the atom lost? (Note that in using $h = 6.625 \times 10^{-34}$ J. sec, all units must be in meters, kilograms, seconds.)

Answers to Exercises

8.1 Among other things one has to mention the physical limitations of the black-box material.

8.2 2070 Å.

8.3 Rigel about 11,000°K; Capella, 6,500°K; Antares, 3,000°K.

8.4 "The investigation which I instituted of the influence of electric light upon electrically charged bodies was carried out mostly in the following way. A brightly polished circular zinc plate of about 8 cm diameter hung from an insulated stand and was connected by means of a wire to a gold-leaf electroscope. In front of the zinc plate and parallel with it stood a large screen of galvanized zinc about 70 cm wide and 60 cm high. In the middle of this was a mica window through which the beams of a Siemens's arc lamp, set up on the other side of it, could fall upon the plate. The system of plate and gold leaves was well isolated: since over a day or so the deflection decreased by only about one-fourth, there was no detectable amount of change during the period of an experiment. Further, if while the arc lamp were lit the mica window were covered with a suitable substance, the isolation was completely preserved.

If the plate, together with the electroscope, which latter could not be struck by the light, were charged electrically negative, then, the moment that the light struck on the plate, the gold leaves rapidly fell together; with a positive charge, the gold leavesdid not fall together at all, and upon accurate investigation, only after a long time was a change first noticeable.

8.5 "Studies on the 'Atomization of substances by ultraviolet light' have pointed to the possibility that there might be ponderable material substance carried off from the irradiated substance. Since that investigation hardly yielded a definite conclusion on the matter, I have carried out special studies concerning which a brief report is given here. A flat surface of sodium amalgam, under hydrogen gas, was irradiated sufficiently long that 2.9×10^{-6} Coulombs of negative electricity leaked off of it, and the electric field above the amalgam was so arranged that all of the electricity must flow to a clean platinum wire. Were it the case that sodium atoms were carriers here, then the corresponding electrochemically equivalent mass of sodium would lead to the collection of a total of 0.7×10^{-6} mg; were a larger proportion of sodium involved, still more. Now Bunsen had shown that as little as 0.3×10^{-6} mg of a sodium salt can be recognized in a flame-test with the greatest distinctness. [The flame is colored yellow.] However, if the platinum wire which has received the carriers is given a flame-test, it yields no trace of sodium reaction."

8.6 $V_t e = (1/2)mv^2$. Therefore v is close to 4×10^5 m/sec.

8.7 $V_t e = h\nu - h\nu_0$. Therefore $(\nu - \nu_0) = 1.18 \times 10^{14}$ per sec. The value found by Millikan from a plot of data is given in the text.

8.8 a) Yes. The correction factor merely displaces the curve upwards or downwards, being additive, and does not change the slope. b) Millikan calculated the value of the slope in volt-frequencies as 4.124×10^{-15}. c) Yes, since the slope is the ratio of two universal constants. d) Millikan determined ν_0 for sodium as 43.9×10^{13} per second.

8.9 Wavelengths shorter than about 2700 Å with this particular trap.

8.10 In the Compton effect the scattered radiation is less energetic than the primary radiation hence with respect to the problem in Exercise 12, as long as wavelengths longer than 2700 Å were being used, there would be no back-reflection from the copper oxide that could cause trouble. However, the question is probably best answered by calling attention to the observation that the number of Compton-effect electrons is quite small relative to the number of photoelectrons under these conditions, and so no effect would be observed, and if it had been observed it would only have changed the shape of the curve (Figure 8-6) and not the position of the threshold value.

8.11 Conservation of energy requires that $h\nu - h\nu_0 = (1/2)mv^2$ (of the electron). This calculates to be 2.7×10^{-13} joules.

8.12 a) $m = h\nu/c^2$. $\nu = c/\lambda = 3 \times 10^8/5 \times 10^{-7} = 6 \times 10^{14}$ per sec.
$m = (6.625 \times 10^{-34} \times 6 \times 10^{14})/9 \times 10^{16} = 4.4 \times 10^{-36}$ kg.
b) $(4.4 \times 10^{-36})/(1.66 \times 10^{-24}) = 2.6 \times 10^{-10}$ in percent.

Bibliography

The books referred to in Chapter 6, list No. 1, are all interesting in connection with this Chapter. Martin J. Klein "Einstein, Specific Heats, and the Early Quantum Theory," *Science 148*, 173 (1965) has given an interesting account of the resistances and the excitement produced by the quantum concept. A sprightly discussion of the wave-particle problem of the nature of light is given by Karl K. Darrow, "The Quantum Theory," *Scientific American*, March, 1952. (W. H. Freeman & Company, reprint #205)

Morris H. Shamos, *Great Experiments in Physics*, Henry Holt & Co., New York, 1960. This little paperback contains an excellent discussion of Planck's and Einstein's work.

Part IV

Chapter 9. Constraint and Variety. An Overall View of the Nature of Matter

Chapter 10. Atomic Structure.

Chapter 11. Atomic Properties

Chapter 12. Nuclear Investigations. The Great Machines

Chapter 13. Nuclear Properties and Processes

Chapter 14. Molecular Properties and Reactions

Chapter 15. Molecular Structure

The Modern View of Matter ~ PART IV

The binding together of parts into a whole is a major theme of these chapters. We meet this theme in the discussion of binding energy, bonding energy, energy of interaction, and intermolecular forces—names given to the energy or the forces (as one looks at the question) which hold together nuclear particles, atomic constituents, molecular constituents, and the molecular aggregates that comprise bulk matter. Two overlapping patterns may be discerned in the constructs that fill this part of the field of constructs. There is an hierarchy of energies: the binding energies that hold the nucleus together are in the range of many millions of electron volts. The innermost electrons of a large atom may be bound by energies in the range of many thousands of electron volts—the outermost electrons with only a few electron volts. The forces that bond atoms into molecules lie in the range of those that hold outer electrons in atoms, and range to small values of magnitude overlapping that of the forces that hold molecules together in liquids and solids. Operating throughout this hierarchy of intensities of binding forces is the idea of energy-levels: each of the bonds described, whatever the entities bound, can be visualized as the state wherein energy has been lost. The bound entities are held together because of this missing energy. If energy is supplied the parts vibrate, rotate, and so on, with ever greater amplitudes, until the energy deficit is met and they fly apart as separated entities. Thus an electron in the lowest stable orbit of an atom, say **n** = 1, may be given energy. A small amount of energy, if absorbed, may "dislocate its probability distribution," that is, cause it to wobble more. A larger amount might, if adequate to the quantum jump, raise it to a higher level, say **n** = 2. If it does not receive more energy, raising it to higher levels or expelling it entirely out of the atom, it will fall back to the lower, more stable level. The spontaneous drive seems to be toward the energetically stable state, whether in the nucleus, the atom, the molecule, or bulk matter. This drive leads to the bonds between atoms that form molecules. It results in the associations between molecules that produce phases. It explains the forces that hold nuclei together.

We begin, in Chapter 9, to show these patterns in the nature of matter. We start at first with the interplay of variety and constraint to show how the rich variety of material things depends on constraints that enable them to have their various forms and to perform their diverse functions.

Next, in Chapters 10 and 11, we study atomic properties. It might be considered logical to start with the nucleus, but the logical must give way to the psychological when the latter dictates content and order. The reason we start with atomic structure and properties is that much of the language used in discussing the nucleus was taken from atomic science, which preceded it in development. We then can discuss the nucleus, in Chapters 12 and 13. One might wonder why these submicroscopic entities, atoms, and nuclei are treated in detail. The reason is that exciting discoveries are being made in this area. Moreover, some rather novel thinking is going on which may influence our culture in due course. So we prepare the ground, and suggest some of the ideas to look for. Further, the subjects studied in previous chapters tie in and unify these.

Molecular properties, reactions and structure, Chapters 14 and 15, complete this part. We try to convey some of the charm of organic chemistry, with its molecular models. And we show in some detail a chain of reasoning of the kind that enables the chemist, operating subtly between P-plane and C-field, to derive structures of substances and confirm them by synthesis.

We do not, in this Part, do more than suggest the fascinating aspects of chemistry and physical chemistry. The student who has already had a course in chemistry will find only a few of the customary topics treated. This is partly because of space limitations; it is chiefly because of our intention to treat whatever we include in some depth and to tie it into what went before.

Chapter 9

9.1 Definitions
9.2 The nucleus
9.3 The atom
9.4 The molecule
9.5 Polymers
9.6 Phases
9.7 Constraints and balance
9.8 Summary

Constraint and Variety ~ 9

An Overall View of the Nature of Matter

Definitions 9.1

We have already remarked on the fact that the extraordinary variety of material things rests on the combinations of relatively few fundamental entities. We have now looked at some of the classical notions about the behavior of matter and seen them all compact in the great laws: Newton's three laws of motion; the laws of gravitational, electrical, and magnetic attraction; the conservation laws, that in an isolated system the total mass does not change; the total momentum is conserved; the total energy does not change.

The Scientists who enunciated these laws were not unconcerned with the behavior of microscopic and sub-microscopic phenomena. But they lacked the discoveries and the tools to make these kinds of studies, and could only speculate from their knowledge of macroscopic phenomena. But, as we saw, when more subtle investigations of matter became possible, anomalies appeared. Some of these laws seemed to have limited scope—powerful though they are in their proper realms. And we have seen that out of the new investigations came new notions that made the older ones "classical": special relativity, which showed Newton's laws of motion to be limiting cases, exactly obeyed only when velocities are slow with respect to the velocity of light; and which showed that the laws of conservation of mass and of energy could be combined into one law, the conservation of mass-energy; and quantum theory, which clarified some aspects of the behavior of light.

In the following six chapters we shall make use of all these constructs and invariant-relations as we develop modern ideas about the nature of material things. In this chapter we take a broad look at what is to come, under the unifying theme of the interplay of variety and constraint.

What makes possible the fantastic variety of things can be thought of as the interplay of variety and constraint: the varieties of possible combinations of protons, neutrons, and electrons and the constraints that limit the naturally occurring chemical elements to ninety-two. (A chemical element is a substance all of which consists of atoms with the same chemical behavior. We shall define these terms and introduce subtleties later.) The thinkable combinations of these ninety-two with each other, and the constraints that make only certain combinations stable; the varieties of size and shape of atomic combinations (molecules), and the constraints that introduce new phenomena at each increase in complexity, continue our theme.

By variety we mean the number of distinct possibilities that exist or may be conceived; by constraint we mean the relation between the variety of what is conceivable or what exists under some given condition, and the variety that exists under some other condition. If the latter is a smaller number than the former, then we say that constraint occurs.

Variety and constraint can be measured in favorable cases. Any physical object, say the leg of a stool, has 6 'degrees of freedom of motion;' that is to say, it requires six numbers to describe its motion: it may undergo translatory motion (from here to there) in the X, Y, Z directions; and in addition it might rotate in the X, Y; X, Z; and Y, Z planes. (Other degrees of freedom may be visualized also, such as spinning about its axis, clockwise or counterclockwise, vibrating a little, and so on. We ignore these complications.) When three legs and a seat are united firmly to produce a stool, with 6 degrees of freedom, the 24 degrees of the four separate parts have been reduced by this constraint to 6. The conservation laws imply constraints—certain behaviors are not possible. Every invariant-relation implies constraint. Were it not for constraints, with attendant orderliness, learning would be useless. Chaos would supervene. It is balance of constraint with variety that we are concerned with—broadly speaking. We proceed, then, to a broad view of the nature of matter.

9.2 The nucleus

Nucleus is a general name for the structure, its diameter of the order of 10^{-15} m, that spins at the center of the atom. A specific nucleus is called a 'nuclide.' It is composed of 'nucleons' (the general name) of which there are two kinds: protons and neutrons. Protons and neutrons are called nucleons only when they are discussed as nuclear constituents, for in the nucleus they seem to have lost their separate identities and to have become interconvertible to some extent. As free particles they are clearly distinguished from each other: the proton has unit positive charge and a mass some eighteen hundred times that of the electron; the neutron is electrically neutral, and slightly more massive than the proton.

Nuclides are characterized by a letter symbol and two numbers. The letter is the chemical symbol of the parent atom, for example, H for hydrogen. At the lower left corner is placed the atomic number, Z, which is the number of protons in the nuclide. At the upper right is placed the mass number, A, which is the number of nucleons in the nucleus. The difference between these two numbers is the number of neutrons in the nuclide. Any nuclide is therefore characterized by the chemical symbol, say X in general, and sub- and super-scripts:

$$_{(\text{atomic number})Z}X^{A\,(\text{mass number})}$$

The chemical symbols of naturally occurring atoms were given long before anything was known about atomic structure. Some of them are easy to remember: hydrogen, H*; oxygen,* O*; nitrogen,* N*; carbon,* C*; fluorine,* F*; lithium,* Li*; neon,* Ne*. Others are not as obvious, being derived from the Latin: sodium,* Na *(natrium); potassium,* K *(kalium); iron,* Fe *(ferrum); mercury,* Hg *(hydrargyrum); gold,* Au *(aurum); lead,* Pb *(plumbum).*

The numbers that replace *A* and *Z* in the type formula in specific cases have to be discovered by actual examination of the substances (we shall see how in some detail later). It is found, for example, that the common hydrogen nuclide consists of a single proton. Its symbol is therefore written $_1H^1$. The rarer 'heavy hydrogen,' named 'deuterium' is twice as massive as common hydrogen. Its nuclide contains a proton and a neutron. Its symbol might be written $_1H^2$ because deuterium is chemically like hydrogen, but in this case it is given the special name and designation $_1D^2$ because in a few of its chemical properties it differs from hydrogen. Similarly for the very rare $_1H^3$, which is named tritium, and given the symbol $_1T^3$. This nuclide contains one proton and two neutrons, and is unstable. These are all hydrogens, chemically speaking, because their Z number is the same, and it is the Z number that characterizes a chemical element. Each chemical element has its own Z number. Atoms with the same Z number but different atomic numbers are 'isotopes' of each other. For example, common uranium $_{92}U^{238}$ is an isotope of the easily fissionable $_{92}U^{235}$. Some hundred-and-three chemical elements are known, and over a thousand nuclides because of isotopy. The elements beyond 92 are synthetic.

Not all nuclides are stable. By stable is meant that they are unchanged over a period of geologic time. The unstable ones tend to break up, or fission radioactively, into smaller particles plus electromagnetic radiation. Of the ninety-two naturally occurring types of atoms, the commonest nuclides of some eighty-three are stable. The other seven are radioactive. All the artificial atoms are radioactive.

It follows, then, that not all ratios of nucleons are stable. What constraints limit the possible variety? A proton by itself is stable; a proton with one neutron forms the stable deuteron. But one proton and two neutrons, as in the triton, is unstable: tritium is spontaneously radioactive. The isolated neutron itself is unstable, and decays to a proton, an electron, and another particle (called an antineutrino); it appears to be quite stable in the stable nuclides, but not in the unstable ones. In some unstable nuclides radioactive decay (as we shall see) seems to stem from protons, too.

Thus there is a kind of intrinsic instability in these nucleons under certain conditions. Many common stable nuclides consist of about equal numbers of the two nucleons. But it does not follow that having equal numbers leads to stability: an uncommon form of beryllium $_4Be^8$, is quite unstable. One has to admit that the constraints that produce nuclear stability and instability are not understood.

One evidence of constraint is the amount of energy that is (or would be) required to separate a nuclide into all its separated nucleons. This is the binding energy of the nuclide. This energy is related by $\mathbf{E} = mc^2$ to the mass difference between the mass of the nuclide and the sum of the masses of the separate nucleons. For example, the mass of the nuclide $_2He^4$ (an α-particle, the bare helium nucleus) is 4.00260361 units on an atomic scale. The total mass of the two protons and two neutrons that might be thought to comprise it is 4.03298132 units. The binding energy is the difference, 0.03037771 units of mass. This is the amount of mass that would be lost were two protons and two neutrons to combine into an α-particle. It is equivalent to a little over 28×10^6 electron volts of energy per nuclide and is the amount of energy that would be given off in such a reaction. It is the amount of energy that would have to be supplied to split the nuclide into the four separate parts. That relativity theory has predicted these mass-energy relations is one of its greatest triumphs.

9.3 The atom

The atom of the chemist and physicist consists in its neutral, elementary state (as distinguished from its combined states) of a nucleus surrounded by a number of electrons equal to the atomic number (Z).

At least three major constraints limit the variety of chemical atoms and prevent all thinkable permutations and combinations of protons, neutrons, and electrons from existing. One constraint is nuclear stability. As the nucleus becomes larger it eventually reaches a region of instability. The nuclei with atomic numbers 90 and above have no stable isotopes at all, and there may be an upper limit to the synthesizable atomic number. Thus, as previously stated, we have some ninety naturally occurring atoms, and some eighty-three of these are stable.

Another constraint is that of electroneutrality. This means that given the naturally occurring atomic numbers, there are only these chemical atoms. One cannot invoke just any combination of elementary particles: a neutron-electron pair or a proton-neutron pair is not an atom. The third constraint is a principle to be discussed later which limits the regions of space about a nucleus which can be occupied by the electrons of a stable atom.

Within these constraints we have a variety of some hundred-and-three or so natural and artificial atom species. Most of these have isotopic forms. The effects of these constraints are beautifully summarized in the Periodic System of the Elements (Section 11-5). The relations in this table were discovered before atomic structure was understood. The atoms can be arranged in periods and groups such that along a given period chemical properties change in a regular way; within groups they tend to show many likenesses. For example, the table begins with the simplest atom, hydrogen. With atomic number one, it contains one electron outside the nucleus, and this may be symbolized by the chemist with a dot after the chemical symbol: H·. The second atom in sequence is helium, He:. This ends a period, because the third constraint (referred to above) says that only two electrons (opposite in spin) may occupy this first shell. All of the atoms after atomic number two have this inner shell filled with two electrons. (A shell is a group of electrons with about the same energy and distance from the nucleus.)

The pattern continues with the third atom, $_3$Li, starting a second shell outside of the first, with its third electron. The chemist is not deeply concerned with the inner filled shell of two (the helium shell) but very much so with that line electron starting a new shell, so he writes: Li , and he notes that lithium has many chemical properties similar to hydrogen. The fourth element $_4$Be is written: Be:. The second electron pairs with the first by constraint number three, but this atom is not at all like helium, He:, for this second shell may contain eight electrons and it is only at $:\underset{\cdot\cdot}{\overset{\cdot\cdot}{\mathrm{Ne}}}:$ that one comes upon a filled shell, and an atom like helium in its chemical behavior. And so on, in a pattern of constraints that will be discussed in a later chapter, and that limit the possible variety in the combinations of protons, electrons, and neutrons.

9.4 The molecule

A chemical element, when the word refers to bulk matter rather than the single atom, is composed of atoms all of the same kind; a compound is bulk matter composed of molecules; a pure compound is composed of molecules all of the same kind. A molecule is composed of atoms of the same or different kinds held together by forces strong enough that the molecule is stable under specified conditions. The designation is arbitrary, but convenient. The constraints that make molecules possible and at the same time limit their variety (though the variety is still prodigious) are few in nature but far-reaching in their effects. We are still, in this section, in the area of sub-microscopic particles: we are discussing molecules, and not bulk matter except in the cases where we are driven by the nature of things into the level of bulk matter.

The simplest molecule is the hydrogen molecule. It is composed of two hydrogen atoms, so tightly bonded to each other that it requires some 4 1/2 electron volts of energy to pull them apart. But three hydrogen atoms do not form a stable molecule; only two. So here is a constraint: the chemical bond-forming property saturates at two; it is used up at two for this atom. The two atoms are said to be chemically bonded when they form a (stable) molecule. The bonding constraint is related to constraint number three in the previous section: each hydrogen atom contributes its single electron, and the bond that forms comprises both, paired, and filling a molecular shell that weaves the two positively charged nuclei together by the behavior of the two negative electrons. The molecule is written H : H, or more simply H—H, or most simply H_2. The subscript 2 says that there are two atoms of hydrogen in the molecule. A bond of this type is called a shared-electron bond, or a covalent bond.

Another type of molecule is formed when the bonded atoms are quite unalike chemically, so that they do not tend to share electrons, but one takes an electron from the other, becoming negative, while the other is left with a net positive charge. For example, sodium $Na\cdot$ reacts with chlorine $:\!\ddot{\underset{\cdot\cdot}{Cl}}\!:$ to form $Na^+ :\!\ddot{\underset{\cdot\cdot}{Cl}}\!:^-$. The outside electron of sodium is given up. This leaves the sodium with a (+) charge because the remaining electrons are not enough in number to neutralize the (+) charge on the nucleus. Such a charged atom is called an ion (Section 3.4). Chlorine, with seven electrons, is neutral. It takes on the electron from the sodium atom, thus filling out a shell of 8 that is a stable arrangement. This converts it to a negative ion. The two ions are held together by the Coulomb attraction between the two opposite charges. The bond is called an electrovalent bond. There is another kind of bond in which one atom shares two electrons with another, but we shall not pursue this further since in principle we have indicated the most important bonds—the constraints—that lead to the appearance of molecules. In the loss or gain of electrons there seems to be a drive toward symmetrical arrangements.

One branch of chemistry is the study of molecular transformations: reactions in which molecules are made or decomposed; in which one material is changed to another. For example, water, H_2O, may be made by burning hydrogen gas H_2 in oxygen gas O_2; it is also a product when caustic sodium hydroxide, NaOH, reacts with acidic hydrogen chloride, HCl. A number of constraints—invariants—were discovered by early chemists to limit the variety of possible products of such reactions, and these have become embodied as chemical laws. It was demonstrated by Lavoisier that mass is neither created nor destroyed in a chemical reaction. (It is known today that the chemical energy changes are so small that relativistic mass changes would not have been noticeable even in the most refined work.) Hard on this demonstration came the discovery that elements form compounds in which the reacting weight ratios are always exactly the same. For example, water contains hydrogen and oxygen in the ratio one to eight. This is the law of definite proportions. It limits the variety of possible compounds when two or more elements combine. It was shown, too, that some elements can combine in more than one ratio (for example, hydrogen and oxygen in the ratio one to sixteen form hydrogen peroxide), but in every case the ratios of the proportions in the related compounds are small whole numbers: in the two compounds water and hydrogen peroxide the oxygen ratios for the same weight of hydrogen are 1:2. This is the law of multiple proportions. Still another constraint of great importance is that equal volumes of gases at the same temperature and pressure contain equal numbers of molecules. The discovery of this law of definite volume proportions led to a rational system of atomic weights and, just after the middle of the nineteenth century, to the development of the periodic table of elements. Shortly thereafter it became possible to think in experimental terms of the shapes of molecules, and while this enlarged enormously the conceptual variety it put definite limits to the variety of possible structures. We pursue these subjects in other ways in a subsequent chapter.

Polymers 9.5

Because atoms of the same kind can combine chemically with each other, and ions with opposite charges attract, an extraordinary variety of stable molecules can exist: over a million are known and indexed in reference books; and chemists can write down general formulas which predict millions more. There is a constraint, however: the number of atoms available in and on the Earth is finite (every breath we take may well contain a few nitrogen molecules breathed by Anthony and by Cleopatra), and the number of possible compounds of carbon, for example, is so great that not all possible carbon structures can exist at any given time.

Very large molecules with a repetitive structure are known as polymers, from the roots 'many parts.' Consider simple table salt. A single crystal may be a giant polymeric molecule, if it is perfect. This is so because table salt, sodium chloride, is composed of equal numbers of sodium and chloride ions. Each sodium ion Na^+ inside the crystal is surrounded by chloride ions Cl^-, for the electrostatic lines of force extend out in all directions from the plus-charged sodium. But not any number of chloride ions may be held to the sodium: there is room for a maximum of six only. This constraint is mirrored in the cubic shape of the sodium chloride crystal: a cube has six faces. It explains that orderly inward structure that older crystallographers postulated as responsible for the orderly outward form. At the surfaces, edges, and corners of the crystal of table salt, a pattern of lines of force is present due to the mosaic of exposed sides of Na^+ and Cl^- ions. Many mineral substances, crystalline or not, are giant molecules composed of billions of billions of atoms. A perfect diamond is pure carbon, covalently linked. But here the constraint of covalent bond formation requires a tetrahedral structure: each carbon surrounded by four others.

Polymers are of great industrial importance: natural and synthetic rubber, polyethylene, wood, natural and artificial fibers are all polymeric. As a very simple example, pure polyethylene consists of thousands of carbon atoms linked to each other covalently, and to hydrogen covalently to fill out the octet. The formula of one 'part' is $\left[\begin{matrix} \mathrm{H} & \mathrm{H} \\ -\ddot{\underset{..}{\mathrm{C}}} & : \ddot{\underset{..}{\mathrm{C}}}- \\ \mathrm{H} & \mathrm{H} \end{matrix}\right]$,
and the polymer would be written $X(—CH_2—CH_2)_nY$, where n might be hundreds or thousands, and X and Y might be H's to complete the ends of the polymer chain, or other atoms depending on how the polymer was made.

Polymers are of the greatest biological importance also. Hair, nails, skin are polymeric protection and support structures; bone is polymeric; proteins are all polymeric, as are enzymes and DNA and RNA; genes, chromosomes, cell membranes—and so on in unknown detail and number of different entities. Here, particularly, constraints of size and shape—very subtle ones—serve to limit variety. The existence of the living creature depends on constraints; the evolution of the creature depends on the balance of initiating variety and viability-giving constraint.

Phases 9.6

Matter exists in one of four states: solid, liquid, gas, and plasma. We shall leave the last of these to a section of a subsequent chapter. (It may be that a fifth state—the living state—should be added. Whether or not it is added, we shall omit discussion of biological phenomena here.) The existence of a given state reflects the operation of constraints, studies of which are largely the province of the physical chemist: the properties of macroscopic systems (bulk matter) are studied as reflections of molecular- and atomic-level properties.

The constraints that lead to the characteristic behavior of a solid are attraction interactions between atoms or molecules. These are usually not strong enough to be called 'chemical,' except that in some cases, as with the sodium chloride crystal, they are chemical. But they are strong enough to give the characteristic rigidity of solid matter. A solid is bulk matter that has and holds a definite shape for a reasonable time. Table salt and table sugar are nicely crystalline solids with permanent shapes until they are dissolved in water; glass is a non-crystalline solid—and it, too, may be dissolved, and lose its shape, but in a special kind of solvent. The binding forces of solids are usually designated physical forces, to distinguish them from the usually much stronger chemical bonding forces.

In liquids the constraints between molecules are less than in solids at the same temperature. The molecules may slip over each other. Thus the ordinary bulk liquid will take the shape of a container: it does not have a characteristic or definable shape. At the same time, the constraints are strong enough that a given weight of liquid at a given temperature will have a definite volume. This is recognized in the property that a liquid has: its density, or weight per unit volume, at a given temperature. One milliliter of water, for example, weighs one gram at 15°C: its density is 1. Another evidence of these constraints is that there is a tendency of the liquid to take a shape under certain conditions: a fine droplet of water in fog or mist is spherical; the tear drop form of a falling larger drop is the result of balance between the attractions of water molecules for each other and the frictional forces against air molecules; the drop standing on a leaf after it has rained, or found as dew in the morning, shows that when the liquid does not wet the surface, it takes on this particular form, with the minimum surface for the volume of the droplet.

Another commonly observed evidence of these constraints is that liquids do not necessarily mix completely. For example, oil does not mix with water: the two, stirred in a container and allowed to stand undisturbed, will eventually separate, with the less dense oil on top, the more dense water below, and a clear line of demarcation between.

Gases are characterized by minimal constraints between molecules (or atoms). They have no preferred shape, nor will a body or gas without a container retain a shape. Gases tend to fill a container uniformly and to intermingle freely (provided that they do not react chemically to form liquid or solid products).

There are always borderline cases. There are 'solids' which are soft enough that they slowly flow, and take the form of the container. They might as well be called quasi-liquid; or they might equally well be thought quasi-solid.

9.7 Constraints and balance

Wherever we have seen a property which reflects a constraint we have also seen the results of balance: balance between motion and constraint, would be one way of putting it. We have discussed the nucleus, the atom, the molecule, and phases as they are found under conditions of relative stability. But in every case each constraint-producing force can be overcome by a greater, oppositely-directed force. Thus the constraints that give a nucleus form and internal structure may be overcome: the nucleus may be smashed. The force that holds electrons to nucleus in the atom may be overcome: the atom may be ionized, and even stripped of its electrons so that all that is left is a bare nuclide. The bonds that hold atoms together to form molecules may be broken by high temperature or electromagnetic radiation or other means: the molecule may be dissociated. The solid, on heating, may melt to a liquid; and this on heating to a higher temperature may vaporize to a gas. This on still further heating may dissociate and ionize, and become plasma. Each constraint has its limits. Physical science is largely the science of finding and making use of constraints.

9.8 Summary

All matter contains atoms, and all atoms contain nuclei. All material states are the result of balance, temporary or longer-lived, between the universal tendency toward motion and the constraints that enforce form. In the nucleus these constraints are reflected in binding energies that are of the order of millions of electron volts. The energies with which electrons are held to nuclei in atoms range from tens of thousands of electron volts downward. The constraints that bond atoms in molecules are of the order of a few electron volts per pair of atoms. The energies with which molecules are held together in bulk matter range from a few electron volts per pair of molecules down to essentially zero in a gas. (These constraints cumulate: along a polymer chain they add, part by part. This is why fibers and metals can be strong.) The next six chapters are devoted to detailed examination of atomic, nuclear, and molecular structures and properties; to the constraints that limit possible variety while at the same time making orderly process, change, and motion possible.

Exercises

Note to Exercises. Sometimes it is desirable to have a quantitative *estimate* in answer to some question where either actual, precise information is lacking, or there is no need for a more precise statement, or there is no time to collect the data necessary for a precise answer. Such situations may arise around the table at a sales-meeting or a board-meeting, a planning session or committee meeting. During the Second World War, such questions arose frequently in the course of the decision-making procedures called Operations Analysis (or Research). The head of one such planning group, a mathematician, told the Author after the War that the two best men in his group were a lawyer and a professor of history. In making these informal guesses, where detailed technical information is not required, qualities of mind become more apparent: ingenuity; the ability to follow a line of reasoning; a hard-nosed attitude towards facts; sheer hedonistic pleasure (in Professor John Weichsel's phrase) in tackling a problem. In answering problems of this kind give a briefly stated, stepwise account of your final process of getting an answer.

9-1 How many drugstores are there in the Continental United States?

9-2 How many blades of grass are there in your fresh, well-cared-for football stadium (before the games begin)?

9-3 What is the pressure, in pounds per square inch, exerted on the pavement by an average-size college girl wearing spike-heel shoes? (Note that a ton is 2000 pounds.)

9-4 Make the calculation hinted at in the text: how many nitrogen molecules might well be breathed in by you that had been breathed at some time by Cleopatra—or Anthony. (Recall or know, that 22.4 l of nitrogen at atmospheric pressure and 0°C contain 6×10^{23} molecules of nitrogen.)

9-5 It has been said that if all the molecules of water in a teaspoonful were marked, then mixed uniformly with the seas, a glassful drawn anywhere would contain a large number of the marked molecules. Is this a reasonable statement? The oceans cover 71% of the Earth's surface. In 18 g. of water there are 6×10^{23} molecules of water.

Answers to Exercises

9-1 I asked my wife this question. She thought for a moment, and then replied: "In my home town of Madison, Indiana, with a population of (a certain number) there were (a certain number) drugstores. Now the population of the United States is about 200 million, and if we assume one drugstore for some (a certain number) people, that might be a reasonable estimate. *But* there are some uncertainties about such an estimate. . . ." (P.S. She majored in English and Music.)

9-4 The question could be answered very precisely by going to the literature and getting accurate data. But let's see what we can do, always erring on the conservative side. What we want to know is if the atoms of nitrogen in one breath were mixed thoroughly with the atmosphere, how many would be found in a breath after mixing. We know that the radius of the Earth is about 3900 miles. This is about 6400 km. Volume is $(4/3)\pi r^3$, which works out to volume of Earth approximately $10.990 \times 10''$ cu km. Volume of the atmosphere is tough to estimate. We know that 5 miles up, planes must be pressurized, and already on the top of Mount Everest oxygen masks must be used. So obviously pressure falls off rapidly. The attenuation is rapid; so if we said that if the atmosphere were compressed down to a height of, say, 15 km, it would have a pressure of 1 atm., we would probably be way overestimating: perhaps less than 10 km would be more like it. However, to get the volume of the "atmosphere" at one atmosphere pressure, we add 15 km to the radius, recalculate the volume $11.052 \times 10''$ cu km, and the difference between this and the volume of the Earth is the figure we want: 6.2×10^9 cu km. One km = 1000 m = 10,000 decimeters; 1 cu km = 10^{12} l. Volume of this much gas is 6.2×10^{21} l. One breath is about 1/2 l (Cleopatra was a small woman) and contained 1.3×10^{22} molecules. Thus if these were evenly distributed in 6.2×10^{21} l, there would be 2 molecules/liter.

Notice that we have made a *very* rough calculation. But at least we are in the right ball-park, because we have been conservative. More accurate calculations might yield 20, or 200, or

2000; but the chances, we feel, are that it won't be less than about 2. We choose nitrogen for this purpose because it is not metabolized by animals and most plants; and though fixed by some plants it does get back into the atmosphere. Why not see if you can actually pin the matter down? You see, the clear implication is that the amount of available molecules on this "precious little Space-Ship Earth" is finite. What does this imply about the oxygen, carbon, water, and so on necessary for life, and the number of people that can be supported on the Earth? (Not to mention animals and plants needed for the support of people, and for esthetic enjoyment. . . .)

Bibliography

W. R. Ashby, *An Introduction to Cybernetics*, Wiley, New York, 1958, has given an interesting discussion of variety and constraint. The example in the text is modeled on his.

Chapter 10

10.1 The nuclear atom
10.2 The Bohr atom
10.3 Energy levels
10.4 Difficulties with the theory
10.5 Wave mechanics
10.6 Test of theory: the contribution of Davisson and Germer
10.7 A picture of the hydrogen atom
10.8 Indeterminacy
10.9 Some implications
10.10 Summary

Atomic Structure ~ 10

The nuclear atom 10.1

We saw, in Chapter 9, that atoms are characterized by their nuclei and their shells of electrons. If one started logically with the smallest particle of matter and developed structure and properties "up" through the hierarchy of size and complexity, the sequence would be from nuclei to atoms to molecules. However, it is psychologically better to start with atomic structure and properties, then turn to nuclear structure and properties, then to molecular structure and properties. This is partly because knowledge of the structure and properties of nuclei came after considerable knowledge of atoms had been achieved, and thus made use of concepts (such as energy levels) that had been developed in studying internal atomic structure. For this reason we begin the study of matter and its submicroscopic properties at the atomic level rather than the nuclear. Also, much more that can be visualized and understood without a mathematical background is known about atomic than about nuclear chemistry and physics.

In 1911 Ernest Rutherford came to the conclusion that atoms must have most of their mass concentrated in a very small positively charged nucleus, and that the electrons necessary for electrical neutrality (elementary atoms are neutral) must in some way occupy space outside of the nucleus. It was, however, difficult to understand how electrons could reside stably outside of a highly positive center to which they would be attracted by Coulomb force. One would expect them to fall into or onto the nucleus. An explanation, and tentative solution, of the problem was offered by Niels Bohr in 1913. This turned out to be applicable only to a few atoms and ions; indeed to be so restricted in usefulness that it was soon abandoned. But what it did provide was a tremendous stimulus to thinking. We therefore begin our discussion of atomic structure with this hypothesis of Bohr's.

10.2 The Bohr atom

Consider the simplest atom, common hydrogen. From chemical evidence it must be concluded that the nucleus consists of a single plus charged particle outside of which there is a single electron. Why does not the electron fall into the nucleus? Bohr's hypothesis may be presented in this way. Assume that there are orbits at certain radii ($r_1, r_2 \ldots r_n$) from the nucleus which the electron may occupy without emitting radiation (energy). We don't enquire at present what we mean by 'occupy' but the electron in one of these states is conceived to have a certain potential energy (or, to be less specific, a *total* energy), $\mathbf{E}_1, \mathbf{E}_2 \ldots \mathbf{E}_n$. The state closest to the nucleus, at r_1, is taken as the reference state, and assigned *relative* energy zero. Then if an electron in this state is moved by some agency to the next stable, stationary state at r_2 work will have to be done on it because a negatively charged particle is being moved away from a positively charged one. The agency that can accomplish this change in energy is electromagnetic radiation: Einstein and Planck had developed this theory in connection with radiation and photoelectric phenomena. Putting these ideas together, a change from an energy state $\mathbf{E}_1$ to an energy state $\mathbf{E}_2$ occurs when the electron absorbs a quantum of suitable size:

$$\mathbf{E}_2 - \mathbf{E}_1 = h\nu$$

The new state, with the electron at r_2, and having energy $\mathbf{E}_2$, will not in general be stable if in the atom an orbit with less energy is available, that is, unoccupied in the special way we shall examine in Section 11.6. When this is the case the electron returns to the 'lower' orbit and gives up the energy difference between the two in the form of a quantum $h\nu$, where ν has the same value as it did for the absorption process. That is, it absorbs a quantum, then may shortly emit it. It might be conceived that if the electron in a given state $\mathbf{E}_1$ receives a quantum of energy very slightly smaller or larger than the exact $h\nu$ it may not accept it, not being 'tuned' to this frequency. However, if the energy were enough larger to raise the electron to a higher energy state, say $\mathbf{E}_3$, or more, then it would be accepted (with the tuning restriction). If the quantum were large enough, $h\nu_o$ of Equation (8-3) or larger, the electron is lifted clear out of the atom in a photoelectric effect. At what radii do these stationary states occur?

Consider the physical picture: a relatively massive nucleus (p) (about 1800 times the mass of an electron) with an electron (e) in orbit around it. Looked at from outside, the two are attracting each other with a force $F_e = K\, q^2/r^2$. If they were stationary, and were allowed to accelerate toward each other, they would collide at the point x in Figure 10-1. The displacements each would undergo,

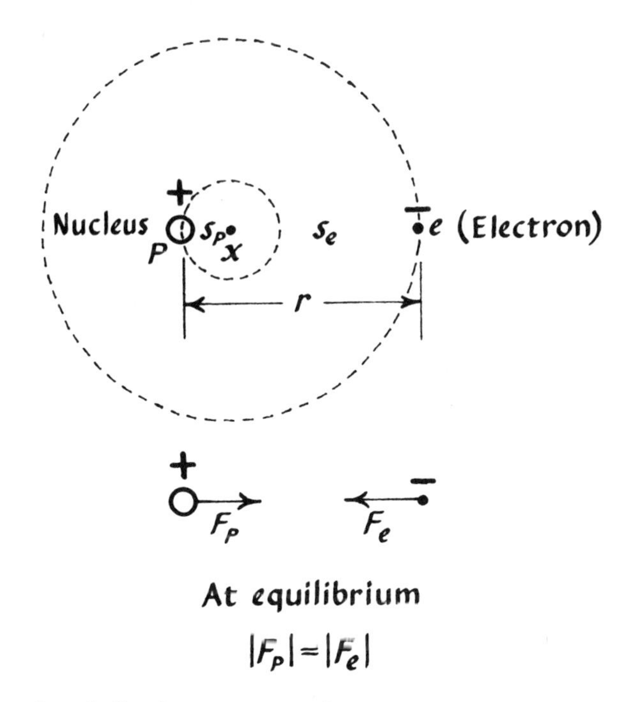

Fig. 10-1 *Rotation of two bodies about a common center.*

namely s_p and s_e, are proportional inversely to the masses. That is, since $F = ma$, and the two particles are exerting equal and opposite forces on each other, $m_p a_p = m_e a_e$; therefore $(a_e/a_p) = (m_p/m_e)$.

Suppose that the particles are revolving at such a velocity about the point x that the Coulomb force just provides the centripetal force to maintain the total distance apart, $r = s_1 + s_2$. The requisite centripetal forces must be $m_p v_p^2/s_1 = m_e v_e^2/s_2$. It is apparent that s is inversely proportional to m, and in the case of the hydrogen atom the point x, which is the 'center of mass' of the system, would be very close to the center of the proton: $s_e = 1800$ times s_p; the figure is far out of scale, of course. But nevertheless the revolution of the two takes place around a center of mass. This detailed picture is given for more complete understanding of the physical situation. In effect, we are always interested in r, the distance of the electron from the nucleus.

The Coulomb force is $F_e = K\,q^2/r^2$ (Equation 3-3). (We can write q^2 instead of $q_e \cdot q_p$ because the electronic and protonic charges are equal in magnitude.) This provides the centripetal acceleration force $\boldsymbol{F}_a = ma_c = m\,v^2/r$ (Section 4.6) and must therefore equal it:

$$Kq^2/r^2 = mv^2/r \tag{10-1}$$

The problem with this formulation is that by this model the electron is undergoing acceleration, and according to Maxwell's theory (Section 7.7) an accelerating charge must radiate energy. But were this to occur; were the electron to lose energy; it would slow down, and consequently F_e would be larger than F_a and the electron would spiral down into or onto the nucleus. This it does not do.

Bohr rationalized the behavior of the electron, assuming this model, by the proposal that those orbits are permitted to the electron wherein the momentum of the electron (mv) multiplied by the distance around the orbit ($2\pi r$) is equal to an integral multiple of Planck's constant, *i.e.*,

$$2\pi r\, mv = n\, h \qquad (10\text{-}2)$$

This is logical: h is the 'quantum of action,' and action has the dimenions of momentum times distance. n Is always a whole (integral) number.

We can eliminate the unknown quantity v between equations (10-1) and (10-2) by algebra. In equation (10-1) v occurs as the square. Rather than work with square roots, we square both sides of Equation (10-2):

$$v^2 = n^2h^2/4\pi^2m^2r^2$$

Substituting for v^2 in Equation (10-1),

$$Kq^2/r^2 = mn^2h^2/4\pi^2m^2r^3$$

This simplifies to:

$$Kq^2 = n^2h^2/4\pi^2mr$$

which can be rearranged to yield the orbital radius r in terms of known quantities:

$$r = n^2h^2/4\pi^2Kmq^2 \qquad (10\text{-}3)$$

Suppose that $n = 1$, that is, by the hypothesis, the electron is in the first orbit, closest to the nucleus. We substitute known values for all the quantities except r, and calculate the radius of the hydrogen atom:

$$r = (6.6 \times 10^{-34})^2/(4)(3.14)^2(9 \times 10^9)(9 \times 10^{-31})(1.6 \times 10^{-19})^2$$
$$= 0.53 \times 10^{-10}m$$

The value comes out just the right magnitude, as determined from experiments with gases and liquids. It was this remarkable discovery that, if it did not solve more detailed problems, at least gave a powerful stimulus to the hope that here was a way out of the dilemmas and anomalies that had beset atomic theory. But much more was present in the hypothesis.

Energy levels 10.3

According to Equation (10-3) r is proportional to n^2h^2 by a constant $1/4\pi^2Kmq^2$; indeed, since h is constant, $r \propto n^2$. By hypothesis, n may take only integral values, 1, 2, 3. . . . Thus r may increase in steps of 1^2, 2^2, 3^2. . . . The cross-sectional diagram of a hydrogen atom is drawn according to this hypothesis in Figure 10-2. The energy

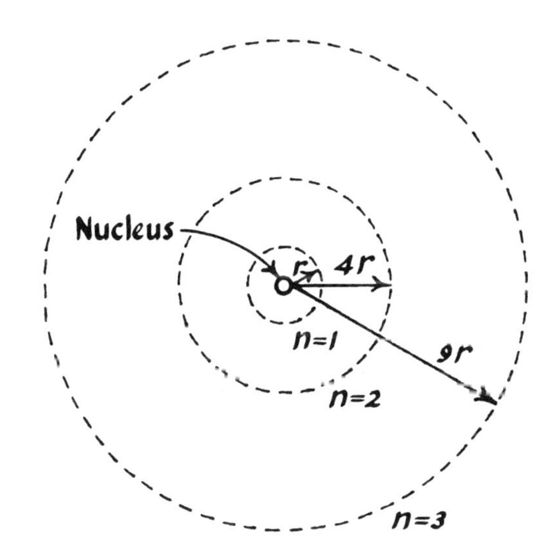

Fig. 10-2 *The Bohr model for the hydrogen atom showing three energy levels at* r, 4r, *and* 9r. *It is drawn to scale (except for the size of the nucleus).*

of an electron with the quantum number n may be obtained from the expression (3-8) KE = $(1/2)mv^2$. The speed of the electron is not measurable, but we can approach the problem algebraically. (As usual, we move around in the C-field along the lines of the invariant-relation equations, from one construct-group to other construct-groups, so as eventually to return to the P-plane at or close to a desired experimentally available point. When this kind of maneuver is done for the first time—in a creative act—it seems to be the result of mathematical intuition. Some people have it; some don't.)

Returning to Equation (10-2),

$$r = nh/2\pi mv$$

This value of r is substituted in Equation (10-1), to eliminate r:

$$Kq^2 = nhv/2\pi$$

We can see that in the definition of energy we need v^2, so we square the above equation:

$$K^2q^4 = n^2h^2v^2/4\pi^2$$

In the energy equation we also need $(1/2)m$, so we multiply both sides of the equation by $(1/2)m$:

$$\text{Energy} = (1/2)mv^2 = 2\pi^2K^2mq^4/n^2h^2 \qquad (10\text{-}4)$$

This rather formidable-appearing relation shows that the energy is a function of n: all the other terms are constants. Thus if we gathered these constants together:

$$k = 2\pi^2K^2mq^4/h^2$$

we see that:

$$\text{Energy} = \text{k}/\text{n}^2$$

The energy of the electron is inversely proportional to the square of n. This expression says that the energy of the electron is quantized (which is not surprising, since quantization was introduced in the original hypothesis). But it also gives details of the functional relation, so that one may calculate the energies for various values of n. The importance of this result will become apparent in the next chapter.

One conceives of the atom in the following way (we are concerned for the present with the 'Bohr atom' but the overall concepts hold in modern atoms). 'n' Is named the 'principal quantum number' of the electron. In its normal state the hydrogen atom consists of a proton nucleus with a single electron at an orbital distance such that $n = 1$. There are available to the orbital electron a set of possible orbits $n = 2, 3. . . .$ When n becomes large enough, the electron has essentially escaped from the Coulomb field of the nucleus. What we are describing is a state of affairs such that in the normal, stable state, only the orbit $n = 1$ is occupied and the 'others,' $n = 2, 3, . . .$ are 'immanent possibilities.' We imagine that if an electron (e^-) were brought up to a proton (H^+) from some distance, which we call infinity, '∞', for convenience since it is so large (relative to atomic distances) that the nucleus exerts essentially no Coulomb force, the electron would 'fall in' to the field of the nucleus and stabilize at the orbital distance for $n = 1$. In the process the electron would give up a certain amount of energy (Section 3.11). This means that to pull the electron back out of the atom, one would have to put in this much energy; one would have to do this much work on the electron. The amount of this energy turns out to be 13.53 eV, for the hydrogen atom, and is named the 'ionization potential' of the hydrogen atom.

We now turn around, so to speak, in our point of view. We think of the electron in its stable state as an energy state 'zero' and we think of the work that has to be done on the electron to pull it away from the nucleus into orbits $n = 2, 3, \ldots$ We speak of 'lifting' or 'raising' the electron to 'higher' energy levels. The analogy is to gravitational experience. We have some object of mass m on a laboratory desk. We lift it to a height s, and calculate that at this height it has gained potential energy $PE = mgs$ due to the work we did on it (Section 3.11). We could get this back by letting it do work as it falls back to the desk-top. At the desk-top the object still has potential energy, for if we pushed it off the edge of the desk it would fall to the ground. However, the desk-top energy is our level of reference, and we call it zero. Then when we raise the object from this level, we do mgs work on it, where s starts at zero at the desk-top. So with the atom. $n = 1$ Corresponds to the desk-top and Coulomb attraction corresponds to the gravitational force. To calculate how much energy is needed to lift the electron to the energy level of the orbit $n = 2$, we substitute in the Equation 10-4, and subtract from this the energy value for $n = 1$. The difference is 10.15 eV. It takes this much energy to lift the electron to the second orbit. By making such calculations we obtain values for the energy at the levels whose quantum numbers are $n = 1, 2, \ldots$ This can be represented in an energy-level diagram, Figure 10-3.

Do not let yourself be subliminally confused by this diagram (Figure 10-3). Electron volts of energy are plotted vertically; these are not radial distances of the electron. Be sure you understand this. The energy, being a function of $1/n^2$ (Equation 10-4), that is, an inverse function of the square of n, the larger is n, the smaller is the energy-difference from neighboring n's, so that the levels converge to the limiting energy value which is 13.53. At the same time, by Equation (10-3), r is directly proportional to n^2, so the larger n is, the larger r is by the square relation (Figure 10-2), and the orbits become much farther from each other, while their energy differences become smaller and smaller.

We need to say a word about energy, here. We conceive of the electron in an orbit as having a certain amount of potential or, at least, 'total' energy, potential and kinetic energy. In $n = 1$, this is our base-line, zero; in $n = 2$, $\mathbf{E} = 10.15$ eV. When the electron is being raised from $n = 1$ to $n = 2$ (we shall soon see how this occurs) we may imagine that it acquires kinetic energy that sums up to 10.15 eV. But at the orbit $n = 2$ this becomes potential energy. The terms potential and kinetic are useful here only for purposes of picturing. It is better to speak only of '*energy*'. We have a situation somewhat like that described above in lifting an object, except that there are no intermediate stages.

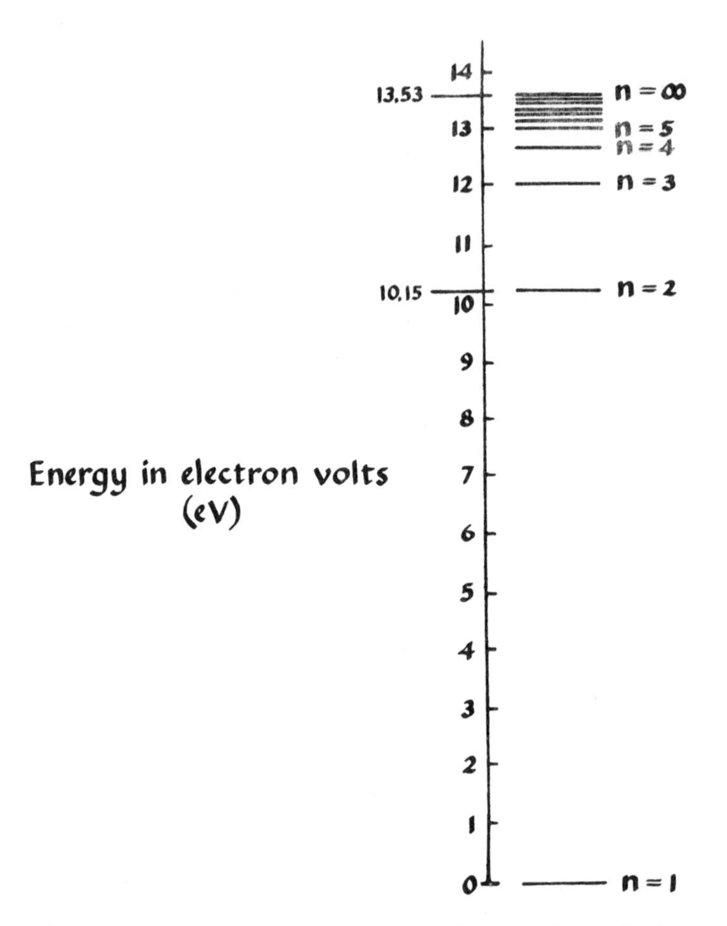

Fig. 10-3 *Energy level diagram (preliminary). A set of energy levels for the hydrogen atom is plotted for* n = *1 taken as zero, the stable level. Potential energy, in electron volts is plotted for* n = *1 to* n = ∞. *The ionization potential is 13.53 eV.*

This is what we meant in Section 8.3 in saying that energy changes along a quantal staircase, not up a smooth classical ramp. Now we see that it is a rather peculiar staircase. The first riser is high, and the steps become smaller as we ascend.

We saw, too, that these higher levels are not stable when lower ones are available. An electron in level $n = 2$ will fall back to $n = 1$ if conditions are right. In doing so it gives up energy in the form of electromagnetic radiation. This is understandable: a charged particle with potential energy at $n = 2$ of 10.15 eV suddenly falls to $n = 1$. It undergoes acceleration, one must suppose. Acceleration of a charge from one stationary state to another produces electromagnetic radiation.

The energy given up is 10.15 eV and by Equation (8-1) this is equal to $h\nu$. Thus the transition from n_2 to n_1 in the hydrogen atom produces a quantum of electromagnetic energy of value 10.15 eV. We shall see in the following chapter what the consequences of this are.

Difficulties with the theory 10.4

Bohr considered his theory preliminary. For one thing, it involved a partial break with classical ideas so that it must certainly be considered provisional and hypothetical pending further thought. The partial break is at the point where the permitted orbits are quantized yet at the same time the orbits are described in classical terms, using Newton's laws for balancing forces, as in Equation (10-1). (Actually it was not until even this vestige of classical notions was erased that atomic theory matured.) The initial Bohr theory did not allow the calculation of the structures of atoms with many electrons; it did not suggest how molecules such as H_2, or other molecules, are formed. The theory, as will be shown in Section 11-3, did predict the positions of lines in the spectrum of hydrogen, but it did not predict their intensities, and it had a serious flaw in that observations of hydrogen showed no magnetic effect such as would be expected of a classical current loop produced by an electron moving in an orbit. Yet it was a remarkable feat, as we shall see in the following chapter, because it fitted exactly to certain optical properties of the hydrogen atom that had been puzzles for years.

Wave mechanics 10.5

If we define quantum mechanics (as we did) as the theories and experiments which invoke Planck's quantum of action h, then wave mechanics is a set of theories and experiments which falls under quantum mechanics. It is an extremely involved set of theories and experimental interpretations, and we can look at it only rather superficially, but it is important to us because of its implications.

It will be recalled that the quantum interpretation of the photoelectric effect employed an equation $\mathbf{E} = h\nu$ (Equation 8-1) which suggested a connection between the particulate quantum and the wave of frequency ν. This dualistic approach was carried over by Louis de Broglie to an hypothesis that freely moving particles would have wave character—matter waves. The relation between the momentum (mv) of the particle and the wavelength (λ) of the associated wave was written (refer to Section 8.7):

$$\lambda = h/mv \qquad (10\text{-}5)$$

This hypothesis was taken up by Erwin Schrödinger and applied to atomic structure as 'wave mechanics': a theory that reconciles particle behavior with wave behavior by quantizing the wave and its associated kinetic energy.

Imagine a string firmly fixed between immovable supports (Figure 10-4). Represent the value of the distance between the fixed ends of the string by d. The amplitude, $\pm A$, gives the maximum distance that a point on the string may move in the Y direction. (You might visualize this as the string on a musical instrument.) Inspection shows that the longest wave that can occur is twice d (Figure 10-4a). When the string is vibrating in this mode (the 'fundamental') it moves up and down as a whole except at the very ends. The maximum value of y occurs at the center of the string for the case $\lambda = 2d$. Other modes of vibration, 'overtones,' are possible. The first occurs when $\lambda = d$. In this mode there is a motionless point, a 'node', at the center of the string, and the string shows two 'loops' (Figure 10-4b). A second overtone arises at $\lambda = (2/3)d$ (Figure 10-4c) and there are two nodes and three loops. And so on. Half-wavelengths occur in whole numbers, n, where $n = 1, 2, 3, \ldots$ for the fundamental and overtones according to the relation:

$$\lambda = 2d/n \qquad (10\text{-}6)$$

Because the ends of the string are fixed only certain wavelengths are physically allowed. Only an integral number of half-wavelengths can occur, not anything in between. The waves are quantized by this restriction, which is the physical basis of this n.

This situation is akin to resonance. If, in the presence of a silent piano, a note C is sounded with a tuning fork, or the voice, or a violin, and then stopped, the C string in the piano will be heard to be singing. Energy has been transferred to the string, but only that string (and overtones) that is tuned to it responds fully. Being tuned to means being able to resonate to. A radio is tuned to a station when its circuit is so set that it will accept (and amplify) the signal impinging on the antenna. This signal is an electromagnetic radiation, carrying energy. Resonance can occur with sound waves (which are longitudinal—compression waves) and with electromagnetic radiation. Essentially, the condition here is that in a distance d only a fixed number of waves will fit.

Now imagine that the two fixed points are brought around together, and the ends of the string are fastened to form a continuous loop (Figure 10-5). Let the waves occur in the string as before, the string moving as a whole; that is, the waves do not travel along it. There may still be no node, one node, two nodes, and so on as in Figure 10-5a. What is not possible, with requirements of this kind, is a situation in which the number of half-wavelengths around the loop is not integral,

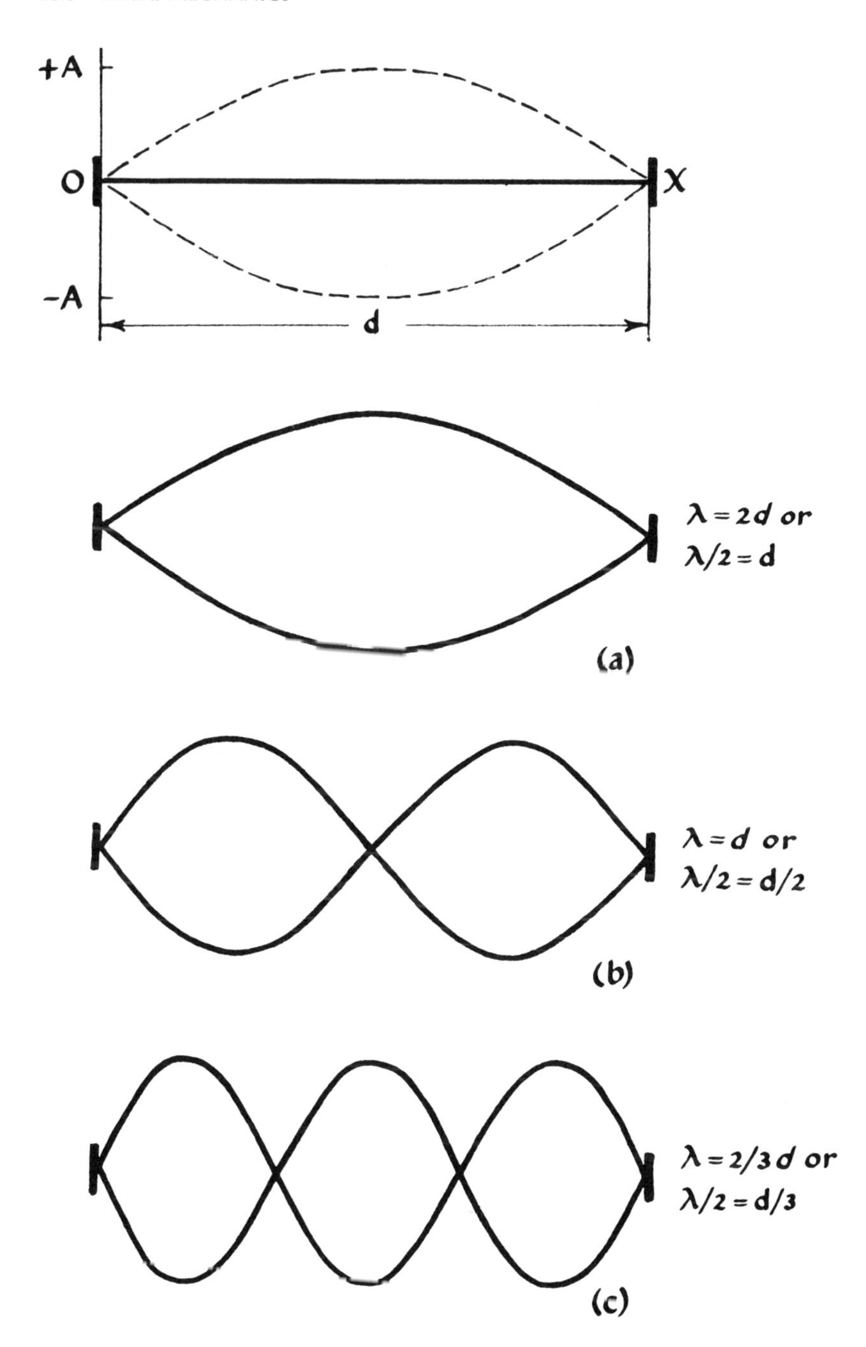

Fig. 10-4 *A string is stretched between two fixed points, 0 and* x = d; y *represents a displacement. The maximum displacement is the amplitude* A. *The fundamental mode of vibration and the first and second overtones are shown at* $\lambda/2$ = d, $\lambda/2$ = d/2, *and* S/2 = d/3.

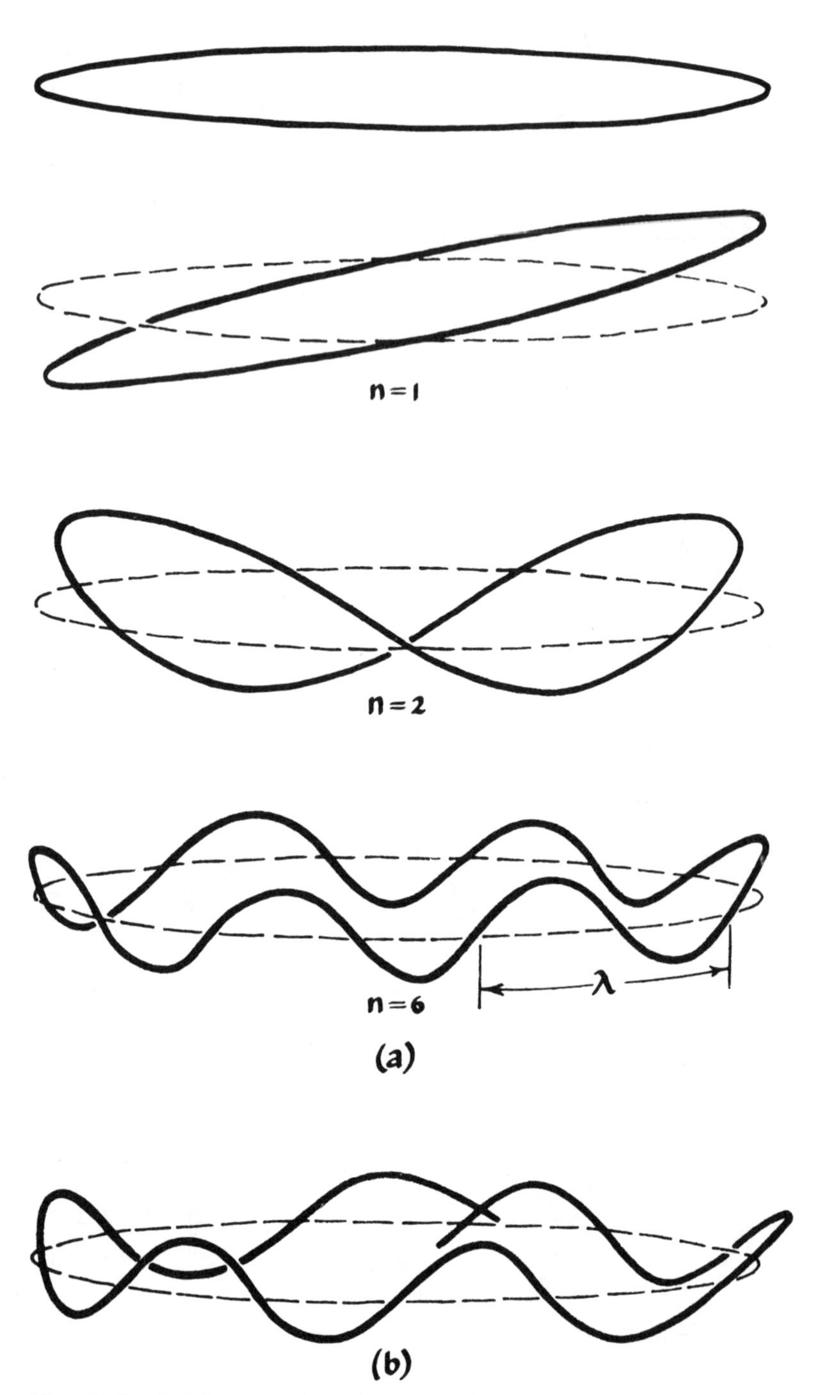

Fig. 10-5 *a) Waves on a loop of string. In* (a) *well-behaved waves for* n = *1, 2, and 6 are shown. The stationary states yield single values of the amplitude for any point on the circumference.* (b) *An ill-behaved wave is shown. The wave is out of phase on each circuit of the circle; a given point on the circumference may show different values of the amplitude at different times. This is not a stationary state.*

as in 10-5b. This would not be stable because overlapping parts of waves would interfere and cancel.

Suppose, now, that the orbit of the electron about a nucleus be likened to a loop of string. The farther the orbit from the nucleus, the larger r is, the larger the loop and the longer the path ($2\pi r$) for accommodating a set of waves. Suppose that an electron in its orbit must have a 'wavelength' such that only an integral number of full waves fit in the orbital path. The orbital path is $d = 2\pi r$. By Equation (10-5) $\lambda = h/mv$, and by Equation (10-6) $\lambda = 2d/n$. But now we wish to have an integral number of whole waves (not half-waves) in the distance $2\pi r$. So we write $\lambda = 2\pi r/n$. We substitute this value of λ in Equation (10-5), and obtain Bohr's equation:

$$2\pi rmv = nh \tag{10-2}$$

In Figure 10-5a are shown schematically four such wave-patterns for $n = 0, 1, 2, 6$. In the last of these, by Equation (10-2) $mvr = 6h/2\pi$, or rearranging, $2\pi r = 6h/mv$. Six waves fit into the distance $2\pi r$ and we are back at the de Broglie relation, Equation (10-5), with h/mv in the role of λ. We have a 'picture' of a standing electron wave (*i.e.*, not a travelling wave) which is overall time-independent and when in this state also non-radiating. One may imagine that radiation is involved as the electron moves from orbit to orbit: if the electron moves to a larger orbit energy is absorbed; if it moves to a smaller orbit energy is given off.

Likening the orbit, as a first approximation, to a loop with waves in two dimensions in it, one may improve the description to a second approximation by considering that the waves have available three dimensions, the third being toward and away from the nucleus. (Further, the waves do not have a two-dimensional loop but the surface of a sphere available to them. This is hard to visualize.) If we remove approximations entirely, the physical picture disappears, and we have only a mathematical expression that gives correct answers to physical experimental questions but is very far removed (in the C field of Figure 1-4) from the plane of perceptions, and from visual representation. We have played fast and loose with some of these concepts in order to provide at least an approximate visualizable representation of the theory.

Test of theory: the contribution of Davisson and Germer 10.6

The Davisson and Germer experiments involved shooting a narrow beam of electrons at the surface of the metal crystal to study how the electrons were reflected. All this was done in a high vacuum to avoid collisions with gas molecules. The narrow beam of electrons (Figure 10-6) is produced by an electron-gun. This is a tungsten fila-

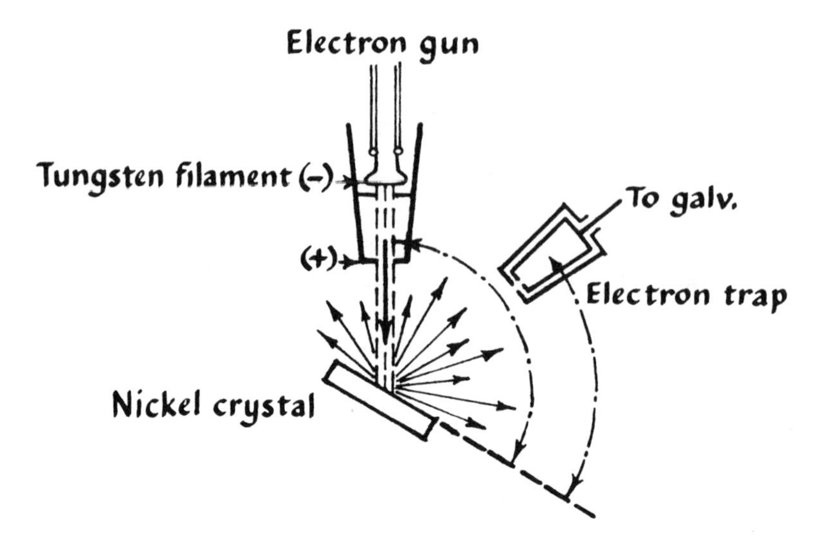

Fig. 10-6 *An electron gun with a tungsten wire filament directs a beam of electrons onto the face of a nickel crystal. The scattered electrons are collected in an electron-trap and the resulting current is measured. The gun, crystal, and collector are shown.* [*Redrawn from C. J. Davisson, "Are electrons waves?"* J. Franklin Institute 206, *597, 1928, Fig. 1, p. 600.*]

ment that is heated white-hot in a tube so arranged that the filament is at a high negative potential with respect to its container. Electrons 'boil' off of the filament in all directions, and are accelerated to the case of the 'gun.' At the front end there is a slit, and the electrons going in this direction have such high velocity that they shoot through as a narrow beam (as in a cathode-ray tube, Figure 3-8). This beam, striking the flat nickel surface, bounces off as shown. The electrons that bounce off in any given direction are measured by means of an electron-trap attached to a sensitive galvanometer. This may be moved through known angles to the plane of the nickel crystal, as shown in Figure 10-6. When measurements made over a range of angles are plotted against the angle relative to the nickel crystal, the pattern in Figure 10-7 is obtained, where each point represents the number of electrons per unit of time (the current) measured by the galvanometer. Evidently a 'peak' of current is found at the angle such that the angle of incidence equals the angle of reflection (Section 5.3). These electrons have the same speed as those shot from the electron gun. The others are slow, low-energy electrons knocked out of the nickel atoms by the electron beam.

It might quite expectedly be supposed that a stream of *particles* would bounce from a surface and exhibit this type of behavior. How-

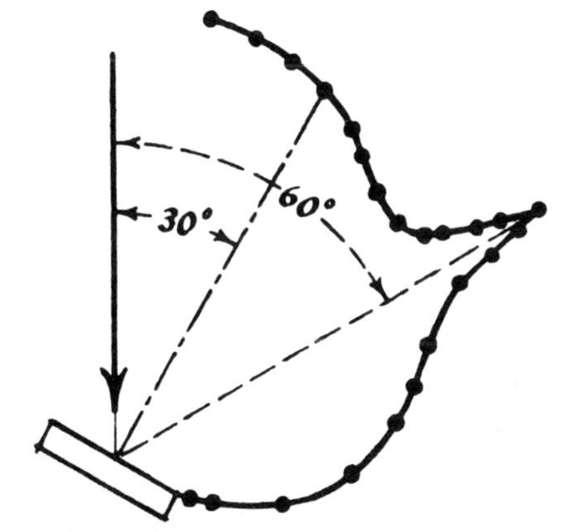

Fig. 10-7 *Typical curve showing a plot of the current measured versus the angle of the collector to the electron beam. The beam strikes the crystal at an angle of incidence 30°. The set-up is that of Figure 10-6. [From C. J. Davisson,* loc. cit., *Fig. 1, p. 600.]*

ever, it is unlikely that the electron is here behaving like a particle. Consider that its diameter is of the order of 10^{-15} m, while the diameter of the nickel atom is about 10^{-10} m, and the smallest distance between atoms in the nickel crystal is 2.48×10^{-10} m. It is like bouncing a stream of bird shot against a pyramid of cannon balls, said Davisson: regular reflection would not be expected, but scattering every which way. Or, more realistically, it is like expecting each electron to swing around a nucleus and fly off at exactly the same angle as every other, a most unlikely behavior. The reflected electrons do come out at a definite angle, and this implies a direction related to the plane of the surface. But it takes at least three points to determine a plane. How, then, can an electron, if its (particle) diameter is 10^{-15} m, be affected by three atoms (minimum) with diameters of 10^{-10} m? Only if it is behaving not as a particle of this small size but as an extended wave. The crystal surface then acts as in a Young's experiment (Section 5.5); the rows of atoms, regularly spaced, act as a set of slits. This wave-interpretation of the electron behavior agreed with the experimental findings.

A picture of the hydrogen atom 10.7

Under certain circumstances, then, a wave behavior can be demonstrated for the electron. The picture one is left with at this point is of an electron in orbit about the hydrogen nucleus. One thinks of the orbit of a planet, say, as being in a plane. The orbit of the electron comprises the entire sphere at the distance r from the nucleus. Further, r must be thought of as the most probable distance for finding the electron. The probability distribution along a radius is shown for the normal hydrogen atom in Figure 10-8. This is the probability that the

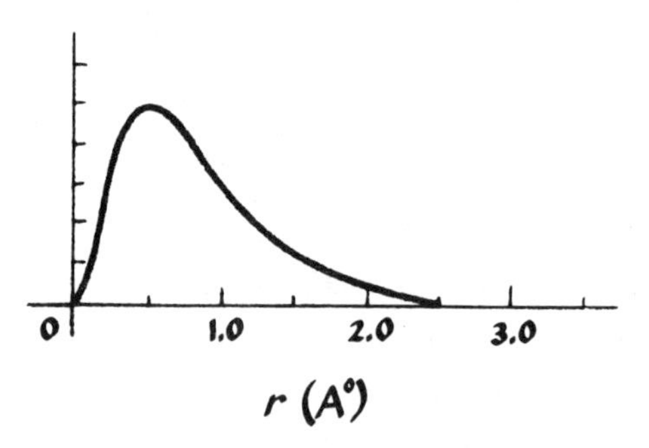

Fig. 10-8 *The probability of finding an electron at various distances* r *from the nucleus of the hydrogen atom when* n = *1.*

electron be found between r and $r + \Delta r$, where Δr is vanishingly small as a limit. There is a small possibility that the electron might be found at the nucleus—and indeed an orbital electron in more massive atoms occasionally falls into the nucleus. Most of the time the electron is about 0.5 Å away from the nucleus in hydrogen, and some of the time it may be as far out as 2.5 Å. But it is very unlikely with a stable atom to find it much farther out. Over a sufficient period of time all the possible regions of this sphere will have been occupied by the electron, but it will have been at the distance 0.5 Å more of the time than at any of the other distances.

10.8 Indeterminacy

The statement of the 'position' of the electron in probability terms is consonant with a principle advanced by Werner Heisenberg in 1927: the *indeterminacy principle* (also called the *uncertainty principle*). This has to do with the general problem of measurement. In physical investigations of these kinds one relies ultimately upon electromagnetic radiation to find where an object is, or to map its dimensions. When the object is of the dimensions of an electron, the radiation used to locate it must be of fairly short wavelength in order to be reflected from it. Suppose, for example, that one had instruments which could be tuned to any wavelength, and an electron gun that would fire out an electron (as Heisenberg imagined in a thought-experiment). The problem is to track the electron and measure its position and velocity at any time. Suppose that the electron is fired along a track which is scanned by two instruments a certain distance apart. By measuring the time taken to traverse the distance one could obtain the velocity of the electron—as is done for automobiles with traffic counters and speed-detectors. But we saw in the Compton effect (Section 8.7) that a short wave-length photon, striking a relatively light particle such as an electron, is reflected with loss of energy (*i.e.*, its wavelength increases) while the particle recoils. If the particle were in motion, the recoil would usually produce a change of course. But the direction of recoil is unpredictable, and therefore the search for the particle at the second instrument would lead to indeterminate results. The precise information that the electron was at *a* certain point under the first instrument, gained by noting the reflected photon, simultaneously prevented determining its velocity. The measuring instrument interacted with the thing measured in a way that introduced an indeterminacy into the result.

Heisenberg showed that the indeterminacy in such a measurement, where Δx is the range in the measurement of position, and $\Delta(mv)$ that in the momentum, has a *minimum* value:

$$\Delta x \cdot \Delta(mv) \text{ at } x \cong h \qquad (10\text{-}8)$$

The smaller Δx is made, the more the uncertainty in mv, and vice versa. The measurement might, of course, be poorly made, with inadequate equipment. This would give a greater uncertainty to the result than were it well made with the most suitable and precise instruments. But even the best measurements cannot do better than the principle allows. George Gamow has given a nice illustration of these ideas. Figure 10-9 shows the trajectory of a moving particle as observed with quanta of three different wavelengths. The broken line indicates a trajectory.

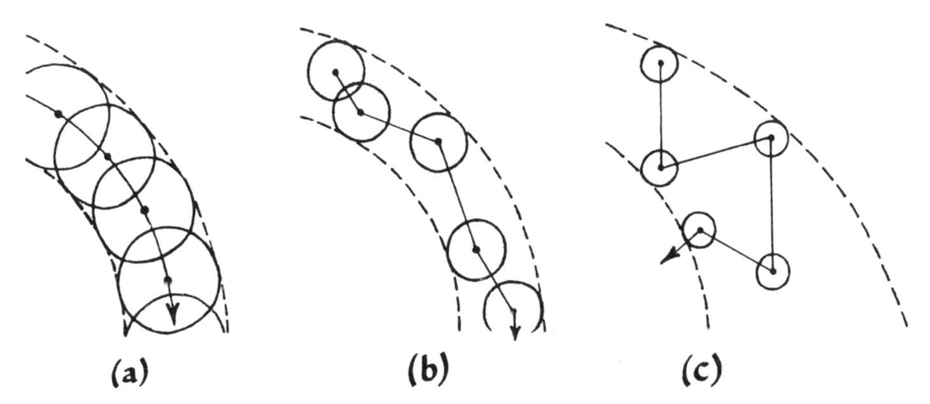

Fig. 10-9 *Trajectory of a moving particle as observed with quanta of three wave-lengths:* (a) *longest;* (b) *medium;* (c) *shortest.* [*From G. Gamow,* Biography of Physics. *Harper & Row, Fig. VII-21, p. 257.*]

In (a) quanta of low energy and long wavelength do not greatly disturb the trajectory, but indicate the position and hence the path of the particle only hazily. In (b) quanta with medium energy narrow the path but somewhat disturb the trajectory. In (c) high energy, short wavelength quanta sharply locate the particle but greatly disturb its trajectory. Thus the trajectory is to an extent indeterminate in every case.

10.9 Some implications

The Heisenberg principle has been discussed a great deal by philosophers in relation to ideas of causality, and rather extreme statements are sometimes encountered in the literature of science and philosophy concerning the meaning or implications of this theory. There are at least two points of view about the meaning of the principle of indeterminacy. One view seems to be that all the data of physical science can be reduced to calculations of the probability distributions of the possible phenomena allowed within operative constraints as, for example, the probability distribution of the electron about the hydrogen nucleus (Figure 10-8). Correlated with this view is the renunciation of causality and continuity at the atomic-nuclear level, and of any way of picturing the phenomena except by pure mathematical symbolism. In other words, probabilistic concepts have replaced deterministic, and now "everything" may be explained in quantum theoretic terms.

The other point of view says, in effect, that this is far too reminiscent of the statements of classical scientists who thought at the end of the nineteenth century that now everything could be explained by Newtonian mechanism. In this view indeterminate mechanism has replaced determinate mechanism as the ultimate theory. This is conceived to be a premature conclusion, for who knows what surprises lie below the level at which indeterminacy is important.

10.10 Summary

In the subject of this chapter we have an example of scientific method. First there was Rutherford's hypothesis of the nuclear atom: vague in many ways, and a marked departure from most thinking of the time. It did, however, fit certain experimental observations, observation that there was something present in matter that was massive and extremely small. Yet it went counter to the classical Coulomb laws because it suggested that electrons close to a positive nucleus did not fall into it. By a combination of classical and quantum arguments Bohr gave so plausible an explanation of why electrons did not fall into the nucleus, and in addition (as we shall see in detail in the following chapter) so added to the experimental data that was fitted, that it was clear that although many experimental findings still remained unexplained, theory was proceeding on a fruitful path. The last vestiges of classical ideas were eliminated from the theory of the atom's structure when the construct of matter waves was invented by de Broglie and experimentally supported by Davisson and Germer. Schrödinger and Heisenberg pioneered theories of atomic structure which have continued to evolve since. Of course, hundreds of scientists contributed to these developments, adding to theory, modifying it, finding new experimental support and new phenomena to be explained. Quantum mechanics is the current theory of physical science of the nuclear-atomic level that has been most generally successful and accepted.

Exercises

10.1 In the Bohr model a) What, in words, is the relation between the kinetic and the potential energy of the electron? b) If the electron (classically) spiraled down to the nucleus, what would happen to its kinetic and potential energies?

10.2 How did Bohr simplify his problem in assuming an electron moving with a speed small compared with that of light?

10.3 Give a free translation of the following passage from de Broglie:

> Bref, j'ai développé des idées nouvelles pouvant peut-être contribuer a hâter la synthèse nécessaire qui, de nouveau, unifiera la physique des radiations aujourd'hui si étrangement scindées en deux domaines où règnent respectivement deux conceptions opposées: la conception corpusculaire et celle des ondes. J'ai pressenti que les principes de la Dynamique du point matériel, si on savait les analyser correctement, se présenteraient sans doute comme exprimant des propagations et des concordances de phases et j'ai cherché, de mon mieux, à tirer de là, l'explication d'un certain nombre d'énigmes posées par la théorie des Quanta. En tentant cet effort je suis parvenu à quelques conclusions intéressantes qui permettent peut-être d'ésperer arriver à des résultats plus complets en poursuivant dans la même voie. Mais il faudrait d'abord constituer une théorie électromagnétique nouvelle conforme naturellement au principe de Relativité, rendant compte de la structure discontinué de l'énergie radiante et de la nature physique des ondes de phase, laissant enfin à la théorie de Maxwell-Lorentz un caractère

d'approximation statistique qui expliquerait la légitimité de son emploi et l'exactitude de ses prévisions dans un très grand nombre de cas.

[Louis de Broglie, *Annales de Physique* Series 10, *3*, 22 (1925), p. 127.]

10.4 Give a free translation of the following passage from Schrödinger:

In dieser Mitteilung möchte ich zunächst an dem einfachsten Fall des (nichtrelativistischen und ungestörten) Wasserstaffatoms zeigen, dass die übliche Quantisierungsvorschrift sich durch eine andere Forderung ersetzen lässt, in der kein Wort von "ganzen Zahlen" mehr workommt. Vielmehr ergibt sich die Ganzzahligkeit auf dieselbe natürliche Art, wie etwa die Ganzzahligkeit der *Knotenzahl* einer schwingenden Saite. Die neue Auffassung ist verallgemeinerungsfähig und rührt, wie ich glaube, sehr tief an das wahre Wesen der Quantenvorschriften.

10.5 Would you expect a radio wave of 1 megacycle frequency to exhibit an effect like the Compton effect?

10.6 What momentum would be calculated for a radio wave of 1 megacycle frequency?

10.7 A bullet of 20 g mass and a length of 2 cm travels with a speed of 300 m/sec. If its speed is measured to 1 part in 10^5, and the Heisenberg principle in this instance is written $\Delta v \cdot \Delta x = h/m$, what is the calculated uncertainty in its position? Is this measurable?

Answers to Exercises

10.1 a) The potential energy is measured by the orbit level, being constant when the electron is in a fixed orbit; the kinetic energy is the energy of motion in the orbit. ($\mathbf{E}_k = Ke^2/2r$; $\mathbf{E}_p = -Ke^2/r$).
b) As it gets closer to the nucleus, it goes faster; but it also radiates energy as the negative potential energy becomes numerically larger: *i.e.*, as it loses energy.

10.2 He could thereby ignore relativistic effects.

10.3 In brief, I have developed new ideas which may perhaps contribute to speed up the necessary synthesis which once more will unify the physics of radiation today so strangely cleaved into two regions in which there reign, respectively, two opposed conceptions: the particle conception and that of waves. I have felt that the dynamic principles of a material particle, if one is able to analyze them correctly, show themselves without doubt as expressing propagations and constructive interferences of phases, and I have sought, as best I could, to extract from thence the explanation of a certain number of puzzles presented by the theory of quanta. In pursuing this effort, I have arrived at several interesting conclusions which perhaps permit the hope of reaching more complete results by pursuing the same path. But it will first be necessary to develop a new electromagnetic theory which fits in a natural way to the principle of relativity, taking account of the discontinuous structure of radiant energy and of the physical nature of phase waves, and finally, permitting to the Maxwell-Lorentz theory a character of statistical approximation which explains the legitimacy of its use and of the exactness of its predictions in a very large number of cases.

10.4 In this communication I show first for the simplest case of the (non-relativistic and non-perturbed) hydrogen atom that the conventional quantization rules may be replaced by another formulation in which no word of 'whole numbers' appears. Rather, the whole-number-ness turns out to be of the same kind as the whole-number ness of the nodes of a vibrating string. The new concept is capable of generalization and strikes, as I hope, very deep into the true essentials of quantum theory. [E. Schrödinger, "Quantisierung als Eigenwertproblem," *Annalen der Physik*, *79*, 361 (1926).

10.5 One megacycle frequency corresponds to a wavelength of $3 \times 10^8/10^6 = 3 \times 10^2$ m. It would probably pass right over the electron.

10.6 $\lambda = h/mv$; $mv = h/\lambda = 3.3 \times 10^{-36}$ kg m/sec.

10.7 $\Delta x = h/m\,\Delta v = 6.6 \times 10^{-34}/0.02 \times 0.003 = 1.1 \times 10^{-29}$ m. This is not a measurable quantity.

Bibliography

C. J. Davisson, "Are Electrons Waves?" *Journal of the Franklin Institute*, *206*, 597 (1928). This is an account, written for the layman, of his and L. H. Germer's work.

George Gamow, "The Principle of Uncertainty," *Scientific American*, January, 1958. (W. H. Freeman & Company Reprint, No. 212) This is a delightful account of the matter. The student with philosophic leanings might enjoy David Bohm, *Causality and Chance in Modern Physics*, Harper Torchbooks, Harper & Brothers, 1961 (Van Nostrand, New York, 1957), *cf.* pp. 102–103. Bohm takes the view that we have arrived at no ultimates as yet.

Chapter 11

11.1 Review
11.2 One consequence
11.3 The wave-mechanical model
11.4 The Pauli exclusion principle
11.5 The Periodic Classification of the elements embodied in the Periodic Table
11.6 Interpretation of the Table
11.7 Summary

Atomic Properties ~ 11

In this chapter we look at some of the implications of atomic structure: at the properties of atoms that allow them to form molecules (Chapter 14).

Review 11.1

We have the picture of an atom of an element as a (+)-charged nucleus surrounded by a number of electrons equal to the (+) charges. The electrons occupy energy levels (Figure 10-3). They are moved out to higher levels—or entirely out of the atom—when they absorb energy. When they fall from a higher to a lower level they give off energy ($\Delta\mathbf{E}$) in the form of radiation with a frequency such that $\Delta\mathbf{E} = h\nu$. We saw in the photoelectric effect evidence that light of sufficient energy can expel electrons from atoms. This behavior explains, qualitatively, certain behaviors merely described previously. For example, in a Crookes tube, a stream of electrons flows at high velocity from cathode to anode, and its path is marked by a ribbon of light. What is happening is that the electrons, behaving as projectiles, hit gas atoms or molecules and knock electrons out of them, ionizing them. When these or other electrons fall back into the ions light of the appropriate wavelength is given off. When the cathode stream in J. J. Thomson's tube strikes the glass, a patch of light appears. This is because electrons in the glass molecules are knocked out of orbits. Falling back in, they radiate light. One might think of these gas atoms and glass molecules as machines for converting electrical energy into light.

One consequence 11.2

We are now able to explain the behavior discussed in Section 8.2. Think of a bar of metal, say, iron. It is composed of atoms of atomic number 26. The 26 electrons are arranged in shells, as we shall see below. Those closest to the nucleus are held tightly and those farther away less tightly. Farthest away are electrons so loosely held that they can be pushed around: they will flow as an electric current when the metal is placed in an electric field—say, connecting the poles of a battery.

We heat the *solid* metal bar, putting in energy. This increases the jittering of the atoms (which even in the solid metal are not absolutely motionless). As they jitter they radiate electromagnetic energy, for as we saw (Section 7.7) an oscillating charge broadcasts electromagnetic radiation. There is a range of energies present because some atoms have hit others, losing energy to them, and these with more energy may have it increased or decreased by further collisions. So the swarm of dancing atoms gives off a range of wavelengths of radiation. As heating continues, this range which was in the low-energy, infrared region of the spectrum extends more and more towards and into the more energetic, higher frequency, shorter wavelength visible. Now the jittering is so violent that electrons are being knocked out of orbit, and are falling back in, cascading down the ladder of energy levels, and giving out characteristic radiation at each energy-fall. The result is the emission of a continuous spectrum of wavelengths.

When a *gas* at low pressure is heated in this way—say, by passing an electric discharge through it as in a Crookes tube—it is found that the spectrum is not continuous, but consists of a set of bright lines, an 'emission spectrum' (Figure 11-1). When hydrogen is examined, its

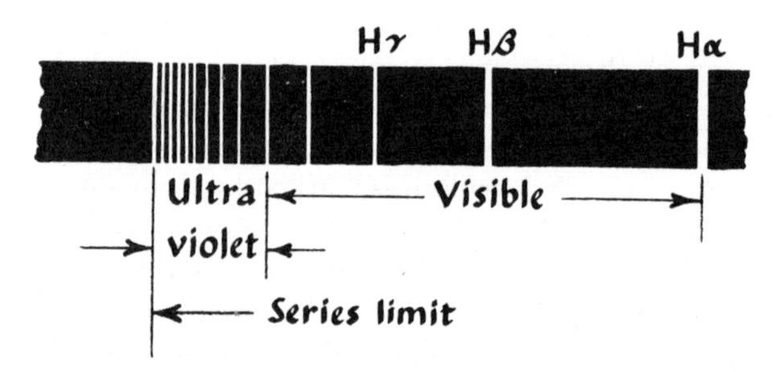

Fig. 11-1 *Part of the line spectrum of glowing hydrogen gas. The Balmer series.*

emission spectrum is found to consist of sets of these lines which can be correlated to transitions between energy levels (Figure 11-2). The sets, or series, are given names of the men who first discovered them. The orderly sets of lines led these experimenters to conclude that within the atom electrons must be arranged in an orderly manner. One of

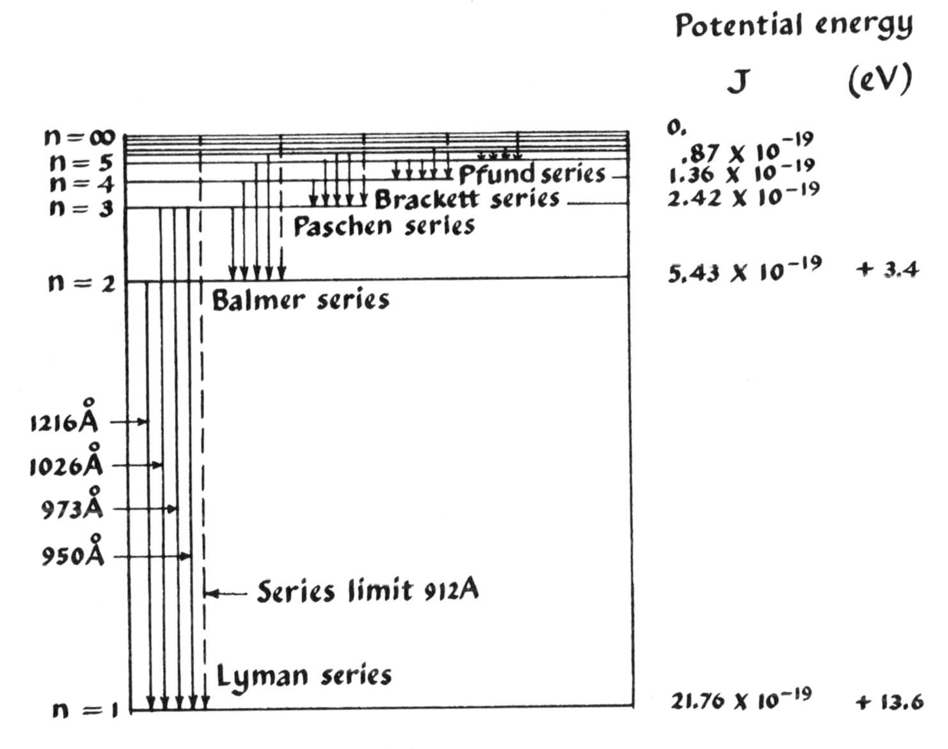

Fig. 11-2 *Energy level diagram for hydrogen.*

the triumphs of Bohr's theory is that it enables the calculation of the wavelength of every one of these lines. That it failed with other atoms was one of the anomalies we spoke of that led eventually to the development of wave-mechanics and more modern notions of atomic structure.

The wave-mechanical model 11.3

In Section 10.5 we discussed the essence of the wave-mechanical approach. The Schrödinger wave equations that describe the probabilities that electrons will be found in certain positions in shells about any nucleus are mathematically highly complex. They allow the radii of various orbits to be correlated to transitions between one orbit and another. In favorable cases they also allow the calculation of the intensity at a given wavelength. In all emission spectra there are found lines of different intensity: some very bright, some less bright, and some very faint relative to each other. Wave-mechanics interprets these differences in terms of probabilities. Some transitions from one orbital to another are more probable than others, and thus as there occur more of these per second in a given mass of radiating atoms, the brightness of these lines will be greater. In this way experimental data are excellently fitted.

A given electron, as we saw, is conceived to occupy a region of space about the nucleus such that at any given instant there is a finite probability of finding it at some particular position in that region. For example, the electron of a hydrogen atom is found in a spherical region about the nucleus a radial section through which is shown in Figure 10-11. As we saw, the greatest probability of finding the electron is at $r = 0.52$ Å, corresponding to the Bohr radius. The physical picture is vague, and indeterminacy ensures this. It can, however, be stated that the $n = 1$ region in which the electron is most likely to be found is spherically symmetrical. There is no preferred radial direction, though there is a preferred radial distance, which accounts for the concept of an orbit. From this point on, however, the old term 'orbit' will be dropped and the term 'orbital' will be used as a noun to designate the region in space about the nucleus in which the electron will probably be found. Then we would say that the orbital of the $n = 1$ electron of hydrogen is spherical. We shall see that at levels where n is larger than 1, orbital structure becomes more and more complex with increase in the value of n. A set of orbitals with the same value of n is named a 'shell.'

The wave equation for calculating the shape and position for an orbital contains four important numbers, the 'quantum numbers' as they are called, and we can readily grasp the numerology without having to calculate the equation itself. The reason we must become involved with this numerology is that we shall make good use of it when we look at how molecules are formed, and why they have the shapes they do (Chapters 14, 15). The number n is the principal quantum number. It corresponds with Bohr's orbital level n, and may take values 1, 2, 3, . . . The highest value of n is 7, found in the atoms with Z numbers 87 and above. The number n sets the most probable distance of the shell from the nucleus. A number l, which may take values of 0, 1, 2, . . . up to $n - 1$, is associated with the angle at which the highest probability of finding the electron lies with respect to a coordinate system centered on the nucleus. For example, when $l = 0$, which is the case for $n = 1$, since l is limited to numbers up to $n - 1$, the orbital is spherically symmetrical and has no preferred direction. When $l = 1$, there are three orbitals possible, as shown in Figure 11-3. These are symmetrically arranged double-spheres lined up at right angles to each other. They are distinguished by a quantum number m_l which may take values of 0, ± 1, ± 2, . . . up to $\pm l$. Thus when $n = 1$, $l = 0$ and $m = 0$. These are the only possibilities which allow the wave

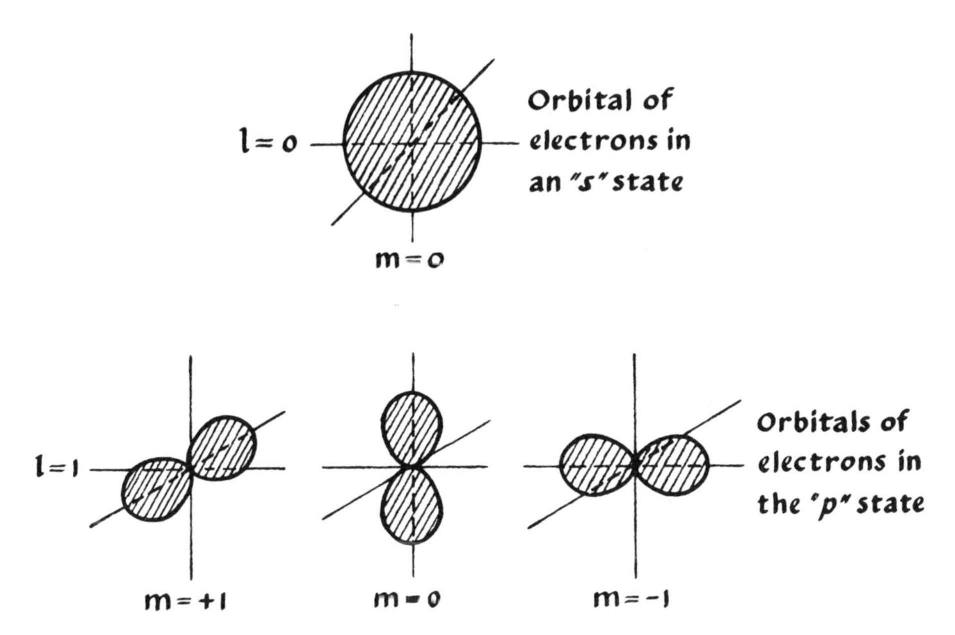

Fig. 11-3 *Sketches of approximate shapes of the* s *and* p *orbitals.*

equation to be solved for $n = 1$ electrons. When $n = 2$, l may take values of 0 or 1, and m_l may take values of $+1$, 0, and -1. This is visualized as follows. The shell designated by $n = 2$ has two kinds of orbitals, a spherical one designated $l = 0$, $m = 0$, and a set of three bi-lobed orbitals as shown in Figure 11-3. It is these orbitals that are distinguished by the three m values $+1$, 0, -1. It has become conventional to refer to electrons for which $l = 0$ as being in an s state. Those electrons with $l = 1$ are said to be in a p state. The designations, which we shall find very convenient, continue: for $l = 2$, d state, for $l = 3$, f state. (And so on through g and h, though these are of no practical value to us.) Following the rules given so far, when $l = 2$, five orbital configurations become possible, namely $m = +2, +1, 0, -1, -2$. These are sketched in Figure 11-4.

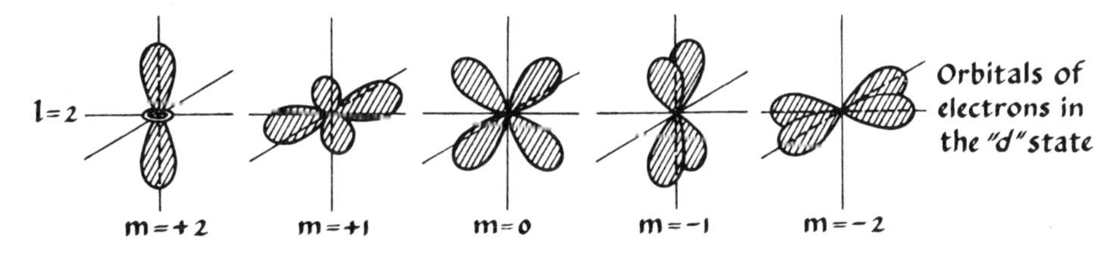

Fig. 11-4 *Sketches of* d *orbitals.*

11.4 The Pauli exclusion principle

The fourth quantum number is m_s. It can take one of only two values, $+1/2$ or $-1/2$. These values are correlated with whether thc electron is spinning in a clockwise or a counter-clockwise direction (Figure 4-7). By this point, however, it is evident that the concept of spin attached to a matter-wave is not a visualizable one. However, the electrons behave in a magnetic field *as though* they have such spins (further splitting of lines) and this sanctions the convenient image.

In 1925 Wolfgang Pauli introduced the assumption that in any given atom no two electrons may have the same set of values for the four quantum numbers n, l, m_l, and m_s. This has been a most powerful tool for rationalizing physical and chemical knowledge. For when the Bohr theory was advanced it was not at all clear why, as atomic numbers increased, the orbital electrons should not all crowd into the lowest shell. It was experimentally quite definite that they did not. For one thing, if they were crowded together as the atomic number increased and the central plus charge on the nucleus increased, the electrons should be drawn ever closer to the nucleus. Thus the volume of the atom should not change very much with atomic number. Yet it does (Figure 11-5), in what is certainly a periodic fashion. Moreover, the steady accumulation of electrons in an inner shell can in no way account for chemical periodicity, discussed below. The Pauli principle, together with wave mechanics, brought the picture into focus.

In calculating the orbital—the probability of finding an electron at a certain distance and with a certain orientation in the atom—the Pauli principle allows every orbital, designated by n, l, and m_l numbers to contain two electrons, designated $+1/2$ and $-1/2$. (In polar coordinates every point requires a distance correlated to n and two angles correlated to l and m_l to specify it.) These electrons can be imagined to be coupled, and so stabilized. For example, helium $_2$He has, as we saw in Section 9.3 and now explain, two electrons in the first shell. These fill the first shell, and their quantum numbers are (1, 0, 0, $+1/2$) and (1, 0, 0, $-1/2$). It is common practice among chemists to suggest this pairing by means of two little arrows, implying directions of spin of the electrons. Then helium might be written He: or He $\downarrow\uparrow$. The Pauli principle explains the filling of shells. The first is filled by two electrons. The second is filled by two s and six p electrons. The p orbitals are commonly designated p_x, p_y, p_z; each may contain a maximum of two electrons. The next shell may contain two s, six p, and ten d electrons, for a total of eighteen. The structures of the first ten atoms are shown in Table 11-1, where the orderly build-up is clear. As the Z number increases stepwise by units of one, so a new electron goes into the orbital structure; into the lowest energy orbital available. When

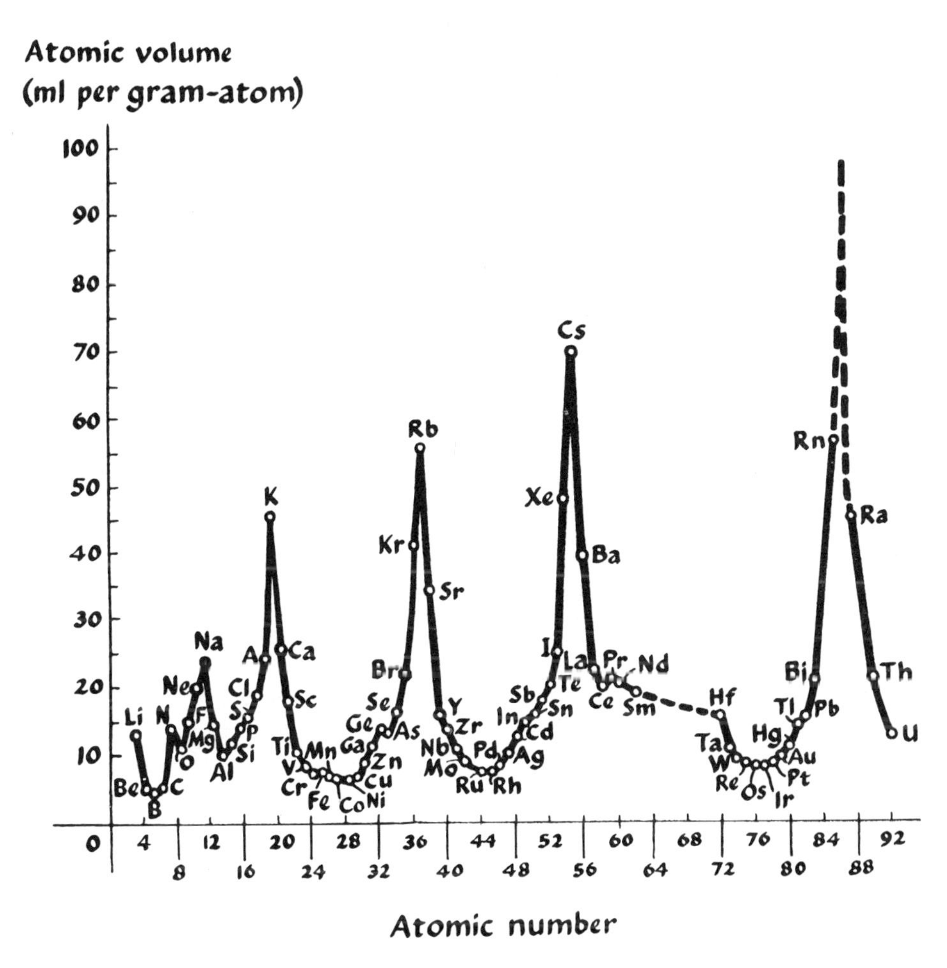

Fig. 11-5 *Atomic volume and its change with atomic number.*

a level n is filled, a new level starts. With heavier atoms (not shown in the table) there is some irregularity in filling orbitals and shells because some higher-numbered energy levels may fall below lower-numbered ones.

We shall use the term 'ground state' to describe an electron in the lowest energy level that is available to it. The electron of the hydrogen atom is in the ground state when its n, l, m numbers are 1, 0, 0. If it has absorbed energy and been lifted to some other level, say 2, 0, 0, it is said to be in an 'excited state.' The ground state is stable, an excited state is not. The 2, 0, 0 electron of lithium is in the ground state when the $n = 1$ shell is filled. In Table 11-1 the convention is to write the state of an electron in shorthand: an electron in $n = 1$ can only be a

Table 11-1

Structures of Shells of Some Atoms

Principal Quantum No.:	$n = 1$	$n = 2$				Element	Z
Orbitals	1s	2s	$2p_x$	$2p_y$	$2p_z$		
	↑					H·	1
	↑ ↓					He	2
	↑ ↓	↑				Li·	3
	↑ ↓	↑ ↓				Be:	4
	↑ ↓	↑ ↓	↑			·B:	5
	↑ ↓	↑ ↓	↑	↑		·Ċ:	6
	↑ ↓	↑ ↓	↑	↑	↑	·N:	7
	↑ ↓	↑ ↓	↑ ↓	↑	↑	·O:	8
	↑ ↓	↑ ↓	↑ ↓	↑ ↓	↑	:F:	9
	↑ ↓	↑ ↓	↑ ↓	↑ ↓	↑ ↓	Ne	10

Note: Each electron is represented by an arrow showing its relative direction of 'spin.'

1 *s* electron. Every level has *s* electrons, thus there are 2 *s*, 3 *s*, 4 *s* . . . electrons. Starting with shell 2, all shells have *p* electrons: 2 *p*, 3 *p*, and so forth. Starting with shell 3, *d* electrons become possible in the ground state.

11.5 The Periodic Classification of the elements embodied in the Periodic Table

Almost fifty years before Bohr proposed his model of the atom, and almost sixty years before wave mechanics came into existence, a remarkable classification scheme for all the then known chemical elements had been proposed. Dimitri Mendeleev and Lothar Meyer independently published schemes, but Mendeleev's was superior because he not only classified all the known elements but he left gaps in his classification scheme at points where, he felt, elements were missing —as yet undiscovered. He then, on the basis of the classified facts, predicted the properties of the then undiscovered elements and suggested in what ores they might be found. His predictions were strikingly confirmed.

The Periodic Classification was a purely empirical discovery, made by classifying elements according to their properties: color, melting temperature, boiling temperature, solubility of compounds in water, alcohol, and other solvents, formulas of compounds, and so on. As knowledge of atomic structure and properties accumulated, so accumulated a more fundamental basis for the Classification. Today this scheme is considered a powerful aid to chemistry and a philosophically important statement in chemical symbolism. Basically, Mendeleev's table still stands.

The Periodic Table (Table 11-2) classifies the elements in Groups and Periods. The period is correlated to the highest *n* number at which electrons are found in the ground state. Thus, H· and He: comprise period 1. Li· to Ne (Table 11-1) comprise period 2. The Periods run horizontally across the table, and there are seven of them. The Groups are somewhat more complicated. They comprise the vertical columns of the table.

The Periodic Classification of the elements is a 'natural classification.' A natural classification, according to the philosopher J. S. Mill, is a scheme that classifies a property so basic that all other properties of the entities are thereby classified.[1] In Mendeleev's Periodic Table atomic weight is classified, and the chemical properties of the elements turn out to be, to a close approximation, periodic functions of atomic weight. Nowadays atomic number is the basis of classification, and anomalies that Mendeleev had to disregard or claim to be errors are explained. For example, if the order of atomic weight contravened the evidence of chemical behavior, Mendeleev followed the chemical and predicted that the analytical measurement was in error. In other cases he found that by putting chemically similar elements in groups he would have to skip one or more places. He suggested the properties and atomic weights of the elements which he said were missing. Today we can see, classifying on the basis of atomic number, how right he was, and how the existence of isotopes helps to explain anomalies. For example, in the modern Table, tellurium and iodine, with atomic numbers 52 and 53, would be out of order if classed strictly on a mass number basis: tellurium has two isotopes with mass numbers 128 and 130, each accounting for about 1/3 of the total atoms in the element, and its atomic weight is 127.61; the atomic weight of iodine is 126.91, and it is composed of a single isotope. Iodine would be out of order on a strict atomic weight classification. But its atomic number shows that iodine does indeed follow tellurium logically—a fact that its chemistry clearly indicates. The Periodic Classification presents us with a picture of periodicity, and an hierarchical arrangement based on complexity and organization. In one sense, Science is a search for basic properties by which phenomena may be classified.

The chemical behavior of an atom, as we saw in a preliminary

---

[1] Edward F. Haskell called my attention to J. S. Mill, and these relationships. See his Scientia Generalis for a more basic scheme.

Table II-2

Periodic Table of the Elements

Atomic weights are based on the most recent values adopted by the International Union of Chemistry. (For artificially produced elements, the approximate atomic weight of the most stable isotope is given in brackets.)

Group→	I	II	III	IV	V	VI	VII		VIII		O
Period 1				1 H 1.0080							2 He 4.003
2	3 Li 6.939	4 Be 9.012	5 B 10.81	6 C 12.011	7 N 14.007	8 O 15.999	9 F 19.00				10 Ne 20.183
3	11 Na 22.99	12 Mg 24.31	13 Al 26.98	14 Si 28.09	15 P 30.974	16 S 32.064	17 Cl 35.453				18 A 39.948
4	19 K 39.102	20 Ca 40.08	21 Sc 44.96	22 Ti 47.90	23 V 50.94	24 Cr 52.00	25 Mn 54.94	26 Fe 55.85	27 Co 58.93	28 Ni 58.71	
	29 Cu 63.54	30 Zn 65.37	31 Ga 69.72	32 Ge 72.59	33 As 74.92	34 Se 78.96	35 Br 79.91				36 Kr 83.80
5	37 Rb 85.47	38 Sr 87.62	39 Y 88.91	40 Zr 91.22	41 Nb 92.91	42 Mo 95.94	43 Tc [99]	44 Ru 101.1	45 Rh 102.91	46 Pd 106.4	
	47 Ag 107.870	48 Cd 112.40	49 In 114.82	50 Sn 118.69	51 Sb 121.75	52 Te 127.60	53 I 126.90				54 Xe 131.30
6	55 Cs 132.91	56 Ba 137.34	57–71 Lanthanide series*	72 Hf 178.49	73 Ta 180.95	74 W 183.85	75 Re 186.22	76 Os 190.2	77 Ir 192.2	78 Pt 195.09	
	79 Au 197.0	80 Hg 200.60	81 Tl 204.37	82 Pb 207.19	83 Bi 208.98	84 Po [210]	85 At [210]				86 Rn [222]
7	87 Fr [223]	88 Ra 226.05	89— Actinide series**								

*Lanthanide series:	57 La 138.91	58 Ce 140.12	59 Pr 140.91	60 Nd 144.24	61 Pm [147]	62 Sm 150.35	63 Eu 151.96	64 Gd 157.25	65Tb 158.92	66 Dy 162.50	67 Ho 164.93	68 Er 167.26	69 Tm 168.93	70 Yb 173.04	71 Lu 174.97
**Actinide series:	89 Ac [227]	90 Th 232.04	91 Pa [231]	92 U 238.03	93 Np [237]	94 Pu [242]	95 Am [243]	96 Cm [245]	97 Bk [249]	98 Cf [249]	99 Es [253]	100 Fm [255]	101 Md [256]	102 No	103

way in Section 9.4, is almost entirely a reflection of the states of its outer electrons, that is, of the electronic state of the outermost ground-state shell. With H·, and He:, this is the first shell; throughout, the ground-state shell gives the period its number. The Group comprises elements that are similar in chemical behavior. The atoms in a Group will therefore have similar outer-shell electron structures, and it is on these that the chemist normally concentrates. If a shell is filled, it is of little chemical interest, and is not explicitly shown in the symbol. Thus lithium is written Li· as we saw; the Li implying the nucleus *and* filled first shell. Na· implies two filled shells and a nucleus; the electron attached to the symbol is the start of a new (outer) shell. The nucleus with its filled or stable shells is often referred to as a 'kernel'; the stable, filled structure of electrons is a 'core.' Thus the kernel of lithium has a helium core; the kernel of sodium has a sodium nuclide and a neon core. By combining the data of Tables 11-1 and 11-2 one can see how beautifully the whole scheme fits together.

Interpretation of the Table 11.6

We defer discussing hydrogen since it has special properties on account of being a unit (+) combined with a unit (−) charge. Consider the elements in Group IA. They form a family. All are metallic, and all react with water to form caustic, alkaline solutions. The family is, indeed, known as the alkali metals. All the members of this family behave similarly in compound formation. For example, if we let 'M' represent any of these atoms from Li to Fr in Group IA, each reacts with hydrogen to give MH, an unstable hydride; all react with chlorine to form stable, water-soluble crystalline chlorides with the formula MCl. (Table salt is NaCl, sodium chloride.) All form oxides of the formula M_2O, that is, two atoms of the metal fastened to one oxygen. All form nitrates, $MONO_2$; all form sulfates $MOSO_2OM$; all form hydroxides MOH, and so on. Another family of elements that was recognized early enough to receive a special name is the alkaline earths, IIA; another the halogens, VIIIA.

Chemists divide elements into three groups: metals, non-metals, and borderline cases. Metals are characterized by an ability to conduct electricity. This reflects the presence of electrons that are easily detached from an atom in the presence of other atoms. In the Periodic Table the most metallic elements are found in the lower left-hand part. The whole left part of the table up to Group IIIA is comprised of metals. These two observations are interpreted by noticing that (for example) the elements in group IA and IIA have respectively one and two outer shell electrons in the *s* state. These, and particularly a lone one as in the alkali family of metals, tend to be readily lost. Strip off the outer *s* electron from a Group IA atom and there remains a kernel with a filled, very stable core. Moreover, as the Z number goes up, the outermost electrons find themselves further and further from the nucleus. It is true that the nuclear charge goes up, but also the concomitantly filling intermediate shells, filled with negative electrons, serve to shield the outer electrons. Thus as Z increases in a family, metallic property increases. This is why cesium and francium are among the most metallic elements, and are more metallic than, say, lithium and sodium. This atomic property is correlated with photoelectric behavior. Cesium loses electrons when irradiated with visible light of a longer wavelength (*i.e.* with less energy) than that which expels electrons from any other common metal. Ionization potentials for a number of representative elements are shown by families in Table 11-3. The elements on the left lose electrons more readily than

Table 11-3

Ionization Potentials for the First Electron Removed from Some Atoms (in Electron Volts) *

Li 5.390	Be 9.321	C 11.265	F 17.422	Ne 21.559
Na 5.138	Mg 7.645	Si 8.149	Cl 12.959	Ar 15.756
K 4.340	Ca 6.112	Ge 8.126	Br 11.844	Kr 13.996
Rb 4.176	Sr 5.693	Sn 7.332	I 10.44	Xe 12.127
Cs 3.893	Ba 5.2097	Pb 7.415		Rn 10.746

* From Gerhard Herzberg, *Atomic Spectra and Atomic Structure*, Trans. J. W. T. Spinks. Dover, New York (1944). Reprinted in paperback from the Prentice-Hall edition of 1937, with some corrections and additions. Table 18, pp. 200, 201.

the others—they have lower ionization potentials—and going down any family, the ionization potential decreases with increase in Z number. This does not mean that iodine, say, is metallic, but it is 'more metallic' than fluorine and the other halogens (and it does have a metallic glance to the crystals).

Non-metallic elements tend to hold on to the electrons they have, and to accept electrons from other elements to fill out their outer shells. The most non-metallic element is fluorine, and other non-metals are found in the upper right of the table from Group VA on. Understandably, then, relative non-metallic behavior increases up the table, for as the outer electrons are closer to the nucleus and less shielded by underlying shells of electrons, they are held more tightly. Also, toward the right the filling out of shells requires fewer electrons than would have to be lost to go down to the lower stable core. For example, a halogen can fill its outer shell to the stable configuration of a noble or inert gas by taking on one electron. Fluorine would have to lose seven electrons to go down to the helium core. The difficulty of losing electrons is shown by the magnitude of the ionization potential. This increases toward the right of the table and up in each family. The chemically rather inert noble gases are in a sense the ultimate non-metals. They have very stable filled shells, with no loose electrons around and no unoccupied niches in the ground state. The difficulty of removing an outer electron is reflected in the high first ionization potentials of these atoms (Table 11-3).

The in-between elements in this classification are at the top of Group IV: hydrogen and carbon. Silicon already shows metallic properties, and these become more pronounced downward. Hydrogen has the property that it may either gain or lose an electron—a symmetrical situation which seems of some importance and which makes us place hydrogen in Group IVA. Sometimes it is placed in Group IA, because it has only one *s* type of electron; other classifiers give it a special place outside of the table, as we have implied. Carbon is also difficult to classify. It is clearly non-metallic because unwilling to give up an electron easily; yet it forms unusual structures such as graphite which conduct electricity. Normally when anomalous behavior (which means behavior that does not fit preconceived categories) is encountered, it is best to accept it and not strain to rationalize it until a good basis for rationalization appears. In the case of graphite there is a good basis, but we will be prepared for it only by Chapter 15.

The chemist classifies atoms and molecules (elements and compounds when in the bulk) in many additional ways besides as metals and non-metals, and the Periodic Table bears these interpretations, too. For example, the regular pattern of orbital structure implies a regular spectral pattern for possible transitions, and this implies regularities in the colors of the elements and compounds. The shapes of the orbitals imply allowed shapes of molecules, which is reflected in the shapes of their crystals. Most importantly, perhaps, the regularities we have discussed manifest themselves in chemical reactions. Some of this will appear in Chapters 14 and 15. All of this is the burden of a pre-professional course in Chemistry.

Summary 11.7

The properties of atoms reflect their structure in the most intimate possible way. The mass of an atom lies largely in its nucleus; it is the Z number that affects its chemistry. As the Z number increases from 1, by units of 1, so the extranuclear electrons increase in number from 1 in units of 1. They become arranged in a definite pattern, the larger units of which are shells. These are numbered from $n = 1$ to $n = 7$. As the electrons increase in number, they fill the shells in a regular pattern: the first shell is filled with 2 electrons, the second with 8, the third with 18, the fourth with 32, and the fifth would be filled with 50. The formula which gives the full shell complement is $2n^2$. At lower Z numbers the process proceeds by filling first the s level, then the p level if there is one, then the d level. However, the guiding principle to which the added electron responds is to take the orbital of lowest energy available to it. In some of the more complicated atoms, levels with higher *numbers* may for reasons of shielding and other phenomena fall below levels with lower numbers. In any atom, no two electrons may have the same four quantum numbers. This is the principle which guides atom building. When electrons absorb electromagnetic energy they are raised to higher orbitals (higher energy levels) and may be expelled from the atom in a photoelectric effect. When they fall into an atom, and when they fall from a level of higher to one of lower energy, energy is radiated as a photon. A macroscopic body of radiating atoms produces a spectrum which is made up of wavelengths characteristic of the electron transitions, and hence of the pattern of energy levels, and thus of the atom itself. Wave mechanics enables the calculation of orbitals (at least in the majority of cases and at least approximately). Four quantum numbers serve to characterize any electron in any atom in the ground or stable state. The picture of atomic structures of the chemical elements explains and amplifies the Periodic Classification that embodies the chemical properties of all the elements. Note that the number of electrons in the outermost shell filled in each Period goes in the order 2, 8, 8, 18, 18, 32 . . . .

Exercises

11-1 Give a free translation of the following note found by Henry Guerlac among Lavoisier's papers in the archives of the Academy of Sciences in Paris. (You may have to guess at some of the old words.)

Sur la matiere du feu

Tous les metaux exposés au feu et calcines augmentent de poids tres sensiblement.

Les auteurs anciens pretendoient qu'on combinoit du feu

avec ces corps dans la calcination et que c'étoit à l'addition de cette Substance pésante qu'on devoit l'augmentation du poids.

Sthal a pretendu que la calcination enlevoit la matiere du feu aux corps qu'on calcinoit mais lui et ses sectateurs sont tombés dans un labirinte de difficultés comment concevoir en effet qu'on augmente le poids d'un corps en lui enlevant un partie de sa substance.

Quoi qu'il en soit de l'explication, le fait n'en est pas moins constant. Tout les metaux augmentent de poids par la calcination. M. de Morvaux le demontre complettement dans Ses digressions académiques page 72 jusqu'à 88. [Henry Guerlac, *Lavoisier—The Crucial Year. The Background and Origin of His First Experiments on Combustion in 1772.* Cornell University Press, Ithaca, New York, 1961.]

11-2 a) Write the expected electronic constitution of an atom with $n = 5$, and all shells filled. b) What is its atomic number expected to be? c) On the assumption of regular behavior what is the electronic constitution of the atom with $Z = 16$? (Answer before you look up the name, and then state what Group it should be in.)

11-3 What would the electronic constitution of the state *g* be?

11-4 Suppose two electrons (*e.g.*, Figure 4-7) spinning in the plane of the paper, with their magnetic poles above and below the plane, and 'swinging' together in clockwise orbit in the plane of the paper about a nucleus. Suppose that a magnetic field perpendicular to the paper is directed into it. What effect(s) would it have? Indicate by drawing a diagram and inserting arrows. This is, of course, a highly artificial thought experiment. What makes it so artificial?

11-5 We have indicated that the electron in the $2p$ state of boron is in the p_x orbital. How do we know this? Might it not be in the p_y orbital (say)?

11-6 How do you suppose Empedocles and his colleagues arrived at the concept of atomicity?

11-7 What is the threshold wavelength in the photoelectric effect for calcium? For lead?

11-8 What is meant by the statement, in connection with ionization potentials that are determined spectroscopically, 'the conversion factor is 8067.5 cm^{-1} = 1 volt?' Here the term with the dimension cm^{-1} is the wave number $1/\lambda \cdot$ the number of waves in 1 cm.

11-9 What would you expect to happen to the electrons in the filled core of a series of atoms as the atomic number increases along the series? Would the core remain unaffected?

Answers to Exercises

11-1 *On the subject of fire*

All metals when exposed to fire, and calcined, gain in weight very noticeably. Early authors supposed that the fire combined with these bodies during calcination, and that it is to the addition of this heavy substance that one owes the increase in weight. Sthal has supposed that calcination removes the material of the fire from the body that one calcines, but he and his followers have fallen into a labyrinth of difficulties in trying to imagine how one increases the weight of a body by taking away part of its substance. Whatever the explanation may be, the fact is no less definite. All metals increase in weight upon çalcination. Monsieur de Morvaux showed this fully in his digressions academiques, pp. 72 to 88.

11-2 a) 1 shell, $1s^2$; 2 shell, $2s^2$, $2p^6$; 3 shell, $3s^2$, $3p^6$, $3d^{10}$; 4 shell, $4s^2$, $4p^6$, $4d^{10}$, $4f^{14}$; 5 shell, $5s^2$, $5p^6$, $5d^{10}$, $5f^{14}$, $5g^{18}$. b) 110. c) 1 shell, $1s^2$; 2 shell, $2s^2$, $2p^6$; 3 shell, $3s^2$, $3p_x^2$, $3p_y^1$, $3p_z^1$.

11-3 In the state g, where $n = 5$, $l = 4$, then $m_l = -4$ to $+4$. Since each of the 9 states may by the Pauli Principle accommodate 2 electrons ($m_s \pm 1/2$), 18 g electrons are possible.

11-4 a) The electron with its poles in the same direction as the applied field would tend to turn over, or to be expelled from the field; the + current, counterclockwise, due to the orbital motion, will produce a force toward the nucleus.

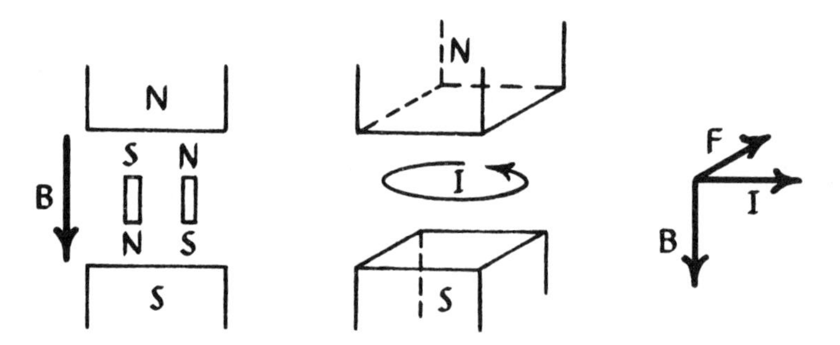

11-5 The arbitrary frame of reference must be alluded to here.

11-6 For one thing, food obviously underwent change when eaten. There were many other evidences.

11-7 The energy required of the radiation quantum must be equal to or more than the ionization energy.

11-8 Since $\nu = c/\lambda$, $\nu = c$ times the wave-number. Then $\mathbf{E} = h\nu = hc/\lambda$, in eV.

11-9 No, it would be expected to draw inward more and more as the positive charge on the nucleus increased.

Chapter 12

12.1 Recapitulation
12.2 Radioactivity and its detection
12.3 Detectors of radiation
12.4 Mass detectors and measurers. The mass spectrograph
12.5 Atomic weight and mass units
12.6 The great machines: atom-smashers
12.7 Nuclei
12.8 Nuclear 'fragments'
12.9 Summary

Nuclear Investigations. The Great Machines ~ 12

Recapitulation 12.1

The charge on the nucleus of an atom determines the number of electrons that will be stably held by that elementary atom. The arrangement of these electrons is described by the quantum numbers that refer to orbital shapes and orientations, and is subject to the Pauli principle. The total number of electrons is closely related to the number and disposition of the outermost electrons that are chemically reactive. It is these that determine what kinds of molecules will be stable, and what kinds of bonds may form between atoms. Further, they determine the associations between molecules that make up bulk matter. Thus it is apparent that a chain of connected properties links the nucleus to bulk matter. This chain may be traced from the construct 'nucleus' lying quite far from the Plane of Perceptions through constructs tied to atomic, molecular, and bulk properties, and thus to the Plane of Perceptions.

Radioactivity and its detection 12.2

Radioactivity was discovered in 1896 by Henri Becquerel. He found that certain materials—ores and pure chemical compounds of uranium, for example—were able to fog a nearby photographic plate (films were not in use in those days) even though the plate was wrapped in black paper or enclosed in a light-proof cardboard box. X-rays had been discovered quite recently (1895) and it was, naturally enough, supposed that the radioactive materials gave off the same new type of penetrating radiation that could easily go through paper and cardboard. Radioactive substances, then, were such as gave off this type of radiation.

The radiation can be detected and measured by photographic means—that is, by its ability to expose a photographic plate, the density of the image being related to the intensity of the radiation. Also, as X-rays do, it causes a fluorescent screen to glow wherever it strikes it (see Section 11.1 for the interpretation). A fluorescent screen is a metal plate coated with zinc sulfide or some other fluorescent material. This is mounted in a light-tight tube. The observer who wishes to test a material for radioactivity looks into the tube and after waiting for his eyes to adjust to the dark, brings the material up to the screen. It is possible, by using a magnifier, to see scintillations: flashes where single fluorescent events occur. These may be counted, so that the intensity of the radiation in terms of scintillations per minute may be measured.

At first, photography and scintillation counting were the only methods available for recognizing radioactivity, but later many more refined tools were devised for this purpose. These are described below, for it is important to realize that objective means are available for detecting and measuring these rays.

It was very soon found that radioactive materials might give off three kinds of rays, α, β, and γ. The α 'radiation' turned out to be composed of doubly charged helium nuclei, ${}_2He^4$. The helium nuclide is now commonly called an α-particle. The β 'radiation' turned out to consist of high-speed negative electrons, whence a single electron ${}_{-1}e^0$ is often named a β-particle. Gamma 'radiation' is actually electromagnetic radiation of very short wavelength—of the order of X-rays, or shorter. These rays could be differentiated by allowing a radioactive source to expel the rays in a narrow pencil, by putting it at the bottom of a well in a lead block (Figure 12-1). The rays are passed between the poles of a strong magnet: α-particles are deflected oppositely to β-particles, while γ-rays pass on in a straight line. These three beams can then be detected separately (see below).

12.3 Detectors of radiation

The detectable properties of nuclei and of nuclear particles such as protons, neutrons, α- and β-particles, γ-rays, and others that we shall encounter, are mass, velocity, charge, wavelength, as these manifest themselves in interactions with matter. Usually mass and velocity are handled together in terms of energy, in electron volts (Section 4.9). For example, the α-particles given off by naturally radioactive materials range in energy from one to about ten MeV (million electron volts). Beta-particles are not massive—less than

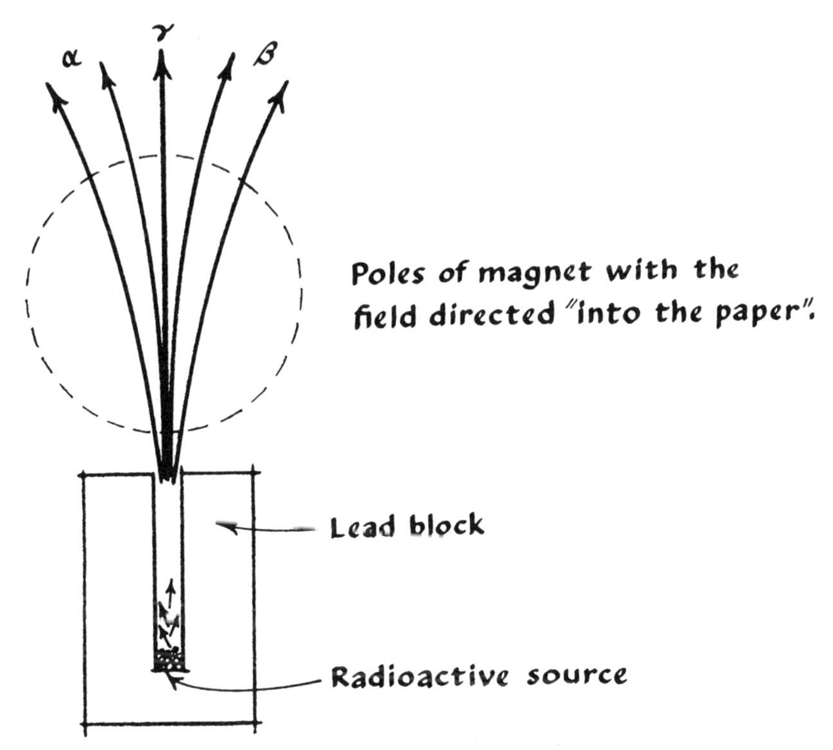

Fig. 12-1 *Rays from a radioactive source at the bottom of a narrow hole in a lead block emerge as a nearly parallel stream, or pencil. (All other rays, given off in every direction by the source, are absorbed in the lead.) When this passes between the poles of a magnet, shown in outline, with the field directed "into the paper," the rays are affected as shown.*

1/10,000 of an α-particle—but they generally move with such high speed that their mass-energy properties have to be calculated relativistically: in general, they may range over a wide gamut of energies, up to several MeV. Gamma photons may show energies up to five or ten or more MeV.

When energetic particles, including photons, interact with matter they produce ionization, by which their presence is recognized. Ionization is the production of charge, as we saw in Section 3.4, and occurs when an atom loses one or more electrons:

$$\text{Zinc sulfide (ZnS)} \xrightarrow[\text{Collision}]{\text{Energetic}} \text{ZnS}^+ + e^-$$

When γ photons produce ionization the process is photoelectric (Chapter 8).

The mechanism of interaction may be visualized. Here is an atom (perhaps in a molecule). Along comes a high energy α-particle. If it passes at some distance from the atom it may be effective—through its charge 2(+)—only in exciting some of the electrons, raising them to higher energy levels from which they cascade down to stability, with fluorescence carrying away the extra energy. The α-particle loses some energy in the interaction. It slows down: the energy it loses is quantitatively accounted for in the fluorescence, and any vibration or other motion given to the molecule (or atom). If the α-particle should pass somewhat closer to the atom, enough energy may be transferred actually to expel one or more electrons from the atom, ionizing it. Again energy is conserved. Or, the α-particle may collide with a molecule and break a molecular bond, as in the above example. If the particle collides head-on with the atom, or glancingly, it will scatter electrons and recoil. Transmutation may even occur if it collides with a nucleus.

Much the same picture may be drawn for the β-particle and the γ-ray. In all these interactions the quantity of energy needed to produce the particular kind of interaction depends on the nature of the atom or molecule. Qualitative effects enter too. Thus electrons are knocked out of the atom by the helium nucleus but none stay with it to form a stable helium atom until near the end of the string of interactions when the residual energy of the α-particle and that of the electrons are in tune (Section 10.5).

There is a rule of thumb which says that on the average an energetic particle loses about 32 eV at each ionization. Thus a one MeV particle might produce 30,000 ionizations. The α-particle plowing into a mass of atoms produces these ionizations over a relatively short path, so that the tracks of ionized particles which it produces, when they are visible, are seen to be dense and stubby (Figure 12-2). Also, because α-particles are relatively large (in terms of atoms) they do not have great penetrating power, and are stopped by metal foils, and by skin (which they may burn because of their high energy). Beta-particles of the same energy, being much smaller and moving much faster, will with the same initial energy as an α-particle produce the same number of ions, but these are strung out over a longer path, leaving a less dense trace (Figure 12-3). Beta radiation penetrates thicker pieces of metal, and when falling on the skin can produce deeper burns. Gamma rays have the highest penetrating power of all. With a given energy the track of the photon is longer and more tenuously marked than that of a β-particle. It is the gamma radiation from radioactive material that is most hazardous, because of its high penetrating ability: not being charged, it is not repelled by electrons or nuclei. (The reason that we must be vitally concerned about radioactivity is that living cells are sensitive to it. The sudden ionization of a protein molecule in a cell,

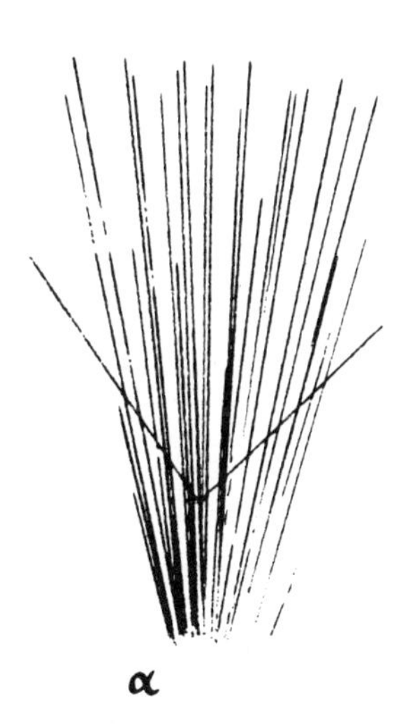

Fig. 12-2 *Cloud-chamber tracks, drawn from a photograph. The tracks are observed as white on a black background. The α tracks are dense and relatively short. (From P. M. S. Blackett, Proceedings of the Royal Society of London* (A) 107, *349*, (*192d*).]

cleaving the molecule, may well kill the cell or produce some chemical damage that is pathological in its results.)

The ionizing ability of radiations has been used as a means of detecting them. In the earliest days the researcher used a low-power magnifier and actually counted individual scintillations occurring on a screen on which the radiation impinged. The number per unit of time gave a measure of the intensity of the radiation. Since then, machines have been devised that not only count faster and more accurately (being not subject to fatigue) but that record and total the counts. By interposing different thicknesses of metal, α-, β-, and γ-rays can be counted: count all the radiation; filter out the α- and count β- and γ-; filter out the β- and the remainder is γ-. (There is always a background count due to cosmic radiation and radioactive contamination which shows up in any of these measurements.) The filters always absorb some of any kind of radiation that falls on them, and this has to be allowed for in an analysis. It is better to separate the rays as in Figure 12-1 before analysis.

One interesting counter uses the photoelectric effect. The particles to be counted are made to strike a scintillation screen, or a crystal that fluoresces (Figure 12-4). The light so produced, striking a photoelectrically active metal, releases a pulse of one or more photoelectrons. These accelerate to an electrode nearby which is charged (+). As they accelerate they gain energy so that, striking the electrode, they expel numerically more electrons—a larger pulse. This group of electrons is

Fig. 12-3 *Cloud-chamber tracks, drawn from a photograph. The tracks are observed as white on a black background. The β tracks, of which only about half the length is shown, are attenuated.* [*From P.. I. Dee, Proceedings of the Royal Society of London* (A) 136, *727* (*1932*).]

accelerated to another electrode, with further multiplication of the effect. In a tube of this kind, a photomultiplier tube, the original pulse might be amplified a millionfold, yielding at the end enough electrons, as a pulse of current, to activate an amplifying and recording machine. The successive electrodes are so shaped that the pulse from that electrode is focused on the following one.

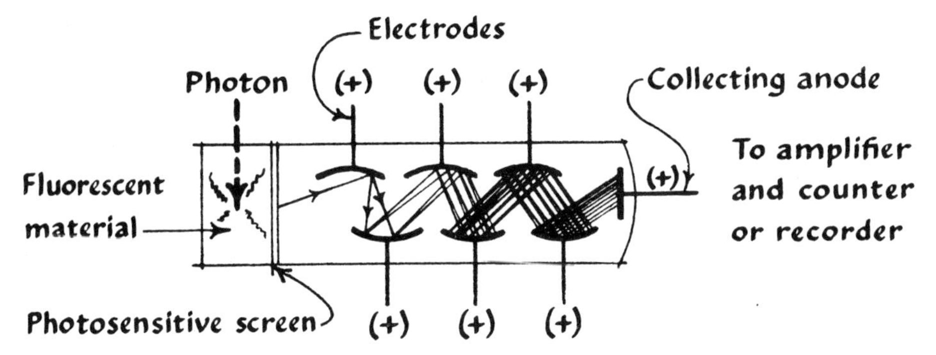

Fig. 12-4 *A photomultiplier tube of six stages attached to a scintillation screen. A photon entering the fluorescent material produces light quanta which expel electrons from the screen. These accelerate to the (+) metal plate where they produce more electrons (because of their high energy). These accelerate to the next plate, and so on, cascading down successive stages to the collecting anode which transmits the resulting pulse of (−) current to a counter and recorder. The plates at each stage are curved to focus the electron-shower on the following plate. The increasing density of fine lines in the drawing is meant to imply the increasing numbers (the "multiplication") of electrons in the cascading stream.*

In another type of detector, the Geiger tube, or Geiger-Müller tube, the arrangement consists of a central wire electrode in a tube containing some suitable gas (Figure 12-5). The central electrode is charged (+) some 1000 to 5000 V with respect to a cylindrical electrode through the axis of which it runs. Suppose a high-energy particle, or photon, enters the tube. It ionizes the gas, producing heavy (+) ions and electrons (−). The former move toward the (−) electrode; the electrons accelerate toward the central electrode. In both cases the moving ions ionize gas particles along the way and the result is a veritable cascade of particles to the central electrode and to the

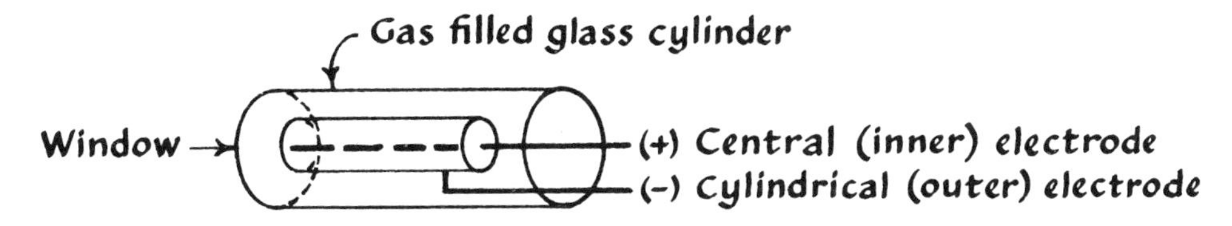

Fig. 12-5 *Schematic of a Geiger-Müller counter. A thin window gives entry to the radiation to be counted. It may be constructed to screen out particular types or energies of radiation. Inside the glass tube are the electrodes, an inner wire (+) and a coaxial cylinder (−). The electrodes are at a high potential difference. When a radiation enters the tube and forms ions in the gas in the tube these form a conducting path between the electrodes and a pulse of current flows between the electrodes. This passes to a pulse counter and recorder (not shown).*

cylindrical electrode. As in the photomultiplier tube this ion-multiplier produces a pulse of current that is enough to activate a counting and recording machine. Thus by this device a single photon, entering the Geiger tube, or a single β-particle produces a magnification effect that allows it to be counted.

The track of ionized atoms may be visualized macroscopically by means of the cloud-chamber or the bubble-chamber. The first *cloud-chamber* was described in 1911 by C. T. R. Wilson. It has been an invaluable tool in nuclear research ever since. Essentially, it consists of a glass chamber containing an air space above a black background of inky liquid (Figure 12-6). The whole is arranged with a piston so that

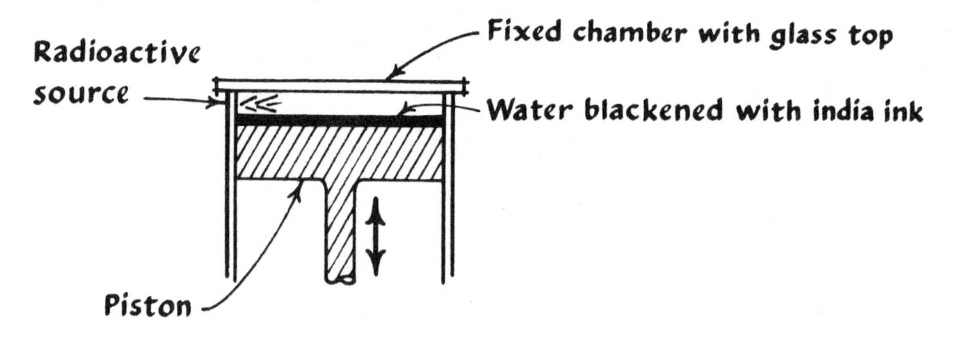

Fig. 12-6 *Diagram of a Wilson cloud-chamber. A piston may move up and down in a fixed chamber with a glass top. Above the piston is a solution of aqueous india ink, and the inside walls of the chamber are blackened. A radioactive source (on the left) gives off radiations which ionize the air in the chamber, and produce fog tracks along the path of ionization. Individual atomic events are thus made visible, as in Figures 12-2 and 12-3.*

the air can be compressed by pushing up on the piston, and then suddenly cooled by releasing the piston. The air is always saturated with water vapor above the liquid. Upon sudden cooling it becomes supersaturated because there is more water vapor present than would saturate the now cooler air, cooled by expansion. This excess moisture readily condenses around ions. Then as an energetic particle traverses the air space, as for example from the 'source' in the diagram, it leaves a track of fog-particles condensed about the ions it has produced. The path of the invisible particle is thus made visible, and can be photographed (Figures 12-2, 3). Once again, as in the photomultiplier tube, a single process of sub-atomic magnitude is amplified to the point where it can be detected macroscopically.

Mass detectors and measurers. The mass spectrograph 12.4

In Chapters 3 and 4 we developed the laws that were made use of by J. J. Thomson (Section 4.7) in his machine for measuring the ratio of charge to mass of the electrons in a cathode-ray tube. In this tube, as we noted, a stream of heavy (+)-charged particles moves in an opposite direction to the electron stream. When he investigated the properties of these particles J. J. Thomson discovered isotopes.

The principle is to some extent like that described in Section 4.7. A hole in a cathode (Figure 12-7) allows a stream of (+)-charged

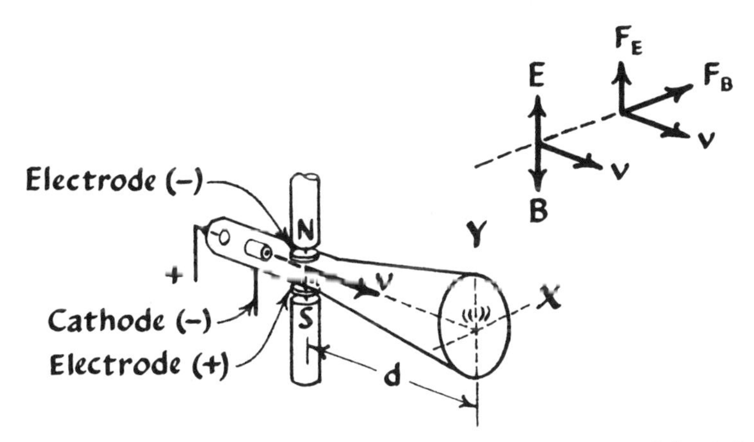

Fig. 12-7 *A simple mass spectroscope of early design to show arrangement of electrical and magnetic poles. On the right, at the window of the tube, is shown an example of the parabolic curves observed for species of ions with different masses (the separate lines) and different velocities (distributed along the lines). The symmetrical arrangement is obtained by reversing the direction of the magnetic field. This makes it easier to obtain measurements of* x *and* y. [*After Richard F. Humphreys and Robert Beringer,* First Principles of Atomic Physics. *Harper & Row, New York, 1950. Figure 18.1, p. 213.*]

particles of the gas in the tube to pass through into a part of the tube that is placed between two electrodes. These electrodes can be charged electrostatically (+) and (−). There are also arranged a pair of magnetic poles oriented so as to give a magnetic field along the same axis as, but oppositely directed to, the electric field.

When the machine is turned on, there are observed fluorescent streaks on the face of the tube. These occur as a group in one quadrant of the X, Y coordinate system on the face of the tube. If the polarity of the magnet is reversed the streaks appear in the neighboring quadrant (Figure 12-8). The streaks are parabolic in form. A parabola is a type

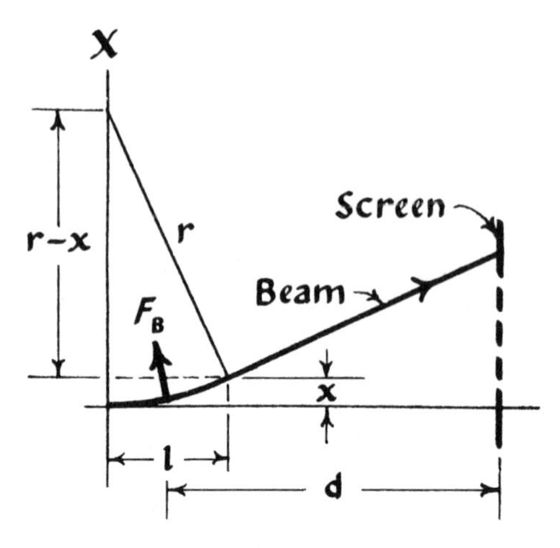

Fig. 12-8 *Analysis of the geometry of the simple mass spectroscope in Figure 12-7. While in the field of the electrodes, the* (+) *ions experience a force upward* (y) *over the extent of the field, a distance* l. *The magnetic field produces a force that swings the particles in an arc (see Section 4.4) as shown. When the ions leave the fields they travel in a straight line to the screen. Slowly moving ions will be deflected in the X and Y directions more than fast ions.*

of curve obtained when a function of the form $x^2 = ky$ is plotted. (k Is a constant.) It is instructive to see how an excursion into the field of constructs yields understanding of why these streaks are parabolic, and not some other shape. It turned out to be an important matter, in J. J. Thomson's hands. Reasoning plus algebra does the job (Figure 12-9).

Suppose that a particle of mass m and (+)-charge q is produced between cathode and anode. According to Equation (3-5) it will experience a force $\mathbf{F} = qE$ in the direction of the cathode and will

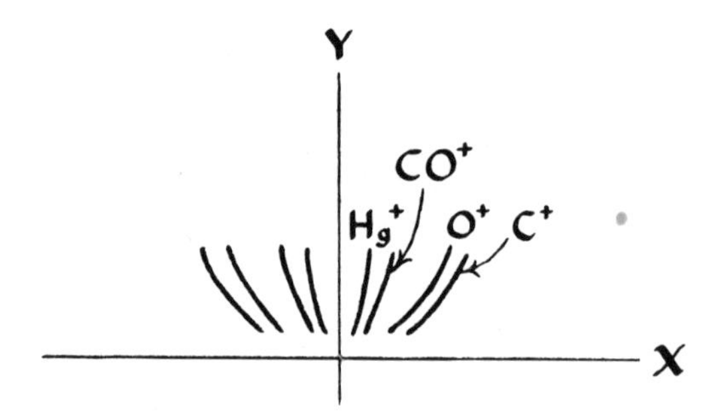

Fig. 12-9 *A set of parabolas of ions: singly charged ions of mercury, carbon monoxide, oxygen, and carbon. [After Humphreys and Beringer,* loc. cit., *Fig. 18.3, p. 215.]*

accelerate towards it. Most of these particles will strike the cathode, pick up electrons, and revert to the state of neutral gas molecules or atoms. But some, travelling fast, will pass through the hole in the cathode and shoot along the axis of the tube toward an origin of X, Y axes placed on the face of the tube. If the electrodes and magnetic poles are not operating, a bright, fluorescent spot appears at this origin. The particles are accelerated according to the laws previously studied:

$$a = \mathbf{F}/m = qE/m \tag{12-1}$$

The particle, now moving at high velocity, passes between the poles charged (+) and (−), and is between them and under the influence of their field for a period of time t. During this period of time it is accelerated upward, in the Y direction, moving a distance y (Section 2.7 and Equation 12-1):

$$y = (1/2)at^2 = (1/2)qEt^2/m$$

(This accelerating force is produced by the (+) and (−) electrodes.) The electric field spans a distance l. Thus by the velocity equation the period of time that the ion spends within the field, travelling with velocity v, is $t = l/v$, and we can write:

$$y = (1/2)qEl^2/mv^2$$

The field between the electrodes is uniform and vertical, except at the very edges (Figure 3-13). It produces vertical deflection of the particle.

The magnetic poles, with lines of force down from N to S in the figure exert a force on the particle tending to drive it in the X direction (Section 4.4). This force extends also over the distance l, beyond which point on leaving the field the particle proceeds in a straight line to the screen. This is shown for the magnetic effect in Figure 12-7. The magnetic force $\mathbf{F}_B$ is $q\overset{\curvearrowright}{vB}$ (Section 4.4). Being centripetal, $\mathbf{F}_B = mv^2/r$. It is permissible to approximate r (because the arc of the massive particle is not highly curved) from the construction, Figure 12-8.

$$r^2 = (r - x)^2 + l^2$$

which yields to good approximation,

$$r = l^2/2x$$

The reasoning goes thus:

$$(r - x)^2 = r^2 - 2rx + x^2 \text{ (multiplying out).}$$

Thus $r^2 = r^2 + x^2 - 2rx + l^2$

Subtracting r^2 from both sides, and rearranging,

$$2rx = x^2 + l^2$$

But x is quite small because the curvature is small, and therefore x^2 will be so small as to be negligible compared with l^2. It may be set equal to zero. Then, $r = l^2/2x$.

It should be noticed that we are looking at what is happening to the particle between the electrodes and the poles. Whatever happens here is projected onto the screen because the particle takes a straight-line path when it leaves this zone. (Newton's first law.)

The magnetic force on the particle, with the above value of r substituted, becomes, on further algebraic manipulation, $x = qBl^2/2mv$. The curves result from a force in an X direction and a force in a Y direction operating together. Here we have x's and y's projected on the coordinates of the tube, their ratios being preserved by the straight-line projection. We do not know v, but this can be eliminated by squaring the equation for x and dividing it by the equation for y:

$$x^2/y = [B^2q^2l^4/4m^2v^2]/[qel^2/2mv^2]$$
$$x^2 = [B^2ql^2/2mE]y \qquad (12\text{-}2)$$

All the terms in the brackets are constant in a given system that produces a single parabolic streak, hence we have the condition for the parabola: $x^2 = (\text{constant})y$. We know now why x turns out to be squared with respect to y. It is because v for any particle is an inverse function of x and the corresponding inverse function for y involves v^2.

The ions produced in the tube between anode and cathode do not all start at a particular spot, but all along the way. Therefore they are accelerated to different extents. Thus the stream of ions that issues from the hole in the cathode is composed of particles with a range of velocities. The slower ones, dwelling between the electrodes and poles for a longer period, are deflected to a greater extent than the faster ones. Since there is a maximum velocity, determined by the accelerating potential difference between anode and cathode, the parabolas are cut off at the lower, least deflected end. At the upper end they are cut off because if a particle moves too slowly it will not get through the hole in the cathode but will fall onto its wall. The above equation shows that each parabola will be a function of q and m. Thus a particle of mass m and charge e will give a single parabola. If its charge is double ($q = 2e$) a separate parabola will be produced. For a fixed charge, those particles with the same mass will fall on the same parabola.

From the equation it is apparent that x and y are functions of e/m. B, e, and E, are known, and x and y can be measured from the images of the parabolas. Thus m can be calculated. One of the discoveries made by Thomson in these investigations was that neon is a mixture of at least two atoms of mass about 20 and about 22. (Today we recognize Ne^{20}, occurring to about 90.9%; Ne^{22}, occurring to about 8.8%; and Ne^{21}, occurring to about 0.3% in the natural neon.) This discovery of the existence of neon isotopes led promptly to the examination of other elements, and resulted in the discovery of large numbers of isotopes. The chemist's definition of 'element' had to be refined: atoms of a chemical element all have the same Z number. Since then, there has been improvement of Thomson's original instruments. The products, under the name of mass spectrographs, have become important analytical tools in physics and chemistry laboratories. Thomson foresaw, and remarked on, the analytical potentialities of this tool.

Atomic weight and mass units 12.5

The machine developed by J. J. Thomson has been redesigned and refined until nowadays there are available mass spectroscopes that will separate a mixture of atoms and give extremely accurate measurements of their relative masses. A problem arises for chemists because the elementary substances isolated from natural sources are mixtures of all the isotopes going. Hydrogen gas such as might be obtained from pure water is almost entirely $_1H^1$ (99.985%) but contains a small amount of deuterium $_1H^2$ (0.015%). At the other extreme, tin is a mixture of some ten isotopes, three of which are present in larger abundance than the others. Chemists had long decided to take oxygen as their reference element for calculating relative atomic weights from combining weights because it combines with so many other elements that numerous direct comparisons of combining weight ratios were available. If the lightest element, hydrogen, were assigned relative weight 1, oxygen had to be assigned very close to 16. That is, if the same number of atoms is taken in each case, the relative weights are close to 1:16 for hydrogen and oxygen. It was decided by international agreement to set the atomic weight of oxygen by definition at 16.0, exactly, as a basis for all atomic weights. This (16.0 g) in grams is the weight of one 'mole' of atoms. On this basis most of the other elements come out with relative weights close to whole numbers. For example the carbon value is 12.01, and hydrogen 1.008. When it was discovered that the chemist's oxygen is actually a mixture of three isotopes with atomic numbers 16, 17, 18, it seemed desirable to re-evaluate the situation, even though the O^{16} isotope occurs with an abundance of 99.759% in the mixture. In 1960–61 it was internationally agreed by physicists and chemists to set up a new scale. This is based on carbon = 12 exactly, and specifically the carbon isotope $_6C^{12}$. On this basis oxygen $_8O^{16}$ has the relative mass 15.99491494 (Appendix 3).

The chemist, working with materials from natural sources, uses atomic weights calculated (today) on the basis of $_6C^{12}$, with account taken of the isotopes. He knows that if he takes 1.00797 g of hydrogen this will contain 6.02252×10^{23} atoms of hydrogen (Avogadro's number), and that the same number of *atoms* will be present in 15.9994 g oxygen, 118.69 g tin, and so on. This is the basis for writing quantitative chemical equations (Section 14.7).

The nuclear physicist speaks in terms of nuclides. The relative mass of carbon being taken as 12, an atomic mass unit, 'u' is defined as 1/12 the mass of a $_6C^{12}$ *atom*. This is the unit of measure in nuclear physics: 'one atomic mass unit.' Relative to the base $C^{12} = 12$, all atomic masses come close to integral numbers. The atomic number A in the symbol $_ZX^A$ (Section 9.2) is the number of nucleons, protons (Z), and neutrons (A-Z). It, of course, must be a whole number. The relative masses of the nuclides are very close to these numbers.

12.6 The great machines: atom-smashers

In previous sections we have reviewed how radiations can be detected, and how charge and mass of nuclear particles can be measured. The existence of natural radioactivity (discussed in detail in the following chapter) implied, when it was realized that α, β, and γ rays originate in the nucleus, that this entity must have a structure. The problem was how to get at it.

The use of α-particles as projectiles to investigate the nature of matter had led Rutherford to recognize how small the nucleus must be (Section 10.1). In the course of his investigations he found that hydrogen was produced when nitrogen was bombarded with α-particles, and he proposed that in the reaction of an α-particle with a nitrogen nucleus, the α-particle unites with the nucleus and a proton is ejected. This, on picking up an electron, yields the observed hydrogen. This explanation was in line with an hypothesis advanced by Rutherford and Frederick Soddy in 1902. In radioactive decay, they suggested, the nucleus of an element changes to that of another element. In the transformation mass, energy, and momentum are conserved. It was apparent quite early in these researches that it would be most desirable to obtain particles with high velocity and, possibly, large mass, with which to probe the atoms of matter. The first machines to produce such particles appeared during 1932.[1] They are called accelerators and may be divided into two classes, the cyclic and the linear. We have already discussed the cyclotron, an exemplar of the cyclic type of accelerator. The linear types use essentially the same principles. The particle to be accelerated is ionized. It is accelerated through a 'drift tube' in a magnetic field that holds it in a straight line. As it comes out of this tube the polarity is switched and the particle is attracted to another close-by tube, accelerating into it (Figure 12-10). By repetition of this process the

---

[1] Ernest O. Lawrence and M. Stanley Livingston, "The Production of High Speed Light Ions Without the Use of High Voltages," *Physical Review*, Sci. 2, *40*, 19 (1932).

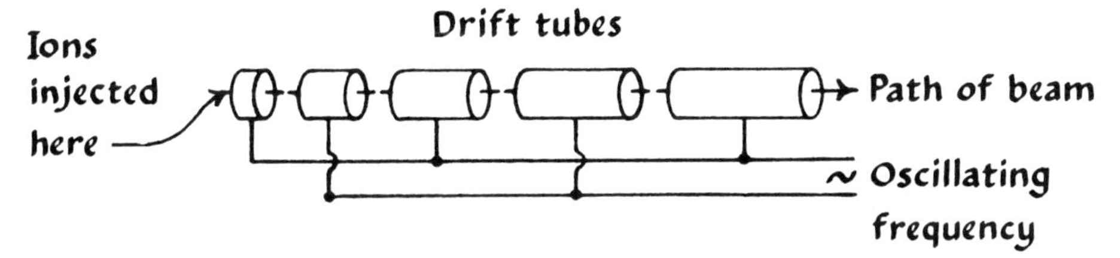

Fig. 12-10 *A scheme of a linear accelerator. Ions enter the "drift tubes," from a source on the left. As a (+) ion approaches the first tube this is charged (−), thus accelerating the ion into it. Within the tube there is no field force on the ion and it "drifts" through at its entering velocity. As it emerges, this tube has changed polarity to (+), while tube 2 has become (−), so the (+) ion, repelled by tube 1 and attracted to tube 2, accelerates across the gap. It is thus given an electrical "kick." Its velocity is now increased, so that drift tube 2 must be longer by the right amount so that the faster ion spends the same time in it as it spent in tube 1. This is so that the polarities of the drift tubes may be changed in regular sequence. The ion emerging from one tube experiences an attraction to the next, and while in the tube the polarity is changed. Each drift tube, then, is longer than the preceding one. The beam of ions is kept compact by specially designed magnets inside the drift tubes, and it emerges as a high-energy, accurately directed* stream *of projectiles. (Redrawn from Humphreys and Beringer*, loc. cit., *Fig. 28.2a, p. 343.)*

particle receives a succession of 'kicks,' and can be accelerated to high energy. As it goes faster, the length of the tube must be increased so that the alternation of polarity of all the tubes can be synchronized. At the same time this allows for adjustments to take account of relativistic effects, such as mass increase. Energies in the billion electron volt (BeV) range are now commonplace.

Nuclei 12.7

There are two ways that we shall look at nuclei. One is a gross accounting method, where the nucleus is thought of as comprising protons and neutrons and the nuclide is given a designation ${}_Z X^A$ to show, as we saw briefly in Section 9.2, the number of protons (Z) and the number of neutrons (A-Z) that are present. This approach will be very convenient when radioactive sequences and certain energy relations are dealt with in the next chapter.

The other method of thinking about the nucleus must be used when the hundred or more artifacts that arise from the activities of atom smashers are considered. The picture of the nucleus as a droplet that is spattered by the impact of a high-energy particle has value in certain respects, but we shall not pursue it in this chapter. Instead, we shall think of the nucleus as describable in terms of energy levels (Section 10.3). It would appear that we do not as yet know which, if any, of the particles that are the result of nuclear breakdown are elementary—using the word in some ultimate sense. Indeed, there is considerable opinion that we have not as yet reached anything elementary but are on the ladder of some kind of an hierarchy stretching upward into the macroscopic world and downward to as yet unsounded depths.

Suppose, however, that we think of the nucleus in terms of energy levels, analogous to those discussed in Chapter 10. There we pictured an electron at the lowest energy level of the hydrogen atom as having *given up* some 13.6 eV of energy, in falling from outside of the range of the nucleus into this stable, lowest orbit. The hydrogen atom is in the ground state under these circumstances, but it can become excited if the electron is 'raised' to a higher energy level by absorbing energy. The energy that is absorbed most efficiently is that which corresponds to one of the transition energies, namely, to one of the lines in the emission spectrum (most of which correspond with lines in the absorption spectrum). One says that there is resonance between the energy of the radiation and the electron state. The analogy is to resonance in the physics of sound. To recall the previous example, if a tuning fork is set vibrating then another tuning fork nearby, tuned to the same key, will respond to the sound vibrations carried in the air by absorbing them and itself vibrating even though it is made of rather heavy metal. Another fork a little off key does not pick up the vibrations at all efficiently. Further, in this picture of the atom we saw that it may become ionized by losing its electron. This would occur when the electron is given enough energy to be lifted out of the range of the nucleus. The minimum amount of this energy for the hydrogen atom is 13.6 eV. It may be transmitted to the atom as X-ray radiation, or it may result from the close approach of ('collision' with) a fast-moving particle, or it may be electrical in origin, being given by a sufficiently intense field. Several of these mechanisms may occur together, of course.

In applying these ideas to the nucleus the convention is to think of the particles in the nucleus as bound together by a *binding energy* which is equivalent to that which would be required to pull them apart. The analogy to the hydrogen atom would be to say that the binding energy between proton and electron in the ground state is 13.6 eV. But we usually try not to carry the analogy too far. There are good reasons for this. An analogy is useful because it can suggest new ideas; but it can also become a limitation by trapping thought and preventing the discovery of *non*-analogous aspects in the new phenomenon. We shall use analogy freely where it helps, but shall at the same time be wary of this use. Thus when we speak of binding energy in the nucleus, we are concerned with energies a millionfold larger than the electron potential energies. The helium nucleus, for example, shows a binding energy of about 28 MeV. Also, when we discuss energy levels we do not necessarily carry the analogy to the point of visualizing orbits about some core.

Everything about the nucleus is exaggerated compared with the atom: binding energies a millionfold greater; distances a hundred thousandfold smaller, and times of reaction incredibly shorter. Certain interactions in the nucleus may occur during a period of some 10^{-23} second, while atomic reactions require periods of about 10^{-8} second: the length of the period required is correlated to the forces that operate. A high energy particle, travelling at nearly the speed of light (or about 10^8 m per second) and impinging upon a nucleus the diameter of which is some 10^{-15} m, must interact within about 10^{-23} second (for the nuclear force does not extend outward very far from the nucleus: it is in this respect not at all analogous to a Coulomb force). Interactions that can occur within 10^{-23} second are called 'strong interactions.' Further, there seem to be a variety of magnitudes of the interactions that can occur in the nucleus, as well as a variety of artifacts (particles, radiations, 'resonances') that may emerge as the result of bombarding the nucleus with high energy particles. If we looked for a broad analogy here, we might conceive of the particles given off by a heavy nuclide as being compounds of some fundamental building blocks, much as atoms above hydrogen are comprised of a compound nucleus and of electrons. Some of them are stable (as are some elements) and some are unstable (as are the radioactive elements). But this analogy is intended only for purposes of aiding visualization and will be modified in the following section.

The energy-level model of the nucleus may be schematized (in analogy to the model of atomic energy levels, discussed in Chapter 10) as shown in Figure 12-11. Imagine a positively charged particle approaching the nucleus, which is also positively charged. The closer the particle approaches to the nucleus, the greater the Coulomb repulsion force. The net repulsion increases up to a short distance from the nucleus. Suppose, for definiteness, we let a proton approach another proton. At a distance of about four times the radius of the nucleus, the electric interaction of repulsion is just about balanced by a new kind of force, a 'nuclear interaction' which is attractive in nature. At closer range, this nuclear force of attraction rises to about forty times the calculated repulsion force. But still closer a nuclear repulsion force supervenes. This is shown schematically in Figure 12-12. The net force on the approaching particle is plotted on the Y axis, and units of distance on the X axis. Closer than about 4 units the nuclear attraction force becomes larger than the Coulomb repulsion force, and increases rapidly as the particles come closer. This assists the particle in surmounting the energy barrier suggested in Figure 12-11. At one unit the particle 'falls' into the potential energy well and occupies one of the

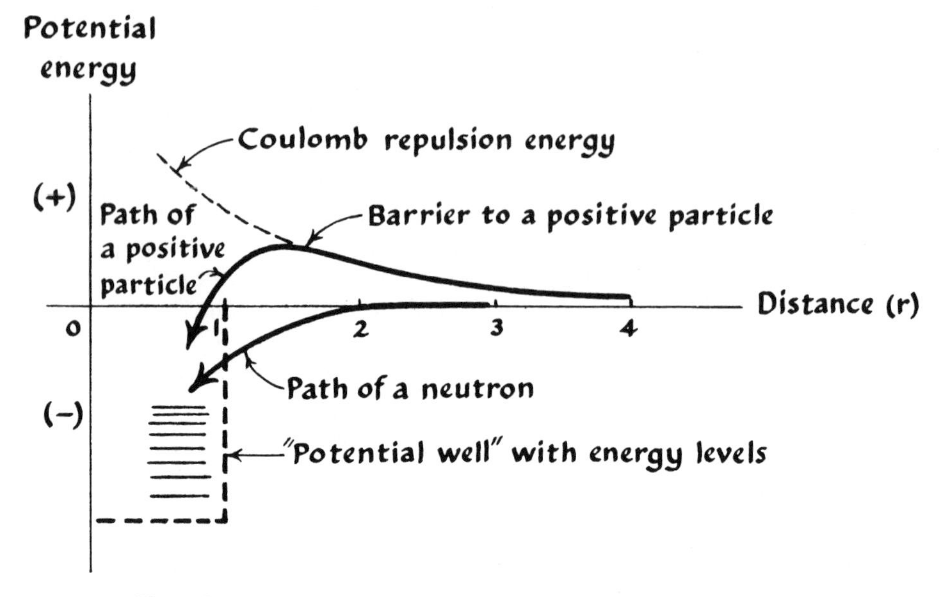

Fig. 12-11 *A mixed metaphor of the energy-level picture of a nucleus. From the energy standpoint the (+) nucleus is a well, surrounded by an electrostatic hill as far as (+) projectiles are concerned, but having no hill to neutrons. The hill is not extremely high, as the dashed curve of Coulomb repulsion energy would make it because in a certain region close to the nucleus the nuclear (strong) attractive force begins to compete. A positive ion with enough energy to surmount the hill "falls into" the well and takes some position consonant with its energy (energy-levels are sketched). If the projectile carries enough energy it can disrupt the nucleus.*

energy levels available there. But the nucleons may not approach much closer than about 1/2 the unit of distance. This is shown by the steep rise of the force-curve into repulsion interaction. The particle (in whatever form it assumes *in* the nucleus) is stable in the potential wells over the 'distance' shown in Figure 12-12. The nuclear force operates over an extremely short distance. It falls off so rapidly that at a distance of twenty-five radii, the electric repulsion force may be a million times larger than the nuclear attraction force even though at this distance the electric force itself has fallen off to the extent predicted by the inverse square law. For a particle to escape from the nucleus it must have enough energy to surmount the energy barrier. In some instances the requisite energy is not available, yet the particle does escape the nucleus. Here it is thought that there is a finite probability that, given a sufficiently long period of time, a particle may find its way through the potential barrier. A neutron finds less of a barrier, or none, as it approaches a nucleus; since it has zero net charge, the Coulomb force does not affect it (Figure 12-11).

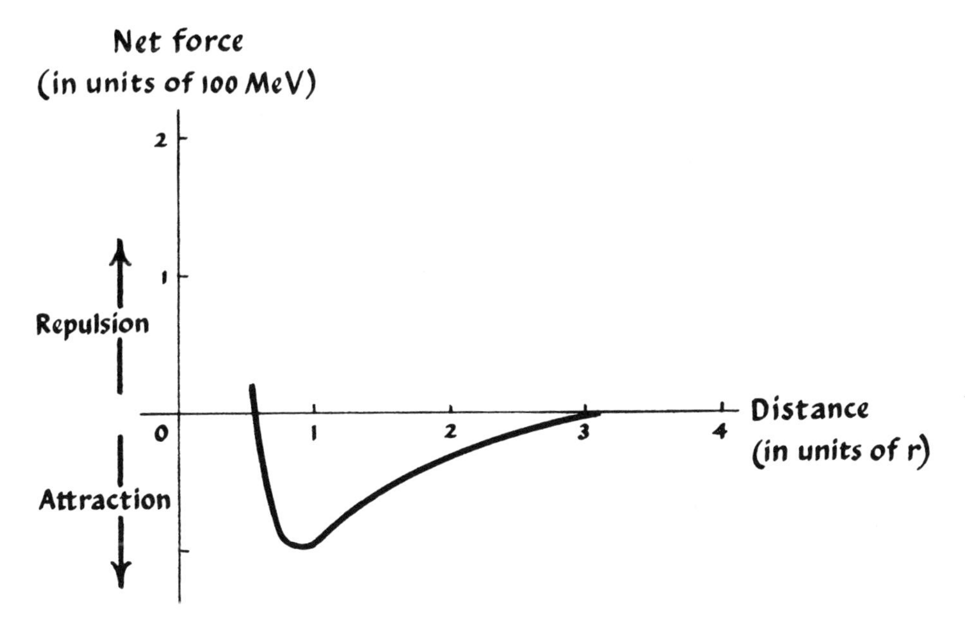

Fig. 12-12 *Plot of nuclear (strong interaction) force against distance between interacting charged particles. The proton of Figure 12-11 approaching along an axis marked 0 experiences essentially no force until it comes close to the nucleus, when Coulomb repulsion occurs (not shown) but is soon overcome, with closer approach, by the far stronger nuclear force. The particle approaches to an equilibrium distance ~r. Here it has given up the maximum energy. Powerful repulsion forces then supervene, opposing closer approach of the two particles.*

The idea of energy levels supplies a picture of radioactive schemes (which will be discussed in Chapter 13). For example, radium $_{88}Ra^{226}$ decays, by giving off an alpha particle, to emanation $_{86}Em^{222}$ (which itself later decays). Close analysis of the process discloses that in the decay of the radium α-particles of two different energies are given off, one of 4.80 MeV, the other of 4.61 MeV. The emission of the lower energy α-particle is accompanied by emission of a γ-ray of energy 0.19 MeV. The interpretation is that the nuclide $_{88}Ra^{226}$ goes to $_{86}Em^{222}$ + $_2He^4$ in a single step in which the more energetic α-particle is released, and that when the less energetic particle is released, the resulting Em remains in an excited state. The release of the gamma-ray allows the nucleus to achieve the more stable state of the Em in the first reaction. This is shown schematically in Figure 12-13. The energy accounting fits this interpretation:

$$4.80 \text{ MeV} = 4.61 + 0.19 \text{ MeV}$$

A nuclear particle may escape 'through' instead of 'over' the

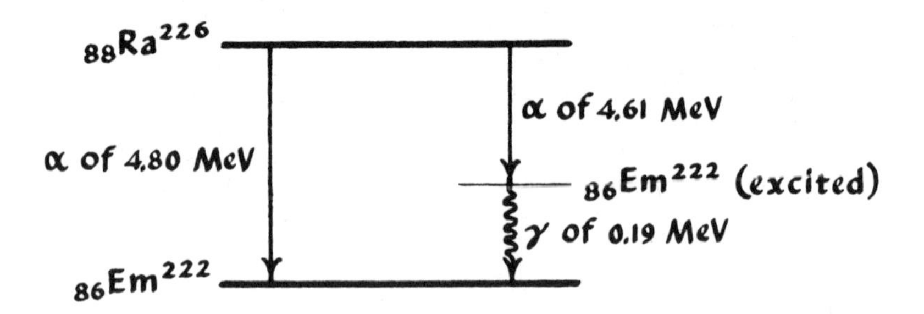

Fig. 12-13 *Representation of two energy levels and a transition between them. Unstable $_{88}Ra^{226}$ can decay, by releasing an energetic α-particle (4.80 MeV) directly to unstable $_{86}Em^{222}$ by one path. Alternatively, it can give off a less energetic α-particle (4.61 MeV) and fall to the level of excited $_{86}Em^{222}$. This then releases energy in the form of a 0.19 MeV γ-ray to become the ordinary unstable $_{86}Em^{222}$.*

energy barrier (or, conversely, may get into the nucleus by a reverse process). This is correlated with the 'width' of the barrier, and the 'number of impacts' of the particle 'against' the barrier. It is known that the potential barrier around the uranium nucleus is 27 MeV, but the α-particles given off when uranium 238 decays have only a little over 4 MeV energy. If they had gone over this barrier, they should have had to achieve at least 27 MeV potential energy which would have shown up as kinetic energy of the expelled particle. Therefore, they must have passed through the barrier. The phenomenon is labeled the 'tunnel effect.'

12.8 Nuclear 'fragments'

Some hundred or more fragments have been produced by bombarding atoms with high-energy particles, and the number of new ones will undoubtedly increase. The information obtained is so complicated and changing that we shall not pursue it here, but refer instead in the bibliography to discussions by experts. We only mention, as confirming the lawfulness that has been found throughout, that energy-and-momentum (mass-energy) of any reacting system is preserved, that is, the total energy and momentum of the products equals that of the reactants. Atomic number is conserved, which implies that the total number of nucleons in the universe is fixed. Charge is conserved. This implies that the ratio of (+) to (−) in the universe is fixed. When a gamma-ray of energy 1.02 MeV or larger passes close to a massive nucleus it may transform into two particles, a (+) electron, a *positron*, and a (−) electron, each with 0.51 MeV or more energy, to make up the initial energy. (The massive nucleus takes up any recoil energy.) But equal (+) and (−) are formed in this process of 'pair-production.' When the electron and positron (an 'anti-electron') recombine, matter is 'annihilated,' and two gamma-rays are produced, firing off in opposite directions to conserve momentum, and having the same total energy ($h\nu$) of the two particles, according to the relation $\mathbf{E} = mc^2$.

Summary 12.9

There are two major sources of information about the interior structures of nuclei. One is radioactive behavior (to be studied in detail in the following chapter) and the other is the pattern of properties of nuclear fragments produced by atom-smashers—more properly, nucleus-smashers. By firing high energy nuclear-size projectiles at nuclei it is possible to break them up, expelling fragments of many different masses, energies, and characteristic behaviors. These can be interpreted in terms of a nucleus with various energy levels from which particles may be expelled by the expenditure of sufficient energy or from which particles may escape by tunneling if given enough time. The patterns of behavior in these nuclear disintegrations show the presence of a short range specifically nuclear force of great magnitude named the strong interaction. It is some hundred times as strong as the Coulomb force. The patterns also obey certain conservation laws: the atomic number is conserved; angular and linear momentum are conserved; charge is conserved. The recognition of these patterns gives confidence that there is a lawful nuclear structure to be discovered.

Exercises

12-1 Suppose that a certain photomultiplier tube (Fig. 12-3) has 12 electrodes before the collecting anode, and at each, one electron produces four electrons. If a single electron impinges on the first electrode, how many will leave the 12th? How many coulombs of electricity will this represent?

12-2 In Figure 12-7, if the magnetic field were directed into the page, assuming the positive rays to be moving from left to right along the page, what effect would the field have on the rays?

12-3 In a measurement of parabolas obtained for singly charged (+) particles, at a certain value of y, x was found to be, respectively, 1.13, 3.15, 4.00, 4.6 mm. If the line at 4.00 was known to be oxygen, O^{16+}, what are the masses of the ions forming the other lines, and what is the name of each?

12-4 How could you demonstrate that a second tuning fork was set vibrating in resonance with a first one to which it was tuned?

12-5 Calculate the repulsion force between two protons brought the diameter of a nucleus apart—say 4×10^{-14} m. What does this value imply about nuclear attractive force? What assumption(s) is being made?

12-6 If a proton and an anti-proton were to annihilate each other, how much energy would be expected to be released?

12-7 The law of conservation of *A* says that the total number of nucleons in the universe is fixed. What if one should suddenly appear out of some very energetic radiation?

12-8 Given the rest mass of an electron as 0.51 MeV, what is the minimum energy of each gamma-ray produced when an electron and a positron annihilate, and to what wavelength does this correspond?

12-9 Suppose a nucleus were to radiate a gamma-ray of 0.5 MeV. How would this affect its mass number and its atomic number?

Answers to Exercises

12-1 $4^{12} = 1.68 \times 10^7$; $1.68 \times 10^7 \times 1.6 \times 10^{-19} = 2.69 \times 10^{-12}$ coul.

12-2 The effect would be to bend them in an arc upward. (The box is drawn to aid three-dimensional visualization.)

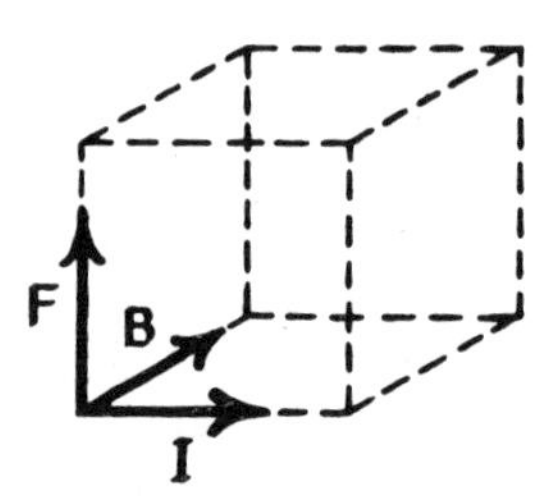

12-3 The masses, in order, are 200, 28, 16, 12. This is calculated from the relation $mx^2 = ky = m'x'^2$. The ions are, respectively, Hg^+, CO^+, O^+, and C^+, as in Figure 12-8.

12-4 Set up the forks on boards or hollow boxes. Start one vibrating. After a short time, silence it by grasping the tines in the hand. Then listen for the other fork.

12-5 $\mathbf{F} = Kq_1q_2/s^2 = 1.43 \times 10^{-1}\text{J} = 8.9 \times 10^{17}$ eV. This implies that the nuclear attractive force must be very great, provided of course that Coulomb's law holds in these extreme conditions.

12-6 $\mathbf{E} = mc^2 = 1.86 \times 10^9$ eV.

12-7 It would be impossible to say that another had not disappeared elsewhere. A law of this kind is not operationally testable. It is more like an extreme act of faith.

12-8 Each gamma-ray would have an energy of 0.51 MeV, since two are produced. By $\mathbf{E} = h\nu$, this corresponds to $\lambda = 0.024$ Å.

12-9 0.5 MeV is equivalent to 5.37×10^{-4} u. This would have negligible effect on A, which is stated in whole units. There is no effect on Z because the gamma-ray is uncharged.

Bibliography

A list of the great particle accelerators that achieve energies of 1 BeV or more and where they are is to be found in John H. Martin, "Smashing the Atom," *Industrial Research*, March 1965, p. 45.

Excellent accounts of recent developments in high-energy physics and chemistry, and of the structure of nuclei, are given in Geoffrey F. Chew, Murray Gell-Mann, and Arthur H. Rosenfeld, "Strongly Interacting Particles," *Scientific American 210*, No. 2 (February), 1964, p. 74, and Victor F. Weisskopf, "The Three Spectroscopies," *Scientific American 218*, No. 5 (May), 1968, p. 15.

Parity and other difficult matters are discussed authoritatively in his well-known sprightly style that combines readability with accuracy and authenticity (he was a nuclear physicist of renown) by George Gamow, *The Atom and Its Nucleus*. A Spectrum Book, Prentice-Hall, Englewood Cliffs, N. J., 1961. See also Ralph T. Overman, *Basic Concepts of Nuclear Chemistry*, Reinhold, New York, 1963, a little paperback in the series Selected Topics in Modern Chemistry, written by the Chairman of the Special Training Division of the Oak Ridge Institute of Nuclear Studies; and Irving Adler, *Inside the Nucleus*, a Signet Science Library Book, The New American Library, New York, 1963, a popular exposition.

Other useful sources are:

M. A. Preston, "Resource Letter NS-1 on Nuclear Structure," *American Journal of Physics*, *32*, No. 11 (November) 1964, p. 1. This is an annotated bibliography. It is available from American Institute of Physics, 335 East 45 Street, New York, 10017. (Enclose stamped return envelope.)

George Gamow, *Thirty Years that Shook Physics. The Story of Quantum Theory*, Doubleday, Garden City, New York, 1966.

Robert R. Wilson and Raphael Littauer, *Accelerators. Machines of Nuclear Physics*. Anchor Books, Doubleday, Garden City, New York, 1960.

John H. Martin, "Smashing the Atom," *Industrial Research*, March, 1965, p. 42.

Hans A. Bethe, "What Holds the Nucleus Together?" *Scientific American* (September) 1953. (W. H. Freeman & Company Reprint No. 201)

Russell R. Williams, *Principles of Nuclear Chemistry*, Van Nostrand, New York, 1950.

Rogers D. Rusk, *Introduction to Atomic and Nuclear Physics*, ed. 2. Appleton-Century-Crofts, New York, 1964.

Henry A. Boorse and Lloyd Motz, *The World of the Atom.* Volumes I and II. Basic Books, New York, 1966. This is an excellent compendium of articles chosen from the literature, or especially composed. It is recommended for source material.

Henry A. Boorse and Lloyd Motz, "Elementary Particles," in *The World of the Atom*, by these authors. Reference 3, Volume II, p. 1799. This is a well-written and sophisticated review which gives a balanced overall picture.

Chapter 13

13.1 Introduction
13.2 Radioactivity
13.3 Half-life
13.4 Nuclear energy
13.5 Mass-energy relations
13.6 Induced radioactivity
13.7 Induced nuclear fission
13.8 Fusion reactions
13.9 Plasma
13.10 Stellar energy
13.11 Radioactive dating
13.12 Summary

Nuclear Properties and Processes ~ 13

Introduction 13.1

In this chapter we turn to macroscopic nuclear properties and processes. Of course, these behaviors are presumed to occur because of the properties of the nuclei themselves—the internal properties, so to speak, that were hinted at in discussing energy levels and conservation laws in the previous chapter. These two chapters are supplementary. Here we are chiefly concerned with bulk behavior, with accounting procedures by which the masses and energies of reacting nuclei, or of radioactive materials consisting of quantities of nuclei, are kept track of.

Radioactivity 13.2

Quite soon after Becquerel discovered radioactive substances it became evident that radioactivity is a property of the nucleus of the atom. Nuclear physics and chemistry were born of this discovery. It came about when it was found that chemical treatment such as forming compounds, precipitating, boiling with the strongest acids, or the most caustic alkalies, has absolutely no effect upon the radioactive behavior of these substances. Also, heating them to the highest temperatures available had no observable effect. Thus the seat of radioactive behavior must be in a region of the atom quite untouched by ordinarily available energetic treatment.

Natural radioactive substances show two kinds of radioactive behavior—of spontaneous nuclear fission. In one type the nucleus expels an α-particle $_2He^4$ along with a γ-ray. In the other type of fission the nucleus expels a β-particle $_{-1}e^0$. Here, also, a γ-ray is given off. In both cases the nuclide is said to have undergone radioactive decay. Other fragments may also be given off (Section 13.3). We shall pay little attention to these.

In the first case, α-dccay, the parent nuclide transmutes to one with a mass number 4 units less, and an atomic number 2 units less. For example (Figure 13-1):

$$\underset{\text{Uranium}}{_{92}U^{238}} \rightarrow \underset{\alpha\text{-particle}}{_2He^4} + \underset{\text{thorium}}{_{90}Th^{234}} + \underset{\text{gamma-ray}}{\gamma}$$

That is, the uranium isotope 238, upon α-decay (the fact of α-decay being an experimental observation), transmutes to the thorium isotope 234 (loss in mass number of 4, loss in nuclear charge of 2) and simultaneously gives off an α-particle and a gamma-ray. In an equation for radioactive decay, the sum of the subscripts must be the same on both sides of the arrow, as must the sum of the superscripts. Charge-number is conserved and mass number is conserved: both Z and A are conserved.

In the second case, β-decay, the parent nuclide transmutes to one with an atomic number larger by one, because the departure of a negative electron leaves behind one (+) charge. But the loss of the electron has negligible effect on the mass of nuclide, so the mass number does not change. For example (Figure 13-1):

$$_{90}Th^{234} \rightarrow {_{91}Pa^{234}} + {_{-1}e^0} + \gamma$$

Upon β-decay, the thorium isotope 234 yields the protoactinium isotope 234 plus a β-particle and a γ-ray. (Of course, chemists have to look—with a mass spectroscope—to see what is produced. They give it a name.)

Three radioactive series have been found in nature, and a fourth has been found among synthetic elements. In each series a high atomic number element begins the series, and successive decays lead eventually to a stable terminus. The three naturally occurring series terminate in stable isotopes of lead (Pb); the fourth terminates in a stable isotope of bismuth (Bi).

13.3 Half-life

The rate of decay of a radioactive element is governed by chance. The image given by Gamow is of the nucleons in the nuclide fluctuating in energy as they 'collide' with each other and as they come up against the 'barrier' produced by the binding forces. Every once in a while one escapes, as an α- or β-particle, and the parent nucleus, now transmuted into a different element, 'settles down' from its excited

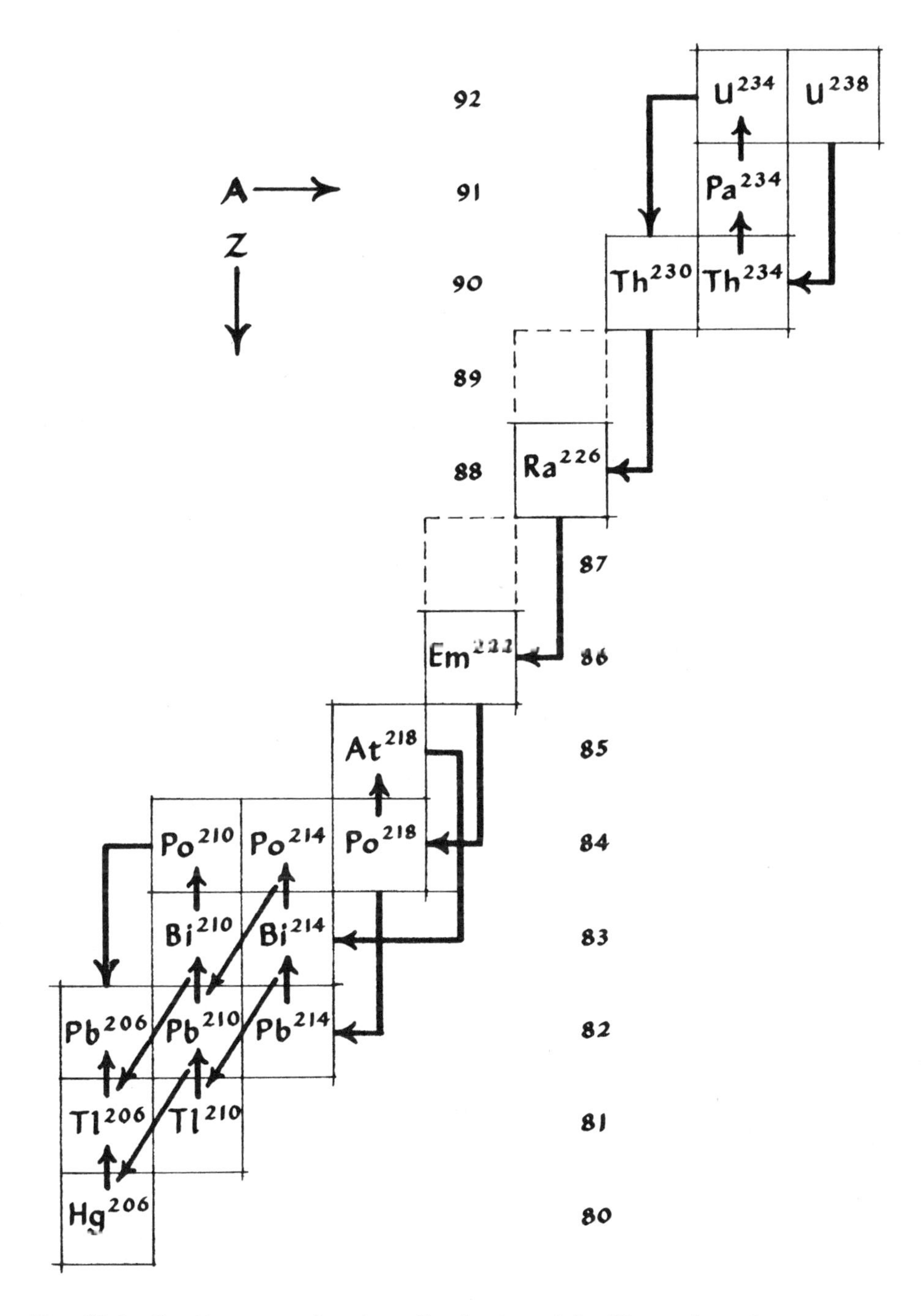

Fig. 13-1 *Graphic representation of a radioactive decay-chain. The atomic numbers are plotted* vs. *mass numbers. The data are from Table 13-1.*

state by giving off excess energy as a γ-photon as discussed in Section 12.7. An anti-neutrino, or some other particle about which we will not be further concerned, may be given off.

If the decay rate is governed by chance, then probability laws can be applied. Suppose we have a nucleus that is capable of α-emission (just to be definite). We consider a time interval starting now, t_0, and ending after a period Δt, at t_1. What is the probability that it will emit an α-particle? (For that is what we measure; moreover, we must know that the substance is radioactive and that it is an α-emitter, and the only way to find this out surely is to examine the evidence given by the radioactive substance itself.) But given the data that the substance is radioactive, and the hypothesis that the process is a chance one, then, letting the probability that it will occur be p, one can reason thus:

$$p \propto \Delta t$$
$$p = \lambda(\Delta t)$$

Here λ is a constant for this particular isotope, the 'decay constant.' (It has nothing to do with wavelength. Eventually we run out of letters and other symbols, so some do double duty.) Because the sum of the probabilities of decay and survival must equal unity, since nothing else can occur, the probability of *survival* of the nucleus is

$$1 - p = 1 - \lambda(\Delta t)$$

This states the probability of survival of a nucleus during one period, Δt. Now a chance process is not governed by its previous history, and if a nucleus survives, the probability of its decay remains the same during the next period. The situation is somewhat like that of flipping a coin, or throwing a die (Section 16.2, 3). Given two intervals in succession, the probability of survival becomes (by probability theory).

$$[1 - \lambda(\Delta t)]\cdot[1 - \lambda(\Delta t)] \text{ or } [1 - \lambda\Delta t]^2$$

The calculation of probability becomes increasingly reliable the larger the number of intervals taken. We can either make the number of intervals, n, large, or Δt, small, with the same result, getting a large number of intervals.

Going through the calculations for $[1 - \lambda\Delta t]^n$ where Δt is very small and n very large, one finds that $\lambda = 0.693\ T$, where T is the time that it takes from when we start measuring for half of the material present to decay. T is the half-life of the radioactive material.

As an example, radium $_{88}Ra^{226}$ has been found to decay at such a rate that its half-life is 1600 years. It decays to emanation $_{86}Em^{222}$ with emission of an α-particle and of some γ-radiation (Section 12.7). We can therefore calculate the data for a chart showing the rate of decay of radium (Figure 13-2). In determining half-lives of radioactive substances, the amount of remaining material is measured after

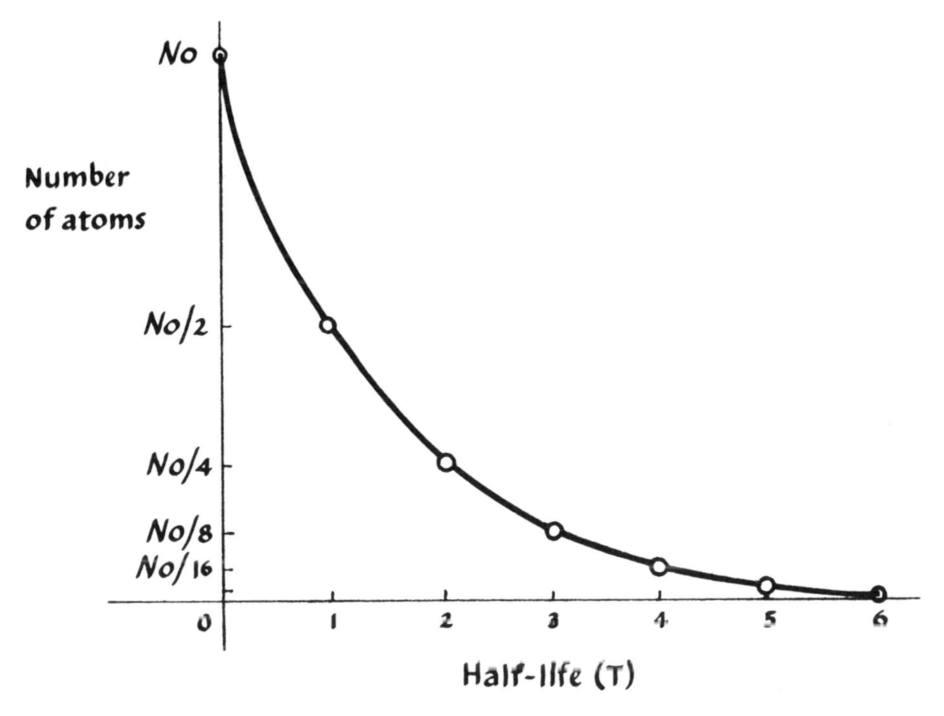

Fig. 13-2 *Generalized scheme for the decay of a radioactive substance in terms of the half-life* T *and the number of atoms* N_0 *present at the start of the period of interest.*

periods of time following some initial analysis at $t = 0$. From such data, T and λ can be calculated. You do not have to wait 1600 years to find out radium's half-life. Careful determination of the rate of the decay over a practicable period of time is sufficient, given the theory.

Returning to radioactive series, a typical example that illustrates decay modes and the nuclear bookkeeping as far as A and Z are concerned is shown in Table 13-1. The old names were given by analysts before the various isotopes were definitely identified. The new name can always be found by suitable bookkeeping with respect to the particles emitted, via the atomic number, and the table of elements (see Appendix 4). Most of the analytic work of determining mass numbers was done with the mass spectroscope. The table is shown graphically in Figure 13-1.

Nuclear energy 13.4

The energy of nuclear processes is in a category by itself compared with ordinary chemical or electrical energy processes (Table 13-2).

Table 13-1

The Uranium-Radium Radioactive Series

Old name	New symbol	Element	Half-life and branch ratio	Emission
Uranium I	$_{92}U^{238}$	Uranium	4.5×10^9 years	α
Uranium X_1	$_{90}Th^{234}$	Thorium	24 days	β, γ
Uranium X_2	$_{91}Pa^{234}$	Protoactinium	1.14 minutes	β, γ
Utanium II	$_{92}U^{234}$	Uranium	3×10^5 years	α
Ionium	$_{90}Th^{230}$	Thorium	8.3×10^4 years	α, γ
Radium	$_{88}Ra^{226}$	Radium	1600 years	α, γ
Radium emanation	$_{86}Em^{222}$	Emanation	3.8 days	α
Radium A (*1)	$_{84}Po^{218}$	Polonium	3.05 minutes	α
↓				
Radium B	$_{82}Pb^{214}$	Lead	26.8 minutes	β, γ
Radium C (*2)	$_{83}Bi^{214}$	Bismuth	19.7 minutes	β, γ
↓				
Radium C′	$_{84}Po^{214}$	Polonium	10^{-5} seconds	α, γ
Radium D (*3)	$_{82}Pb^{210}$	Lead	22 years	β
↓				
Radium E (*4)	$_{83}Bi^{210}$	Bismuth	5 days	β, γ
Radium F	$_{84}Po^{210}$	Polonium	140 days	α
Radium G	$_{82}Pb^{206}$	Lead	Stable	—
↓				
(*1) Radium A			(0.02%)	β
Astatine-218	$_{85}At^{218}$	Astatine	1.3 seconds	α
↓				
(*2) Radium C			(0.04%)	α
Radium C″	$_{81}Tl^{210}$	Thallium	1.32 minutes	β, γ
↓				
(*3) Radium D			(2×10^{-6}%)	α
Mercury	$_{80}Hg^{206}$	Mercury	8.5 minutes	β, γ
Radium E″	$_{81}Tl^{206}$	Thallium	4.3 minutes	β
Radium G		Lead	Stable	
(*4) Radium E				α
Thallium				

* The asterisk indicates 'branching,' where the reaction may go more than one way.

For example, one of the hottest conventional flames is that of the oxy-hydrogen blow-torch. Hydrogen is burned in oxygen with the production of water and with the release of energy, as shown by the equation:

$$\underset{(4\text{ g})}{2\,H_2} + \underset{(32\text{ g})}{O_2} \rightarrow \underset{(36\text{ g})}{2\,H_2O} + 116\text{ kcal of energy}$$

Table 13-2

Energies per Particle *

Kinetic energy of one water molecule 450-foot waterfall	0.00025	eV
Average kinetic energy of a gas molecule at room temperature	0.025	eV
Carbon atom oxidized to CO_2	4.0	eV
Visible photon	2.0	eV
Ultraviolet photon	3.0 to 100.0	eV
Hard X-ray photon	0.1 to 1.0	MeV
Gamma-ray photon	1.0 to 3.0	MeV
Radium disintegration	4.8	MeV
Fission disintegration, uranium	200.0	MeV
Cosmic ray particle	1.0 to 10.0	BeV

* From J. A. Richards, Jr., F. W. Sears, M. R. Wehr, and M. W. Zemansky, Addison-Wesley Publishing Co., Reading, Mass. 1962, p. 924, Table 44-1.

This equation, typical of chemical equations, is written in moles. One mole of oxygen molecules contains 6.02×10^{23} molecules. If we calculate on the basis of atoms of hydrogen, of which $4 \times 6 \times 10^{23}$ are reacted to yield 116 kcal, we find that the energy production is about 1.25 electron volts per atom of hydrogen.[1] The disintegration of one atom of radium yields nearly four million times as much energy.

Mass-energy relations 13.5

That mass and energy are interconvertible under certain conditions harks us back to Section 6.14. Einstein's relativistic equation $\mathbf{E} = mc^2$ is abundantly confirmed in nuclear science. This equation relates two concepts: binding energy and mass deficit. Binding energy (Section 12.7) is the energy that would be needed to pry a nucleus apart into its separated protons and neutrons. (It is essentially a calculated value, since no reactions are available to dissociate any large nucleus in this manner.) Mass deficit is the difference in mass number between the calculated sum of the masses of protons and neutrons in the nuclide and the actually observed mass of the nuclide itself. This latter is experimentally determined—for example, with a mass spec-

[1] The conversion factor from kcal to joules is $1/2.389 \times 10^{-4}$; also, 1 eV = 1.6×10^{-19} J. Combining this with Avogadro's number of particles in a mole, namely, 6.02×10^{23}, gives the conversion from kcal per mole to eV per particle: eV $\times$ 23.053 = kcal/mole. Since $4 \times 6 \times 10^{23}$ hydrogen atoms are involved, this calculates to about 1.25 eV per hydrogen atom, or 2.5 eV per atom of oxygen.

trograph (Section 12.4). An alpha-particle is a very stable nuclide. From its symbol, $_2He^4$, it is apparent that it contains two protons (Z = 2) and two neutrons (A-Z = 2). Summing their masses (Appendix 3) and finding the difference between the calculated value and the actually observed value of the α-particle (using rest masses in all cases, of course, so as to avoid additional energy-mass) yields the mass-deficit:

mass of 2 protons = 2 × 1.00782522 =	2.01565044 u
mass of 2 neutrons = 2 × 1.00866544 =	2.01733088 u
Total, calculated	4.03298132 u
Found rest mass of α-particle	4.00260361 u
Difference (mass-deficit)	0.03037771 u

It is supposed that if two protons and two neutrons were made to combine into the α-particle, 0.03037771 u of mass would be given off in the form of energy. This would amount to 0.03037771 × 931.478 = 28.3 MeV per α-particle formed, or a little over 7 MeV per nucleon. The α-particle is stable by the amount of mass-energy lost—that is, by the binding energy—the energy that is, so to speak, "not there;" that would have to be supplied as energy, and/or mass, to go back to the original four separated nucleons.

To find the binding energy of a nuclide, calculate the sum of proton and neutron masses and subtract the actual observed rest mass of the nuclide. This gives the mass deficit. Multiply by 931.478, or a rounded-off value, to obtain the binding energy in MeV. To find the energy per nucleon divide by the number of nucleons.

When calculations of this kind are made, an important fact emerges. The binding energy per nucleon for stable nuclei is relatively low for the light nuclides (except helium) and increases with atomic mass number up to about 50, after which it declines slowly (Figure 13-3). This means that elements near the middle of the Periodic Table are most stable. That is, if radioactive elements decompose they tend to end up as stable heavy nuclides. If heavy nuclides are forced to decompose (Sections 13.6, 7) they do not set up a chain reaction that continues down to small fragments, but instead there result stable products in the mass number 50 region. If light nuclides are fused (Section 13.8) they do not produce a run away reaction. The binding energy per nucleon seems to reflect a balance between the constraints of nuclear forces holding the nucleons together, and the Coulomb repulsion of the (+) charges.

13.6 Induced radioactivity

When a normally stable element is made radioactive by artificial means, it is said that radioactivity is induced in it. We can picture the

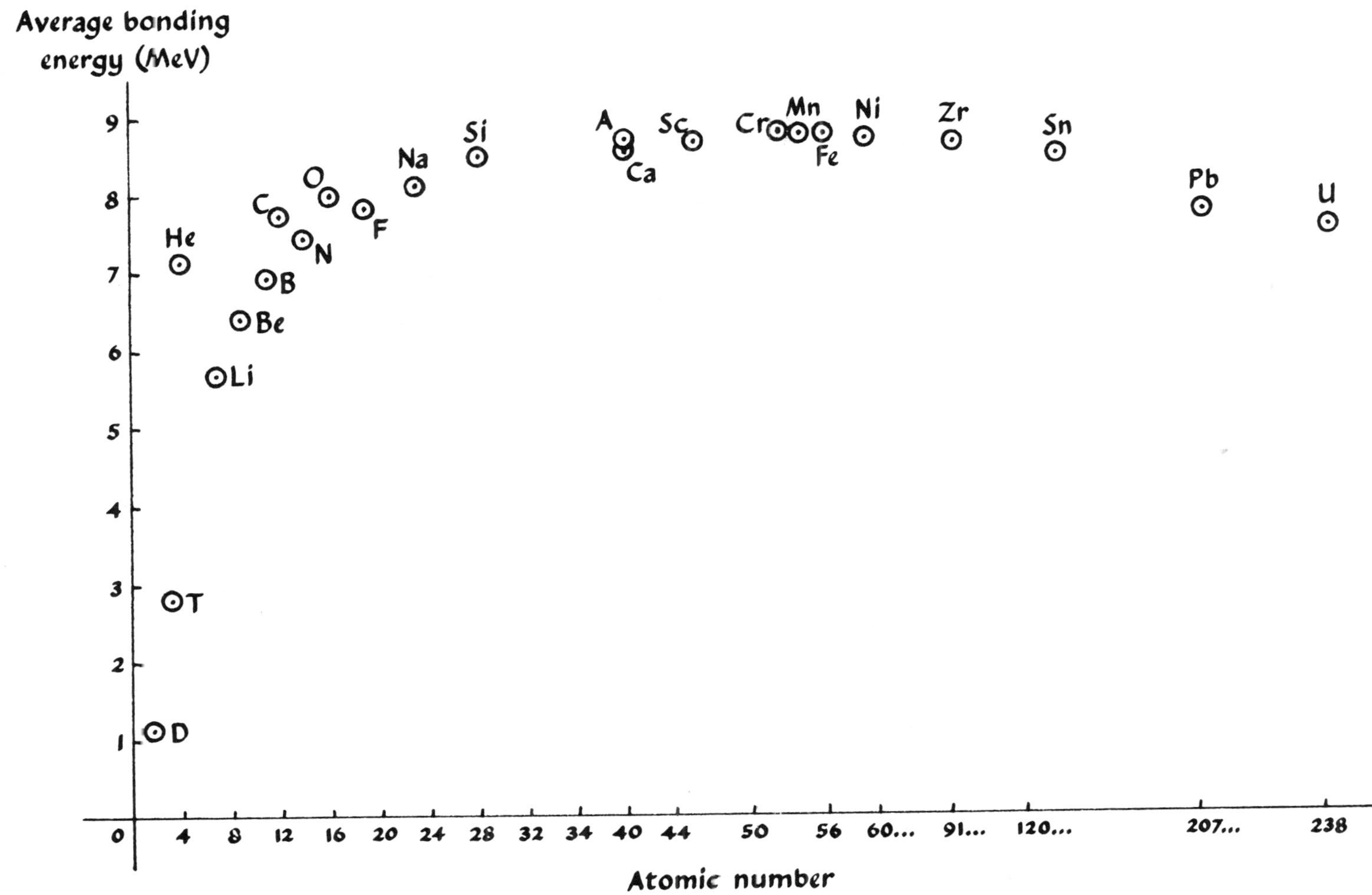

Fig. 13-3 *Average bonding energy per nucleon is plotted against atomic number.*

process in terms of energy levels (Section 12.7). Suppose a nucleus $_ZX^A$ is in a stable state, with low potential energy, its nucleons down in the bottom of its energy-well. A slow neutron, falling (or wandering) into this nucleus, converts it to $_ZX^{A+1}$ which may now be in an unstable state because the ratio of neutrons to protons has been altered (Figure 13-4). The atom is in an excited state $[_ZX^{A+1}]^*$ from which it returns to a new stable state with mass number increased by 1, by emitting the excess energy as a γ-ray (Figure 13-5). (Excited states are often designated with an asterisk.) Suppose, instead, the neutron

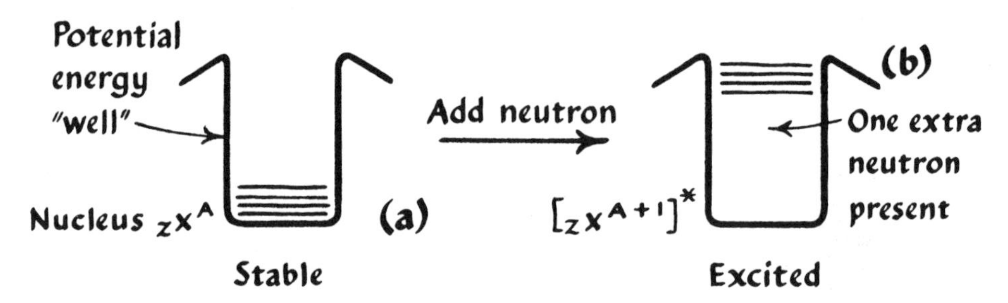

Fig. 13-4 *In the ground state* (a) *the energy levels occupied by nucleons are near the bottom of the potential energy "well;" the entering neutron raises some at least of the nucleons to higher energy states. The nuclide is excited* (b).

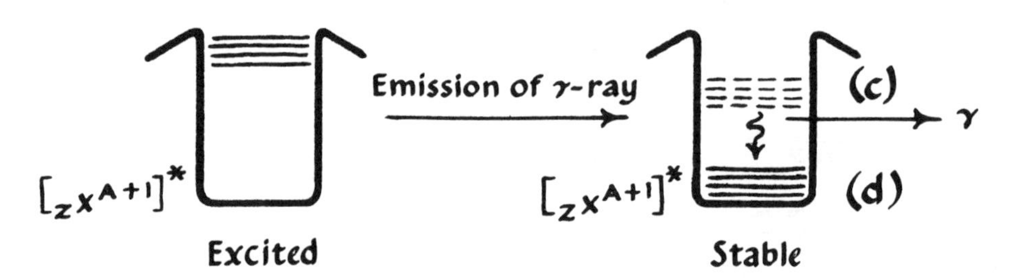

Fig. 13-5 *The excited nuclide gives off a gamma photon, falling to a less excited state* (c) (*compare Figure 12-12*). *By giving off one or more gamma photons it converts to a new stable state of a nuclide with* Z *unchanged, and* A *increased by one mass unit.*

brings in more energy (it might enter with higher velocity). Then, the energy of the product may exceed that (negative energy) of the potential barrier that gave the atom its stability (Figure 13-6). It might have gained enough energy to decompose by a mechanism such as emission of an α-particle, or of a proton or neutron, that brings it back to a stable state. Or it may fission into fragments of approximately the same size. In each case mass energy is conserved.

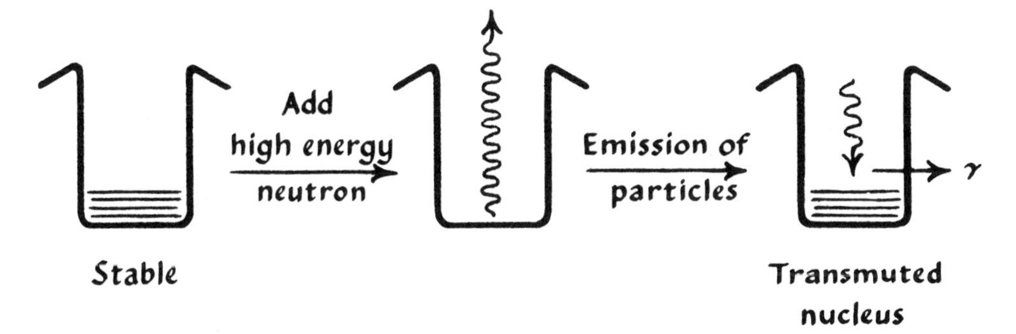

Fig. 13-6 *A stable nuclide, receiving a high-energy neutron, is excited to the extent that a particle (α-, β-, or some other) is emitted. The resulting nuclide then settles down to a stable state by giving off excess energy in the form of gamma radiation.*

For example, if a stable $_{13}Al^{27}$ nuclide reacts with a slow neutron, there is formed the excited isotope $_{13}Al^{28}$ which is radioactive, gives off a γ-ray and becomes stable $_{13}Al^{28}$.

$$\underset{(26.981535\ u)}{_{13}Al^{27}} + \underset{(1.008665\ u)}{n} \rightarrow [_{13}Al^{28}]^* \rightarrow \underset{(27.98190\ u)}{_{13}Al^{28}} + \underset{\substack{(0.008292\ u)\\(7.7\ MeV)}}{\gamma}$$

Should the neutron come into the $_{13}Al^{27}$ nucleus with high energy, the following reactions may occur. But since the products show more mass than the reactants, the entering neutron must make up the balance with the energy it carries:

$$_{13}Al^{27} + {_0n^1} + \text{energy} \rightarrow {_{12}Mg^{27}} + {_1H^1}$$
$$_{13}Al^{27} + {_0n^1} + \text{energy} \rightarrow {_{11}Na^{24}} + {_2He^4}$$
$$_{13}Al^{27} + {_0n^1} + \text{energy} \rightarrow {_{13}Al^{26}} + 2\,{_0n^1}$$

Cf. Ralph T. Overman, *Basic Concepts of Nuclear Chemistry*, Reinhold, New York, 1963.

Induced nuclear fission 13.7

Of course, any radioactive decay involves some nuclear decomposition, or fission. This might be spontaneous, or it might be induced by bombarding the nucleus with some particle or high energy radiation. Out of studies of the production of artificially radioactive substances there came a suggestion in 1939 by Lise Meitner and Otto R. Frisch. Results obtained by Enrico Fermi and other nuclear physicists, they thought, showed that when uranium is bombarded with neutrons the nuclide breaks up into two more or less equal parts with the release of some 200 MeV per atom and with the additional production of two or more neutrons. It was this last observation that lent great excitement to the suggestion for it meant that if a single neutron would cause an atom to fission and produce two or more neutrons, then these might produce four or more, so that the reaction would cascade, or escalate. Since each reaction takes only a short time but gives off much energy, the cascade could lead to an explosion.

The suggestion was immediately taken up. The results of further investigation were shrouded in secrecy. By now enough information has been released that the following description of the reaction may be given. It is found that the reactive isotope is U^{235}. When this absorbs a slow-moving neutron, reaction occurs:

$$U^{235} + n \rightarrow$$
$$\text{(smaller nuclides)} + \gamma + 2.5\,n \text{ (on the average)} + 200 \text{ MeV of energy}$$

This is not a clean reaction such as we shall find in some chemical reactions of molecules. When the uranium isotope of mass number 235 absorbs a slowly-moving neutron, it transiently forms an unstable complex which promptly fissions to produce pairs of nuclei. These may range in mass numbers from the low seventies to some 160. Mass number is always conserved. These fission products are, in general, unstable. At the same time there are produced two neutrons. Some of the fission products are themselves neutron emitters, so that on the average about 2.5 neutrons result promptly per reaction of U^{235} with one neutron. In principle, then, other things being equal, this reaction could sustain itself, because at each reaction step there are produced more neutrons than are required to initiate the step.

But the first fission reaction that was devised using natural uranium was not self-sustaining. The important reasons turned out to be that the neutrons appeared with considerable kinetic energy (of the order of 2 MeV) and many of them escaped from the uranium mass without reacting. In the second place, naturally occurring uranium consists very largely of U^{238}, and this nucleus is able to capture neutrons without fission:

$$_{92}U^{238} + {_0}n^1 \rightarrow {_{92}}U^{239} + \gamma \rightarrow {_{93}}Np^{239} + {_{-1}}e^0$$
$$^{93}Np^{239} \rightarrow {_{94}}Pu^{239} + {_{-1}}e^0 + \gamma \text{ (half-life } 2.44 \times 10^4 \text{ years)}$$

That is, U^{238} absorbs a neutron to form neptunium-239, which has a very long half-life.

Neutrons that become slowed down may be captured by U^{235}, with resultant fission. But there are too few of these events to sustain a chain reaction in natural uranium. This is because the isotope U^{235} is present to only 0.7% in natural uranium. Fast neutrons also bring about fission of U^{235}, but they are captured with one-hundredth or less the probability with which slow neutrons are captured. Thus at least three factors having to do with neutrons conspire to prevent a piece of ordinary uranium metal from blowing up: escape from the metal of fast neutrons, absorption of neutrons by U^{238} without fission, and the low ability of U^{235} to capture fast neutrons. (Additionally there is the low concentration of U^{235}.)

To obtain a self-sustaining reaction (as in a nuclear-energy plant) or a self-accelerating reaction (as in a nuclear bomb) these factors have to be manipulated. The first requirement is to separate out the U^{235}, or at least to enrich the mixture by removing U^{238} so that any neutrons present have an improved chance of colliding with a 235 nuclide. This can be done in various ways, and it was a technological triumph of the first order that between 1939 and 1942, when on December 2 the first self-sustaining "atomic pile" was brought into operation, the necessary physical and chemical methods were worked out and the necessary huge plant was built to concentrate U^{235}. Thus one of the factors that stood in the way of producing a self-sustaining reaction was brought under control: U^{235} was concentrated. The concentrate took the form of metal slugs or rods.

A second factor is the low ability of U^{235} to capture fast-moving neutrons. This difficulty is overcome by slowing down the neutrons produced in the pile by fission. Atoms with light nuclei are provided in the pile. Neutrons collide with these, and they recoil, taking up energy from the neutron without absorbing it. The atoms that serve this purpose are provided by water, particularly heavy water, and by highly purified graphite (a form of carbon). By constructing a pile with enriched uranium metal slugs interspersed in a regular pattern among blocks of graphite it becomes possible to slow down neutrons. These are emitted in the fission process with 2 MeV energy. They are slowed to 0.025 eV, to 'thermal neutrons,' see Table 13-2. These are readily captured by U^{235}, causing it to fission and to give off more fast neutrons.

The ability of a nucleus to capture a particle that impinges on it is measured in terms of 'cross-section.' This measures the probability of reaction between the nucleus and the particle. It varies from nucleus to nucleus, and it varies with the nature of the bombarding particle. For example, a particle may be charged (+) and hence be repelled by the nucleus. It varies with the speed of the particle. U^{235} has a larger cross-section for absorption of slow neutrons with fission than for faster neutrons. By subjecting the given substance to bombardment by a known number of neutrons per second, and measuring the number captured without fission, the number captured with fission, the number scattered by non-capture collisions, and the number that pass right through the substance, one can determine cross-sections that enable proper bookkeeping, and calculation for a nuclear process. (The unit of cross-section is the 'barn'—an area of 10^{-24} cm^2. The name is related to the well-known problem of 'hitting a barn door.') The self-sustaining nuclear reaction, once it became controllable through manipulation of the variables described above, could be utilized as a source of sustained energy production (as in a power-plant) or of explosive violence (as in a bomb). The same fission reaction occurs in either case but the method of control differs (Section 17.8). Other energy-producing fission reactions have been studied and are being employed. We do not pursue them here, since few new principles are involved.

In the fission of U^{235} or in any fission process initiated by a neutron, the barrier to the reaction is low because it is relatively easy for even a slow neutron, being neutral in charge, to come close enough to a nucleus for the short-range nuclear force to operate on it (Figure 12-10). When fission occurs, the major part of the energy appears as kinetic energy of the fission fragments; some is lost as gamma radiation. But by the use of a heat-exchanger which might be, for example, a stream of water or other fluid which acts as a cooling agent flowing through pipes in the energy-producing zone of the reactor, the heat energy can be drawn off and converted to other forms of energy. For example, the water might be heated to high pressure steam. The reactor might drive a steam turbine which in turn generates electricity. The most serious problem is the disposal of the waste products of fission, which are radioactive. This is one of the sources of concern with the process, together with the always-present but actually vanishingly small possibility of a runaway reaction due to failure of the control system. All atomic energy reactors are provided with multiple safety devices to prevent explosive behavior.

13.8 Fusion reactions

Fission reactions occur with nuclides at the high atomic number end of the Periodic Table, and lead to products near the middle of the Table, where the binding energy per nuclide is higher for products than reactants. Fusion reactions are applied to low atomic number nuclides which upon combination yield energy, and produce products the binding energies of which per nuclide are also higher for products than for reactants. In the former case the production of nuclides of highest binding energy is approached from the high atomic mass number side; in the latter from the low. In both cases energy is released.

In the case of fusion, the problem basically is to bring two nuclei which are positively charged close enough together—with their centers about 10^{-14} m apart—for the short-range nuclear forces to take effect. The Coulomb law (Section 3.8) states that the force of repulsion between like-charged particles increases inversely as the square of the distance apart of their centers of charge, and directly as the product of their charges. The charge on each nucleus is Ze, and thus the higher the atomic number, the more force is required to bring the nuclei together; also, the closer together they are brought, the more force is required. It follows that the former difficulty can be minimized by using low mass-number nuclides in the fusion process. This is part of the reason for investigating nuclides such as ${}_1H^1$, ${}_1H^2$, ${}_1H^3$, and ${}_3Li^7$ as potential fuel. The latter difficulty cannot be avoided: the nuclei must be brought within range of nuclear forces.

The major problem, then, is to cause the nuclei that comprise the fuel of the process to approach each other on a collision course with sufficient velocity to bring them within reacting distance. This requires very high velocity. Particles moving at high velocity in a closed region and in a random manner with respect to each other are said to be 'hot.' The molecules of the air at room temperature move on the average at about the speed of a rifle bullet. Some move faster, some more slowly. Room temperature is about 293°K. (20°C) To have some nuclei move with a velocity great enough to have them come into contact they would have to be heated to about 100 million 'degrees.' And to have the fusion reaction go at an appreciable rate would require some 300 million 'degrees of temperature.' This is a good deal hotter than the temperature estimated for the interior of the Sun. The fusion reactions that take place there are aided by the tremendous gravitational force of the Sun, which helps to compress the nuclei. The fusion temperature of a hydrogen bomb is achieved by an atom bomb, the explosion of which generates for an instant, locally, sufficient heat and pressure to set off the fusion reaction.

Plasma 13.9

When matter is raised to the high temperature mentioned in the previous paragraph, it is called plasma. This special name is given because no molecules or even atoms can be present for any length of time: it is a special state of matter. The violence of the collisions—at 100 million degrees[1] a particle would have an energy of about 10,000 eV —breaks bonds and strips electrons from nuclides.

---

[1] What is called the kinetic temperature is calculated for a particle in motion by the relations below, where the energy referred to is average energy:

1 eV corresponds to 1.16×10^4 °K
1 °K " " 8.62×10^{-5} eV
1 °K " " 1.38×10^{-22} J

The average energy of a particle at room temperature thus calculates to 0.025 eV (Table 13-2).

The pressure of a gas on its container is related directly to the square of the velocity of the gas particles, to their mass, and to the number of them bombarding the walls of the container. Plasma, with its high-velocity particles, obviously exerts pressure, and to keep this within practical limits—with respect to the 'container'—the density of the plasma is kept low. In most instances, it is kept in the neighborhood of 10^{15} particles per milliter, or about 10^{-4} atmosphere: usually it falls within the range from 10^{14} to 10^{18} particles per ml.

The problem of a 'container' for plasma is obviously an exigent one. At 300,000,000°K no material can serve as a container—every known substance would be decomposed and vaporized. However, because all particles in plasma are ionized, it is possible to contain them in a *suitably shaped magnetic field.* We saw in Section 3.12 that a flow of charges constitutes a current, and that with it is associated a magnetic field (Figure 4-6). If the particle moves in a magnetic field, it is affected by a force at right angles to both the field direction and the direction of motion (Figures 4-8, 4-9). Under suitable conditions the particle may thus be made to take a circular path (as in Figure 16-20), and a mass of plasma between the poles of a properly shaped magnet may be constrained to a doughnut shape (Figure 13-7). Indeed, if in the plasma moving in this way there is induced a current (Section 4-10) this will generate its own magnetic field at right angles to the current, and then the charged particles will respond to this field by being "pinched" in toward the central part of the field, to give a denser, and therefore more compressed, hotter plasma. This device provides electromagnetic "walls" to constrain the plasma. Magnetic constraint, however designed, appears to be the most promising method of plasma confinement, although so far (1969) both technical and conceptual difficulties of great complexity have prevented any sustained reaction from being achieved. The plasma is said to be held in a 'magnetic bottle.' In the Sun and other stars the enormous gravitational forces that are present serve to confine the particles of plasma to the hot region and to compress them.

The advantages of fusion power, if it could be controlled, are that the fuel is cheap and plentiful, and especially that the products of the reaction are not radioactive and so would not present the ferocious problems of waste disposal that are already plaguing fission power plants.

13.10 Stellar energy

Hans Bethe and Carl von Weizsacher have suggested, on a principle proposed by Robert Atkinson and Fritz Houtermans, that the energy released by the Sun and other stars might arise from a sequence

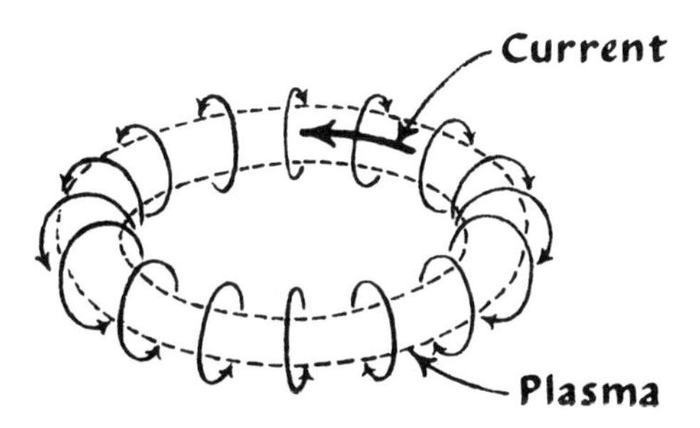

Fig. 13-7 "*The Pinch Effect. As current enters the tube under the influence of the voltage induced in it, an encircling magnetic field begins to form. This field exerts an inward force on the moving charged particles of plasma, pinching them to the center of the tube and thereby tending to 'insulate' the plasma from the walls of the tube. This compression also raises the plasma temperature.*" [*From A. S. Bishop,* Project Sherwood. The U.S. Program in Controlled Fusion. *Addison-Wesley, Reading, Massachusetts, 1958, Figure 3-1, p. 23. Reprinted by permission from the United States Atomic Energy Commission.*]

of reactions—a chain-reaction—which is started when a hydrogen nucleus (a proton) and a carbon nucleus interact. The net result of the process is the conversion of four hydrogen nuclei to one helium nucleus, He^4, thus:

$$
\begin{aligned}
H^1 + C^{12} &\rightarrow N^{13} + \text{energy} + \gamma \\
N^{13} &\rightarrow C^{13} + e^+ \\
H^1 + C^{13} &\rightarrow N^{14} + \text{energy} + \gamma \\
H^1 + N^{14} &\rightarrow O^{15} + \text{energy} + \gamma \\
O^{15} &\rightarrow N^{15} + e^+ \\
H^1 + N^{15} &\rightarrow C^{12} + He^4 \\
\hline
4H^1 &\rightarrow He^4 + 2e^+ + 26.7\ \text{MeV}
\end{aligned}
$$

The cycle is a slow one, with a total period of some 6 million years.

Another fusion process was proposed by Charles Critchfield. It is the proton-proton cycle:

$$
\begin{array}{l}
2\,(H^1 + H^1 \rightarrow D^2 + e^+ + \nu) \\
2\,(D^2 + H^1 \rightarrow He^3 + \gamma) \\
\quad He^3 + He^3 \rightarrow He^4 + 2\,H^1 \\
\hline
4\,H^1 \rightarrow He^+ + 2\,e^+ + 26.7\text{ MeV}
\end{array}
$$

This cycle is estimated to take 3×10^9 years, and appears to predominate in the Sun. It operates at some 20 million degrees K. In some stars the carbon cycle seems to predominate. Both of these cycles are too slow to serve for Earthly energy-production; but they are quite able to account for the energy-production of stars because of the colossal mass of these luminaries.

13.11 Radioactive dating

Naturally occurring radioactive disintegration-series such as the uranium-radium series, Table 13-1, and Figure 13-1, and the Thorium and Actinium series, Exercises 13-8, 13-9, end in stable isotopes. Thus, if a rock were found to contain an atomic ratio of uranium 238 to lead 206 of 1:1, it would imply that approximately half of the original amount of U^{238}, laid down when the rock was formed, had had time to decay, provided that no Pb^{206} had been deposited with the U^{238} in the beginning, and that none of the radioactive daughter products of the decay process, particularly the gaseous emanation (radon), had been leached out of the rock and thus lost to later accounting, and that the half-lives of the daughter products may be neglected as being very short in comparison with the half-life of the parent. There are, of course, accurate means of taking into account all the daughter products. Detailed calculations of these kinds, based on analyses of samples of the Earth's crust, and on the known half-lives of U^{238}, U^{235}, Th^{232}, and K^{40}, namely 4.5×10^9, 7.1×10^8, 1.4×10^{10}, and 1.3×10^9 years, have led to an estimate of $(4.55 \pm 0.07) \times 10^9$ years elapsed time since the Earth became an isolated body, and some 3×10^9 years since the solid crust of the Earth was formed.

Estimates of this kind, made in a preliminary way as long ago as 1907, using these naturally occurring 'radioactive clocks' have drastically changed previous estimates of the age of the Earth. They have shown that a very long time has been available for biopoesis (the coming into existence of living things) to take place, and for evolutionary processes to continue as they appear to have. The data from the uranium-lead series has been consonant with those obtained in other ways as, for example, from the helium that has accumulated due to trapped α-particles; from thorium-lead and actinium-lead decay chains; as well as potassium-argon ratios (for Ar^{40} is a product of K^{40}

decay, which occurs to about 0.110 of all fissions) and others. It is interesting that when these methods are applied to stony meteorites, the data suggest that they solidified (whatever their origin) some 4.5×10^9 years ago.

Radioactive C^{14} is a naturally-occurring isotope formed constantly in the upper atmosphere by a reaction of neutrons produced by cosmic rays. These, slowed down, react as follows:

$${}_7N^{14} + {}_0N^1 \rightarrow {}_6C^{14} + {}_1H^1$$

This carbon, as a component of carbon dioxide becomes mixed with the free carbon dioxide in the air. It then finds its way into the oceans, into living bodies, and as these die into dead organic matter. The ratio of C^{14} to C^{12} seems to have remained constant overall for thousands of years. There have been some fluctuations due to suddenly increased cosmic ray activity—perhaps from solar flares—which have shown up in examinations of tree rings. But on the whole it can be assumed that a piece of wood, bone, cloth, and so forth, which has been preserved, as in burial pits or mummified remains, originally contained the normal C^{14}/C^{12} ratio, so that the present ratio in the material gives information about its age. This has been checked back to about 5000 years ago by means of well authenticated specimens.

Radioactive isotopes with useful half-lives, such as C^{14}, can be used to 'label' organic molecules so that metabolic pathways may be followed. An animal is fed the labeled material; then after a period the animal is sacrificed and the presence of the label is sought in specific chemical substances, such as cholesterol, or a fat constituent, or a particular amino acid, and so forth. If there are no suitable radioactive tracers, a heavier or lighter isotope may be used—for example, C^{13} instead of C^{12}, or O^{18} may be used in place of some of the O^{16} in the particular molecule; or N^{15} instead of N^{14}; or H^2 instead of H. From studies with these substances many sequences of reactions on which life processes depend have been elucidated. These and other applications of isotopes are so widely practiced that a quite specialized literature has grown up. The uses spread into applications such as historical validation in history, geology, and anthropology; studies of reaction paths in biophysics, biochemistry, molecular biology, chemistry, geology, ecology; diagnostic and curative uses in medicine; applications in technology such as determining the wear of piston rings by measuring the amount of radioactive iron rubbed off during operation of the engine and appearing in the lubricating oil.

Summary 13.12

In a period of about two generations, from the discovery of radioactivity to the development of the hydrogen bomb and the construction of nuclear power plants, knowledge of nuclear properties has exploded in a manner to rival any other increase in knowledge. Rutherford and his students had produced transmutation of elements (an early alchemist dream) in some of their first experiments. Then the development of atom-smashers made transmutation commonplace. (Some transmutations build nuclides to higher mass numbers. That is how the artificial, trans-uranium elements have been made; most produce isotopes, or nuclides of smaller mass number.) In investigating these nuclear processes it was discovered that in some reactions more neutrons are produced than are required to initiate the reaction, so that a reaction once started might become self-sustaining, even explosive. Einstein's equation relating mass and energy was confirmed to the smallest detail and it became possible to develop vast quantities of energy by nuclear fission processes. The possibility of nuclear fusion as an energy source was demonstrated by the hydrogen bomb, but this source has been refractory to such control as would allow it to be a source of industrial power.

At the same time other uses for the knowledge of nuclear processes have continued to be developed. For the first time speculation about the source of the energies of stars has been given a reasonable basis. The vestiges of radioactive decay products found in meteorites and in rocks of the Earth have enabled estimates of the age of the Earth. Radioactive tracers as well as isotopic tracers have become powerful tools in scientific research and industrial development, and artificially radioactive elements have been used for the treatment of disease.

This pattern of development presents a picture that deserves this additional comment, not out of place in a text that is committed to the holistic view. Whatever men may accomplish, whether in the Sciences, Humanities, Philosophies, may be used for good or evil purposes. Our knowledge of anything is neither good nor evil in itself. It is the application of the knowledge that makes the difference. One must, however, have some working understanding of the knowledge that is applied so as to be able to judge consequences; otherwise a good purpose may lead to a greater evil. The teachings of history, the rationale of morals, the philosophy of human behavior, and the psychology of individual human beings all need to be joined with Science when it is to be put to use.

Exercises

13.1 Give a free translation of the following passage:

En effectuant ces diverses opérations, on obtient des produits de plus en plus actifs. Finalement nous avons obtenu une substance dont l'activité est environ 400 fois plus grande que celle de l'uranium.

Nous avons recherché, parmi les corps actuellement connus, s'il en est d'actifs. Nous avons examiné des composés de presque tous les corps simples; grâce à la grande obligeance de plusieurs chimistes, nous avons eu des échantillons des substances les plus rares. L'uranium et le thorium sont seuls franchement actifs, le tantale l'est peut-être très faiblement.

Nous croyons donc que la substance que nous avons retirée de la pechblende contient un métal non encore signalé, voisin du bismuth par ses propriétés analytiques. Si l'existence de ce nouveau métal se confirme, nous proposons de l'appeler *polonium*, du nom du pays d'origine de l'un de nous.

[P. Curie and S. Curie, *Comptes Rendus 127*, 175 (1898), p. 177.]

13.2 Give a free translation of the following passages:

Les diverses raisons que nous venons d'énumérer nous portent à croire que la nouvelle substance radio-active renferme un élément nouveau, auquel nous proposons de donner le nom de *radium*. . . .

La nouvelle substance radio-active renferme certainement une très forte proportion de baryum; malgré cela, la radio-activité est considérable. La radio-activité du radium doit donc être énorme.

L'uranium, le thorium, le polonium, le radium et leurs composés rendent l'air conducteur de l'électricité et agissent photographiquement sur les plaques sensibles. A ces deux points de vue, le polonium et le radium sont considérablement plus actifs que l'uranium et le thorium. Sur les plaques photographiques on obtient de bonnes impressions avec le radium et le polonium en une demi-minute de pose; il faut plusieurs heures pour obtenir le même résultat avec l'uranium et le thorium.

[P. Curie, Mrs. P. Curie, and G. Bémont. *Comptes Rendus 127*, 1215 (1898), p. 1217.]

13.3 Suppose a nucleus were to radiate a gamma-ray of 0.5 MeV. How would this affect its mass number and its atomic number?

13.4 Show that the energy-releases from the following reactions are as indicated:

Number	Reaction	Yield
1	$H^2 + H^3 \rightarrow He^4 + n$	17.6 MeV
2	$H^2 + H^2 \rightarrow He^3 + n$	3.2 MeV
3	$H^2 + H^2 \rightarrow H^3 + H$	4.0 MeV

13.5 Estimate quantitatively the kinetic temperature of an apple that has fallen from a tree just before it lands on a dry leaf on the ground. Will it burn the leaf? Well, then?

13.6 In a certain uranium ore deposit there is found Pb^{206}. The deposit contains 0.865 g. Pb^{206} for each gram of U^{238}. a) What is the minimum age of the deposit? b) What assumptions have been made to yield this estimate?

13.7 In Figure 12-13 an excited state of Em^{222} is shown. What is its mass-energy before it returns to the ground state?

Answers to Exercises

13.1 "In the course of carrying out these different operations one obtains products that are more and more active. Ultimately we have obtained a substance with an activity some 400 times greater than that of uranium.

"We have sought, among known substances, whether there is any as active as this. We have examined compounds of nearly all the elements; thanks to the kindness of many chemists, we have had samples of the rarest substances. Uranium and thorium alone are indubitably active; tantalum is perhaps very feebly active.

"We think, then, that the substance which we have recovered from pitchblend contains a hitherto unknown metal, related to bismuth according to its analytical properties. If the existence of this new metal is confirmed, we propose to name it *polonium*, from the name of the country of origin of one of us." [*Note:* the S. Curie is Madame Sklodowska Curie. Later she signed herself Mrs. P. Curie.]

13.2 "The various considerations which we have enumerated lead us to believe that the new radioactive substance contains a new element, for which we propose the name *radium*.

"The new radioactive substance contains a very large proportion of barium; in spite of that, however, the radioactivity is considerable. The radioactivity of radium, then, should be enormous.

"Uranium, thorium, polonium, and radium, and their compounds cause air to conduct electricity and expose sensitized photographic plates. From these two points of view polonium and radium are considerably more active than uranium and thorium. Good exposures on photographic plates are obtained with radium and polonium in a half-minute exposure; many hours are necessary to obtain the same result with uranium and thorium."

13.3 0.5 MeV is equivalent to 5.37×10^{-4} u. This would have negligible effect on *A*, which is stated in whole units. There is no effect at all on *Z*, because the gamma-ray is uncharged.

13.4 For example, in reaction 3, $2.01410219 \times 2 - (3.01604940 + 1.00782522) = 0.00432976 = 40.5 \times 10^5$ eV.

13.5 Assume the apple, weighing about 1/4 lb. (0.15 kg) has fallen about 21 ft. (7 m). It will have reached a velocity obtained from $v^2 = 2gs$. $mv^2 = 20.58$ kg m^2 sec^{-2}. $T° = 2.48 \times 10^{23}$°K. Kinetic temperature is not the same as thermal at the macroscopic level.

13.6 a) The *atomic* ratio is 1:1; thus half the original U^{238} has decayed. The deposit has been undisturbed for some 4.5×10^9 years. b) The assumptions are enumerated in the text.

13.7 It is greater by the mass equivalent of 0.19 MeV over its ground state mass (see Appendix 3).

Chapter 14

14.1 Introduction
14.2 Chemical bonds
14.3 Covalent bonds
14.4 Ionic bonds
14.5 Coordinate covalent bonds
14.6 Metallic bonds
14.7 Writing ionic chemical equations
14.8 Chemical reaction rates
14.9 Simple acids and bases
14.10 Oxidation and reduction
14.11 Molecular aggregates. Phases
14.12 Effects of heat on reactions
14.13 Le Chatelier's theorem
14.14 Summary

Molecular Properties and Reactions ~ 14

Introduction 14.1

The material world may be divided, from the chemist's point of view, into natural products and artifacts. Natural products comprise the naturally-occurring constituents of the atmosphere, the lithosphere, the bathysphere, and the biosphere: the gases, rocks, great waters, and thin living layers of the Earth. Artifacts are those artificial products made by man through the transformation of natural products: building and paving materials, papers and materials of communication and transportation, synthetic fibers and drugs, and so on, including a wide variety of pollutants. We are concerned, in this chapter and the next, with the molecular constitution of the simpler of these natural products and artifacts. We use them to demonstrate the reasoning that has guided knowledge in this area.

Chemistry is the science of the transformations of matter, whether naturally or artificially produced. It arose out of alchemy, and came into flower when symbols and concepts capable of rational manipulation were devised. We mentioned some of these early discoveries and inventions in Chapter 9.

Another basis of classification is to divide substances into mixtures, compounds, and elements. Elements, to the chemist, are substances not capable of being broken down into simpler parts by chemical means. To the nuclear physicist or nuclear chemist (nearly indistinguishable in what they do) a chemical element is a substance all of whose atoms have the same *Z* number. It may or may not be isotopically pure. Compounds are substances capable of being broken down by chemical means into elementary parts (atoms). Mixtures can be separated into simpler parts by relatively mild mechanical or physical means such as sorting, filtering, evaporating.

In this chapter we are concerned with compounds, and with their behavior in chemical reactions that transform them into other substances. We are also concerned with the physical interactions that occur when molecules are gathered into bulk aggregates or are mixed with other molecules. We are interested in molecular-level properties and in bulk properties. Molecules are made up of atoms chemically bonded to each other. In bulk they form liquid, solid, or gaseous 'phases' in which there is more or less physical interaction between the molecules, but this interaction is not strong enough to be called a bond between any two of the molecules in the bulk phase.

14.2 Chemical bonds

Chemical compounds may be divided, on the basis of their properties, into two broad classes: the covalently bonded and the ionic. The two classes are not mutually exclusive. Indeed, they are arbitrarily divided to help discuss a range of behaviors that extends from pure covalent behavior at one extreme to pure ionic at another. Covalent behavior is shown by covalently bonded compounds; ionic behavior by molecules the parts of which are held together by Coulomb interactions. Some molecules show both types of behavior in different bonds. Ionically bonded molecules when dissolved in water, or when molten or gaseous, conduct electricity (Section 3.4). Covalently bonded molecules in solution do not. This is one of the chief distinctions between them.

Within the two broad classes, we recognize four types of chemical bonds: covalent, ionic, coordinate covalent, and metallic. These types are not mutually exclusive. They represent constraints on the chemical interactions of atoms (Section 9.4) and are responsible for an extraordinary variety of molecular properties. We shall discuss first the simplest molecule, H_2, the atoms of which are covalently linked by a shared pair of electrons.

14.3 Covalent bonds

The bond between two hydrogen atoms is explained in terms of a molecular orbital, analogous to an atomic orbital (Section 11.5). Consider two hydrogen atoms $H_A^{\bullet}$ and $H_B^{\bullet}$, where *A* and *B* are labels given for convenience. The atoms themselves are identical as atoms. Each electron is in a 1*s* spherical orbital about its nucleus (Table 11-1). We indicate this orbital by a circle (Figure 14-1) and the relatively

Fig. 14-1 *Two isolated hydrogen atoms.*

massive nucleus by H. We conceive the electron to occupy this orbital, scribing out the orbital sphere. The nucleus, too, is in motion since it ponderously follows the rapid electron under the drag of a Coulomb force (Section 10.2). Suppose the two *atoms* are at a relatively large distance from each other. We let them approach each other, the internuclear distance r becoming smaller and smaller. At some point they begin to attract each other, as is shown by the appearance of a force between them, and by a release of potential energy as they come closer. At a certain distance apart, r_o (Figure 14-2, compare Figure 12-12), they have given off all the energy that will be released, the bonding energy $\mathbf{E_b}$. Now the bond is formed. To push the atoms closer together requires great energy; to pull them apart the bonding energy E_b must be exceeded. The energy release is a little over 4 eV per bond formed, that is, per two H· atoms.

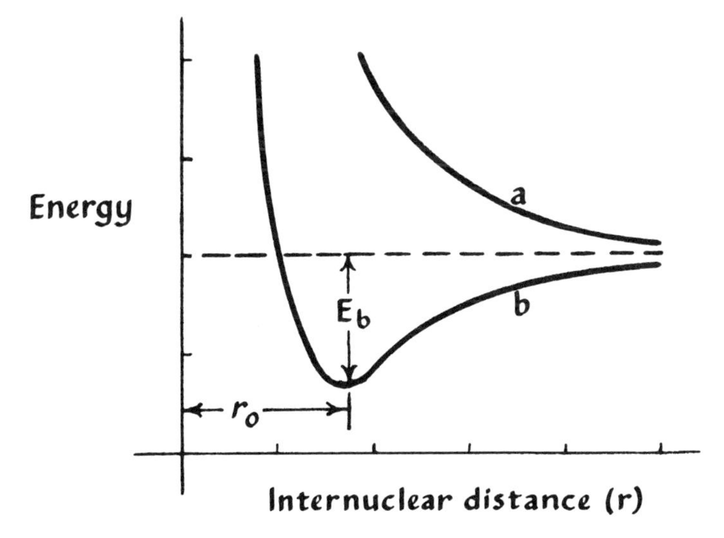

Fig. 14-2 *Calculated curves for the energy* E *of the system of two hydrogen atoms at various internuclear distances* r. a. *A curve for repulsion and non-bonding (See Exercise 14-1).* b. *A curve for attraction, with a minimum at* r_0. *This distance is the bond length. Energy* E_b *has been* released *in bond formation. This is the bond energy: the energy that would be required to dissociate the two atoms.*

The radius of the 1*s* orbital of a hydrogen atom is 0.53 Å. The bonded nuclei are found to be 0.74 Å apart on the average (they jitter a little relative to each other). Therefore, the orbitals must overlap (Figure 14-3). Through this formulation we imply a new kind of orbital

Fig. 14-3 *A shared-electron bond, conventionally written.*

which comprehends *both* nuclei—a 'molecular orbital.' Here the electrons may be with one or the other nucleus or with both nuclei. Both pictures have special conveniences. We shall use both. In either case, the Pauli Principle (Section 11.4) holds: the orbital is filled by two electrons. Each nucleus shares the electrons part of the time. The state of the molecule is dynamic. The electrons occupy positions (that is, the probability of finding them is high) sometimes about one, sometimes about another nucleus.

The consequences of the behavior of the electrons are these. Let one electron spin be indicated ↑ and the other ↓ (Section 11.4). Then, as in Figure 14-4, part of the time the electrons will be unpaired and far from each other. The unshielded nuclei repel, and begin to move apart. But they are massive, relative to the electrons, and move slowly, while the electrons move rapidly. Before the nuclei have moved very far apart the electrons are back holding them together. The nuclei thus oscillate in bond-stretching and compression motion. Further, the bond is contributed to by an ionic component. Suppose at one moment both electrons are closer to one nucleus than the other. The charge on this

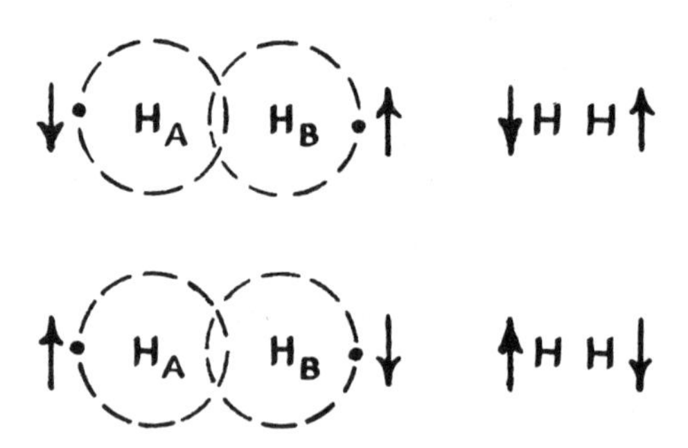

Fig. 14-4 *The electrons are unpaired for some of the time.*

nucleus is 1(+) and it has 2(−) close to it; then the net charge must be (−) on that side of the molecule while it is net (+) on the other side (Figure 14-5). This would help the nuclei to move toward each other,

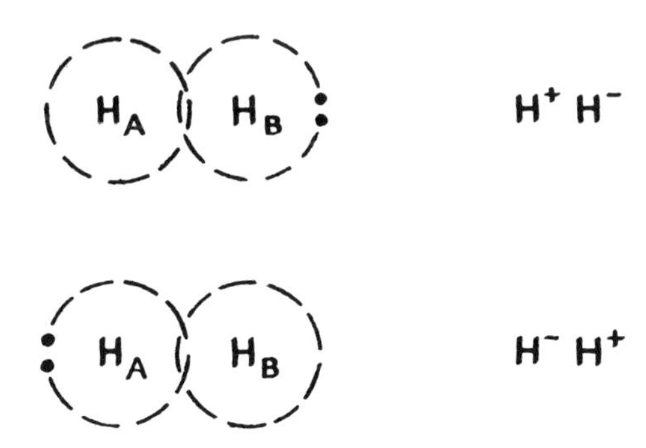

Fig. 14-5 *The bond is ionic for some of the time.*

but again their relatively ponderous motion is countered by the electron agility. Notice that although the two atoms in this shared-electron bond attract their electrons equally, yet because of the electron motion even this equal sharing displays some ionic character. The two bonded atoms in this covalent bond show about 5% ionic character—one imagines that about 5% of the time the two electrons are enough closer to one nucleus that the two may be designated ions as in Figure 14-5. We shall write the covalent bond as H—H or H : H, whichever is more convenient, and for any bond so written the symbol implies the properties we have described.

When a bond is formed between two atoms, as in H—H, the shapes of the orbitals of each atom must change. With the 'overlapping' must also go an adjustment to the presence of the other nucleus and the other electron. When the bonded atoms are of the same kind, the sharing must be equal, and we would expect to find (and do find) such equal sharing not only in H_2, but in N_2, O_2, F_2, and so on, though the orbitals that overlap are not *s* orbitals in these cases. When, however, the bonded atoms are of different kinds—differing in their affinities for electrons—it must be expected that the sharing will not be equal. For example, the simplest bond between C and any H in, say,

```
   H
H : C : H
   H
```

has about 5% ionic character; in

```
   ..
H : N : H
    H
```

the N : H bond is more ionic: N has a greater affinity for electrons than C. Continuing across the Periodic Table, the bonds in

```
   ..
H : O : H
   ..
```

are still more ionic in character,

and the ionic character reaches in $H:\ddot{\underset{\cdot\cdot}{F}}:$ a maximum of 45%. This is in agreement with what we saw about the electron affinity of nonmetals in Section 11.6.

Suppose we look down the Periodic Table in the halogen group. The percent ionic characters of the different bonds are shown in Table 14-1. Here we see that the bond becomes more covalent as the affinity

Table 14-1

Estimated Percent Ionic Character of Some Bonds

Bond	% Ionic character
H—H	5
H—F	45
H—Cl	17
H—Br	12
H—I	5

for the outer electrons decreases with their distance from the nucleus (Table 11-3).

Proceeding in the other direction from C in the Periodic Table, we expect the bonds between hydrogen and boron, beryllium, and lithium in that order to become more ionic, with the Li-H bond being most ionic, the H now being (−) with respect to Li since Li is metallic and releases its outer-shell electron readily (Section 11.6).

In order to emphasize that some bonds show partial ionic character above the minimum of 5% in the pure covalent bond, one may write small $\delta+$ and $\delta-$ to show the direction of displacement of electrons. $\overset{\delta+}{H} - \overset{\delta-}{F}$ and $\overset{\delta+}{Li} - \overset{\delta-}{H}$ show this convention in use. These are said to be polar molecules: they have a little 'dipole.' The consequences of polar property will turn out to be important.

14.4 Ionic bonds

Returning to the Periodic Table, it can be expected, on the basis of the range of types of bonds from Li—H to H—F, that the compound LiF would be highly ionic. It is. The lithium atom has lost its outer-shell (2*s*) electron to the fluorine, and now carries a full (+) charge while the fluorine is (−). The two ions are held together by Coulomb force, and the pair would be written Li^+F^-. Compounds of this kind are salts. Coulomb force is not directed in space but extends out in all directions around the ion (Section 9.4). For this reason, when the oppositely charged ions aggregate into a solid, they form giant "molecules": crystals all cross-lined by Coulomb force. This is suggested for sodium chloride (common table salt) in Figure 14-6.

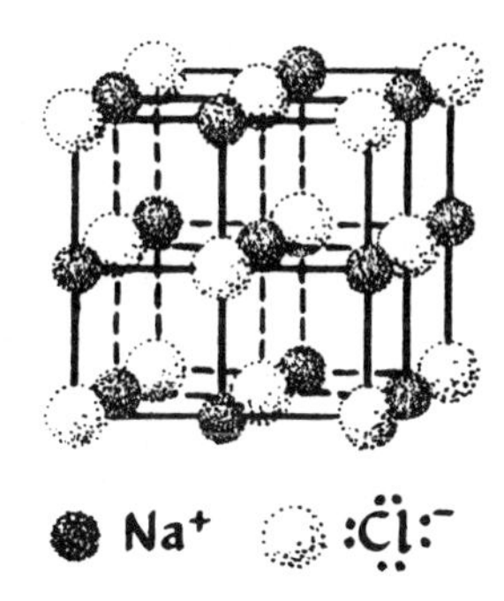

Fig. 14-6 *Arrangement of ions in part of a sodium chloride crystal. At the faces, edges, and corners of the crystal the ions are exposed—more so at the corners, less so in the faces—because not completely surrounded by other ions. Thus there are residual static fields due to these ions. The dissolution of the crystal in water is aided by interaction of the water molecule dipoles with these residual fields. [From Linus Pauling,* The Nature of the Chemical Bond. *Ed. 3, Cornell University Press, 1960, Fig. 13-4, p. 519, by permission.]*

Coordinate covalent bonds 14.5

In some molecules the bonded atoms may share two electrons both of which are contributed (no matter how the molecule was actually formed) by only one of the atoms. Such bonds are quite common. They are polar bonds. Suppose, for example, a neutral ammonia molecule reacts with an oxygen atom:

$$\begin{array}{c}\mathrm{H}\\ \mathrm{H\,:\underset{..}{\ddot{N}}:}\\ \mathrm{H}\end{array} + \mathrm{\underset{..}{\ddot{O}}:} \rightarrow \mathrm{H:}\begin{array}{c}\mathrm{H}^{\delta+}\ {}^{\delta-}\\ \mathrm{\underset{..}{\ddot{N}}\ :\underset{..}{\ddot{O}}:}\\ \mathrm{H}\end{array} \text{ or } \mathrm{H-}\begin{array}{c}\mathrm{H}^{\delta+}\ {}^{\delta-}\\ |\\ \mathrm{N}\\ |\\ \mathrm{H}\end{array}\!\!\rightarrow \mathrm{O}$$

Evidently the orbitals of the oxygen are now filled, at least part of the time, by the *pair* of electrons shared with and coming from the nitrogen. The bond is sometimes designated by the arrow that points in the direction of the sharing, so as to accentuate the very polar nature of the bond.

Metallic bonds 14.6

These are of a special kind. Outer-shell electrons, relatively weakly held (Table 11-3), are freely shared among the atoms in a metal. They can be made to move as a current of electricity under the gradient of a potential difference. One might think of a metal as a structure overall electrically neutral that is filled with mobile electrons and that is porous to electrons. We shall not be concerned with metallic bonds—this special kind of sharing—any further.

14.7 Writing ionic chemical equations

Given the types of bonds that may hold atoms together, the question arises how bonds are formed. What is, in other words, the course, or mechanism of a chemical reaction in which substances are transformed and bonds broken and made? We approach this subject, central to this chapter, by describing the conventions of chemical language, the grammar and syntax, used in writing about reactions; by presenting the code used to conceptualize chemical transformations.

Basically, of course, one has to find out what actually happens in a chemical reaction. If a water solution of silver nitrate is mixed with sodium chloride solution it is observed that a white material precipitates out—insoluble in the solution. The liquid, supernatant to the precipitate, is clear. There has been a chemical reaction. When it was first studied, someone had to determine by chemical analysis that what precipitates is silver chloride, and that its molecule is composed of one silver atom, Ag, and one chlorine atom, Cl. (And he was carefully checked by several other people.) The convention is that the formula of the compound will be written AgCl, with the symbol for the metal written before that for the non-metal. Likewise the reagents that produced this reaction had to be known: silver nitrate is $AgNO_3$, where NO_3 is a group of atoms called 'nitrate' that tends to go around as a whole; and sodium chloride is NaCl.

Furthermore, as the reaction is studied it is found that when equal numbers of molecules of $AgNO_3$ and NaCl are made to react in this way in solution, practically all the silver precipitates as the chloride AgCl, leaving only traces in the supernatant. This contains all of the $NaNO_3$ that is formed. The reaction of these two salts is carried out in aqueous solution, and examination of each reagent solution shows that the salt in it is highly ionized. (We shall see later why this is.) Also the resulting $NaNO_3$ supernatant contains sodium and nitrate ions. All of this information is packed into an equation for the reaction. The convention is that the formula (e.g., AgCl) represents one mole, Avogadro's number, of molecules, and not just a single molecule. This makes accurate bookkeeping possible because in any given reaction at equilibrium some molecules are reacting and others just forming at any instant. If we consider a very large number of molecules (recall that Avogadro's number is 6×10^{23}), the statistical fluctuations that must occur when by chance a large number of molecules happen to collide in the same way, and form many product molecules locally in the solution, are ironed out. The result is an orderly equation. For even if one were to take a very small amount of a given reagent, say 10^{-5} mole of $AgNO_3$ (this would weigh about 0.0017 g) there would be present 6×10^{18} molecules of $AgNO_3$—quite a large enough number to hide fluctuations of millions.

The equation may be written to show the ionization of the reagents (the water is understood and is usually left out) and the interaction to form products. The arrows indicate the direction of reaction; their length and heaviness suggest the quantitative findings, and the whole equation represents what is found at equilibrium. At equilibrium, in this case, some of all species of molecules and ions are present, and they are colliding, ionizing and recombining, precipitating and redissolving, in a way described by this overall pattern. The pattern is that of dynamic equilibrium:

$$\begin{array}{lcccc} AgNO_3 & \rightleftharpoons & Ag^+ & + & NO_3^- \\ NaCl & \rightleftharpoons & Cl^- & + & Na^+ \\ & & \downarrow\uparrow & & \downarrow\uparrow \\ & & \underline{AgCl} & & NaNO_3 \end{array} \qquad (1)$$

A line under a molecular formula indicates that the substance is present as a solid—usually as a precipitate. A line over the formula, or an upward arrow beside it, indicates that the substance is gaseous. The overall reaction may be written:

$$AgNO_3 + NaCl \rightarrow \underline{AgCl} + NaNO_3$$

The reaction is balanced when each atom on the left is present in the same amount on the right. The subscript 3 in NO_3 means that in one nitrate ion, or group of atoms, there are three oxygens for each nitrogen atom, a fact that has to be pried out of Nature by chemical analysis.

Turning to a table of chemical atomic weights, or better yet, a handbook containing tables of formula weights, one can determine how many grams of silver nitrate of any other pure substance will contain one mole of molecules. These data may then be written under the symbols in the equation:

$$\begin{array}{ccccccc} AgNO_3 & + & NaCl & \rightleftharpoons & AgCl & + & NaNO_3 \\ (6 \times 10^{23} & & (6 \times 10^{23} & & (6 \times 10^{23} & & (6 \times 10^{23} \\ \text{molecules)} & & \text{molecules)} & & \text{molecules)} & & \text{molecules)} \\ 169.89\text{ g} & : & 58.45\text{ g} & :: & 143.34\text{ g} & : & 85.01\text{ g} \end{array}$$

These figures give the combining ratios by weight of the molecules in the equation. Thus to get a 1:1 ratio of silver nitrate to sodium chloride *molecules*, one must take a ratio of 169.89 g of the former to 58.45 g of the latter. One can see then that with this information if one wished to precipitate virtually all the silver in a solution containing 17 g silver nitrate (knowing that silver chloride is insoluble) one would have to add a solution of sodium chloride containing x g NaCl:

$$\begin{array}{c} 169.89:58.45 \\ 17 \ : \ x \end{array}$$

or at least 5.9 g (in round numbers) of NaCl. One could also calculate how much AgCl would be formed:

$AgNO_3$, 169.89 g forms 143.34 g AgCl

Thus 17 g $AgNO_3$ will form y g AgCl in the relation $y/17 = 143.34/169.89$.

$$y = 17 \times 143.34/169.89 = 14.3 \text{ g AgCl}$$

Similarly for the sodium nitrate. The equation therefore implies rigorously quantitative data. A good deal of a course in elementary chemistry consists in learning the vocabulary: the names and formulas of substances; and in learning what reactions will occur; what the physical properties of reactants and products are—their colors, solubilities—and in writing equations and making calculations.

We shall not pursue the matter further, for with what is given in this and the following chapter a student can work his way through a text in elementary chemistry. It suffices to add that if the formula weight of the silver nitrate were dissolved in water, and the solution made up to a total volume of one liter, and the same were done for the sodium chloride, then since equal volumes of liquid now contain equal numbers of molecules, one could write:

$$\begin{array}{cccc} AgNO_3 + & NaCl & \rightleftharpoons \underline{AgCl} + & NaNO_3 \\ \text{1 liter} & \text{1 liter} & \text{ppt} & \text{2 liters} \end{array}$$

The $NaNO_3$ would be half as dilute, on a molecular basis, as the reagents, since now one mole is present in 2 liters of solution. A solution of one mole of substance in one liter of solution is a molar solution. Clearly, 10 ml of molar $AgNO_3$ would react with 10 ml molar NaCl to yield 20 ml of half-molar $NaNO_3$ and 0.01 mole of AgCl. Also, if an excess of one reagent were present over what would react according to this equation, the excess, calculable by the above reasoning, would be left over, and appear in the products. (Note that one l of one molar $AgNO_3$ contains 169.89 g $AgNO_3$, so that the weight of silver nitrate in any volume of such a solution is known.)

The equation rests on the conservation law of Lavoisier: what goes in as reagents must be fully accounted for in the products.

14.8 Chemical reaction rates

Consider a reaction between hydrochloric acid, HCl, and a base, sodium hydroxide, NaOH (Section 14.10). Each is dissolved separately in water. HCl is a gas, and in water it ionizes. NaOH is a white crystalline ionic solid, also very soluble in water. The equilibrium reaction would be written for the water solution of the mixture, based on observation and analysis of *what actually happens:*

$$\begin{array}{lcll} HOH + \overline{HCl} & \rightleftharpoons & H_3O^+ & + \; Cl^- \\ Na^+ \; \bar{O}H & \rightleftharpoons & OH^- & + \; Na^+ \\ & & \;\;\updownarrow & \;\;\;\updownarrow \\ & & HOH & \;\; Na^+ \, Cl^- \end{array} \qquad (2)$$

Rates of reactions are influenced by: the states of the reactants and the products; the concentrations of the reactants and the products; the temperature of the system (Section 14.12); the presence of catalysts.

The states of reactants and products are important because for molecules or ions to react they must collide and stick—at least for a time. Suppose we make HCl solution by bubbling HCl gas (from a tank) rapidly through a small tube that dips below the surface of water in a flask. When the $H:\ddot{\underset{..}{Cl}}:$ molecules collide with the $H:\ddot{\underset{..}{O}}:H$ molecules a competitive reaction takes place. Water forms a stronger bond with H^+ than Cl^- does, and most of the protons end up as $H:\overset{H}{\ddot{\underset{..}{O}}}:H^+$, hydronium ion ($H_3O^+$). The first few HCl molecules entering the water have full opportunity to react. One may visualize the $H:\ddot{\underset{..}{Cl}}:$ colliding with $H:\ddot{\underset{..}{O}}:H$. If the Cl side of the molecule strikes the oxygen no reaction, only repulsion by the electron clouds, would be expected.

$$H:\ddot{\underset{..}{Cl}}::\overset{H}{\underset{H}{\ddot{\underset{..}{O}}}}: \rightleftharpoons H:\ddot{\underset{..}{Cl}}: + :\overset{H}{\underset{H}{\ddot{\underset{..}{O}}}}: \text{ (no reaction)}$$

But if the $H:\ddot{\underset{..}{Cl}}:$ collided H-first, then reaction might occur:

$$:\ddot{\underset{..}{Cl}}:H:\overset{H}{\underset{H}{\ddot{\underset{..}{O}}}}: \rightleftharpoons :\ddot{\underset{..}{Cl}}:^- + H_3O^+ \text{ (ionization)}$$

Here is an important case of the effect of the relative orientation states of the reactants. The two molecules must collide with the correct orientation in space relative to each other for the reaction to be favored.

Gradually, as more and more HCl reacts with water, Cl^- and H_3O^+ accumulate, and the relative amount of free H_2O in the solution (the concentration of H_2O) decreases.

The effect of concentration on a reaction is this: the rate of a reaction (moles reacted per second) increases with increase in the concentration of reactants. This is the Law of Mass Action.

We understand, therefore, that the rate of the reaction $HCl + H_2O \rightarrow Cl^- + H_3O^+$ slows down as the concentrations of HCl and H_2O fall. Concomitantly, the concentrations of Cl^- and H_3O^+ are building up, and the rate of the reaction $Cl^- + H_3O^+ \rightarrow HCl + H_2O$ increases. There comes a point when the two rates exactly equal. At this point, the point of equilibrium, we have a solution of the reactants and products the concentrations of which remain constant while the opposing reactions continue to occur. This is why the state is named 'dynamic' equilibrium.

Similarly, when $Na^+ \bar{O}H$ is dissolved in water, it ionizes because Na^+ ions become surrounded by a cloud of water molecules which shield it electrically from OH^-. The OH^- ions are also surrounded, but not by as dense a cloud. The ions are said to be hydrated, but the water molecules are not written into the equation because they are not held as tightly as the single H_2O to H^+. H_3O^+ is also hydrated as in Cl^-.

The reason for the hydration is that H_2O has a small dipole. The four pairs of electrons of oxygen in $H:\ddot{O}:H$ form bonds and potential bonds oriented in space as far apart as possible. It is found that the two H atoms in H_2O are at an angle of 104.5° to each other. The O—H bond has about 39% ionic character. The (+) character of the two H nuclei cause them to repel each other somewhat. Thus H_2O may be written:

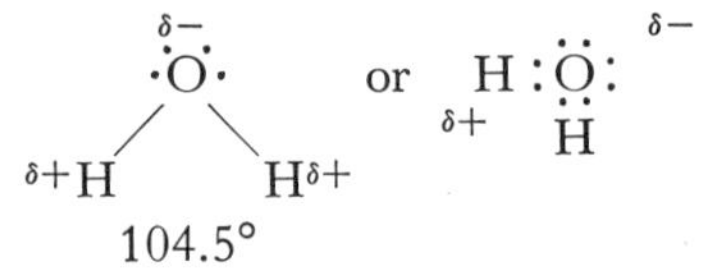

```
    δ−                      δ−
   ·Ö·          or   H:Ö:
   /  \             δ+  H
δ+H    Hδ+
  104.5°
```

This means that one side of the molecule is (−) relative to the other, giving existence to a dipole H—Ø(↗) with H below. The negative end is attracted toward (+) ions, the plus toward (−) ions, and thus the hydration clouds about ions in aqueous solution are composed of more or less oriented water molecules. There is a great deal of coming and going out of the cloud, and jittering about in it.

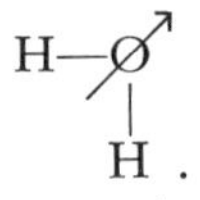

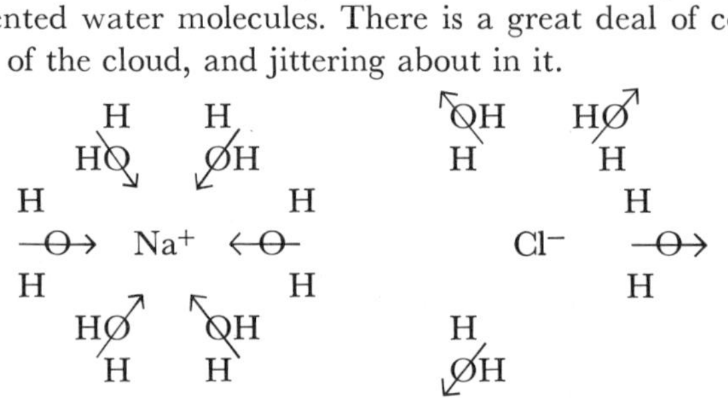

```
    H     H                    ↖ØH    HØ↗
   HØ↘   ↙ØH                   H       H
H             H                          H
—Ø→   Na⁺   ←Ø—             Cl⁻        —Ø→
H             H                          H
   HØ↗   ↖ØH                   H
    H     H                   ↙ØH
```

When the acid and base solutions are mixed, nothing much happens to the Na^+ and Cl^- ions, since the product, Na^+Cl^-, is highly ionized in water. However, H_3O^+ and OH^- react strongly to form un-ionized water:

$$H_3O^+ + OH^- \rightleftharpoons 2\ H_2O$$

This drives the reaction toward formation of H_2O.

In each reaction, the reagents and products come into equilibrium following the course described above. The overall reaction may be written at equilibrium as:

$$HCl + NaOH \rightleftharpoons H_2O + NaCl$$

This is a homogeneous reaction: reagents and products remain in the same phase—a liquid one. (In a homogeneous gas reaction, reagents and products would be gaseous.) The reaction between $AgNO_3$ and NaCl (Section 14.7) is heterogeneous. Two phases are present: solid AgCl and solution containing all the ions and molecules.

Suppose that we generalize the above reaction:

$$A + B \rightleftharpoons C + D \quad (3)$$

The law of mass action states that the rate of a reaction depends directly on (is proportional to) the concentrations of the reactants. Concentration is measured in moles per liter of total solution, and the concentration of A is written [A]. Then, by the law, for the reaction toward the right:

$$\text{Rate} = k[A][B]$$

k is the rate constant for the reaction. For the reverse reaction:

$$\text{Rate}' = k'[C][D]$$

Here k' is the rate constant for the reverse reaction. At equilibrium, Rate = Rate′, thus $k[A][B] = k'[C][D]$, and (by convention)

$$\frac{[C][D]}{[A][B]} = \frac{k}{k'} = \boldsymbol{K} \quad (4)$$

$\boldsymbol{K}$ is named the equilibrium "constant," and must be determined by experiment. It usually varies with temperature. It is determined by measuring the concentrations of reactants and products at equilibrium.

When several reactions occur simultaneously, as in the equations (1) and (2), the constants at equilibrium can be equated. If the reactions go essentially to completion, as in the precipitation of silver chloride in (1) or in the formation of water in (2), the mathematical treatment can be simplified. Chemists are concerned with calculations of these kinds, for the purpose of predicting and controlling reactions.

Simple acids and bases 14.9

Classically, acids are substances which when dissolved in water yield hydrogen ion. Many are highly corrosive, and some are highly toxic. Those that can safely be tasted, such as acetic acid in vinegar and citric acid in citrus juice, taste acid to most people. When an acid such as hydrochloric (HCl), nitric (HNO_3), sulfuric (H_2SO_4), phosphoric (H_3PO_4), acetic ($HC_2H_3O_2$) is dissolved in water ionization occurs, as we saw in Section 14.8, in the example of the ionization of hydrochloric acid. This reaction can be thought of very helpfully as the result of a competition for the proton H^+ between $:\ddot{\underset{\cdot\cdot}{Cl}}:^-$ and $H:\ddot{\underset{\cdot\cdot}{O}}:H$ (Section 14.8), and as is implied by the first line of equation (2).

In the reaction the oxygen shares a pair of electrons with the proton. Protons are almost never found free in solution. Substances that share electrons in this way are named 'bases.' As a class they are non-metals with available unshared electrons. Classically, the name base, or alkali, was used for substances that ionize in water to yield hydroxide ion $H:\ddot{\underset{\cdot\cdot}{O}}:^-$. When caustic or other alkalies react with acids to yield water, each is said to be 'neutralized' by the other. This was shown in equation (2). The name base has been given more general meaning, as described above. The same broadening, as we saw, has occurred in the meaning of 'acid.' (Generalized) acids, then, are substances that form coordinate bonds with (generalized) bases. In the reaction of $H:\ddot{\underset{\cdot\cdot}{Cl}}:$ with $H:\ddot{\underset{\cdot\cdot}{O}}:H$, the Cl^- ion is a base, as is the $H:\ddot{\underset{\cdot\cdot}{O}}:H$. Because the reaction goes largely to the formation of H_3O^+, with very little HCl present at equilibrium, we have evidence that HOH is a stronger base than Cl^-. In general strong acids are those in which H^+ is bound to weak bases (ions or groups that hold electrons strongly) such as NO_3^-, $SO_4^=$, Cl^-. Weak acids are those wherein the base moiety is strong: HCN (hydrocyanic acid), HOH. $HC_2H_3O_2$.

Some acids have more than one available ionizable proton. For example, H_2SO_4 has two, and H_3PO_4 has three. In these cases, the first H^+ may come off easily, as in

$$H_2SO_4 + H_2O \rightleftharpoons HSO_4^- + H_3O^+$$
$$H_3PO_4 + H_2O \rightleftharpoons H_2PO_4^- + H_3O^+$$

The second proton ionizes less readily since it must separate from an ion that is already negative. The same applies even more strongly to the third hydrogen in phosphoric acid. In a large class of acids containing carbon and hydrogen, as for example acetic acid, $HC_2H_3O_2$, some of the hydrogens are so firmly bound that they are not available for ionization. Only the one set apart in the formula for acetic acid may ionize. We shall see the reason for this in the following chapter.

14.10 Oxidation and reduction

In another large class of chemical reactions the classifying behavior is the transfer of one or more electrons. We have already seen an example of this:

$$2\,Na\cdot + :\ddot{\underset{\cdot\cdot}{Cl}}:\ddot{\underset{\cdot\cdot}{Cl}} \rightarrow 2\,Na^+ + 2:\ddot{\underset{\cdot\cdot}{Cl}}:^-$$

The $1s$ electron in the outer shell of $Na\cdot$ is transferred to the $:\ddot{\underset{\cdot\cdot}{Cl}}:$ atom, where it pairs with a lone $3p$ electron. The sodium atom has lost an electron: it has been 'oxidized.' Chlorine is the oxidizing agent, or electron acceptor. A chlorine atom in the process has taken on the electron: it has been 'reduced.' Sodium is the reducing agent.

It is often not easy to write and balance equations for oxidation and reduction because the electron may not appear explicitly in the overall reaction:

$$2\ Na + Cl_2 \rightarrow 2\ NaCl$$

The procedure is to write down the changes of oxidant and reductant separately, as though they were independent. To do this one has to know that elementary atoms have charge zero. The reaction would be written:

$$Na^0 \rightarrow Na^+ + e^-$$
$$e^- + Cl^0 \rightarrow Cl^-$$

Clearly, then, for charge conservation, one Na^+ reacts with one Cl^-. In more complicated situations one observes that the Periodic Table gives the possible charge on an atom in the reduced form (for atoms on the right half of the table) and for oxidized atoms throughout the table. Thus Na, with one outer electron, will have the formal charge (+) when in the oxidized form. Be, with 2 electrons, will be oxidized to Be^{++}; B, with three electrons, may take on the formal charge 3(+) as in BF_3, where the F will have the formal charge 1(−). Carbon in CO_2 has the formal charge +4. Nitrogen, a non-metal, tends to accept electrons, and can be reduced to the formal charge 3(−) as in ammonia, NH_3. Oxygen is always 2(−) in its combinations (as far as we are concerned). Some of the non-metals may also take on (+) formal charges, when combined with a strong oxidizing agent. This is the state of N in the nitrate ion, NO_3^-, where the formal state is 5(+). In the brown gas NO_2 the formal charge on N is 4(+). In the nitrite ion NO_2^- it is 3(+); in colorless nitric oxide, NO, it is 2(+); in nitrous oxide (laughing gas) N_2O, it is 1(+); in nitrogen gas, N_2, it is zero; in other compounds it may be 1(−) or 2(−); and in ammonia, NH_3, it is 3(−) as noted above. Here is a case where the structure of the atom, and the position in the Periodic Table which is related to structure, says that up to three electrons may enter the outer shell—when the formal charge becomes 3(−)—or electrons may successively be shared with a strong acceptor up to the point when all 5 are withdrawn to some extent, as in NO_3^-. Not many other atoms show the entire range of oxidation and reduction states, and part of Chemistry involves learning the stable states and their behaviors.

Molecular aggregates. Phases 14.11

The substance water exists in three possible states: gaseous, liquid, and solid. Since these are commonly experienced, our language is full of names for forms and states of water: ice, snow, frost; river, stream, creek, trace; water vapor. Mist, fog, dew, and steam are all finely divided forms of liquid water. These states were surveyed in Section 9.6. Here we are concerned with the interactions that produce them.

We saw, in Section 14.4, that ionic molecules tend to form solids held together by Coulomb force. These materials tend to have a structure in which the ions fall in a regularly repeated pattern: they are crystalline. Also, because the bonding is ubiquitous throughout the crystal, they tend to have high melting points (see next section). Materials that are not ionic may also be solid and high-melting if they are composed of very large molecules or if the individual atoms are highly cross-linked by covalent bonds, as is the diamond. But all substances, if cooled to a sufficiently low temperature, become solid. Some solids are crystalline, showing extensive order in the arrangement of the molecules —as a snowflake, or a crystal of table sugar. Others become amorphous solids—the domains of order are tiny, and randomly arranged—as for example ordinary glass.

The solid state results when the forces between molecules are stronger than disrupting thermal motion. Consider ice. We saw in Section 14.8 that the water molecule is dipolar. There are two pairs of unshared electrons capable of interacting with hydrogen ion:

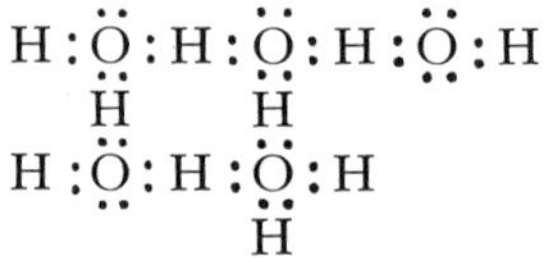

```
   ..    ..    ..
H :O: H :O: H :O: H
   ..    ..    ..
   H     H
   ..    ..
H :O: H :O: H              etc. in three dimensions.
   ..    ..
         H
```

The bond that is formed, $:\ddot{\underset{\cdot\cdot}{O}}:H\text{---}:\ddot{\underset{\cdot\cdot}{O}}:$, is a hydrogen bond. In ice, the two oxygen atoms are about 2.76 Å apart, and the hydrogen that "bonds" them is 1.00 Å from one oxygen atom and 1.76 Å from the other. It is better to think of the hydrogen-bonded water molecules as being physically associated rather than bonded because the energy of the hydrogen bond is about 0.5 kilocalories/mole (0.02 eV per bond) while a chemical bond, as that between H and O in HOH, has an energy of some 110.6 kilocalories/mole (4.8 eV per bond). At 0°C the vast number of hydrogen bonds in the mass of water molecules add up to sufficient constraint that the substance solidifies. At lower temperatures, or under pressure, the pattern of the crystalline arrangement may change.

When ice is heated it melts. In this process one mole of crystalline ice absorbs 1.437 kilocalories. This input of thermal energy breaks enough of the hydrogen bonds (about 15%) that the solid structure breaks down, and small groups of oriented molecules can slide and tumble over others: the substance becomes a liquid. When this liquid freezes, at 0°, it gives up 1.437 kcal/mole of heat. It might be added that even in the frozen state individual water molecules have some freedom of motion and may jitter in a constrained way.

If liquid water at 0°C is heated, the energy put in raises its temperature, increasing the velocity of motion of the molecules and aggregates, and gradually breaking more hydrogen bonds. The calorie is defined as the quantity of heat needed to raise the temperature of one gram of water from 15° to 16°C. As water is heated further the temperature increases, viscosity decreases because the internal structure is loosened by breaking of hydrogen bonds so that the molecules, and decreasingly large aggregates, can roll over each other more easily, and for the same reason the increased molecular motion causes the liquid to expand. At 100°C the liquid absorbs 9.73 kcal/mole and turns into water vapor. In this vapor there is virtually no structure. The molecules move randomly and independently except that there may be some temporary sticking on collision.

The description given for water holds in all its essentials for other substances, though the range between melting and boiling, and the temperatures, will be different.

Effects of heat on reactions 14.12

The constraints that hold molecules together against thermal motion in phases, and that cement atoms into molecules can be broken by heat energy. When a substance is heated, the motions of its molecules increase: translatory motions, rotatory motions. When a crystalline or other solid material is heated, the molecules jitter about centers, each gradually increasing the range of its excursions on the average as the energy input increases. This causes the substance to expand.[1] Finally, as the heat energy is increased (reflected in the temperature of the substance) it becomes high enough to overcome the associative forces, and the solid melts. A crystal starts to melt on the corners and edges where molecules are less firmly held—they have fewest near neighbors. These slide over onto the faces, and the sharp corners and edges disappear, the crystal taking a drop-like form and fusing with neighboring drops.

If heating is continued the molecular motion in the liquid increases —viscosity decreases, density decreases—and more and more of the molecules that happen to be at the surface of the liquid escape into the gas phase, giving increasing vapor pressure above the liquid. When this pressure reaches atmospheric the liquid boils and goes over into the gas phase. If heating is continued the molecules may actually decompose.

---

[1] Some substances contract on heating because over this range of temperature the molecules rearrange into more compact form. This occurs with water which, at 0°, and as ice, is less dense than liquid water at 4°, the temperature of maximum density. At 4° 1 g water has a volume of 1 ml; its density is 1.000 g/ml. That ice floats prevents the beds of oceans and lakes from being permanently coated with ice.

The molecular motion due to heat energy includes, besides the motions of translation and rotation of whole molecules, distortion of the molecules due to collision: bending, stretching, and spinning of atoms and groups of atoms about the bonds. If this becomes sufficiently energetic, the molecule may dissociate into fragments.

Chemical reactions may give off heat when they occur in one direction and absorb heat when occurring in the opposite direction. For example, in the extremely important Haber process for fixing nitrogen of the air in ammonia (for making fertilizers and for other uses) the overall reaction is:

$$N_2 + 3\ H_2 \rightleftharpoons 2\ NH_3 + \text{heat}$$

The reaction is extremely slow at ordinary temperatures. It is not practically feasible unless the mixture of nitrogen and hydrogen is heated to a high temperature. But this input of heat tends to reverse the reaction, as the equation shows, for heat is a "reagent" on the right, tending to drive the reaction to the left. This will be further discussed in the following section.

14.13 Le Chatelier's theorem

This principle states that if a stress is placed on an equilibrium system, the equilibrium will shift in the direction that relieves the stress. Consider the equation of the Haber process, Section 14.12. It is found that even with a catalyst a temperature of some 400°C is needed to obtain a practical reaction rate, but at this temperature the amount of ammonia formed is only about 4% of what it would be were there complete conversion. At 200°C the rate is far too slow to be practical even though the equilibrium is much more favorable, the yield of ammonia being a little over 50%. By the Le Chatelier principle, the increased heat drives the reaction to the left, the direction that absorbs heat. Decreasing the heat of the system displaces the equilibrium toward the right, which supplies heat.

Further examination of the equation shows that when the reaction proceeds to the right the volume of the system decreases. One mole of any permanent gas, especially at such a temperature as this, has the same volume at a given pressure as any other. Thus by the equation:

$$\begin{array}{c} N_2 + 3\ H_2 \rightleftharpoons 2\ NH_3 + \text{heat} \\ 1\ \text{vol.} + 3\ \text{vol.} \rightleftharpoons 2\ \text{vol} \end{array}$$

The volume of the reagents on the left is twice that of the products on the right. If, therefore, the pressure on the system were increased the equilibrium would shift so as to relieve that stress: the volume would decrease; the equilibrium would be displaced to the right. The figures given above were for a pressure 10 times atmospheric. At 400° and 100 times atmospheric 25% of ammonia is produced at equilibrium; at 600 atmospheres about 65%, and at 1000 atmospheres nearly 80% ammonia.

Le Chatelier's principle is used to obtain the best yield for the process. The need to have the reaction proceed at a fast rate requires the decreased yield that temperature increase alone would produce, and this is counterbalanced by raising the pressure. Other factors arise to limit what is practical. For example, the powerful pumps and the very heavy and expensive vessels needed to produce and contain high pressure at high temperature, as well as the corrosive property of hot ammonia, place an economic constraint on the process. The designer of a chemical process and the plant to carry it out in must take all these factors into account.

Summary 14.14

In this chapter we have been chiefly concerned with chemical reactions and the types of molecules that produce them. The classifications served to bring out orderly relations inherent in the fantastic variety of molecules and reactions. Several themes were played upon and illustrated. The constraints that bind molecules together into phases are much weaker than those that bond atoms into molecules, but are conceptualized in much the same way. The state of the system, whether phase or molecule, is the result of competition between the bonding forces, ionic or covalent or both, and the disrupting presence of thermal energy—related to work, measured in calories or their equivalent in joules—and ultimately traceable to fundamental properties of charges, forces, masses, and accelerations. We saw how the rates of equilibrium reactions are conceptualized and how at equilibrium the rates of opposing reactions are the same, and how the equilibrium constant may be quantitatively determined. We looked briefly at the factors that affect the position of an equilibrium: state of the reactants and products; temperature, which may actually change the constant; concentrations of the reagents and products. Finally we illustrated the operation of Le Chatelier's principle, a powerful theorem of chemistry. Application of this principle shows how heat and pressure (and though we did not discuss it explicitly, concentration of a component) may shift an equilibrium state.

Exercises

14.1 Under what conditions might the electrons of two $H\cdot$ atoms not participate in bond formation: $2H\cdot \not\rightarrow H:H$?

14.2 How might the bonding energy of two atoms be measured?

14.3 Write a brief statement which shows that you have thought about any connection there might be between the photoelectric effect and electronegativity. (This is a difficult question.)

14.4 Given the following compounds, indicate, where proper, the polarization of the bond by inserting $\delta+$ and $\delta-$.

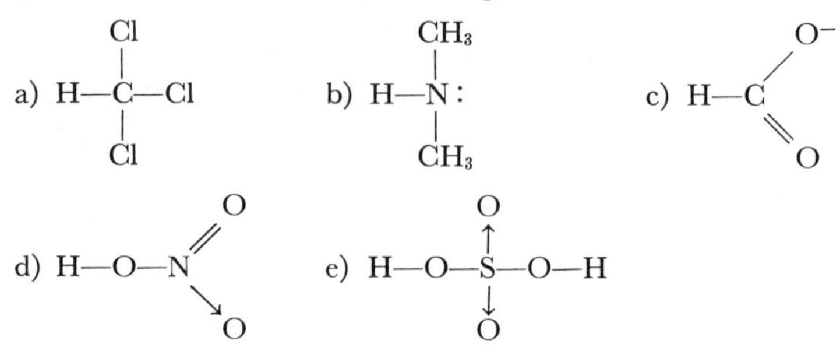

14.5 Which of these molecules would you expect to form hydrogen bonds with water?

```
  H       H          H                          H     O
  ..      ..         ..        ..      ..       |    //
H:B    H:Si:H     H:N:      H:S:    H:I:    H—C—C
  ..      ..         ..        ..      ..       |    \
  H       H          H         H                H     O—H
boro-   silane    ammonia  hydro-   hydri-     acetic acid
hydride                    gen sul- odic
                           fide     acid
```

14.6 Is it in any respect legitimate to say that the dissolution of a solid is analogous to the evaporation of a liquid?

14.7 a) Why does sodium chloride not dissolve in gasoline? b) Sodium chloride is not very soluble in ethyl alcohol. If alcohol is mixed with a concentrated aqueous solution of salt, some salt precipitates in crystalline form. Why is this?

14.8 In successive ionizations of an atom such as Li to give Li^{+}, Li^{++}, Li^{+++}, the ionization potential increases much more for the second ionization (5.39 eV, 75.6 eV, and 122 eV are the respective potentials) than can be accounted for by purely Coulomb attraction. Draw an energy level diagram and explain this.

14.9 a) Balance the equation:

$$C_8H_{18} + O_2 \rightarrow CO_2 + H_2O$$

b) The equation a) represents an idealized formulation of what occurs in an automobile engine when gasoline is burned. A gallon of gasoline weighs 2.6 kg. How much oxygen, in kg., is needed to burn it in the engine, making the (false) assumption of 100% efficiency? c) At 1 atmosphere pressure and 0°C 32 g oxygen has a volume of 6 gallons. How many gallons of air, of which 1/5 is oxygen, are needed to burn one gallon of gasoline, assuming 100% efficiency? d) How much water is produced?

14.10 In the reaction of these gases, explain what will be the effect of an increase in pressure on the position of equilibrium.

a) $H_2 + I_2 \rightleftharpoons 2\ HI + 3$ kcal

b) $2\ SO_2 + O_2 \rightleftharpoons 2\ SO_3$

c) $N_2O_4 \rightleftharpoons 2\ NO_2$

d) $CO_2 + H_2 \rightleftharpoons CO + H_2O - 10$ kcal

e) $N_2 + O_2 \rightleftharpoons 2\ NO - 43.2$ kcal.

14.11 In Exercise 14.10, explain what is the effect of an increase in temperature on the positions of the equilibria of a), d), e).

14.12 Balance the following equations:

a) $H_3PO_4 + KOH \rightleftharpoons KH_2PO_4 + H_2O$

b) $H_3PO_4 + KOH \rightleftharpoons K_2HPO_4 + H_2O$

c) $H_3PO_4 + Ca(OH)_2 \rightleftharpoons Ca_3(PO_4)_2 + H_2O$

d) $KOH + CO_2 \rightleftharpoons K_2CO_3 + H_2O$

e) $FeSO_4 + Br_2 \rightleftharpoons Fe_2(SO_4)_2 + FeBr_3$

f) $Cu + HNO_3 \rightleftharpoons Cu(NO_3)_2 + NO + H_2O$

14.13 In the equilibrium $\overline{H_2} + \overline{I_2} \rightleftharpoons 2\ \overline{HI}$, suppose you pumped extra hydrogen into the system without changing the total volume. What effect on the position of the equilibrium would it have? Why?

Answers to Exercises

14.1 If both have the same quantum number; the same spin.

14.2 In analogy to other similar situations, by finding the temperature at which the material dissociates; or by finding the wavelength that brings about dissociation and using the relation $E = h\nu$.

14.4 a) The bonds $\overset{\delta+}{C}—\overset{\delta-}{Cl}$; b) $\overset{\delta-}{N}—\overset{\delta+}{C}$;

c) $\overset{\delta+}{C}=\overset{\delta-}{O}$ and $\overset{\delta+}{C}—\overset{\delta-}{O^-}$; d) $\overset{\delta+}{N} \rightarrow \overset{\delta-}{O}$, and $\overset{\delta+}{H}—\overset{\delta-}{O}$;

e) $\overset{\delta+}{H}—\overset{\delta-}{O}—\overset{\delta+}{S}$, and $\overset{\delta+}{S} \rightarrow \overset{\delta-}{O}$.

14.5 Only ammonia and acetic acid. None of the others contain N, O, or F.

14.7 a) Gasoline is non-polar and cannot shield the ions. b) For one thing, the alcohol forms hydrogen bonds with the water and steals it away from its shielding function.

14.8

2s ↑ ↑ ↑

1s ↑ ↓

The first electron is removed from a higher level. The second must be uncoupled from the third, at a lower, 1s, level. Also, this is a filled shell (helium shell) and is very stable. Therefore extra energy is needed.

14.9 a) $2\ C_8H_{18} + 25\ O_2 \rightarrow 16\ CO_2 + 18\ H_2O$

$$\begin{array}{ccc} 2\ C_8H_{18}: & 25\ O_2: & 18\ H_2O \\ 228 & 800 & 324 \end{array}$$

b) 9.1 kg O_2; c) 1,710 gal. O_2, 8,550 gal. air; d) 3.7 kilograms

14.10 a), d), e) No effect because there is no volume change. b) The reaction will proceed toward the right, since this relieves the pressure by decreasing the volume. c) Toward the left, since this decreases the volume.

14.11 An increase in temperature drives the reaction in the direction that takes up heat: a) to the left; d) and e) to the right.

14.12 a) Already balanced

b) $H_3PO_4 + 2\ KOH \rightleftharpoons K_2HPO_4 + H_2O$

c) $2\ H_3PO_4 + 3\ Ca(OH)_2 \rightleftharpoons Ca_3(PO_4)_2 + 6\ H_2O$

d) $2\ KOH + CO_2 \rightleftharpoons K_2CO_3 + H_2O$

e) $6\ FeSO_4 + 3\ Br_2 \rightleftharpoons 2\ Fe_2(SO_4)_3 + 2\ FeBr_3$

f) $3\ Cu + 8\ HNO_3 \rightleftharpoons 3\ Cu(NO_3)_2 + 2\ NO + 4\ H_2O$

14.13 If more H_2 is added, this increases the concentration of this reagent and drives the reaction to the right.

Chapter 15

15.1 Introduction
15.2 Carbon atoms
15.3 Preliminary definitions
15.4 Beginning nomenclature
15.5 Writing equivalent structures
15.6 Some reactions of paraffins
15.7 Mechanism of reaction, I
15.8 Double and triple bonds
15.9 Unsaturation
15.10 Isomers
15.11 Rings
15.12 Aromatic hydrocarbons
15.13 Körner's proof of structure
15.14 Epistemological interlude
15.15 Proof of structure
15.16 Optical isomerism
15.17 Mechanism of reaction, II
15.18 Summary

Molecular Structure ~ 15

Introduction 15.1

The purpose of this chapter is to show how a chemist might set about determining the structure of a molecule. How does one determine which atoms are attached to which, and what their arrangement in space is? We have chosen as examples rather simple molecules. Even so, however, we need to go through some discussion of naming and related matters before we become ready for the main job. At the same time the principles illustrated hold even for very complicated problems. Our purpose, as before, is to show how this aspect of Nature can be grasped conceptually and controlled; how scientists think in dealing with these problems. The concepts of bond structure and reaction rate, developed in the previous chapter, are now applied systematically. For convenience, though these concepts are widely applicable, we discuss only organic molecules. Organic molecules are the class of molecules that contain at least one carbon atom. There are more than a million different members of this class.

We begin with a discussion of carbon atoms. The outstanding property of carbon atoms is their ability to form bonds with other carbon atoms and with a wide variety of other atoms. In their simplest compounds carbon atoms are linked together and to hydrogen. Even in this class—the hydrocarbons—fantastic variety shows itself and we look at chains, branched chains of carbon atoms, rings of several kinds and three-dimensional structures. We then turn to the linkages with other atoms. In each case the variety of possibilities increases.

As these new molecules are introduced we develop the systematic nomenclature which makes it possible for the Student to name ordinary compounds and to invent and name extraordinary ones.

15.2 Carbon atoms

In its ground state a carbon atom has a pair of electrons in a 1*s* orbital, and in the second shell two 2*s* electrons and one each $2p_x$ and $2p_y$ electrons (Section 11.4). When it reacts to form four single bonds, as in CH_4 (methane), the two 2*s* electrons are raised to a higher energy level and become "hybridized" with the two 2*p* electrons to form, sharing with the 1*s* electrons of hydrogen, *four equivalent bonds.* That these four bonds are equivalent is shown by the fact that (for example) only one CH_3Cl is known. If one bond were different from the others, then presumably there would be at least two different CH_3Cl's. Furthermore, only one CH_2Cl_2 is known. If the molecule CH_2Cl_2 were pyrimidal, or planar, it should be possible to find or make two compounds of the type CH_2Cl_2 (Figure 15-1). There is no evidence for two molecules CH_2Cl_2. All the evidence indicates that CH_4 is tetrahedral. All the evidence suggests that the four carbon bonds in similar structures are tetrahedrally arranged, each of the same distance from the others, except when bulky groups may distort the shape a little. Figure 15-2 shows ways of representing methane. We shall write abbreviated

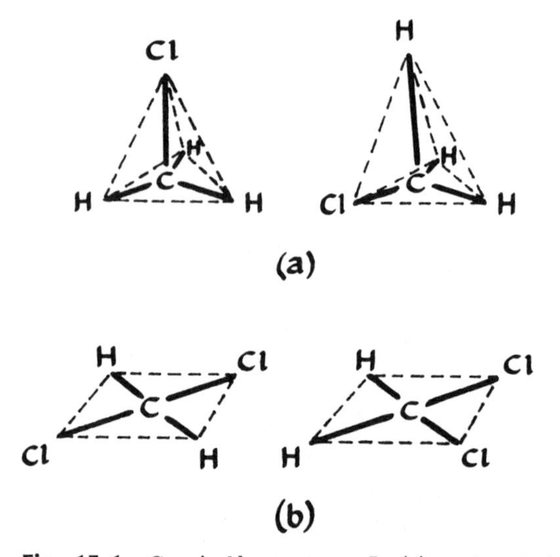

Fig. 15-1 *Conceivable structures. In* (a), *a hypothetical pyramidal structure, there would be two molecules with the formula CH_3Cl: one with the Cl at the apex of the pyramid, and one with the Cl at the base. These would not be identical because they are not superimposable. In* (b), *a hypothetical planar structure, there would be two different arrangements possible for a compound of the formula CH_2Cl_2.* [*From James English and Harold G. Cassidy,* Principles of Organic Chemistry. *Ed. 3. McGraw-Hill, New York, 1961, Fig. 1-2, p. 7.*]

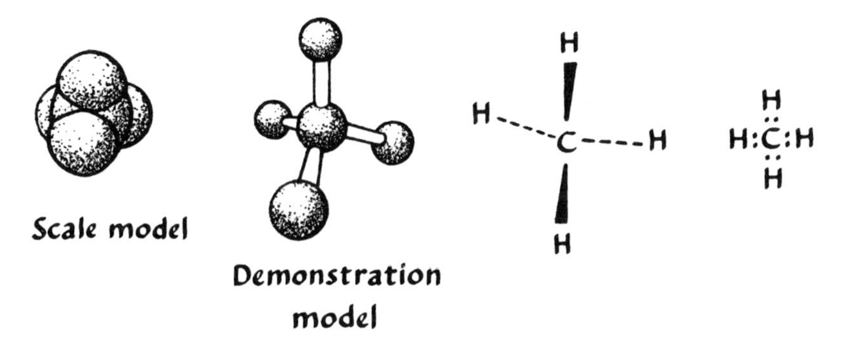

Fig. 15-2 *Symbols for the methane molecule, after J. English and H. G. Cassidy,* loc. cit., *Fig. 1-1, p. 6.*

formulas and let the three-dimensional shape be implied since, because H can form only one single bond with carbon, there can never be any doubt about such arrangements.

The four identical bonds in compounds like CH_4 are formed by overlap of the 1*s* orbitals of the four hydrogens with four sp^3 orbitals formed from hybridization of the *s* and 2*p* orbitals of the reacted carbon atom (Figure 15-3). The electrons, two for each overlapping orbital, are shared by C and H, and form the C—H bond. In some molecules only two *p* orbitals hybridize with the *s*, forming sp^2 hybrid orbitals; in some cases only the *s* and one *p* hybridize, forming an *sp* orbital (Figure 15-3).

Preliminary definitions 15.3

A 'group' is defined as a part of a molecule—it may be an atom or a group of atoms—which is dealt with as a unit, perhaps for convenience in nomenclature, perhaps because it passes through some chemical transformation unchanged. H— is usually not named. Common groups are —Cl, chloro-; or -chloride; —Br, bromo-; or -bromide; —I, iodo-; or -iodide; —OH, hydroxy-; or -hydroxyl; —CH_3, methyl-; or -methyl; and other organic groups which are dealt with in the next section. A 'hydrocarbon' is a molecule containing only carbon and hydrogen.

The 'molecular formula' of a compound shows the actual number of atoms in the molecule. For example, the molecular formula for ethane is C_2H_6. The 'empirical formula' is the simplest correct ratio of atoms in the molecule. For ethane it is CH_3. The 'structural formula' or structural symbol of the molecule is the molecular formula written

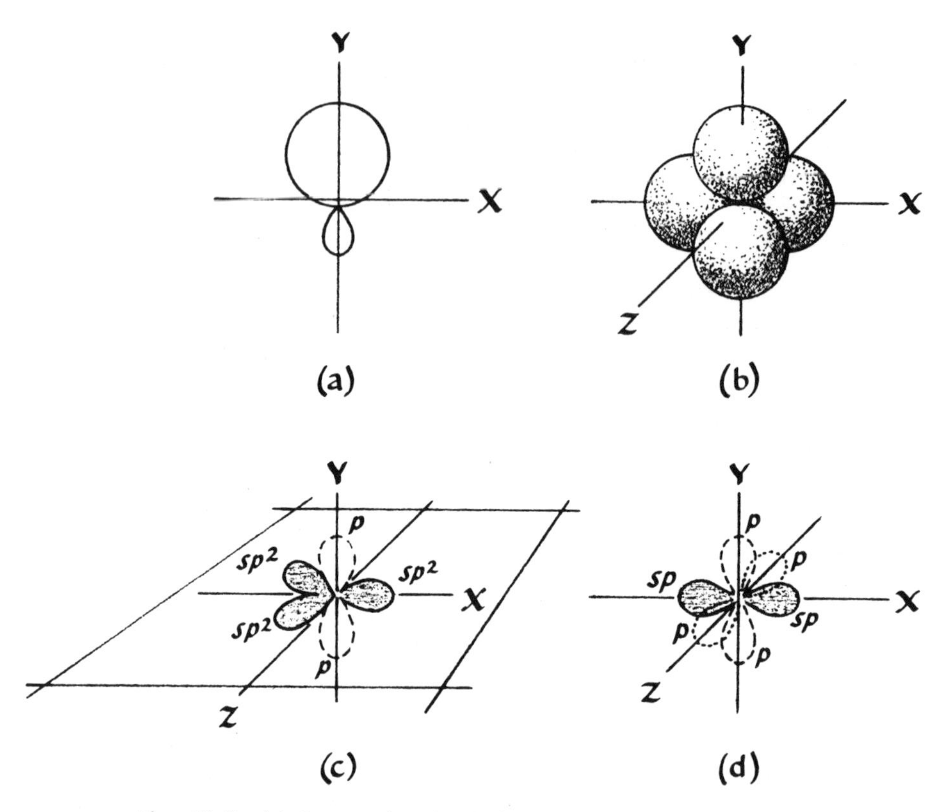

Fig. 15-3 (a) *Cross-section of a single* sp^3 *hybrid atomic orbital, and* (b) *perspective of four tetrahedrally arranged* sp^3 *hybrid orbitals.* (c) *Three* sp^2 *orbitals shaded, with the axis of the remaining* p *orbital perpendicular to the plane of the three.* (d) *Two* sp *orbitals. The axes of the remaining two* p *orbitals are at right angles to each other and to that of the* sp *orbitals.*

out to show the relationship of the atoms (with spatial arrangement implied).

```
  H  H
HC—CH
  H  H
```

15.4 Beginning nomenclature

A series of hydrocarbons is known in which the members have 1, 2, 3, . . . *n* carbon atoms linked to each other in a chain. A few of these are shown in Table 15-1, with names. The first four have

Table 15-1

Some Paraffin Hydrocarbons

Molecule	Name	Molecular formula
CH_4	methane	CH_4
$H_3C—CH_3$	ethane	C_2H_6
$H_3C—CH_2—CH_3$	propane	C_3H_8
$H_3C—CH_2—CH_2—CH_3$	butane	C_4H_{10}
$H_3C—(CH_2)_3—CH_3$	pentane	C_5H_{12}
$H_3C(CH_2)_4—CH_3$	hexane	C_6H_{14}

C_8, octane; C_9, nonane; C_{10}, decane; C_{11}, hendecane.

trivial, or popular, or non-systematic names that were given before systematic naming was agreed on; the rest take Greek-root prefixes familiar to anyone who has studied geometry (*e.g.*, 'penta'; a pentagon has five angles). To the prefix is appended the suffix '-ane,' which signifies that the compound belongs to this family, known as the paraffin hydrocarbons. The family has a 'type formula' C_nH_{2n+2}, where n is the number of carbon atoms present. From each, a group can be derived, for nomenclature purposes, by removing a hydrogen. Then, the suffix '-ane' is replaced by '-yl,' thus: $H_3C—$, methyl; $CH_3CH_2—$, ethyl; and so forth (Table 15-2). The general symbol for an organic group is R.

Table 15-2

Some Groups Derived from Paraffin Hydrocarbons

Group	Name
H HC— H	methyl-
H $H_3C—C—$ H	ethyl-
$C_3H_7—$	propyl-
$C_4H_9—$	butyl-
$C_5H_{11}—$	pentyl-

The International rules for naming paraffin hydrocarbons and their derivatives are:

1. Find the longest continuous chain of carbon atoms. The basic name for the compound is that of the paraffin hydrocarbon with this number of carbon atoms. For example, (I) is named as a derivative of propane, as is (II). (III) Is named as a derivative of pentane.

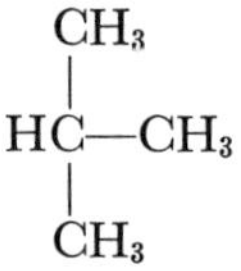

$$\begin{array}{c} CH_3 \\ | \\ HC{-}CH_3 \\ | \\ CH_3 \\ \text{(I)} \end{array} \qquad \begin{array}{c} CH_3 \\ | \\ CH_3C{-}CH_3 \\ | \\ CH_3 \\ \text{(II)} \end{array} \qquad \begin{array}{c} CH_2CH_3 \\ | \\ CH_3{-}CH_2{-}C{-}CH_3 \\ | \\ CH_3{-}CH{-}CH_3 \\ \text{(III)} \end{array}$$

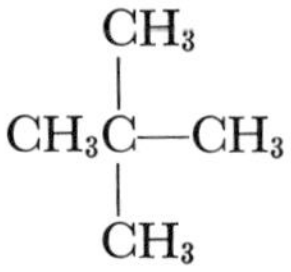

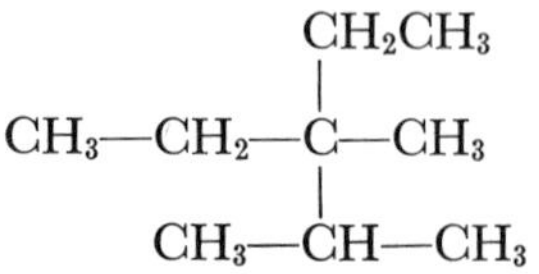

2. Number the carbon atoms of this continuous chain.

3. Each substituent other than hydrogen, which is assumed to fill all places undesignated in the name, is given a number corresponding to the carbon atom of the continuous chain to which it is attached. Each substituent receives both a name *and* a number. Groups that appear more than once receive, if there are two of them, two numbers and the prefix di-; if three of them, three numbers and the prefix tri-, and so forth. Thus (I) is 2-methylpropane; (II) is 2,2-dimethylpropane.

4. The numbering of the chain must normally start from the end of the molecule which allows the smallest numbers to be used in locating substituents. Thus (III) is 2,3-dimethyl-3-ethylpentane.

Compound (III) could without sacrifice of clarity be named 3,4-dimethyl-3-ethylpentane. Rule 4 introduces uniformity in naming. The over riding rule is to use the simplest unambiguous name. These names are sometimes referred to as blue print names, since they map the structure. Commonly, where there is a choice, the most euphonous name is chosen.

15.5 Writing equivalent structures

Since it is not feasible to write out all molecular structures in three dimensions, the conventional procedure is adopted of projecting the molecule, in imagination, onto the paper and writing the structure in two dimensions. This takes a little getting used to, and demands the cultivation of visual imagery. For example, $\begin{array}{c} H\ \ H\ \ H\ \ H \\ |\ \ \ |\ \ \ |\ \ \ | \\ HC{-}C{-}C{-}CH \\ |\ \ \ |\ \ \ |\ \ \ | \\ H\ \ H\ \ H\ \ H \end{array}$ might equally well be written $H_3C{-}CH_2{-}CH_2{-}CH_3$, or $\begin{array}{cc} CH_3 & CH_3 \\ | & | \\ CH_2 & {-}CH_2 \end{array}$.

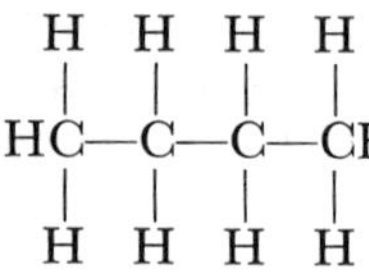

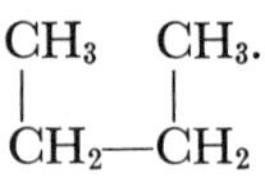

Looking at the molecule from one side, it might look like

```
 H      H
  \    /   H
   C      CH
H/  \   /H
HC     C
 H    / \
     H   H
```

because of the tetrahedral angles of the bonds (about 109°). Further, there is considerable freedom of rotation about a single bond, unless the groups on the bonded atoms are quite bulky. The molecule in gas or solution phase is continually changing its 'conformation,' writhing about under the buffeting of neighboring molecules—unless, of course, it happens to be a rigid molecule. Thus $\begin{array}{c}\mathrm{H\ \ H}\\ \mathrm{HC{-}C{-}Br}\\ \mathrm{H\ \ H}\end{array}$ might equally well be written $\begin{array}{c}\mathrm{H\ \ Br}\\ \mathrm{HC{-}CH}\\ \mathrm{H\ \ H}\end{array}$.

Some reactions of paraffins 15.6

The production of heat (or work) is the chief industrial use of the paraffins: Those in natural gas are largely 1- or 2-carbon paraffins—methane or ethane; those in low pressure (LP) tanks, such as bottled gas, are mainly butanes and pentanes; those in gasoline are in the octane range with about 8 carbon atoms; in kerosene the paraffins have 12 or so carbon atoms; fuel oil and solid paraffins have larger numbers of carbon atoms. The paraffins burn in air (the lower paraffins may form explosive mixtures) according to an equation of the form:

$$2\ CH_3CH_3 + 7\ O_2 \rightarrow 4\ CO_2 + 6\ H_2O + 2 \times 368.4\ \text{kcal}$$

If not enough oxygen is present for complete combustion carbon monoxide, CO, may be formed, or soot, largely elementary carbon. Most organic substances will burn in air, some with more difficulty than others.

Paraffin hydrocarbons react with chlorine or bromine in what is named a substitution reaction: hydrogen is replaced by halogen. Thus

$$\begin{array}{c}\mathrm{H}\\ \mathrm{HCH}\\ \mathrm{H}\end{array} + Cl_2 \rightarrow \begin{array}{c}\mathrm{H}\\ \mathrm{HCCl}\\ \mathrm{H}\end{array} + HCl$$

$$\begin{array}{c}\mathrm{H}\\ \mathrm{HCCl}\\ \mathrm{H}\end{array} + Cl_2 \rightarrow \begin{array}{c}\mathrm{H}\\ \mathrm{HCCl}\\ \mathrm{Cl}\end{array} + HCl$$

and so forth to $CHCl_3$ and CCl_4. The reaction goes slowly in the dark, rapidly in the light; the build-up of the more and more chlorinated products proceeds on a statistical basis, the result of chance collisions.

15.7 Mechanism of reaction, I

By mechanism we mean a detailed, preferably quantitative description of the reaction. Consider the above reaction. What is its mechanism, and why does light make the reaction go faster? When light of suitable energy ($\boldsymbol{E} = h\nu$) is absorbed by a chlorine molecule, it dissociates it into two atoms:

$$:\ddot{\underset{..}{Cl}}:\ddot{\underset{..}{Cl}}: + h\nu \rightarrow :\ddot{\underset{..}{Cl}}\cdot + \cdot\ddot{\underset{..}{Cl}}: \qquad (1)$$

The chlorine atom is now energy-rich, and attacks a paraffin molecule by abstracting a hydrogen atom, leaving a paraffin radical (in this example a methyl radical):

$$\begin{matrix} H \\ HCH \\ H \end{matrix} + \cdot\ddot{\underset{..}{Cl}}: \rightarrow \begin{matrix} H \\ HC\cdot \\ H \end{matrix} + H:\ddot{\underset{..}{Cl}}: \qquad (2)$$

(A radical is a group with a single electron.) The radical is reactive and attacks Cl_2 which may collide with it:

$$\begin{matrix} H \\ H\text{—}C\cdot \\ H \end{matrix} + :\ddot{\underset{..}{Cl}}:\ddot{\underset{..}{Cl}}: \rightarrow \begin{matrix} H \\ HC \\ H \end{matrix}:\ddot{\underset{..}{Cl}}: + \cdot\ddot{\underset{..}{Cl}}: \qquad (3)$$

This regenerates a chlorine atom, and the last two steps alternate. Thus one quantum may start a 'chain reaction' that continues for many steps. If the steps give off energy (are exothermic) the system becomes hotter, and an explosion may occur. (Methane mixed with chlorine explodes violently when a light is shined on it.) The reaction stops when the reagents are used up, or when some reaction such as the combination of radicals terminates it by not forming a new radical, *e.g.*,

$$:\ddot{\underset{..}{Cl}}\cdot + :\ddot{\underset{..}{Cl}}\cdot \rightarrow Cl_2, \text{ or } :\ddot{\underset{..}{Cl}}\cdot + \cdot CH_3 \rightarrow ClCH_3,$$
$$\text{or } H_3C\cdot + \cdot CH_3 \rightarrow H_3C\text{—}CH_3 \qquad (4)$$

The mechanism of the chain reaction involves the initiation step (1), the propagation steps (2) alternating with (3), and the termination step or steps (4).

That the reaction goes at all in the dark is because there is always some slight dissociation of the halogen molecule into atoms—after all, everything is being bombarded by α-, β-, and γ-rays from residual radioactivity and by cosmic rays.

$$CH_4 + Cl_2 + \quad \rightarrow CH_3Cl + CH_2Cl_2 + CHCl_3$$
$$+ CCl_4 + \text{traces of } C_2H_6$$

Double and triple bonds 15.8

Carbon forms not only single bonds with carbon, but also double and triple bonds. These are written:

```
H           H   H       H
 \         /     .     .
  C  =  C     or   C::C    and H—C≡C—H or H:C:::C:H
 /         \     .     .
H           H   H       H
```

ethene, or ethylene — ethyne or acetylene

The -ene (as in sc*ene*) compounds belong to the family of olefins, the -yne (as in l*ine*) to the family of acetylenes. The type formula for the mono-olefins is C_nH_{2n}; that for the mono-acetylenes is C_nH_{2n-2}.

The existence of structures of this kind requires additional naming rules. In general (at least for the compounds we shall be concerned with) the numbering now is arranged to run through the carbons of the double or triple bond, and the position of the bond is indicated by the smaller of the numbers of the two bonded atoms. The suffix -ene is used for double and -yne triple bonds. The other rules continue to hold. Examples of the application of these rules are:

```
          CH2CH2CH3
          |
CH3CH = C—CH2CH3
```

3-ethyl-2-hexene

```
CH3CHCH = CH2
   |
   Cl
```

3-chloro-1-butene

$CH_2 = CH—CH = CH—CH_3$

1,3-pentadiene

$H_3C—CH_2C≡CH$

1-butyne

$CH_2 = CH—C≡C—CH_2Cl$

1-chloro-4-penten-2-yne

Unsaturation 15.9

When a compound such as ethylene is treated with bromine, there is an immediate, rapid reaction in which the bromine color disappears, *but no HBr is formed.* The reaction is clearly not substitution because of the lack of appearance of HBr, and also because on analysis the product turns out to be $C_2H_4Br_2$—the bromine 'added' to the molecule. The structure of the product turns out upon examination to be $BrCH_2—CH_2Br$. Similarly, when bromine is added to acetylene, the product may be $BrCH = CHBr$, if only one Br_2 adds, or with more bromine $CHBr_2CHBr_2$. Olefins and acetylenes are 'unsaturated': they undergo addition reactions.

What is the mechanism? It appears to be the case that the double and triple bonds are not simply two or three covalent bonds. At the double bond each of the linked carbon atoms has three sp^2 hybrid bonds in a plane. The third p orbital is positioned symmetrically above and below this plane, and the p orbitals of neighboring carbon atoms overlap to form a molecular orbital, called a π orbital (Figure 15-4). In the triple bond each carbon has two sp hybrid bonds arranged linearly, with bond formation between the carbons. The remaining two p orbitals on each carbon overlap to form π bonds at right angles to each other (Figure 15-5).

It will be noted that the double-bond compound is flat at the

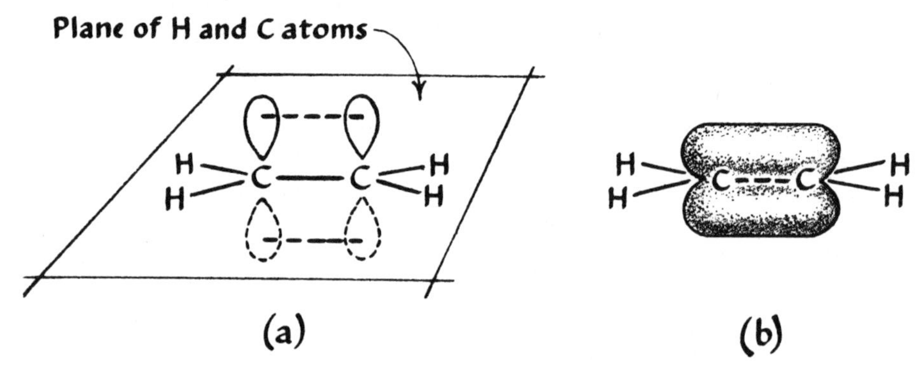

Fig. 15-4 *Schema of molecular orbitals* (a). *The 'overlapping'* p *orbitals, dashed line, are above and below the plane of the H and C atoms of the ethylene molecule.* (b) *The* π *molecular orbital of ethylene in perspective.* [*From J. English and H. G. Cassidy,* loc. cit., *Figs. 3-1 and 3-2, p. 60.*]

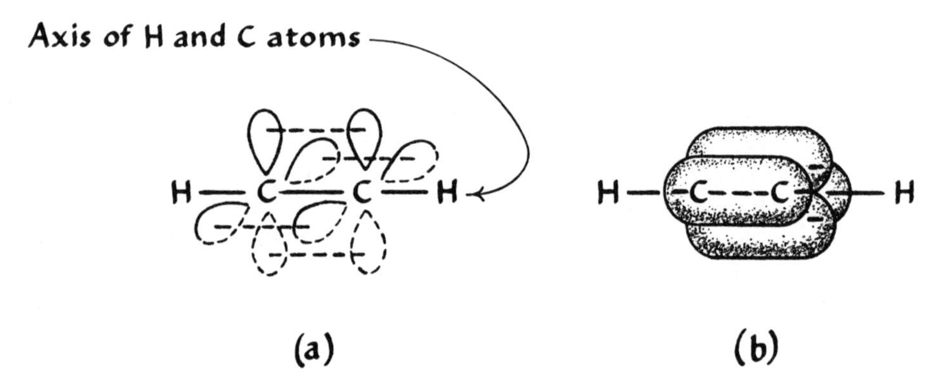

Fig. 15-5 (a) *The two sets of overlapping* p *orbitals of acetylene are shown in outline, connected by dashed lines.* (b) *Perspective representation of the* π *molecular orbitals of acetylene.* [*From J. English and H. G. Cassidy,* loc. cit., *Fig. 5-1, p. 97.*]

bond, with the six groups on the two bonded carbon atoms lying in the same plane. The π orbital, source of the unsaturation, lies above and below this plane. It may be conceived that some of the time the paired electrons of the π bond are above, sometimes below, and sometimes one or the other above and below the plane.

$$\overset{:}{C}—C \leftrightarrow \underset{:}{C—C} \leftrightarrow \overset{\cdot}{C}—\underset{\cdot}{C} \leftrightarrow \underset{\cdot}{C}—\overset{\cdot}{C} \qquad (5)$$

We met a similar situation in Section 14.3.

Now, when bromine adds it may do it by one of two mechanisms: ionic or radical.[1] In a bromine solution there are some ions formed by dissociation:

$$:\ddot{Br}:\ddot{Br}: \rightleftharpoons :\ddot{Br}^{+} + :\ddot{Br}:^{-}.$$

The π bond is a site of negative character due to its pair of electrons (Equation 5). Br^+ lacks a pair of electrons and readily shares this π bond pair to form an unstable ionic intermediate:

$$C\overset{\cdot\cdot}{—}C + \ddot{Br}:^{+} \rightleftharpoons \left[\begin{array}{c} :\ddot{Br}: \\ C—C \end{array}\right]^{+}$$

Now we have a more or less flat underside to the molecule, and this can be attacked by a negative ion, *e.g.*, Br^-, with the product being neutral CH_2BrCH_2Br:

$$\begin{array}{c} Br \\ C—C \\ :\ddot{Br}:^{-} \end{array} \rightarrow \begin{array}{c} Br \\ | \\ C—C \\ | \\ Br \end{array}$$

We shall show, below, how this mechanism was worked out. A similar mechanism applies to acetylenes.[2]

---

[1] In reactions of this kind, the adding reagent is usually dissolved in some inert solvent. Carbon tetrachloride, having no H to be replaced, is quite inert to Br_2. The C—Cl bond is a stronger covalent bond than the C—Br bond, as might be expected (Section 11.6). Bromine in CCl_4 is a red solution. Add to it an olefin and the red color immediately disappears. The brominated hydrocarbons are colorless.

[2] $:\ddot{Br}^{+}$ and similar substances that lack a *pair* of electrons and so will react with available, unshared pairs are spoken of as 'electrophilic,' or electron-loving. They are generalized acids (Section 14.9). The molecule or group with an unshared pair, as $:\ddot{Cl}:^{-}$ or R—O—H, which partakes of the bond-forming sharing is 'nucleophilic,' or nucleus (*i.e.*, (+))-loving. It is a generalized base.

Substances that form ions of a non-metallic kind will add to -ene and to -yne compounds in this way, thus:

$$H_2C{=}CH_2 + H^+ Br^- \rightarrow \underset{\text{ethyl bromide}}{CH_3CH_2Br}$$

$$H_2C{=}CH_2 + H^+O^-SO_2OH \rightarrow \underset{\text{ethyl hydrogensulfate}}{CH_3CH_2OSO_2OH} \qquad (6)$$

In the presence of light, Br_2 yields $Br\cdot$ atoms. These add rapidly to double and triple bonds, as might be expected from Equation (5).

$$>\dot{C}{-}\dot{C}< + 2:\ddot{\underset{..}{Br}}\cdot \rightarrow -\overset{Br}{\underset{|}{\overset{|}{C}}}-\overset{Br}{\underset{|}{\overset{|}{C}}}-$$

This is not a chain reaction, of course.

It is interesting that ethylene (and propene) may be made to add to itself by the use of a reagent (M) that converts it to an electrophilic ion:

$$H_2C = CH_2 + M \rightarrow H_2\overset{\ddot{M}}{C}-\overset{H}{\underset{H}{C^*}}$$

$$H_2\overset{\ddot{M}}{C}-\overset{H}{\underset{H}{C^*}} + H_2C = CH_2 \rightarrow H_2-\overset{\ddot{M}}{C}-\overset{H}{\underset{H}{C}}-\overset{H}{\underset{H}{C}}-\overset{H}{\underset{H}{C^*}}$$

The nucleophilic end of the molecule, attacking another double bond, generates a new nucleophilic end, and so on. The molecule grows by accretion, and becomes a 'polymer' (Section 9.5). In this case hundreds or even thousands of parts may add to form long chains: polyethylene is the product. This is a chain reaction. Similarly, a radical may initiate such a reaction, which then proceeds by a radical mechanism.

15.10 Isomers

Compounds that contain the same numbers of the same kinds of atoms but in different arrangements are 'isomers.' For example, $CH_3CH_2CH_2CH_3$ and $CH_3\underset{CH_3}{\underset{|}{C}}HCH_3$ are chain isomers; 1,1-dibromoethane, CH_3CHBr_2, and 1,2-dibromoethane, CH_2BrCH_2Br, are position isomers. Isomers almost always have different physical and chemical properties.

The olefins display an additional kind of isomerism: 'geometrical isomerism.' The C=C bond is rigid—no easy rotation about it—and the molecule is flat. Now if each bonded carbon has two different substituents on it, two arrangements become possible:

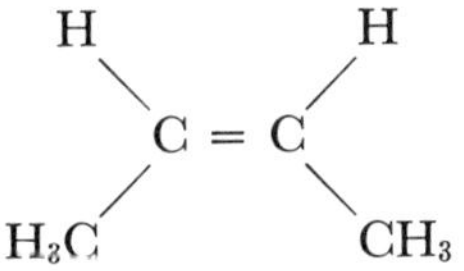

```
H           H            H           CH3
 \         /              \         /
  C   =   C                C   =   C
 /         \              /         \
H3C         CH3          H3C         H
```

cis-2-butene[1] 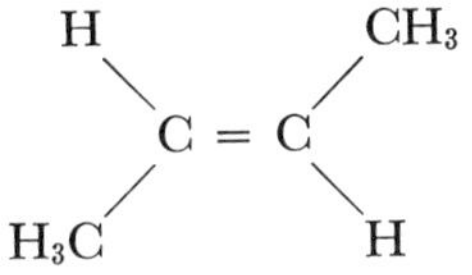 *trans*-2-butene[1]

These two substances are known. They have different physical and chemical properties. X-ray analysis of compounds of this type shows the *cis*-groups closer to each other than the trans-.

Rings 15.11

Hydrocarbon molecules are known in which the carbons are attached to each other to form a ring. For example:

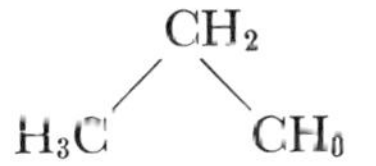

```
      CH2                      CH2
     /   \                    /   \
  H3C     CH3             H2C-------CH2
```

propane, b.p. −42.2° 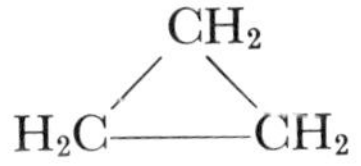 cyclopropane, b.p. −32.9°

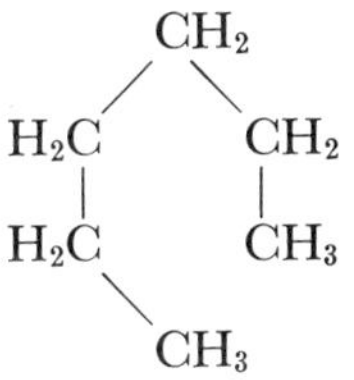

```
     CH2                    CH2
    /   \                  /   \
 H2C     CH2            H2C     CH2
  |       |              |       |
 H2C     CH3            H2C     CH2
    \                      \   /
     CH3                    CH2
```

hexane, b.p. 69° 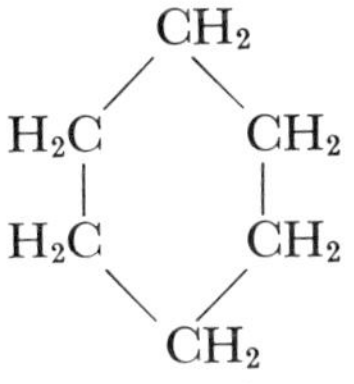 cyclohexane, b.p. 69°

The names of these substances are based on the size of the ring. The carbon atoms are numbered so as to use the smallest numbers for substituents. Examples of naming show the conventions:

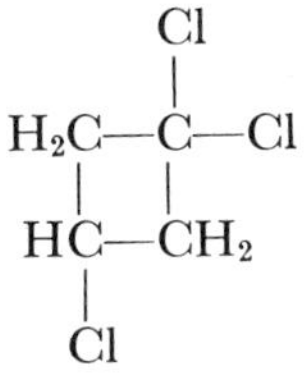

```
       Cl                         Br
       |                          |
 H2C---C---Cl                     CH
  |    |                         /   \
 HC---CH2                     H2C     CH
  |                            |      ||
 Cl                           H2C     CH
                                 \   /
                                  CH
                                  |
                                  C6H13
```

1,1,3-trichlorocyclobutane 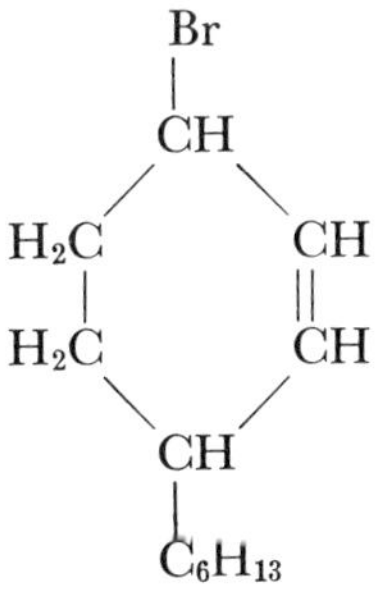 1-bromo-4-hexyl-2-cyclohexene

[1] cis- means "on this side," as in Caesar's 'cisalpine', meaning on the southern side of the Alps, from the viewpoint ('frame of reference') of Rome. Trans- means "on the other side of," as in transatlantic.

```
     CH2  H        CH3
    /     \|        |
CH2        C--------C—CH3
|          |        |
|          |        H
CH2-------CHOH
```

1-hydroxy-2-[2-propyl]-cyclopentane

To produce 3- and 4-member rings, the carbon-to-carbon bonds have to be "bent" or "strained." Whatever the imagery, the fact is that cyclopropane and cyclobutane *add* bromine readily, showing a degree of energy-richness, while higher cyclic compounds substitute for hydrogen, like other paraffins.

When a model of cyclopentane is made, the ring is found to be flat. Cyclohexane and all higher cycloparaffins have puckered rings. The puckered rings may take various conformations. Two assumed by cyclohexane are shown in Figure 15-6. The atoms on the bonds marked with asterisks interfere with each other, and this conformation (the 'boat' form) is less stable than the strain-free 'chair' form.

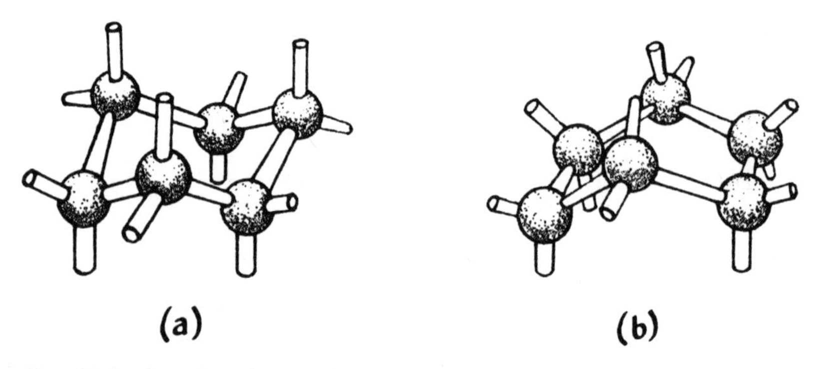

Fig. 15-6 *Demonstration models of unstrained, puckered rings of cyclohexane.* (a) *Chair form.* (b) *Boat form.* [*From J. English and H. G. Cassidy,* loc. cit., *Fig. 4-1, p. 87.*]

Since the ring is a structure with two 'sides' and cannot undergo free rotation about C—C bonds, it can show geometric isomerism:

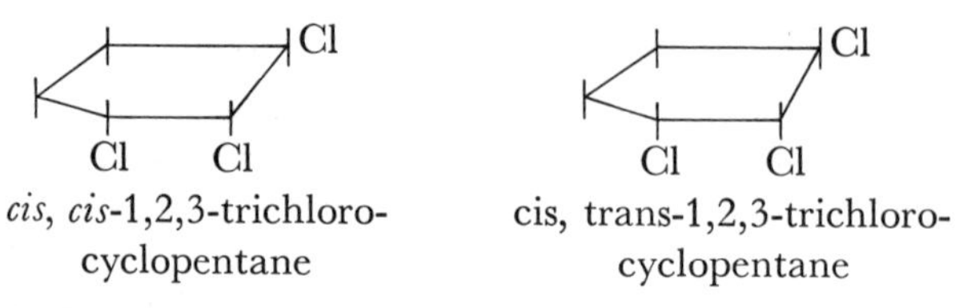

cis, cis-1,2,3-trichloro-cyclopentane   cis, trans-1,2,3-trichloro-cyclopentane

We shall see later how this property of rings allows us to determine mechanism.

Aromatic hydrocarbons 15.12

The last types of hydrocarbons that we shall consider are named aromatic because they were first isolated from aromatic plant gum. The simplest member of the family is benzene, which might be called 1,3,5-cyclohexatriene were it not that the molecule does not easily add bromine, but substitutes it instead.

H
C
HC CH
HC CH $+ Br_2 \rightarrow$ Br (ring) $+$ HBr.
C
H

bromobenzene

The ring is usually written as a simple hexagon, with double-bonds. When it is a substituent group it is designated *phenyl-*.

All the hydrogens in benzene are equivalent. Only one bromobenzene is known. Furthermore, only two 1,2-dibromobenzenes are known, and therefore the three double bonds cannot be as shown in the above structure for then two 1,2-bromobenzenes would be known.

Br Br Br Br
C = C and C − C, and they would be different compounds,

because a C=C bond is shorter than a C—C bond. Also, three and only three dibromobenzenes are known:

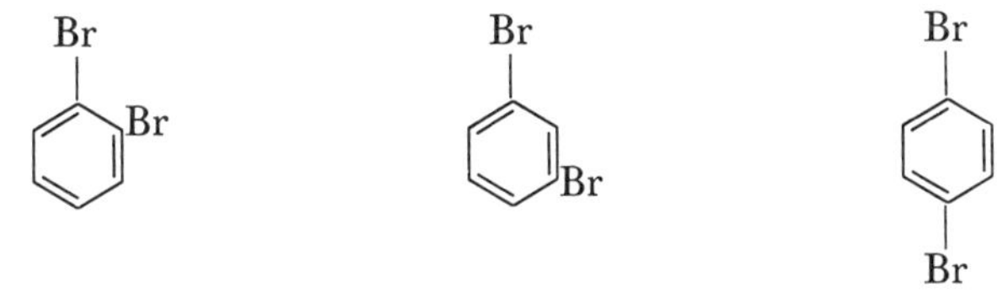

1,2-dibromo-benzene or *ortho*-dibromo- 1,3-dibromo-benzene or *meta*-dibromo- 1,4-dibromo-benzene or *para*-dibromo-

These and a myriad of other findings are rationalized by the molecular orbital picture of benzene. Each carbon, with three sp^2 bonds has a p orbital perpendicular to the ring (as an olefin does to its double bond) but now these overlap uninterruptedly, and the result is an extremely stable molecule (Figure 15-7).

Körner's proof of structure 15.13

How did it first become possible to attach the correct symbols to the different dibromobenzenes? Suppose that you had brominated

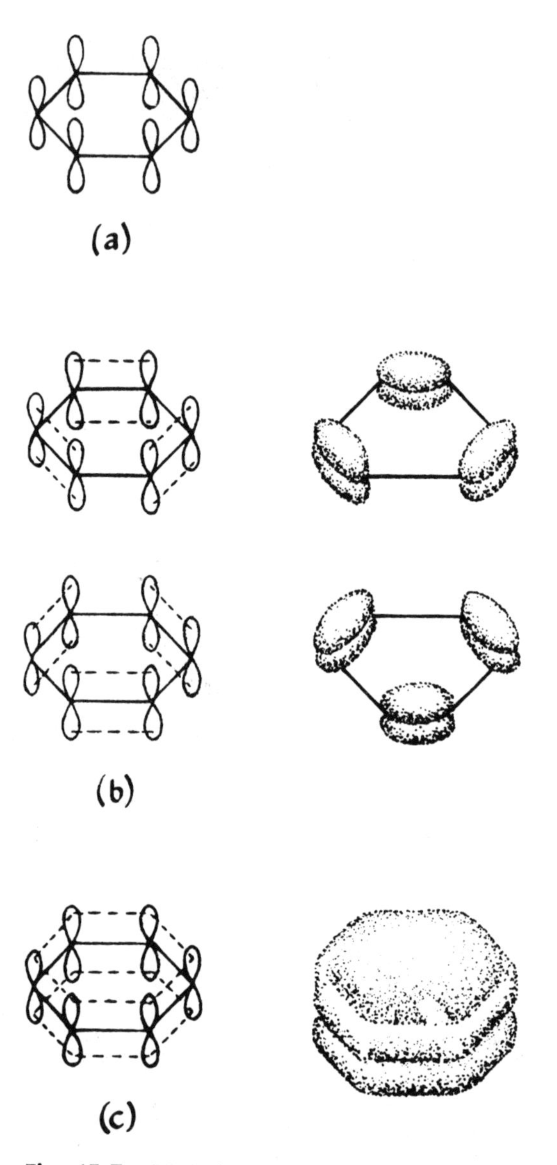

Fig. 15-7 *Schematic representation of π bonds in benzene. The bonds to H are omitted.* (a) *The* p *orbitals at each carbon.* (b) *The representation of fixed double bonds.* (c) *The correct representation as a molecular orbital.* [*From J. English and H. G. Cassidy*, loc. cit., *Fig. 6-1, p. 114.*]

benzene with enough bromine and heat to give you a mixture of dibromobenzenes. (All possible isomers are formed.) You have crystallized the products of the reaction and find that you have three substances, all of which analyze for $C_6H_4Br_2$. One melts at $-6.9°$; one at 2°; one at 89°. There are no other dibromobenzenes present. Logic argues that there are only three ways of arranging two bromines on a benzene ring. These are shown above. You label each bottle with the melting point. Now, which is which: how do you draw the correct structural formula on each label? Körner solved the problem in the following way.

Assume that the ring is symmetrical and that in any reaction *all* possible isomers will be formed. They need not, and do not, form in equal amounts. But if you substitute another bromine on the ring to form all possible tribromobenzenes, and then work up the products of the reaction carefully enough to find all of them, then, according to the logic of the situation, you will obtain different *numbers* of tribromobenzenes for the three isomeric dibromobenzenes. On the basis of these experimental results you can write the correct names on the bottles containing the dibromobenzenes. Further, following the logic, you can also write the correct names on the bottles of tribromobenzenes:

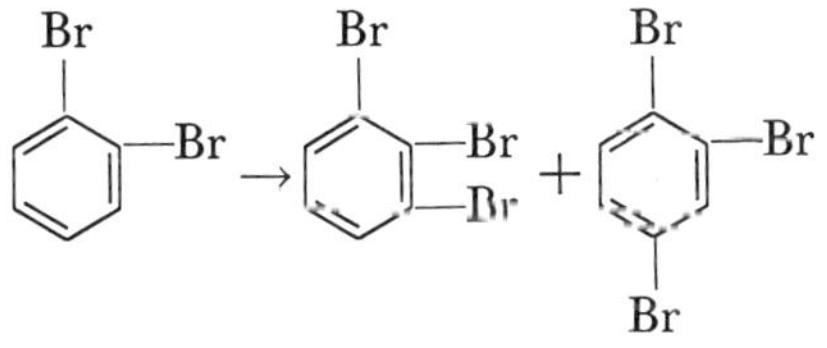

1,2-dibromobenzene (melts at 2°)

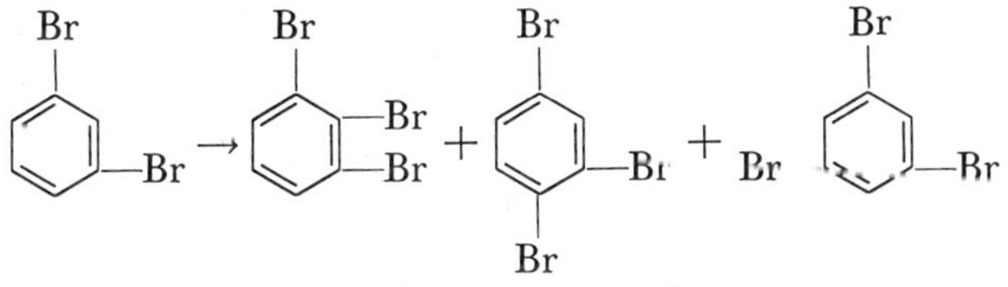

1,3-dibromobenzene (melts at −6.9°)

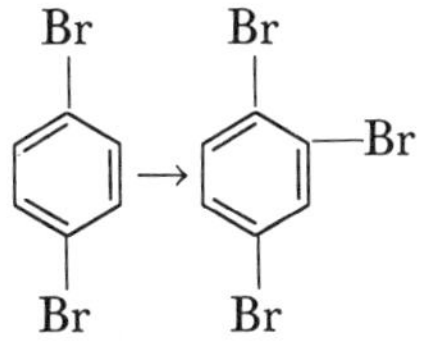

1,4-dibromobenzene (melts at 89°)

Determination of the structure of a molecule follows logic of this kind. Sometimes it is quite straightforward, as here; in other cases, extremely subtle.

Epistemological interlude 15.14

We will be concerned in the next section with proofs of structure: with what groups are attached to what other groups. The conclusions arrived at depend on analytical data such as empirical formula, molecular formula, kinds and number of groups present such as hydroxyl or keto, as found by specific reaction sequences. These data are interpreted through inferences based on logical rules. Put into the idiom of Chapter 1 the experimental observations at and close to the Plane of Perceptions are transformed to symbols and connected by rules. These symbols, signs for Constructs, are manipulated according to rule—new Constructs and rules being invented or discovered from time to time. Return is made to the Plane of Perceptions through experiments.

15.15 Proof of structure[1]

When starches and sugars are fermented, as in the early stages of the production of beer, wines, or hard liquors, there is produced a substance, ethyl alcohol, which may be concentrated by careful distillation in an efficient still. The highest concentration available upon ordinary distillation is 95% ethyl alcohol, 5% water. (This is 190 proof.) By other means the remaining 5% of water may be removed. The pure product is a water-white liquid with a boiling point of 78.4°C, a specific gravity of 0.789 gm per ml, and the empirical and molecular formula C_2H_6O, as determined from analysis: 52.18% carbon, 13.04% hydrogen, 34.78% oxygen.

Our rules allow us to write two structures with this formula:

$$
\begin{array}{ccccccc}
 & \mathrm{H} & & \mathrm{H} & \\
 & | & & | & \\
\mathrm{H}- & \mathrm{C} & -\mathrm{O}- & \mathrm{C} & -\mathrm{H} \\
 & | & & | & \\
 & \mathrm{H} & & \mathrm{H} & \\
 & & \mathrm{I} & &
\end{array}
\quad \text{and} \quad
\begin{array}{cccccc}
 & \mathrm{H} & & \mathrm{H} & \\
 & | & & | & \\
\mathrm{H}- & \mathrm{C} & - & \mathrm{C} & -\mathrm{O}-\mathrm{H} \\
 & | & & | & \\
 & \mathrm{H} & & \mathrm{H} & \\
 & & \mathrm{II} & &
\end{array}
$$

These are quite clearly different substances. If we name them, I is 'dimethyl ether,' for ROR is an ether linkage; the other, II, is hydroxy ethane, or 'ethanol' for —OH is a 'hydroxyl', or '—Ol' group. (There is no need to number the position of the —OH since on either carbon it would be '-1-'.) We now must resort to chemical reactions to determine which of the two structures properly represents our substance whose trivial name is ethyl alcohol. What this amounts to is that we have a bottle of liquid and two labels, dimethyl ether and ethanol: we wish to know which label to affix to the bottle. If we were physicians and were faced with a patient whose disease we had to determine in order to treat, we would follow a similar procedure, looking for symptoms, making tests, and finally coming up with a diagnosis. (The

---

[1] The suggestion for this whole sequence comes from William T. Caldwell, *Organic Chemistry*. Houghton Mifflin, Boston, 1943. See pp. 61 ff.

physician's problem may be made easier to solve, or more difficult, by the fact that the patient usually can talk.)

When ethyl alcohol is treated with metallic sodium, a quiet reaction ensues. Hydrogen gas bubbles off at the surface of the metal, and at the end of the reaction a white, solid product can be isolated. Analysis of the reagents and products allows us to write the reaction:

$$2\ Na + 2\ C_2H_6O \rightarrow \overline{H_2} + 2\ C_2H_5O\ Na$$

The products are clearly quite different from the reactants. Moreover it is found that even in the presence of a large excess of sodium only *one* of the six hydrogens of the ethyl alcohol is replaced. Thus we must single out one of them as being quite different from the other five. Further, if the white product is mixed with water and the mixture is distilled, out comes ethyl alcohol identical in all ways with our original substance, and the residue is a water solution of sodium hydroxide, NaOH.

$$HOH + C_2H_5O\ Na \rightarrow C_2H_6O + NaOH$$

Thus we are assured that in the course of the original reaction with sodium, no deep-seated rearrangement of the structure of the molecule can have occurred, for in the simple, quiet mixing with water, the original substance was regenerated: the -Na was replaced by -H.

Now our analysis of the bonds tells us that only compound II has one and only one of the hydrogens different from the others. We may therefore tentatively assign to the bottle the label II, namely ethanol. But although one reaction may be definitive in its conclusiveness, caution and the knowledge that new things are discovered this way urge us to validate the inferences by other means. However, we may tentatively write for the ethyl alcohol the structure

```
    H  H
    |  |
 H—C—C—OH
    |  |
    H  H
```

A number of analogies come to mind. We know from inorganic chemistry the reaction:

$$2\ Na + 2\ H—O—H \rightarrow 2\ H—O—Na + \overline{H_2}$$

This is a more vigorous reaction than that of sodium with ethyl alcohol. But the analogy is clear, and given the bond numbers of Na as one, and H as one, and O as two, we are satisfied that the structure of NaOH is as written, though if we wished to consider it in more detail we would insert appropriate $\delta+$ and $\delta-$ signs. Thus we feel that the reaction with ethyl alcohol involves the replacement of H on OH by Na.

We have another analogy from inorganic chemistry. When phosphorus tribromide is reacted with water it forms HBr and $P(OH)_3$:

$$2\,H—O—H + PBr_3 \rightarrow 3\,HBr + P(OH)_3$$

Reacting ethyl alcohol with PBr_3 we find that a new substance is formed: a colorless liquid, almost insoluble in water, with boiling point 38.4°C, specific gravity 1.431 g/ml, and the molecular formula C_2H_5Br. Clearly, a —Br has replaced a —OH. Moreover, if we treat C_2H_5Br with NaOH, the products that we get back are ethyl alcohol and sodium bromide. We are therefore justified in writing:

$$3\,C_2H_5OH + PBr_3 \rightarrow 3\,C_2H_5Br + P(OH)_3$$
$$C_2H_5Br + NaOH \rightarrow C_2H_5OH + NaBr$$

We carry our investigation further, for each structure that is determined plays the role of a construct in the field of constructs of organic chemistry and serves to validate, through interconnections, other constructs (Figure 1-4). Therefore we attempt to make connections to other substances. We shall show the reasoning and the reactions because in their very simplicity they imply the kind of elegance that has always intrigued organic chemists. This whole section is, of course, meant only as a prototype. Today chemists merely look up properties of such simple substances in handbooks in order to identify them. However, our account here is for heuristic purposes, and we can afford to be straightforward and naive.

When wood chips are placed in a retort and distilled by the application of heat out of contact with air there is obtained a reddish-brown liquid 'pyroligneous acid' (from Greek *pyros*, a fire, plus Latin *lignum*, wood). This is a mixture of many substances, but the major components are wood alcohol and acetic acid. Acetic acid is also found as the active product of vinegar, obtained by the fermentation, or souring, of wine or cider. When these are separated, purified, and analyzed, one finds for wood alcohol the molecular formula CH_4O. It is a colorless, poisonous liquid, completely miscible with water, with a boiling point of 64.7°C, and a specific gravity of 0.792 g/ml. Acetic acid has the molecular formula $C_2H_4O_2$. It is a colorless, acrid liquid, with a boiling point of 118.1°C and a specific gravity of 1.049 g/ml. We wish to relate the structures to each other and to that of ethanol.

Wood alcohol, tested as we did ethyl alcohol, gives the following reactions, which allow us to write its formula as CH_3OH. We write the reactions using the formula and giving appropriate names:

$$2\,\begin{array}{c}H\\HC—OH\\H\end{array} + 2\,Na \rightarrow 2\,\begin{array}{c}H\\HC—O\,Na\\H\end{array} + \overline{H}_2$$

methanol
(wood alcohol, methyl alcohol) sodium methylate[1]

[1] The '-ate' ending as in sodium methylate signifies a 'through-oxygen-bond;'

$$3\ \mathrm{H\overset{H}{\underset{H}{C}}{-}OH} + PBr_3 \rightarrow 3\ \mathrm{H\overset{H}{\underset{H}{C}}{-}Br} + P(OH)_3$$

$$\mathrm{H\overset{H}{\underset{H}{C}}{-}Br} + NaOH \rightarrow \mathrm{H\overset{H}{\underset{H}{C}}{-}OH} + NaBr$$

Methyl bromide, or bromomethane, formed in the reaction with PBr_3, is a colorless gas only very slightly soluble in water, with a boiling point at about 4.5°C and a density, in the liquid state, of 1.73 g/ml.

We find now that if we react pure $\mathrm{Na\ O{-}\overset{H}{\underset{H}{C}}{-}H}$ with pure $\mathrm{H\overset{H}{\underset{H}{C}}{-}Br}$, a new product is formed which is a colorless gas, somewhat soluble in water, which can be condensed to a liquid at −23.7°C. It has the formula C_2H_6O, and we feel quite secure in assigning it the structure:

$$\mathrm{H{-}\overset{H}{\underset{H}{C}}{-}O{-}\overset{H}{\underset{H}{C}}H}$$

dimethyl ether
methoxy methane

The assignment of this structure rests upon the analogy of the reaction between CH_3Br and NaOH. Since CH_3ONa does not react with itself any more than HONa does, and since the Na of the NaOH may be replaced by CH_3 using CH_3Br, it seems eminently reasonable to write:

$$\mathrm{H\overset{H}{\underset{H}{C}}{-}O{-}Na} + \mathrm{Br{-}\overset{H}{\underset{H}{C}}H} \rightarrow \mathrm{H\overset{H}{\underset{H}{C}}{-}O{-}\overset{H}{\underset{H}{C}}H} + NaBr$$

Furthermore, the product does not react with metallic sodium, nor with phosphorus tribromide, so clearly it has no —OH in it.

$$H_3C{-}O{-}CH_3 + Na \rightarrow \text{N. R. } (\textit{No Reaction})$$
$$H_3C{-}O{-}CH_3 + PBr_3 \rightarrow \text{N. R.}$$

And clearly its physical properties confirm that it has a structure that is radically different from that of ethanol.

Turning now to acetic acid, with the molecular formula $C_2H_4O_2$, we might apply our previous reactions. However, it would be unwise to

---

sodium methylate has the sodium connected to the methyl through oxygen. The same suffix is used with the same meaning in naming carboxylic acid salts, *e.g.*, sodium acetate.

add sodium to the acid (we know this from other reactions of sodium with acids) for the result would be a quite violent explosion. Instead, we treat the acid with a base, sodium hydroxide. Even so, the reaction produces heat. From it we isolate a white, crystalline product:

$$C_2H_4O_2 + NaOH \rightarrow C_2H_3O_2Na + H_2O$$

No matter what excess of sodium hydroxide we use, or whether we now attempt to react the product with sodium, the other hydrogens remain inert. *Only one* of the hydrogens is replaced, and is thus shown to be different from the others. We suspect an —OH group. Treatment of the acid with PBr_3 confirms this supposition:

$$3\ C_2H_4O_2 + PBr_3 \rightarrow 3\ C_2H_3OBr + P(OH)_3$$

We find, further, that when ethyl alcohol is treated with an inorganic oxidizing agent, there is formed acetic acid:

$$CH_3CH_2OH + 2[O] \rightarrow C_2H_4O_2 + H_2O$$

Oxidation is usually a fairly drastic treatment, and it is the case that when a carbon has been partially oxidized—for example, by having a —OH group in place of —H, as in CH_3CH_2OH—the carbon so oxidized is susceptible of further oxidation. We may infer this from the experimental fact that CH_4 and CH_3—CH_3 are quite inert to oxidation of this kind while CH_3OH and CH_3CH_2OH are readily susceptible of oxidation.

$$CH_3OH + 3[O] \rightarrow CO_2 + 2\ H_2O$$

At the same time the carbon-to-carbon bond in ethyl alcohol is stable to oxidation (one may reasonably assume) since acetic acid still has 2 carbons.

Hydrolysis, decomposition brought about by water, is a relatively mild reaction with compounds of this kind, so that the following sequence of reactions bears quite convincingly on the structure of acetic acid:

$$H_3CBr + KC{\equiv}N \rightarrow H_3C{-}C{\equiv}N + KBr$$

$$H_3C{-}C{\equiv}N + 2\ HOH \xrightarrow[\text{acid}]{\text{dilute}} H_3C{-}C(O)OH + NH_3$$

methyl cyanide (acetonitrile) → acetic acid + ammonia

Given these reactions, then, and the evidence for —OH in acetic acid, we write as its structure, adhering to our rules about bond number:

```
     H     O
     |    //
  H—C—C
     |    \
     H     OH
```

acetic acid

We know that the methyl group is present as an intact group since it is stable, and we have used quite mild replacement reactions; thus both oxygens must be on the other carbon. Then the product of reacting acetic acid with NaOH must be

$$H_3C{-}C({=}O){-}O^- \; Na^+$$

sodium acetate

and the product of reacting acetic acid with PBr_3 must be

$$H{-}CH_2{-}C({=}O){-}Br$$

acetyl bromide

If acetyl bromide is treated with water, acetic acid is produced by hydrolysis. If acetyl bromide is reacted with ammonia, acetamide is analogously produced by ammonolysis.

$$H{-}CH_2{-}C({=}O){-}Br + H\ddot{N}H_2 \rightarrow H{-}CH_2{-}C({=}O){-}\ddot{N}H_2$$

acetamide

Now if acetamide is treated with a powerful dehydrating agent, P_2O_5, there is formed methyl cyanide:

$$H{-}CH_2{-}C({=}O){-}NH_2 + P_2O_5 \rightarrow H{-}CH_2{-}C{\equiv}N + 2\,HPO_3$$

acetamide methyl cyanide

We must conclude that these sequences of reactions, involving simple compounds which are reacted with and built into more complex ones, validate themselves, provided we adhere to bond number rules, and base our arguments on reason applied to the results of experiment and elementary analysis.

But the wise scientist is the cautious one: he keeps in mind the old aphorism "seek simplicity—and distrust it." He knows that KCN reacts with CH_3Br to give C_2H_3N which on hydrolysis yields acetic acid. So he tries the behavior of silver cyanide, AgCN.

$$H_3CBr + AgCN \rightarrow C_2H_3N + \underline{AgBr}$$

Silver bromide is a nice, insoluble substance, easily filtered off from the reaction mixture, but the product C_2H_3N turns out *not* to be the same as that from the reaction of methyl bromide with sodium or potassium cyanide. Something is amiss. (This compound, C_2H_3N, has a most unpleasant odor.)

Recourse to hydrolysis gives us the answer. The reaction takes the course shown by the following equation based on analysis of the separated products:

$$C_2H_3N + 2\,H_2O \xrightarrow[\text{acid (HCl)}]{\text{dilute}} \underset{\text{an acid}}{H_2CO_2} + \underset{\text{a salt}}{CH_5N(HCl)}$$

Following the argument already applied to acetic acid, the structure of the acid H_2CO_2 is shown to be:

$$\mathrm{H{-}C}\begin{matrix}\nearrow\!\!\!\!\!\!\!=O \\ \searrow OH\end{matrix}$$

formic acid[1]

The salt, on treatment with sodium hydroxide, releases a volatile substance, CH_5N, which has the odor of rotten fish, and which is produced when methyl bromide is treated with ammonia and the product distilled from sodium hydroxide solution. These behaviors are summarized in the sequence of reactions:

$$CH_5N(HCl) + \underset{\text{excess}}{NaOH} \xrightarrow{\text{distil}} CH_5N + NaCl + H_2O$$

$$H_3CBr + HNH_2 \rightarrow H_3C\overset{\overset{\displaystyle H}{|}}{N}H_2^+Br^-$$

$$H_3C\overset{\overset{\displaystyle H}{|}}{N}H_2^+Br^- + \underset{\text{excess}}{NaOH} \xrightarrow{\text{distil}} \underset{\text{methyl amine}}{H_3CNH_2} + NaBr + H_2O$$

So now we must write for the course of the reaction of AgCN with CH_3Br and for the product of the reaction:

$$\underset{\text{methyl isocyanide}}{H_3C{-}N{=}C} + H_2O \rightarrow \underset{\text{methyl amine}}{H_3CNH_2} + \underset{\text{formic acid}}{HC\begin{matrix}{=}O \\ {-}OH\end{matrix}}$$

[1] This acid is found in the sting of ants: Latin, *formica*, an ant.

Here it is apparent that the terminal carbon is in the same state as carbon in CO, and we are confirmed in the need to remember rules and to distrust simplicity of interpretation even while seeking it.

We have gone into detail in establishing the structures of ethyl alcohol and acetic acid, proving (testing in the sense of a printer's 'proof') each point along the way, to show how the chemist attempts unequivocally to establish structure and to correlate chemical behavior with the inferred arrangement of the atoms in the structure. This is how organic and inorganic molecular structures have been developed by thousands of individuals and teams over the years. Conceptually, then, we have developed a fragment of the field of constructs of organic chemistry as shown in Figure 15-8. If one imagines hundreds of thousands of such interconnections (remembering that some are not well established and may be wrong), and if one connects to them the more abstract constructs such as 'ionic mechanism,' 'radical mechanism,' and so on, one can imagine to some extent the validity and grandeur that the chemist (and the physicist in his realm) can feel for what he considers reality (Figure 1-4), and the sense of security it gives him.

Optical isomerism 15.16

If a carbon atom has four different groups attached to it it is said to be 'asymmetric.' Two different arrangements of the groups are possible, due to the tetrahedral geometry of the carbon atom. They are shown in Figure 15-9 for two models. The figure shows an attempt to superimpose the two forms, matching the four different groups, represented by I to IV. If (III) and (III) are matched, and (IV) and (IV) are also matched, then the two possible arrangements of the two other groups stand out: they do not match. Molecules so related are optical isomers; specifically, they are 'enantiomorphs' (opposite forms) or enantiomers. They bear the relation of image to mirror image, as shown in Figure 15-10. The two bulk substances formed of enantiomorphs have identical melting and boiling points. Most other physical properties, and most chemical ones, are the same. The chief differences are two, one optical and one biological.

If a beam of light, coming out of the page toward the observer in Figure 15-11*a*, is thought to vibrate as a wave with components in all directions at right angles to the direction of propagation, then by passing it through a polarizer (Figure 15-12) one can sort out all the vector components in one direction (Figure 15-11*b*), and one can think of the beam as being plane polarized. When such a beam is passed through a sample of one enantiomer the *plane* of polarization is rotated to the right, as on Figure 15-13. The substance is said to be dextrorotatory. The other enantiomer rotates the plane of polarization to the

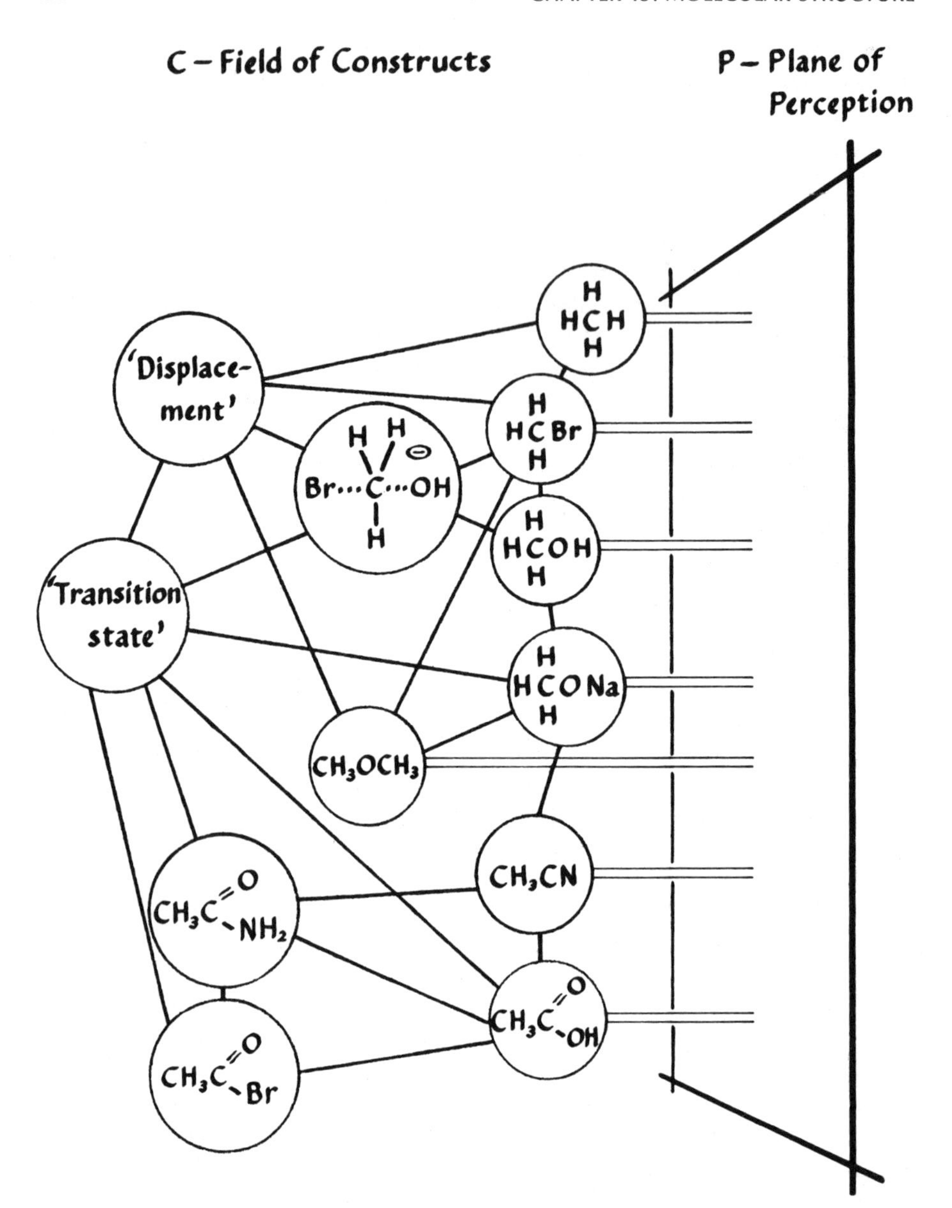

Fig. 15-8 *Partial diagram of the constructs and their connections in the proof of structure of acetic acid. The constructs CH_4, CH_3Br, and so forth, are connected to the perceptions of the bulk substances; the more 'remote' constructs, 'displacement reaction,' 'transition state' are not immediately connected to the plane of perceptions.*

left by exactly the same amount when the same number of molecules is in the path of the light. The substances are said to be 'optically active.'

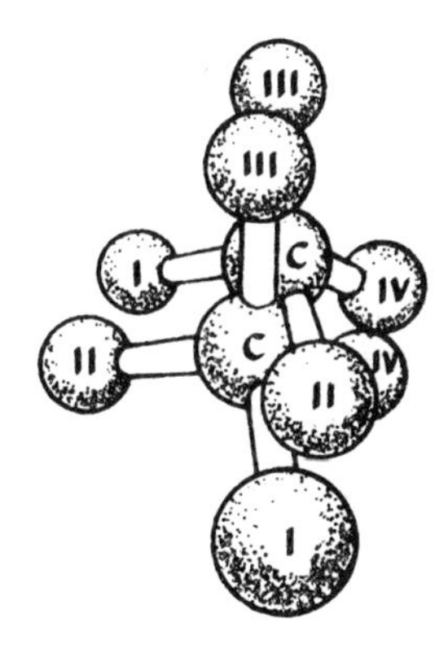

Fig. 15-9 *An attempt to superimpose the two models of Fig. 15-10.* [*From J. English and H. G. Cassidy,* loc. cit., *Fig. 22-4, p. 410.*]

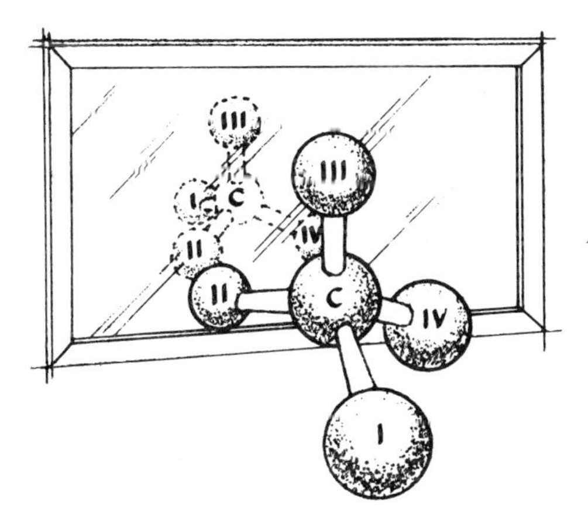

Fig. 15-10 *If a molecule were constructed with four different groups I, II, III, IV about the central carbon atom, as shown, its mirror image, seen in the mirror, could be constructed also, and the two molecules would be found not to be superimposable, as an imaginative consideration of the figure will show. Try imagining taking the two into your hands and fitting them together, as in Figure 15-9.* [*From J. English and H. G. Cassidy,* loc. cit., *Fig. 22-5, p. 411.*]

The substance glyceraldehyde, $CH_2(OH)CH(OH)CHO$, is known to occur in two different optically active forms. A solution of one form rotates plane polarized-light to the right; of the other form, to the left. The two are therefore named *dextro-* (or d−, or (+)) and *levo-* (or *l*− (−)) glyceraldehydes. Theoretical and X-ray investigations have led to the conclusion that the absolute configurations in space of these two molecules are as shown below, where the dotted bond is below the plane of the page; the thin line bond is in the plane of the page, and

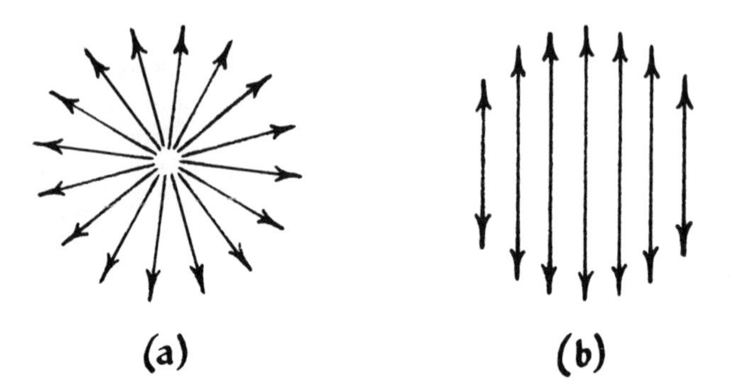

Fig. 15-11 *Diagrammatic representation of the polarization of light.* (a) *Unpolarized light beam coming "up" out of the page, showing vibrations in all planes.* (b) *The beam polarized in a vertical plane. [From J. English and H. G. Cassidy,* loc. cit., *Fig. 22-1, p. 408.]*

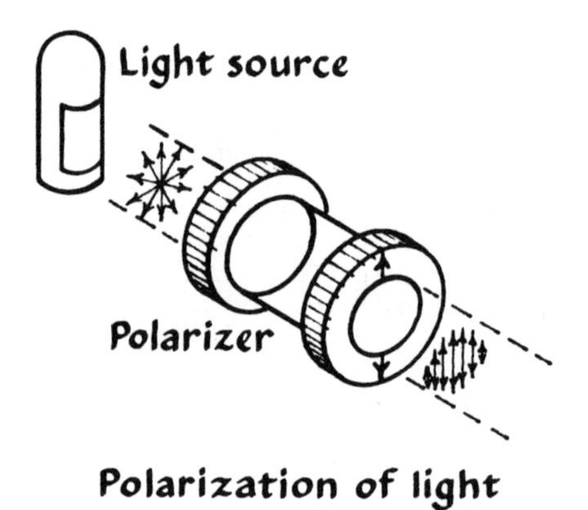

Fig. 15-12 *A polarizer. The direction of the polarization is shown by the arrows. An analyzer is the same instrument arranged to read off the angle of rotation, as in Figure 15-13. [From J. English and H. G. Cassidy,* loc. cit., *Fig. 22-2, p. 408.]*

the wedge-shaped lines indicate bonds extending up out of the page.

CHO	CHO
|	|
H► C ◄OH	HO► C ◄H
⋮	⋮
CH_2OH	CH_2OH
dextro-glyceraldehyde	*levo*-glyceraldehyde
D-glyceraldehyde	L-glyceraldehyde

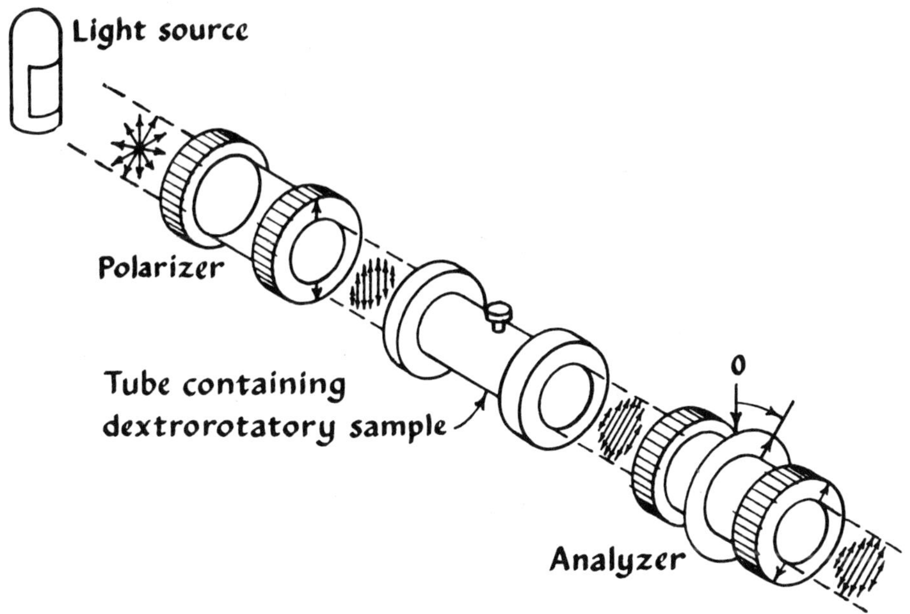

Fig. 15-13 *Diagram of a polariscope. Light is polarized in the vertical plane. On passing through the sample, its plane of polarization is measured by the analyzer, which is rotated to pass the maximum amount of light. In this case the sample rotates to the right.* [*From J. English and H. G. Cassidy,* loc. cit., *Fig. 22-3, p. 408.*]

The observed optical rotation of a substance can vary with the solvent in which it is dissolved and with the temperature. It depends, too, on the four groups that are present and on their absolute configuration. Thus we carefully distinguish between the 'absolute configuration' (designated D or L in certain series of substances such as the carbohydrates and proteins) and the observed rotation of the substances. The configuration of a given structure is a constant; the observed optical rotation can be made to vary by changing the solvent or the temperature.

Any lingering doubts that chemists might have had about the relationships of two asymmetric forms were dispelled in an elegant way by Emil Fisher. He carried out a sequence of reactions, shown below, in which a pure dextrorotatory acid was converted to a pure levorotatory form (with identical, though opposite, molar rotation) by reactions which retained the asymmetry while transforming groups at a distance from the asymmetric carbon. (Ignore the fact that we have not studied the reactions.)

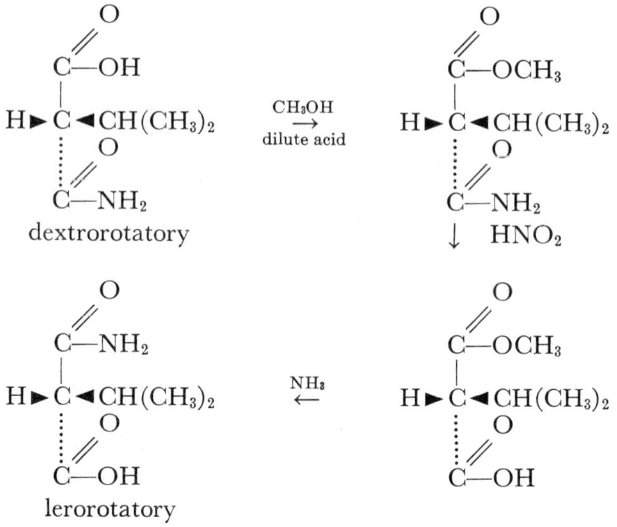

$H \blacktriangleright C \blacktriangleleft CH(CH_3)_2$ with $C(=O)$—OH above and $C(=O)$—NH_2 below
dextrorotatory

$\xrightarrow[\text{dilute acid}]{CH_3OH}$

$H \blacktriangleright C \blacktriangleleft CH(CH_3)_2$ with $C(=O)$—OCH_3 above and $C(=O)$—NH_2 below

↓ HNO_2

$H \blacktriangleright C \blacktriangleleft CH(CH_3)_2$ with $C(=O)$—OCH_3 above and $C(=O)$—OH below

$\xleftarrow{NH_3}$

$H \blacktriangleright C \blacktriangleleft CH(CH_3)_2$ with $C(=O)$—NH_2 above and $C(=O)$—OH below
lerorotatory

15.17 Mechanism of reaction, II

Optical asymmetry has been a powerful aid to the study of reaction mechanisms. Suppose that we look at the addition of bromine to cyclopentene.

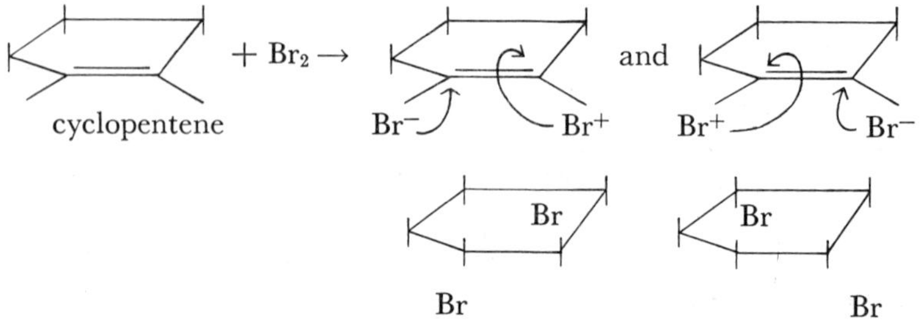

cyclopentene $+ Br_2 \rightarrow$ [Br^-, Br^+] and [Br^+, Br^-]

Br Br
(I)

Br Br
(II)

These two compounds, (I) and (II), are not superimposable. They have two asymmetric carbon atoms. Within each compound the two carbons have the same configuration; but the configurations are opposite in the two compounds. This can be seen by choosing some order of passing from group to group, for example H to C to Br. This requires clockwise movement in (I) and counterclockwise in (II) as is shown by redrawing the structures with the rest of the ring below the plane of the paper and looking down on the two compounds:

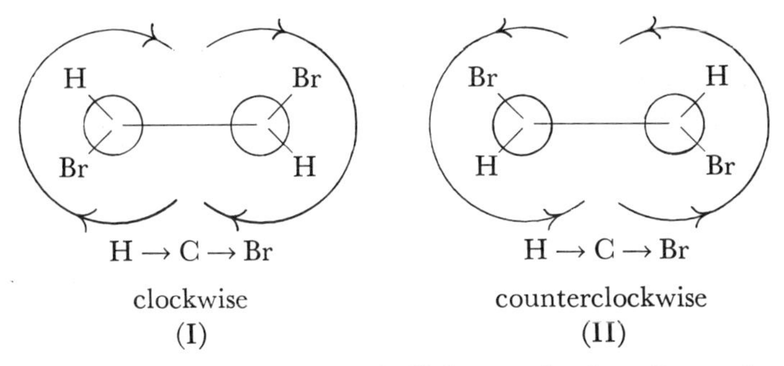

Now this result could occur only if the attack of one Br was from above the plane of the ring and the other from below. For consider the other possibility:

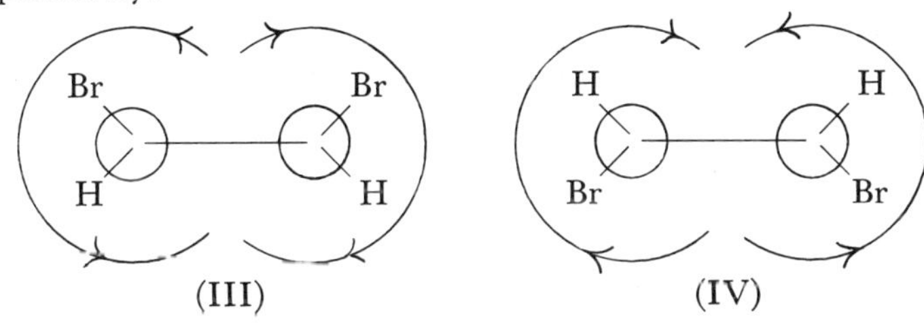

We still have two asymmetric carbon atoms, but now their configurations are opposite, and since they have exactly the same 4 groups that produce the asymmetry on each carbon, one is exactly as much dextrorotatory as the other is levorotatory. Thus they cancel exactly, and the molecule is optically inactive: it has no rotation. Moreover (IV) is identical to (III), for the two can be superimposed. This is seen by rotating (in the imagination) one molecule through 180° *without lifting it out of the paper*, then if we slip it up to the other it becomes obvious that they are identical. Thus we have another way of validating the mechanism. (Some consequences of these arguments are played upon in the exercises.)

Summary 15.18

In this part we have traced the structure of matter, and have shown how nuclear properties control electronic structure of atoms, and how this is related to the ability to form one or many bonds, weak or strong, with other atoms to form molecules. In this chapter the thread was singlemindedly traced to show how atoms may be linked to form three-dimensional structures. The general purpose was to give a glimpse of the extraordinary variety of molecules that may be built out of a few types of atoms, and to lead into some, even if impressionistic, awareness of what is known about molecules. It can truly be said that physicists and chemists have barely penetrated into the mysteries of material structures and functions. At any rate, however, we have presented several rather subtle examples of the reasoning by which structures of molecules and mechanisms of their reactions may be worked out.

Exercises

15.1 a) Name the following compounds:
(1) $CH_3C(CH_3)_2C(CH_3)_2CH_2CH_3$;
(2) $CH_3C(C_2H_5)_2C(C_2H_5)_2C(CH_3)_3$;
(3) $(CH_3CH_2CH_2)_4C$.
b) Write the structure of 2,3,4,5,6,7,8-heptamethyldecane.

15.2 a) Which of the following structures are identical? (Refer to them by number.)

(1) $CH_3CH(OH)CH_2CH_3$

(2)
$$\begin{array}{l} \qquad\qquad\quad\;\; CH_3 \\ \qquad\qquad\quad / \\ (CH_3)_2CHCH \\ \qquad\qquad\quad \backslash \\ \qquad\qquad\quad\;\; C_2H_5 \end{array}$$

(3) $CH_3—CH{=}CH—CH_2CH_3$

(4)
$$\begin{array}{rcl} & CH—CH—CH_3 & \\ & \| \qquad\; | \quad\;\; & \\ CH_3—CH & \;\; CH_3 & \end{array}$$

(5)
$$\begin{array}{c} \qquad\qquad OH \\ \qquad\qquad | \\ CH_3CH_2—CH—CH_3 \end{array}$$

(6)
$$\begin{array}{c} \quad Cl \\ \quad | \\ CH_3CH_2CH—CH_2—CH_3 \end{array}$$

(7)
$$\begin{array}{c} \qquad\qquad CH_3 \\ \qquad\qquad | \\ CH_3—CH—CH—CH_3 \\ | \qquad\qquad \\ CH_2—CH_3 \qquad \end{array}$$

(8) $CH_3CH_2CH{=}CH—CH_3$

(9)
$$\begin{array}{c} \qquad\qquad CH_3 \\ \qquad\qquad | \\ CH_3—CH—CH—CH_2CH_3 \\ | \qquad\qquad\qquad \\ CH_3 \qquad\qquad\qquad \end{array}$$

(10) $(CH_3)_3COH$

(11)
$$\begin{array}{c} Cl \qquad\qquad\quad \\ | \qquad\qquad\quad \\ CH_3CH—CH_2CH_2CH_3 \end{array}$$

(12) $(CH_3)_2CH—CH{=}CH—CH_3$

(13) $(C_2H_5)CHCl$

(14)

$$\begin{array}{l} CH_3 \diagdown \\ \qquad CHCl \\ C_2H_5{-}CH_2 \diagup \end{array}$$

(15)

$$\begin{array}{c} CH_3 \\ | \\ HO{-}C{-}CH_3 \\ | \\ CH_3 \end{array}$$

Or (b) name the compounds.

15.3 a) Write out and name all the isomers with the molecular formula C_6H_{14}. b) Write out and name all the isomers with the molecular formula $C_4H_8Cl_2$.

15.4 A sample of pure hexane is reacted with bromine in the presence of ultraviolet light, and the products are all separated by careful distillation. Assume all possible products will form. a) How many different monobromides will there be? b) How many dibromides? c) In each case, name them.

15.5 Write the structures of a) 1,1,3-trimethyl cyclohexane. b) *trans* 1,4-dimethyl-*cis*-2,3-dihydroxy cyclopentane.

15.6 A certain compound X has an empirical formula corresponding to the acetylenes C_nH_{2n-2}, yet it adds only one molecule of bromine, while acetylenes will add two. What type of structure must be present?

15.7 Which tribromobenzene gives three different tetrabromobenzenes on bromination?

15.8 Assume that C_2H_5Li can be prepared. Would you expect it to react with ethyl bromide? What would be the products? Is this a chain reaction?

15.9 a) Write the structures of all the possible pentenes and name each. b) Compare the number of these with the number of pentanes.

15.10 A certain substance was found to contain 85.7% C and 14.3% H. One gram just decolorized 38.0 g of a 5% solution (by weight) of bromine dissolved in CCl_4. Write a possible structure for the substance.

15.11 It is discovered that there are two isomeric 2-butenes, and only one single 1-butene. Explain this.

15.12 There are found to be two (*cis-trans*) isomeric 1,2-dimethylcyclohexanes. Would there be two isomeric 1,4-dimethylcyclohexanes? Explain with structural diagrams.

15.13 There is only one monobromobenzene, while there are two monobromonaphthalenes and three monobromobiphenyls. Explain these facts. Biphenyl is ⬡—⬡; naphthalene is ⬡⬡.

Answers to Exercises

15.1 a) (1) 2,2,3,3-tetramethylpentane; (2) 2,2,4-trimethyl-3,3,4-triethylhexane; 4,4-dipropylheptane.
b) $CH_3(CHCH_3)_8CH_3$

15.2 (1) = (5); (2) = (7) = (9); (3) = (8); (4) = (12); (6) = (13); (10) = (15); (11) = (14).

15.3 a) There are 5 of these. b) There are 9 of these.

15.4 a) 3 monobromides. b) 12 dibromides.

15.5 a) CH_3 CH_3 CH_3 b) CH_3 H_3C OH OH

15.6 A cyclic olefin.

15.7 1,2,4-tribromobenzene.

15.8 Yes. $C_2H_5C_2H_5$ or C_4H_{10} and LiBr. No. The products are inert.

15.9 a) There are five.

15.10 One of the hexenes.

15.11 *Cis-trans* isomerism is possible with 2-butene; not with 1-butene.

15.12 Yes. CH_3 H_3C H_3C CH_3

The ring has two sides.

15.13 All H's on benzene are equivalent. But in naphthalene there are two kinds, because the second ring marks a position on the first.

In biphenyl, one ring is a substituent on the second, so that there are *o*-, *m*-, and *p*-bromophenylbenzenes.

Part V

Chapter 16. Chance, Probability, and Statistics

Chapter 17. Process. An Introduction to Cybernetics

Chapter 18. Bodies, Particles, and Fields

Chapter 19. Directional Arrows and Universal Laws

Chapter 20. Toward a Natural Philosophy

Larger Issues and Unifying Principles ~ PART V

The intent of this book is holistic; to teach certain aspects of physical science in the most unified, clear, and exposed way. We have chosen those topics that seem central to today's problems. We have therefore omitted vast areas of physical science. This needs no defense, but it is nevertheless of some concern. To offer redress is the function of this Part.

With the substantive knowledge of modern Physics and Chemistry that we have presented, we are able to look over the whole field of Science and see what it has to offer to our culture—to making us civilized. We endeavor to find the grand pattern of our entire educational experience. We do not, of course, abandon a scientific approach. What we do is to move some of the time to a high enough level of abstraction—far enough out in the construct-field—so that we can obtain a vaster panoramic view. We remain warned that thereby many vital details become blurred, vague, or even invisible. Therefore from time to time we return to a closer examination of what is our safeguard on theory, our bond to sanity: the way things are.

Cause-and-effect is an underlying theme of Chapters 16 and 17. The subject hardly arose in our discussion of classical science because there was no problem. If an object is dropped it falls; like charges repel, unlike attract. Cause-effect relations were quite clear, for each mechanical cause was linked to its effect in a one-to-one relation: in a good mechanism the effect follows the cause one hundred percent of the time. Even the behaviors of living creatures were conceived to fit in this scheme. Granted that one did not always observe the expected effect when some causal change occurred, but then this was because of lack of knowledge of all the causative factors. Now, however, this premise has been questioned at a basic level by the hypothesis of indeterminacy. At the same time there clearly are processes in which the effects from some set of causes are distributed over a range, with some results more probable than others, as we shall see. In Chapter 16 we discuss certain aspects of probability theory, and then in Chapter 17 we introduce the theory of processes known as 'cybernetics.' Here it becomes evident that in many kinds of processes the effects retroact (through feedback) upon the causes. This chapter, interesting particularly from a technological point of view, is offered, as is the chapter on probability, to connect the earlier parts of this book with certain subjects outside of physical science, as well as to clarify some aspects of physical science.

In Chapters 18 and 19 we move to a still broader context. Here again, we can only touch on these subjects. However, it seems desirable for the student to be aware of the two "explanatory" premises that are being widely explored—and argued—today, not only in physical science but throughout the whole intellectual sphere. These are 'atomistic' "versus" 'continuum' views. Quantum and wave theories have gradually taught us that these are not necessarily mutually exclusive. After all, what is visible light but a phenomenon that has particle aspects when you look for them with a particle-finding machine, and wave aspects to an interferometer? To see the light in this area of physical science is to think of light as just light.

In Chapter 19 we unify some previously discussed matters under the heading of 'directional arrows.' In the relativistic way of thinking there have to be—or at least one urgently seeks—invariant-relations, preferably in the form of universal laws which translate between the relative states: they are 'absolutes,' which serve as touchstones. What we have called directional arrows are in a sense process absolutes: they indicate irreversible directions. One of these is 'Time's arrow,' one of the few directional absolutes recognizable to us now through the still darkened glass of Science.

Chapter 16

16.1 Introduction
16.2 Coin-tossing
16.3 Conceptualization
16.4 More complicated cases
16.5 A practical example: the Second Law of Thermodynamics
16.6 Time's arrow
16.7 Summary

Chance, Probability, and Statistics ~ 16

Introduction 16.1

From the moment that one particular spermatozoan (and not any other) fertilized the egg that, surviving the vicissitudes of existence, grew to be an adult, to the event that this adult dies in one way and not another, chance appears to pervade his life. The chanciness of existence has been borne in on us especially since the downfall of classical mechanism. In a sense, however, chance has two faces. A particular radium nuclide presumably will ultimately fission, but we have no way of predicting when it will fission. At the same time we know that given a sufficiently large number of radium nuclei—say 10^{18}—we can predict with very high probability that half of them will have decayed in 1600 years (Section 13.3). In arriving at this prediction statistics and probability theory were made use of to supplement observation of the actual behavior of the substance. Statistics has to do with gathering, organizing, and interpreting numerical data. Probability has to do with interpreting and predicting, and is not restricted to numerical data. The essential difference between the two is that probability is the theoretical science, and deals largely with idealized, invented constructs; statistics is the applied science, and is concerned with actual situations in all their complexity, intransigence, and fuzziness: probability lies largely in the C-field; statistics is close to the P-plane (Figure 1-4). In this chapter we examine in an introductory way how people think in probability terms: how a measure of control is gained in the face of chance. Further reasons for including this chapter in this book are mentioned in Section 16.7.

16.2 Coin-tossing

If we flip a coin it will come to rest showing heads (H) *or* tails (T). That it stand on the edge is so unlikely that from a practical view we ignore the possibility. We do have to decide beforehand which face to call H. Then the other is T. If we flip a coin once, the flip is a 'trial' of which the outcome is H or T; the outcome is called an 'event.'

If we flip a coin twice or flip two coins then instead of just two possible outcomes we have four: HH, HT, TH, TT. One of these will be the actual outcome. If we flip the coin three times, the possible outcomes (events) are HHH, HHT, HTH, HTT, THH, THT, TTH, TTT. What actually occurs will be one of these. The possibilities can be written in the form of a table—as a matter of convenience—so as to make it easy to see that we have not missed any (Table 16-1). It is

Table 16-1

Possible Outcomes of Tossing Coins

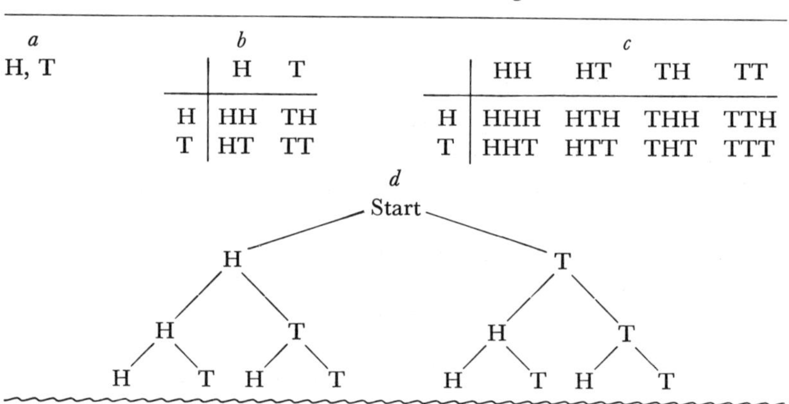

a

H, T

b

	H	T
H	HH	TH
T	HT	TT

c

	HH	HT	TH	TT
H	HHH	HTH	THH	TTH
T	HHT	HTT	THT	TTT

d

apparent that we can be quite certain about the possible outcomes in these simple cases. What we remain uncertain about is which particular outcome will result from a given trial. The theory of probability is concerned with this uncertainty.

The fundamental assumption that is made in probability theory is that there are things that can be called chance events: one speaks of the chance of getting heads or tails when one flips a coin, or of getting seven when one rolls a pair of dice. We know, of course, that under certain circumstances a skillful person can 'make' heads or tails come up, or can throw a 'seven' or 'eleven' at will by manipulating the coin or the dice, and in such cases we say that he has determined the event, and that the outcome has not been a chance one. We look at the statistics of a set of such flips of the coin or of throws of the dice and we see that the events do not show a random distribution and so we look for causative factors that have determined the events; an untrue coin, or loaded dice, or a skilled hand.

Another assumption that is made is that equivalent events must be assigned the same value of probability. Thus a fair coin is expected to come up, on the average, as many times heads as tails; therefore the probability of each is 1/2. The total range of probability values is from 0, or no probability at all, to 1, or complete certainty. In a set of values, such as of the probabilities of H and T, the sum must be 1, since it is certain that one or the other event will ensue. Again, it is reasonable to assign the probability 1/6 to each face of an unbiased die, since there are 6 possibilities, each equally likely. The probability of the die resting on an edge or a corner is zero in each case.

Conceptualization 16.3

We know that the chances of getting H or T on flipping a fair coin are equal—or do we? People have tested the assumption by actually flipping coins, or by equivalent means (see Exercise 16-2). Suppose that a coin is flipped, and for each 100 trials the number of heads is set down. If, for example, there were found for a sequence of such counts 54, 46, 53, 55, 46, 54, 41, 48, 51, 53 heads (this is the first line, 1000 trials, from a table of 10,000 trials given by William Feller, *loc. cit.*) we would be able to say that in this test the chances for obtaining H were 54/100, 46/100, 53/100, and so forth. The numbers 54, 46, 53, and so on, are the 'frequencies' of occurrence of H in samples of 100 numbers. The ratio of frequency to sample size is the proportion, sometimes called the 'chance,' of occurrence, or loosely the 'probability,' and designated by p. If we add up all the frequencies given above, we get 501 H's for 1000 tries, or $p = 0.501$. It follows that the not-H's (that is, the T's) must have "come up" 499 times. Designate the chance for *not* getting the particular occurrence 'q.' It follows that:

$$p + q = 1 \tag{16-1}$$

We try to be careful about symbols. In an experimental test, the chance is symbolized with a lower-case p: "p turned out to be 0.501 for the number of heads in 1000 trials." This is a statistical judgment. From a theoretical point of view, the probability, symbolized by a capital P, is expected to be 0.500, that is, $P(\text{H}) = P(\text{T}) = 0.05$. In the table referred to above, p for 10,000 trials is 0.4979. We can see intuitively how this might be. We can see in Table 16-1 that in any group of trials there is a chance for runs of H or T. Even though these tend to even out when the number of trials is large, it would not take many runs, occurring by chance, to make evening-out difficult. And there is always a finite chance for a long run.

In any record of actual tosses of a coin, or rolls of dice, there are found fluctuations; runs of H, and T, or failure to throw 7, and so forth. It is the phenomenon of fluctuations which allows quite hard-headed people, who know the chances for throwing '7' or '11' at dice, or for the various hands at poker, or even (though there is some doubt about the hard-headedness in this case) for filling an inside straight, to enjoy wooing the Goddess of Chance in gambling.

In simple cases, as those in Table 16-1, we can determine the probability of occurrence of a given event, say HHH in 16-1*c*, by counting: it is one in 8. To use precise language, we call each possible event a 'sample point,' and we designate the aggregate of all thinkable points for that particular physical set-up a 'sample space.' Thus HHH is a point, or sample point, in the sample space of Table 16-1*c*. We already have a fundamental relation, that $P(\mathrm{H}) + P(\mathrm{T}) = 1$. Then it is easy to see, from Table 16-1, letting $P(\mathrm{H})$ be represented by x and $P(\mathrm{T})$ by y for convenience of writing and to link up with algebra:

For one toss $x + y = 1$

For two tosses, or the toss of two coins:

$(x + y)^2 = 1$ or, multiplying out, $x^2 + xy + yx + y^2 = 1$

Each term in the equation represents a sample point. For example x^2 represents HH; xy, HT; yx, TH; and so on. Each point has a 1/4 probability of occurring. In the case of three tosses (or a toss of three coins) the sample space is derivable from:

$$(x + y)^3 = x^3 + x^2y + yx^2 + xyx + yxy + xy^2 + y^2x + y^3 = 1$$

By this kind of calculation any sample space may be worked out, and the probability of any point may be determined. It becomes complicated, of course, when the number of trials is large, and one then has to use special mathematics. But the principle is clear here. Moreover, one can determine the probability of any cluster of points. For example, suppose we are interested in the probability of two T's coming up in any order. We see that by the above equation this may occur at 4 points out of the 8 in the sample space. The probability is thus 1/2. If we call the event 'two T's in any order' 'A,' we can show what is meant by means of Table 16-2, which implies the same conclusion.

Table 16-2

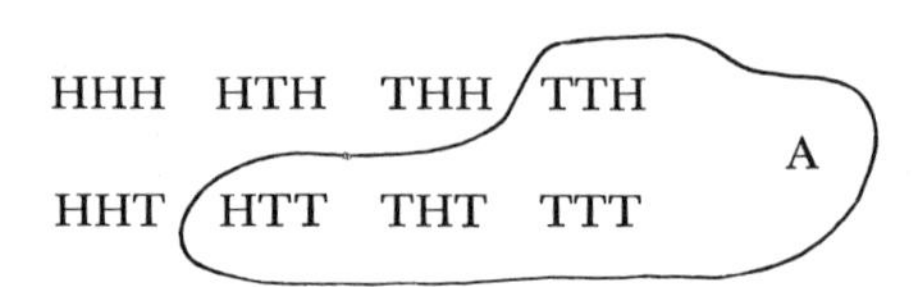

More complicated cases 16.4

In the previous examples all the sample points have the same probability, and any of them may occur in a given trial. In practical situations it may turn out that not all sample points have the same probability (this is the case with loaded dice) and also that the previous history of the event affects the probability of the outcome. This can be shown at a level of utter simplicity by taking three billiard balls, identical except that one is marked '1', another '2', and the third '3'. We put these in a sac, stir around, and then withdraw one. At the beginning, the probability for each is 1/3. After we have withdrawn one of the balls, this changes the probability to 1/2 for each of the other two. Finally, when only one is left, the probability becomes 1, or certainty, when it is withdrawn.

We are all aware of these different kinds of situations. Their discussion cannot be pursued in this text, but a bibliography is appended to this chapter to lead the interested student into the literature.

A practical example: the Second Law of Thermodynamics[1] 16.5

The computations we have described have practical value when we deal with really large numbers of otherwise identical particles which may be in different states—as in the calculation of half-life (Section 13.3). A very practical (and enlightening) example of the application of these ideas and methods involves the possible arrangements of quanta of heat energy distributed among a number of atoms. We know that atoms may possess energy only in definite multiples of quanta (Chapter 8). The total energy of a macroscopic system is the sum of the energies of the elementary entities (the atoms or molecules) of which it is composed. Further, it is found that these elementary entities can exchange energy—perhaps by collision—so that the number of quanta associated with any particular atom or molecule may fluctuate with the passage of time.

Any macroscopic system will contain a truly fantastic number of elementary entities. (Recall that 4 grams of helium gas contains 6×10^{23} atoms of helium.) We simplify drastically and imagine three atoms which we can 'label' α, β, γ. We imagine that they share 5 quanta of energy. For physical flavor we suppose that at the beginning atom α has 3 units of energy and each of the others only one. The particle with more energy is hotter than the particle with less energy.

---

[1] This section and the next section are derived from a treatment in Chalmers W. Sherwin's *Basic Concepts of Physics*, Holt, Rinehart and Winston, New York, 1951, with permission.

The initial state of the system of three atoms is shown in Figure 16-1 where we have plotted numbers of quanta of energy (n) on the left, and on the right units of energy E, for emphasis. We bring the atoms into contact and allow them to exchange energy. What will happen? (Of course, we isolate this little system from any sources of energy so that no additional quanta can enter during the 'experiment.')

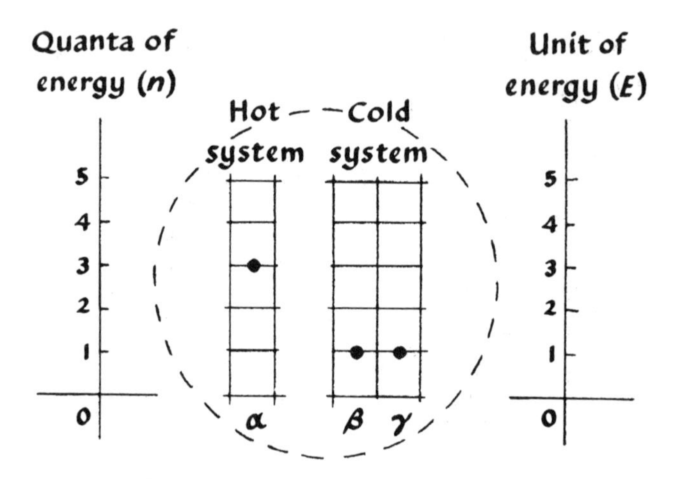

Fig. 16-1 *Three molecules, α, β, and γ share 5 quanta of energy. At the beginning α contains 3 units of energy, and β and γ one each. The three atoms are brought into contact and the system isolated (the dashed line). They may exchange energy but no energy may enter or leave the system.*

In Table 16-3 are enumerated all possible distributions of the five quanta between the three atoms. We make the fundamental postulate that, given time, all possibilities will occur; moreover all are equally likely. (This fits the way physical systems behave.) There are 21 possible arrangements. The sample space comprises the 21 sample points, called microstates to emphasize the physical application. Of these there are 3 (numbers 1, 4, and 5) in which atom α possesses more than 3 quanta of energy, and in such states α is hotter than it was initially. There are 15 microstates in which α possesses less than 3 quanta of energy (numbers 2, 3, 6-9, 12-15, 17-21) and is therefore cooler than initially.

Suppose that this system is examined from time to time. Over a large number of observations the probability of observing any microstate approaches 1/21 of the number of observations, since all microstates occur with equal likelihood. Then the chances of finding our original system, α, hotter than it was when we started (*i.e.*, containing

Table 16-3

Possible Arrangements of 5 Quanta between 3 Atoms

	Atom α	Atom β	Atom γ	
1	5	0	0	one atom has all 5
2	0	5	0	
3	0	0	5	
4	4	1	0	one atom has 4, and the others have either 1 or none
5	4	0	1	
6	1	4	0	
7	0	4	1	
8	1	0	4	
9	0	1	4	
10	3	2	0	one has 3 and the others have fewer than 3
11	3	0	2	
12	2	3	0	
13	0	3	2	
14	2	0	3	
15	0	2	3	
16	3	1	1	
17	1	3	1	
18	1	1	3	
19	2	2	1	no atom contains more than 2
20	2	1	2	
21	1	2	2	

more energy per elementary system) is 3/21, or 1 in 7. The chances of finding it with less energy are 15/21, or 5:7. Thus as a matter of probability, even with this very simple system, we are more likely to find that heat has flowed away from the hotter into the colder atoms than *vice versa*. Instead of observing one system over a long period we could have examined many systems of 3 atoms and 5 quanta all at once, when we would again have found the microstates distributed among the possibilities, with the α's having more energy in 1/7 of the cases having the same energy as initially in 1/7 of the cases, and having less energy than initially in 5/7 of the cases.

Suppose, now, that we do not distinguish between the three atoms. This means that we do not distinguish between arrangements such as 1, 2, 3; or 4 to 9; or 10 to 15; or 16 to 18; or 19 to 21. Each of these is an event such that one atom has a certain amount of energy, another another amount, and the third the rest, but with no enumeration of which atom. Thus the event realized in 1, 2, 3 is "one atom has 5 quanta." The other two must then have no quanta, since only 5 quanta are available for sharing. Each of these events is called a 'configuration,' and under this condition—the only different ways of distributing energy are tabulated—the Table 16-3 reduces to Table 16-4.

Table 16-4

Configurations for 3 Distinguishable Elementary Systems Sharing 5 Units of Energy

	Configuration	*Number of Microstates*
1	One 5, one 0, one 0	3
2	One 4, one 1, one 0	6
3	One 3, one 2, one 0	6
4	One 3, one 1, one 1	3
5	One 2, one 2, one 1	3

This situation is more like a physical one in many ways—chiefly that in a gas or liquid the atoms, or molecules, cannot be individually labeled as in our first table, and also we do not have nearly as much information about individual atoms as would be required for that kind of tabulation. An advantage of the tabulation of Table 16-4 is that the number of microstates in any configuration can be calculated in a simple way. Take, first, number 2, with one 4, one 1, one zero. The 4 units may be in 3 different atoms, α, β, or γ. Once a choice has been made, however, only two possibilities remain for the single unit, and then only one possibility for zero energy. The total number of possible microstates is therefore $3 \times 2 \times 1 = 6$.

When a configuration contains microstates in which the same energy is present in two or more atoms, so that there is no way of telling them apart even by energy content, the calculation has to be adjusted, to take account of this. Such a situation arises in configurations 1, 4, and 5 (Table 16-4). In 1, for example, 5 units being assigned, with 3 choices, the other atoms must both have zero energy. But the zeros are identical numbers, so in effect there are only 3 choices. The methods of calculation of the number of microstates in such cases are slightly more complicated, and we shall not pursue them since they add nothing to our overall objectives.

Returning to Table 16-3 we can see by inspection that the chances of α possessing 3 units of energy are 3/21, the chances for 4 units of energy are 2/21, while the chances for 5 units are 1/21. This suggests that if we have a hot and a cold system in contact the farther apart the initial temperatures of the two systems—the greater the initial difference—the less likely is it that the hot system will become hotter. By calculation it can be shown that when the hotter and colder systems are closer together in initial temperature the chances for heat to flow "uphill" are increased.

What happens as the systems become larger? The calculations are laborious, but not in principle very different from those already made. For the system '22 atoms share 44 units of energy' there are about 2.5×10^{17} microstates. But macroscopic systems contain of the order of 10^{20} or more atoms or molecules sharing a very great many units of energy. The number of calculatable microstates in ordinary macroscopic systems is astronomic. The distribution of microstates has been calculated, and it has been shown that the likelihood of having all the energy in a few atoms is vanishingly small compared with that in which the energy is statistically distributed: the former configurations contain only a few microstates compared with the latter. What this means is that a pan of water heating on the stove will behave normally: never will half the molecules boil out, leaving the other half colder, or frozen.

Our chief object in pursuing this highly simplified model, a model which is simplified both quantitatively and qualitatively, is to show that even with such simplification we are able to understand the statistical and probabilistic basis of the observation that heat flows in only one direction. We have an additional object in view. This is to show how powerful this kind of approach, created by J. Willard Gibbs (1839–1903), Ludwig Boltzmann (1844–1906), and Max Planck can be. The irrelevant appearances in the phenomena are cut away, the unmanageable complexity is subdued by the creative acts of drastic simplification combined with insight. The resulting conceptualized model enables us to explain the phenomenon at a finer-grained level. But a return to the physical situation can always be made. That heat flows from a hot macroscopic body to a colder one, and not the other way around, is one statement of the second law of thermodynamics. Another way of saying this is that in an isolated system not at equilibrium the number of microstates can only increase.

Thermodynamics deals with macroscopic systems at equilibrium. The number of microstates is astronomical, and thus the laws of thermodynamics are firm invariant-relations. Macroscopically speaking, an equilibrium state is one that shows no observable overall change even after a very long period of time. But we know that at the atomic or molecular level there is a chaos of motion in gas or liquid. In a solid, even at room temperature, there may be vibrations and rotations going on, so that the energy states of given molecules are changing all the time. We speak of dynamic equilibrium at this level. Thus while there may be no observable macroscopic change in a system at equilibrium, we know that at the microscopic level change is continuous.

Put another way, the distribution of microstates among configurations fluctuates. But the number of microstates that happen to be in configurations far from equilibrium is negligible compared with the number in the total macroscopic system. In addition, the effects of fluctuations in one direction tend to cancel out those in an opposite direction.

The extraordinary number of microstates in a macroscopic system is a consequence of the small magnitude of Planck's constant. It is because of this fact that the macroscopic system is not sensitive to small fluctuations and that, therefore, the system, seen to be fantastically complicated at the atomic level, may nevertheless be described by means of the values of a few properties. For example, an ideal gas may be described by the values of P (pressure), V (volume), and T (absolute temperature) from which the number of particles may be calculated.

Someone has said that scientists and insurance companies can bet on the statistics, but philosophers must remember the possibility for exceptions. The quantum of energy is of such small magnitude compared with the energies we usually deal with in the macroscopic world accessible to our senses that we can and do bet our lives on the probabilities. In the midst of the tremendous flux of shifting microstates, we yet have a sense of stability and predictability at the macroscopic level. It is at this level that the thermodynamicist quite categorically states, on the basis of a vast amount of experience, that heat flows only from a hot to a cold body, and not the other way around.

16.6 Time's arrow

Suppose a copper bar with many holes into which are inserted a set of identical thermometers. The bar is heated rapidly at one end, photographed (Figure 16-2a), and isolated from loss or gain of heat. The photograph shows a hot region at one end of the bar, as indicated by the readings of the thermometers. After a certain time another photograph is taken (Figure 16-2b) in which the thermometers all show nearly the same temperature which is partway between the extremes of photograph A. If now these two photographs are shown to an intelligent person, and he is convinced that in the period between the two photographs the bar was indeed isolated, he will pick A as the earlier photograph, and bet any amount on it.

The second law defines a unique direction of flow and provides one of the few directional arrows in this World: Sir Arthur Eddington spoke of it as 'Time's Arrow.'

16.7 Summary

In this brief introduction to probability and statistics we have limited our discussion to a few kinds of experiments. Particularly, we

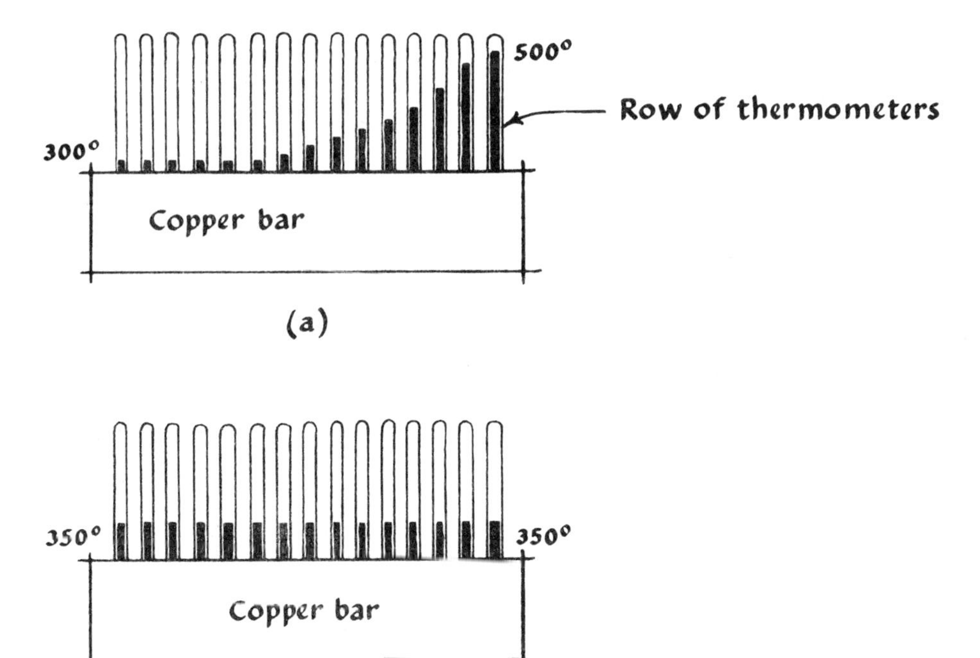

Fig. 16-2 *A copper bar is heated at one end, and this shows in the temperatures read in the row of thermometers,* a. *The bar and thermometers are isolated. After a period of time, inspection shows* b. *The thermometers now all read the same at a temperature between the extremes in* a. [*From Chalmers W. Sherwin,* Basic Concepts of Physics, *Holt, Rinehart & Winston, Inc., New York, 1961, Fig. 7-15, p. 338.*]

wished to develop the idea that the very orderly behaviors of macroscopic systems, the systems of classical chemistry and physics, are orderly because of the statistically large number of events that are involved, and because the individual event, being extremely small, is swamped out by the large number of other events; and in any case, fluctuations away from some mean of behavior tend to be cancelled by equal fluctuations in the other direction. A system, put into an "improbable" state, as is the copper bar heated at one end, tends spontaneously to move in the direction of a more probable distribution of microstates and configurations. Note that this situation is more probable because that is the way Nature is. This gives us a reliable directional arrow between the two states of a system. A physical system, isolated from outside disturbing factors, and found to be at one time in an improbable state and at another time at a more probable state, or at equilibrium, can be certified to have been in the improbable state before, in time, it reached the more probable state.

We develop these ideas, also, because probability thinking is beginning to pervade our culture. One of the lessons of this chapter is that probability considerations cannot predict exactly about the behavior of an individual event. In the humanistic realms and those of the social sciences, a single man or woman may determine the course of development of a country. A great novelist, poet, dramatist, songwriter, a great leader, may sway the mass of men in a way that has no counterpart in physical phenomena (unless these have been specially arranged by an intelligent person or are connected to catalytic processes in a living system). The dilemma of the social scientist is to be a scientist and at the same time to remain true to the social part of his title: for that intransigent datum, the individual person who defies categorization under the rubric "all so-and-so are thus-and-such," keeps obtruding himself. The educated citizen needs to remind himself and others of the powers *and* limitations of probability and statistics.

Exercises

16.1 Write out a table—like 16-1—of possible outcomes for the situation where you have the digits 1, 2, 3, 4, 5, and you choose two in order: one and then another. This is a sample space with 20 points. Associate the same probability with each point (1/20) and find the probability that two odd numbers will be drawn in succession.

16.2 Open the telephone directory of a large city and copy out 100 phone numbers, one after another, taking only the last 4 numbers and ignoring the exchange (first three numbers). Omit numbers for institutions. This should give a reasonably uniform distribution of odd and even numbers (0 is even) and also of digits from 0 to 9. Check this.

16.3 a) Write the sample space for 4 flips of a coin in the form of a compact table.
b) What is the probability of the event A "at least two consecutive heads?"

16.4 In a New Haven hospital a mother of 4 girls was heard to remark after her fifth child turned out to be a boy, "A boy in time saves nine." Assuming equal probabilities of boys and girls a) what is the probability in a family of 5 children that the fifth child will be a girl? b) what is the probability of G G G G B?

16.5 Two dice are thrown once. What is the probability of obtaining a) 7; b) 4; c) at the most 4?

16.6 What is the probability of drawing, from a well-shuffled deck of playing cards, first an ace of spades and then a king of spades, without replacing the ace?

16.7 It is calculated that there are some 6.35×10^{11} possible hands at bridge. Estimate the number of bridge hands dealt in the United States in one year. Is it likely that one of these will be thirteen cards of the same suit?

16.8 A man nervous about flying consults his friend: "What are my chances of getting on a plane that has a bomb on it?" The friend estimates something of the order of 1/1,000,000. As his unhappy questioner turns away, he calls him back and says, "I suggest that you take a bomb along in your suitcase, because I estimate conservatively that the chances of having two bombs on the same plane is more like $1/10^{10}$." Discuss this advice from the point of view of this chapter.

Answers to Exercises

16.1 3/10

16.3 a)

HHHH	HTHH	THHH	TTHH
HHHT	HTHT	THHT	TTHT
HHTH	HTTH	THTH	TTTH
HHTT	HTTT	THTT	TTTT

b) 8/16 = 1/2

16.4 a) 1/2. b) $1/2 \cdot 1/2 \cdot 1/2 \cdot 1/2 \cdot 1/2 = 1/2^5 = 1/32$.

16.5 a) 1/6. b) 1/12 c) 1/6

16.6 $1/52 \times 1/51 = 1/2652$

16.8 Evidently the friend is no student of probability. That the man takes along a bomb has no influence on the possibility of another being present; the chance of a second one being present for him remains as before.

Bibliography

William Feller, *An Introduction to Probability Theory and Its Applications.* John Wiley, New York (1959), Vol. I, ed. 2, Table 1-3, p. 21.

Edmund C. Berkeley, Probability and Statistics—*An Introduction Through Experiments.* Science Materials Center, Inc., New York (1961). A kit of dice, disks, urns, cards, coin-flipping machine, and other tools are available from Science Materials Center, Inc., 59 Fourth Avenue, New York.

Henry L. Alder and Edward B. Roessler, *Introduction to Probability and Statistics*. W. H. Freeman, San Francisco (1960).

Solomon Diamond, *Information and Error. An Introduction to Statistical Analysis*. Basic Books, New York (1959).

Frederick Mosteller, Robert E. K. Rourke, and George B. Thomas, Jr. *Probability with Statistical Applications*. Addison-Wesley, Reading, Mass. (1961).

J. G. Kemeny, H. Mirkil, J. L. Snell, G. G. Thompson, *Finite Mathematical Structures*. Prentice-Hall, Englewood Cliffs, N. J. (1958). This excellent text on modern mathematics is well worth reading through.

J. D. Williams, *The Compleat Strategyst, being a primer on the theory of games of strategy*. McGraw-Hill, New York (1954).

Harold F. Blum, *Time's Arrow and Evolution*. Harper Torchbooks, Harper & Brothers, ed. 2, 1962.

Chapter 17

17.1 What cybernetics is
17.2 Some definitions
17.3 The system
17.4 Feedback; retroaction, I
17.5 Types of material systems
17.6 Interim remark
17.7 The types of effects
17.8 Atomic power plant. Atomic bomb
17.9 Conceptualization
17.10 Effector with different possible transitions
17.11 Feedback, II
17.12 Interim remarks
17.13 The goal
17.14 Summary

Process. An Introduction to Cybernetics ~ 17

What cybernetics is 17.1

A person coming on cybernetics for the first time may feel that it is not new. Nor is it, in its aspect as an art. "The art of steermanship," says W. R. Ashby of cybernetics, and Louis Couffignal defined cybernetics as "the art of ensuring the efficacy of an action." It is surely an old art: the art of getting one's way. A physical scientist may argue convincingly that physicists and chemists have just as good, or better, and certainly more familiar, ways of saying what cybernetics would say in their field. This is quite true. The science of cybernetics, formulated by Norbert Wiener and his colleagues in the mid-1940's, is a new science which like many modern sciences has grown out of ancient arts. He defined it as "the science of control and communication in the animal and the machine." Cybernetics is a unifying science. It is concerned with processes of all kinds, from nuclear to political; that is, up through the whole range of sciences on the right side of Figure 1-1. Moreover, it may be the best way of dealing with very complicated processes.

It is true that looking back into the history of technology one can discern machines which employed principles now known as cybernetic. A thermostatically controlled incubator for eggs was made in the mid-seventeenth century. Seventy-five years before that windmills were in use that fed grain automatically to the grindstones in accord with the wind-power. A little over a century later James Watt invented his 'governor,' which controlled the speed of a steam engine, causing it to change its power output as the load changed on it. All these, as we look back on them now, involved the principle of feedback, or retroaction. Moreover, once Norbert Wiener had called attention to the principles—a creative act of invention on his part—it became evident that exemplars of these principles lay all around us. All living beings owe their stability to processes that can be understood through cybernetic reasoning; so also the societies of such beings; evolutionary behavior, both of man and Nature, embodies cybernetic principles. In industry, the assembly line, a primitive cybernetic arrangement as it is seen to be, has been developed into the highly automated factory of today by virtue of cybernetic theory and 'cybernation,' the engineering application of cybernetics. The difference in behavior between an atomic bomb that explodes and an atomic power plant that produces useful energy—both utilizing the same nuclear reaction—is clearly explained by cybernetic theory. And so on, *ad infinitum.*

Cybernetics deals with the behaviors of 'systems,' which are not treated as things but according to their behavior. Actually, we need two kinds of language, as we have seen before. There is the 'thing-language' so to speak, by which we describe machines of all kinds as well as natural processes (which are not machines but which by a kind of reverse logic are sometimes said to be machine-like). This language has to do with specific things in all their multifariousness. The other language is construct-language. It is a kind of metalanguage used for talking about thing-language and the behavior of things. For example an atomic bomb may be discussed in thing-language. But to show its relation as a process to a forest fire, a malignant tumor, cloud-seeding or a riot, one has to move into the field of constructs. It is in this field that cybernetics resides. It takes a dynamic approach to processes of all kinds and provides a uniform language for discussing them. It can be rigorous and quantitative; but in those applications where mathematics is not appropriate, it may still be rigorous though qualitative.

17.2 Some definitions

We shall refer to the thing, simple or complicated, which is the seat of a process under consideration as an 'effector.' (The effector is often a machine, and there is a tendency to name all effectors machines, but we do not bind ourselves to this restricting dogma.) The effector undergoes changes (the process) from one state to another. Following Ashby's notation, the first state before the change will be named the 'operand.' This undergoes a 'transition' to the 'transform.' A set of transitions is a 'transformation.' The thing which undergoes change is never completely isolated (if it were we could not even observe it). It exists in a habitat. If we think of the effector as an entity, its interaction with its habitat may be symbolized in the diagram entity—habitat→ (17-1)

E. F. Haskell names that which directly affects the entity, and which it affects, the 'habitat.' Everything else is 'environment.' The entity may be a unitary or a complex effector; the habitat is almost always a complex effector; the system entity—habitat may itself be considered an entity; then the environment is its habitat.

The effector undergoes change which may be symbolized:

$$\text{operand} \rightarrow \text{transform} \qquad (17\text{-}2)$$

or, since the transform results from a transition on the operand, one might adopt functional notation (Section 2.5) and describe this process as

$$\text{operand} \rightarrow \text{f(operand)}. \qquad (17\text{-}3)$$

Suppose a very simple effector undergoing a process: a ball rolling with constant velocity in a straight line on a desk-top that has been marked off with a Cartesian grid (Figure 17-1). We describe the posi-

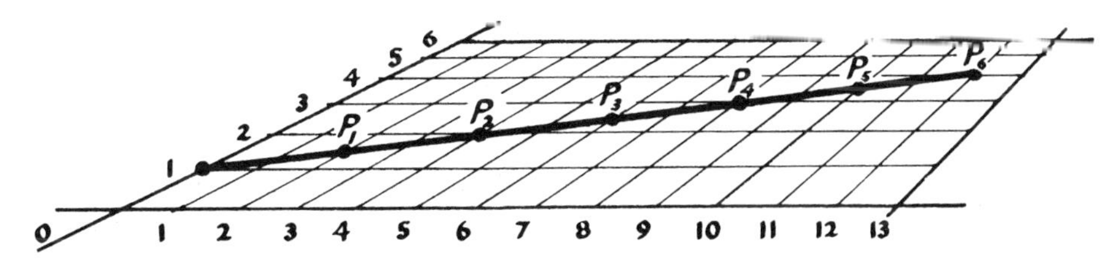

Fig. 17-1 *Transformation. A ball rolls on a table top, marked off in an X, Y grid and its positions are noted at equal time intervals.*

tion of the ball on consecutive periods by reading and recording the X and Y coordinates. This provides us with a set of transitions which take the form of a data-sheet, Table 17-1. The input to this effector was the impulse (Section 2.7) given the ball to start it rolling. Notice that we are optically coupled to the effector, since we observe it, and that this does not have any effect on its behavior. Part of its output ends up in our data sheet. At consecutive times the ball was observed at consecutive positions, $p_0, p_1 \ldots p_6$. The individual transitions, and the whole transformation 'T' may be described, following Ashby, as

$$T: \downarrow \begin{matrix} p_0 & p_1 & \ldots & p_4 & p_5 \\ p_1 & p_2 & \ldots & p_5 & p_6 \end{matrix} \qquad (17\text{-}4)$$

or as a kinematic graph: $p_0 \rightarrow p_1 \rightarrow p_2 \rightarrow p_3 \rightarrow p_4 \rightarrow p_5 \rightarrow p_6$

Table 17-1

Data for Behavior Shown in Figure 17-5

Time in equal units	X coordinate	Y coordinate	'P'
Begin at 0	0	1	0
1	2	1.5	1
2	4	2	2
3	6	2.5	3
4	8	3	4
5	10	3.5	5
6	12	4	6

or

$$\begin{array}{ccccccc} p_0 \rightarrow p_1 & & & p_4 \rightarrow p_5 \\ \downarrow & & \nearrow & \downarrow \\ & p_2 \rightarrow p_3 & & p_6, \text{ etc.} \end{array}$$

In this graph all that counts is the sequence; the spatial arrangement is not relevant.

17.3 The system

If the effector is something we observe, we describe it and its behavior in language suitable to the plane of perceptions. Oftentimes the data-sheet consists of this description, for example: "at 10:02 I mixed the reagents; at 10:15 the mixture was clear and bright blue . . . and so forth." The cybernetic investigation is concerned with constructs and relations between them. The cybernetic 'system' is the variables that are undergoing change: color, and the optical clarity of the solution, *e.g.*, freedom from cloudiness, in the chemical description; X and Y coordinates in the case of the rolling ball. In general, the system is described by a 'state vector.' The transitions may be described verbally preparatory to interpreting or otherwise manipulating them. For example, in the chemical reaction the transformation 'C' might be written:

$$C: \downarrow \begin{array}{lll} \text{(clear, colorless)} & \text{(clear, blue)} & \text{(cloudy, faded)} \ldots\ldots \\ \text{(clear, blue)} & \text{(cloudy, faded)} & \text{(precipitate, water white)} \ldots\ldots \end{array}$$

Each operand and transform is a verbal state vector with two components. It is perfectly definite and precise, and could be recognized if it occurred again, so that the description is rigorous even though not mathematical or quantified. In discussing the rolling ball, we have in the data-sheet the components of mathematical state vectors. The particular positions of the ball are described in Table 17-2. Change

Table 17-2

Behavior of the Effector in System Terms

Position	State vector
p_0 . . .	[0, 1]
p_1 . . .	[2, 1.5]
p_2 . . .	[4, 2]
. . . .	.
. . . .	.
. . . .	.
p_6 . . .	[12, 4]

in a system is recognized when the state vector at one point in time does not have the same value as at another point in time. Further examination of the system would lead to the functional relation between position and time, as discussed in Section 2.5; $\Delta p/\Delta t = \sqrt{(\Delta x)^2 + (\Delta y)^2}/\Delta t$, or $p = 2.06\ t$. The numerical value is the speed of the ball (a behavior that may be changed by changing the input).

We have chosen this simple example because it is familiar. *The (conceptual) transformation corresponds exactly to the (perceived) behavior of the ball.* That this is so follows from the circumstances that each state vector corresponds one-to-one (uniquely) to an observed position of the ball at the linked time. The succession of states corresponds to the succession of positions as they smoothly occur. Ashby points out that in this relationship between the transformation and the physical process, the transformation is the 'canonical representation' of the physical process, and the process 'embodies' the transformation.

It may seem that we have gone the long way 'round to arrive at something we have said before in other ways, but this is only because the illustration we chose is so very simple. We have generalized. If q is distance and t is time, then depending on the functional relation we have a symbolism for representing velocity, acceleration, oscillation, and other complicated motions. But q could be any set of variables. We have shown in principle that a process, observed at discrete times may be given canonical representation. Before taking up examples of processes that might be analyzed in cybernetic terms it is necessary to discuss several related matters.

Feedback; retroaction, I 17.4

When an effect retroacts upon the causes of subsequent effects, that is feedback. Watt's fly-ball governor, which became of tremendous industrial importance, is a good embodiment of feedback. To the shaft of a steam engine is attached through a train of gears a spindle on which are hung two arms with weights on the ends. These are hinged to a collar which slides along the spindle (Figure 17-2). The collar, through a lever, controls the steam valve to the engine. As the collar moves up, the valve closes; as it moves down, the valve opens. The engine spins the spindle, and the balls swing out to a degree that has been set ('programmed' in modern parlance) so that the valve is open enough to maintain the engine at that speed. Suppose a load comes

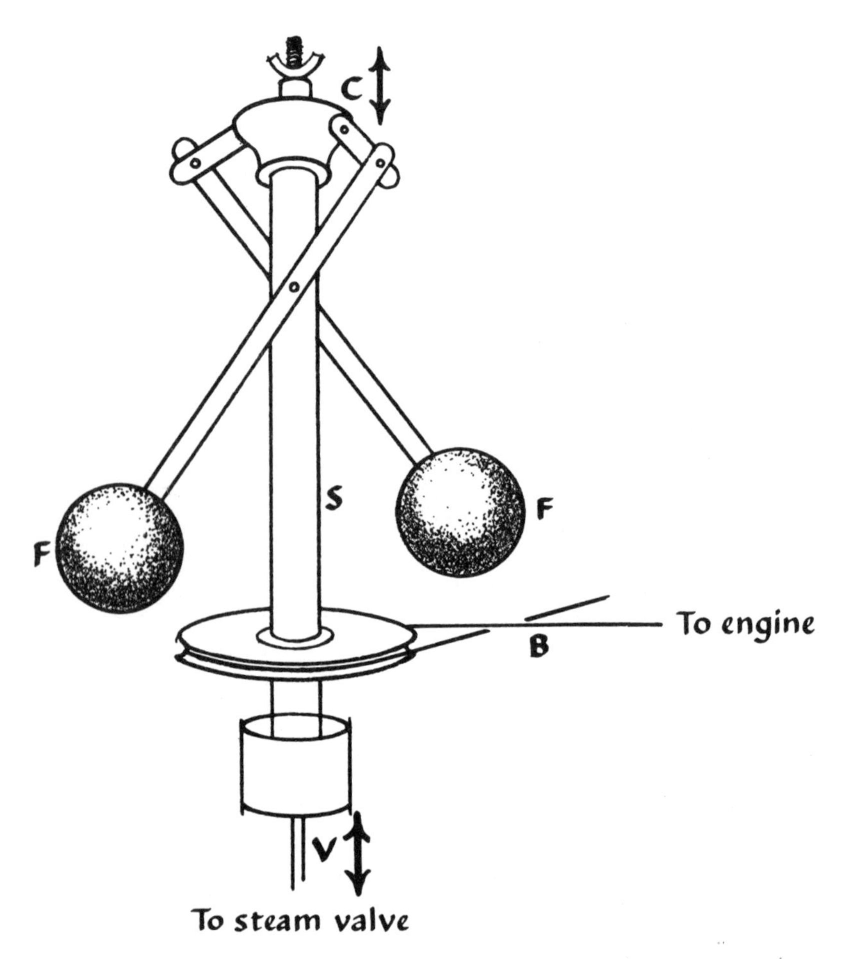

Fig. 17-2 *Watt fly-ball governor. The collar C can slide up and down. It raises or lowers the rod V which controls the steam valve. C rides on the spindle S which is driven by gears, or a belt B from the engine. The fly-balls F are attached to rods that are pivoted to the shaft, and their movement out or in with changes in speed of rotation of the shaft is transmitted to the cap. The relation of V to C is adjustable.*

on the engine. For example, if it is a railroad engine, it starts up a grade; if it is the engine of a saw-mill, a log comes under the saw. The load slows the engine, and the coupled spindle slows. This allows the weights to fall, and in the process, through collar and lever, opens the steam valve, calling for more steam to the engine. Thus the advent of the load calls forth energy to cope with it. With more steam fed to it the engine puts out more power, and as it speeds up the weights swing out. When the railroad engine gets to the top of the hill, or the saw comes to the end of the log, the load lessens, the engine, now receiving unneeded steam, speeds up, throwing the weights outward, and in this process shutting down on the valve. The engine slows down, perhaps overshooting a little, going too slowly, whereupon the governor speeds it a little, and if the governor is well-behaved the engine settles down rapidly with minor oscillation to its set speed.

Here we have an effector, the engine and associated parts, that is doing work. It changes energy of one kind into energy of another kind (basically, heat energy to mechanical energy). A very small quantity of its output is used to spin the spindle, but this is sensitive to the output since directly coupled to it. An adjustable balance between the speed of the spindle and the openness of the steam valve has been set, whereupon the swinging weights hold open the valve to just the extent that keep them spinning fast enough to be raised enough to hold the valve open. The steam valve controls one of the work-factors (the most important one, though there are others, such as lubricant systems, cooling systems, negative factors such as frictions of many kinds). It thus controls the power output of the engine; and the valve is controlled by the governor. A very small amount of the output energy serves to control the entire machine through feedback, making it responsive to demands placed upon it within the limits of its mechanism. Feedback is an arrangement by which a very small effect may produce a very large result.

Types of material systems 17.5

de Latil[1] has classified tools and effectors in a fruitful way based essentially on their degree of retroaction and adaptability to and influence on their habitat. The classification, adopted below, is to some extent arbitrary. Categories are set up in what is really a range of complexity and organization of effectors. However, just because the range is so complex some classification is helpful. Operand and transform are symbolized with arrows, and the transformation might be shown as in Figure 17-3. The two arrows, *w.c.* and *c.c.*, represent work

[1] Pierre de Latil, *Thinking by Machine. A Study of Cybernetics*, Houghton Mifflin, Boston, 1957.

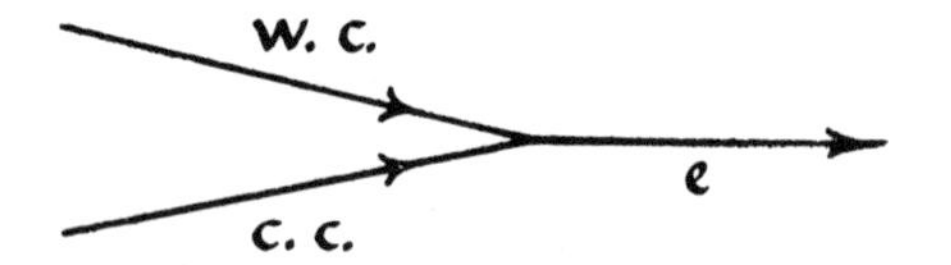

Fig. 17-3 *A simple first-degree effector, or machine, w.c., work component(s); c.c., contingent component(s); e., effect.*

component and contingent component respectively in the state vector. The two represent the state vector, and there may be many such arrows, one for each variable that makes up the state vector. The third arrow in the figure, '*e*', is the effect, the transform. Where operands become transform is the locus of the functional relation $q(t)$. *That* the components come together in this way is contingent. It might not happen. Once it does, the functional relation (natural law) leads to the effect, either deterministically or in a probability relation, as we saw. One might think of the state vector as the existential and the functional relation as the essential components of the process described by the diagram.

The simplest kind of mechanism is a tool such as a hammer or a lever. It is a passive object that requires to be manipulated in a particular way. It is a vestigial machine. A machine is an artificial object constructed by man to accomplish a certain action when energy is supplied in a suitable form together with necessary matter. A stone used as a hammer, or a log used as a lever, would not be classed as machines since not made by man, but a carpenter's claw-hammer might well be classed as a vestigial machine. The stone held in its jaws and used by a mud-wasp to tamp down the mud with which it closes its larva's nest is a tool, as is the thorn carefully selected and held in its bill which one of Darwin's tree finches uses to prise insects out of cracks in barks that are too narrow and deep for its beak to enter. The tool is a material extension which the living creature uses for a purpose. It may be natural in origin, as a twig, pebble, bludgeon, or levering log, or it may be artificial—that is, made for the purpose. A lever may be diagrammed as in Figure 17-4.

The simplest kind of machine, a first-degree effector (de Latil), is shown in Figure 17-3. It belongs to the class of effectors with determined effects. A light-bulb or typewriter key are examples. Sometimes one of the factors is left free to vary, as the bimetallic strip in a thermostatic control which bends and makes or breaks electrical contact under the influence of the ambient temperature. Its actual value at any time is not programmed in beforehand, but its value at any temperature (within a permissible range) is. (Preprogramming is indicated by an x on a factor.)

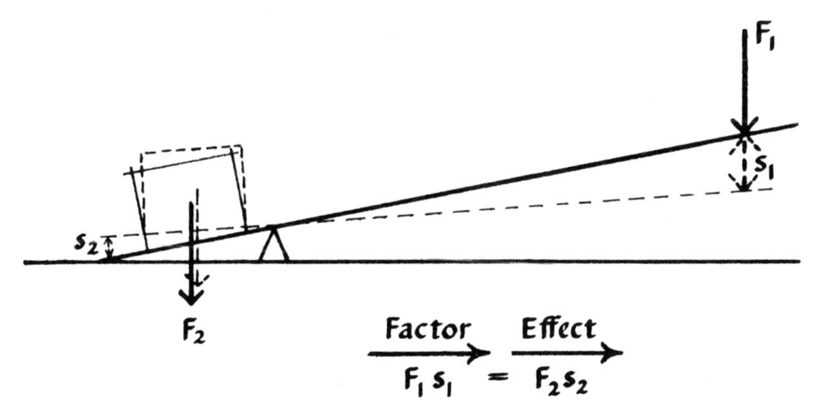

Fig. 17-4 *A simple tool, a lever and its diagram.*

In a second-degree machine (Figure 17-5) several simple actions are coordinated and the factors, preprogrammed, may operate in a variable time sequence so that simple actions occur in phase. As examples, one may take a player piano, or a simple clock. Their behavior is usually repetitive.

Fig. 17-5 *A second-degree effector, programmed (indicated by x) to coordinate several actions, two of which are shown.*

Machines of the third degree are sensitive to external factors that can modify their program or otherwise change their behavior. This external factor, represented by a short arrow impinging on a factor (Figure 17-6), is contingent. A fire-alarm system that does nothing except what is programmed into it—*i.e.*, ring a bell and turn on sprinklers—but does this only in response to contingent events is this

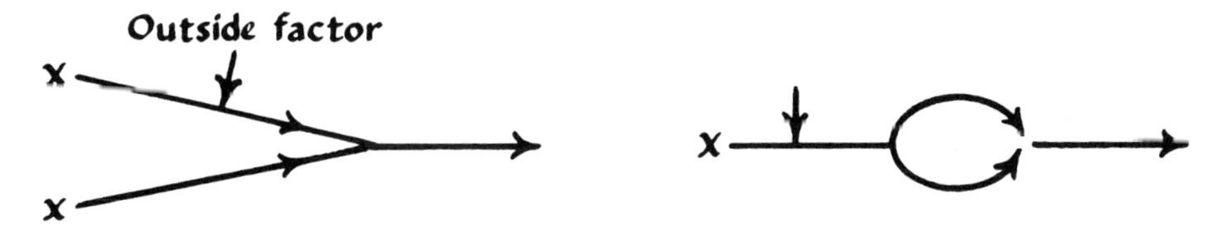

Fig. 17-6 *A third-degree effector. The program may be actuated or changed by an outside factor, symbolized by the short arrow.*

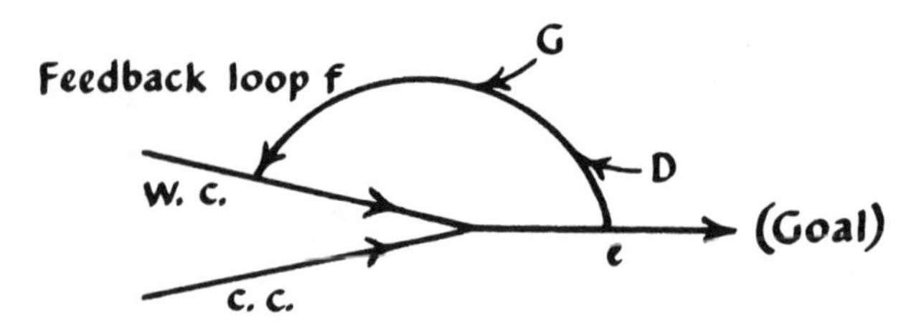

Fig. 17-7 *A fourth-degree effector. f Is the feedback loop, comprising two linked effectors; D, a detector that senses the effect and G, a governor to which information is transmitted from the detector and which controls the work component(s).*

kind of effector, as is an automatic traffic light actuated at a minor access road by an occasional car. The contingent variable has to be dealt with in terms of the probability of its changing.

In machines of the fourth degree the beginnings of effectors with organized effects producing feedback, or retroactive coupling, appear. Several simpler effectors are coupled (Figure 17-7). In the simplest type of fourth-degree machine there is a feedback loop comprising a detector, which is usually a first or second degree effector coupled to a governor which, at this level of complexity, is a third-degree effector. The detector responds, and its effect is transmitted to the governor as a contingent factor. The governor regulates the functioning of the machine by controlling the work-component. The machine has been set to produce a given effect—for example in a household heating system to hold the temperature at or close to 70°. The detector responds to deviations from this setting, and so the machine may be called a deviation-controlled effector. Retroactive coupling closes the circuit

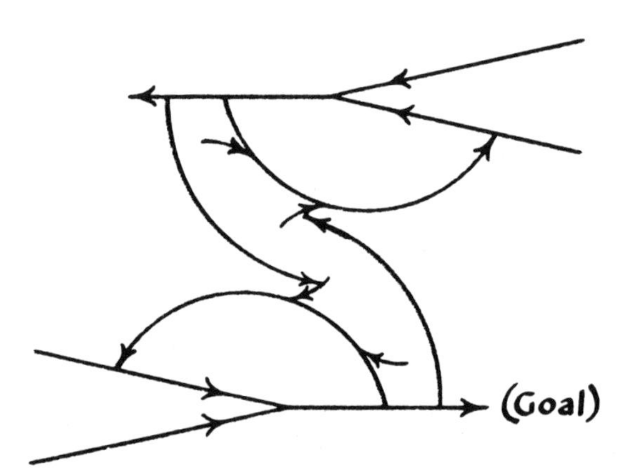

Fig. 17-8 *Fifth-degree effector. Two linked fourth-degree effectors affect the behaviors of each other.*

between effect and cause. This kind of organization has the greatest consequences.

In effectors of the fifth degree (Figure 17-8) we begin to find representations of the simplest living creatures. These effectors are characterized by the ability to escape to some extent from the limitations of simple feedback operation. This is accomplished by linking together a network of two or more fourth-degree effectors. A given fourth-degree effector (such as a Watt governor) seeks a state of equilibrium. But if linked with other effectors which also tend toward equilibrium the two, or more, acting together and upon each other, can reach an overall equilibrium which is compounded of several disequilibrial states of the individual effectors. The effector then gains a kind of 'ultra-stability,' as Ashby names it, and achieves a multiplicity of behaviors. Ashby built a machine, a 'homeostat', that embodies this principle. Some abilities of animals and persons imply the presence of fifth-degree effectors. For example, a person wore special spectacles that inverted everything he saw. After a period of confusion he became able to operate normally, and when the eye-glasses were removed he had to again go through a period of adjustment and confusion. When he put on the disturbing glasses he assaulted the equilibrium relations (invariant-relations) that his brain had developed from the deliverances of all his senses and related to his sight. Adjustments led to new states which compensated for the disequilibrial optical situation. Fifth-degree effector behavior abounds in our experiences.

Sixth-degree effectors are those that are able to change their own programs as well as adapt to new circumstances. Here the state of homeostasis of the effector is achieved by complicated interrelations. Effectors of the fifth degree can modify their behavior in line with inputs that may change in ways unforeseen by the designer; this behavior-modification has the aspect of 'learning.' Those of the sixth degree are in addition able to adapt the means at their disposal; the programs available, and the goals sought (see Section 17.13). The feedback may be not only internal but external, through the operation of sensory mechanisms. These effectors can adapt to their habitat as it is perceived, and can also modify their habitat. Many higher animals show sixth-degree effector behavior.

Effectors of the seventh or eighth degrees display transcendental effects. These have been described by de Latil, and will not be discussed at this point.

Interim remark 17.6

The effectors met with in Physics and Chemistry are mostly low-degree effectors, as we shall show, the degree of the effector depending to some extent on what is the 'physical system' of interest. Our purpose is to show the relationships of a conceptual kind that link these to the more complicated, higher-degree effectors encountered in Biological, Behavioral, Social, and Policy sciences. For this reason our examples will range over the whole gamut.

17.7 The types of effects

Basically, there are only three types of effects. There may be no effect at all, and there may be, in de Latil's terminology, 'tendency' and 'constancy' effects. These may occur in mixtures and sequences and in quite complicated combinations. A constancy effect is shown by the properly connected Watt governor, or thermostat. It maintains the behavior of the system constant, or close to constant, at or near to some set state. The prototype behavior of a constancy effector is shown in Figure 17-9. Constancy effect is obtained when the governor con-

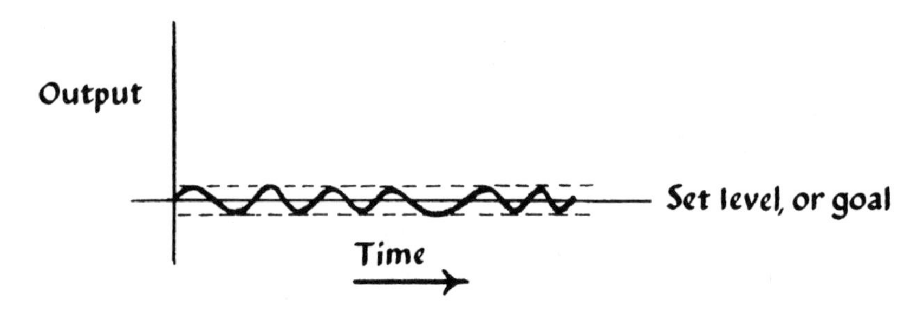

Fig. 17-9 *Behavior in constancy effect. The output is held close to a set goal. Deviations are sensed by the detector and opposed by the governor.*

tinually operates against deviations from the set state: the governor says 'No! no!' When the room temperature rises to a specified point it shuts off the heat; when it falls it turns on the heat. It operates through 'negative feedback,' a name given to suggest deviation control.

Tendency effect is produced when the governor calls for more of the same. It is the result of 'positive feedback.' The system tends to some limit. For example, a forest fire may start with a few smoldering leaves. The heat produced dries out close-by material which catches on fire. This larger fire dries out and heats material at a longer periphery and helps it to ignite more easily. . . . Thus the process escalates. The prototype behaviors are shown in Figure 17-10. An epidemic follows this pattern, as does a riot, or the stopping of a car by pressing on the brake. In each case some variable is plotted against time: acres of wood destroyed or board-feet of lumber; number of persons infected,

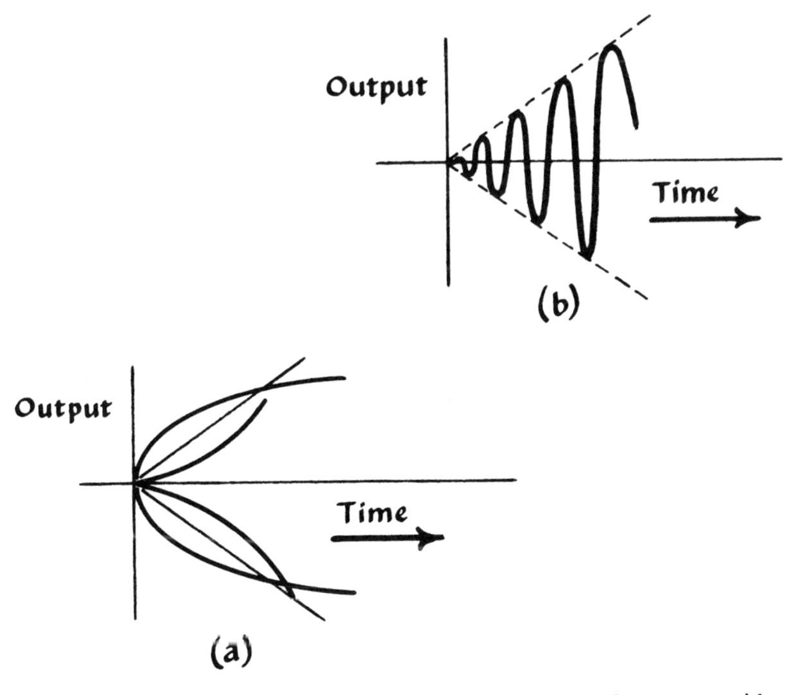

Fig. 17-10 *Behavior in tendency effect. The system tends toward an extreme without* (a) *or with* (b) *accompanying oscillation. Possible shapes of curves of behavior are shown.*

or dead; numbers of persons involved, or dollars of damage done; decrease in speed, or negative speed. Imperfect constancy behavior may lead to oscillation when the system does not damp down. It may lead to destructive oscillation if overshooting occurs and overcorrection, as in Figure 17-10*b*. The increasing amplitude of the oscillations imply a tendency effect, as shown by the dashed lines. On the other hand, a tendency behavior may lead to a stable state when it approaches a limit—as in the example of stopping a car.

The important point is that complicated processes, analyzed in these terms, make possible the setting up of hypotheses and the asking of relevant questions which might otherwise be obscured by the complexity of the unanalyzed system. If you see a tendency or constancy effect in operation you can ask about detectors (there may or may not be any), governor (may or may not exist overtly), and work factors: you can make an hypothesis, and so begin to study it and understand it.

Atomic power plant. Atomic bomb 17.8

We now can appreciate that it is their differences in organization that essentially distinguish an atomic power plant from an atomic bomb. So as to make the picture as simple as possible we shall ignore the modern refinements and technological changes that have been made in both. Both machines use the same type of reaction:

$$_{92}U^{235} + _0n^1 \rightarrow \rightarrow \rightarrow _{Z_1}X_1{}^{A_1} + _{Z_2}X_2{}^{A_2} + 2_0n^1 + \gamma + 200 \text{ MeV}$$

X_1 and X_2 represent fission products ranging in mass number mostly around 95 and 135, which themselves may be radioactive. Z_1 and Z_2 must add to 92, and A_1 and A_2 must add to 234 for material balance, considering that one neutron is on the left and two on the right.

The fission neutrons (somewhat more than two because of the radioactivity of fission-products) are produced with high energy, perhaps up to 2 MeV. To recapitulate briefly, these 'fast neutrons' are slowed down in the pile of a power plant by letting them collide with suitable light nuclides, 'moderators' which have a high cross-section for scattering relative to that for absorbing the neutrons. The cross-section is not a physical cross-section but is a number proportional to the probability of reaction (Section 13.7). The reason that the fast neutrons must be slowed down is that very few of them will be captured (by uranium 238 and 235) and produce fission. On the other hand, if they are slowed down to the energy range of an ordinary gas molecule, becoming 'thermal neutrons' or 'slow neutrons' with about 0.025 eV energy, they are readily captured.

The heart of the reactor is the energy-producing 'pile' shown schematically in Figure 17-11. It is built of blocks or containers of moderator (which may be purified carbon or water or heavy water) with uranium enriched in U^{235} dispersed in it and with channels

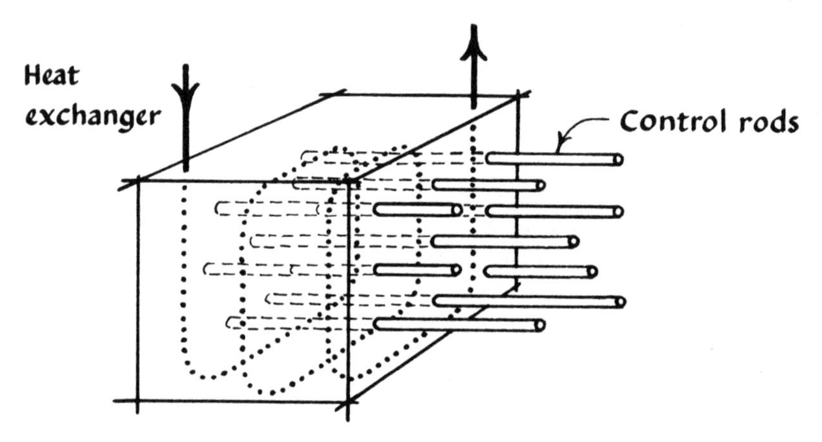

Fig. 17-11 *Scheme of an atomic pile for power production. Control rods may be pushed in or out of the pile to control critical-ness. The heat produced is drawn off by a flowing liquid heat-exchanger.*

containing movable rods, 'control rods,' of cadmium or boron which can be pushed in or withdrawn. The control rods are made of metal which has a high cross-section for absorbing neutrons but does not react with them to produce more neutrons. Through and around the pile flows heat-exchanger, in pipes. This carries off the heat that is produced and in turn produces steam or some other gas to run turbines to generate electricity.

Suppose that there are present at any given instant in the mass of uranium dispersed throughout the pile N_0 fast neutrons. Analysis of what happens to them shows that in one type of pile about two-tenths will escape completely by leaking out of the pile. Another fraction of about one-tenth will be captured by the uranium 238 and 235 without producing fission. A few, less than one percent, will be captured by these isotopes and produce fission. The rest are slowed down by collisions with the moderator. Among these, and continuing our book-keeping on the basis of N_0, about seven and a half percent of the slow neutrons leak out and are lost; two-tenths or so are captured without fission; about two percent are lost by capture by fission products and by support materials in the pile. Thirty-eight percent are captured by U^{235} and produce fission, and another one percent are captured by plutonium Pu^{239}, produced in the pile, and produce fission.

Only four fission reactions yield neutrons that can maintain the chain-reaction: capture by U^{238} and U^{235} of fast neutrons; capture by U^{235} and by Pu^{239} of slow neutrons. These last two account for 98% of the fission neutrons that maintain the chain. If the total neutrons produced by fission is N, and the number producing fission is N_0, for a pile to be self-sustaining the number produced must at least equal the number required to produce them. That is, N/N_0 must at least equal 1. This ratio is named the 'multiplication factor' or 'multiplication constant': $k = N/N_0$. When $k = 1$, the mass, or pile, or reactor, is said to be *critical*. If $k < 1$, it is *subcritical*, and if $k > 1$, it is *supercritical*. At $k = 1$, the reaction just sustains itself; at $k < 1$, it dies out. At $k > 1$ it increases in rate. The difference between a bomb and a power reactor is summed up in the behavior described by k. A bomb is made so that k is as much above 1 as the physical and chemical properties of the material permit. A reactor is made so that k never goes above 1 for long, but hovers as close to 1 as controls permit.[1]

The self-sustaining pile, or reactor, is a constancy effector. As it is

---

[1] The data are from the Westinghouse Electric Corporation, and are adapted from a table given by J. A. Richards, Jr., F. W. Sears, M. Russell Wehr, and M. W. Zemansky, *Modern College Physics*, Addison-Wesley, Reading, Massachusetts, 1962. Analogous data are obtainable from other sources, *e.g.*, General Electric Company. The actual distribution of neutron absorptions, losses, and fission effectiveness varies greatly from design to design.

being assembled, control rods are inserted to absorb neutrons, in this way keeping the value of k less than 1. When all is ready to start up the reactor, rods are withdrawn until k becomes greater than 1. The number of fissions produced per unit of time increases exponentially (Figure 17-12). Heat is produced, and when the power production

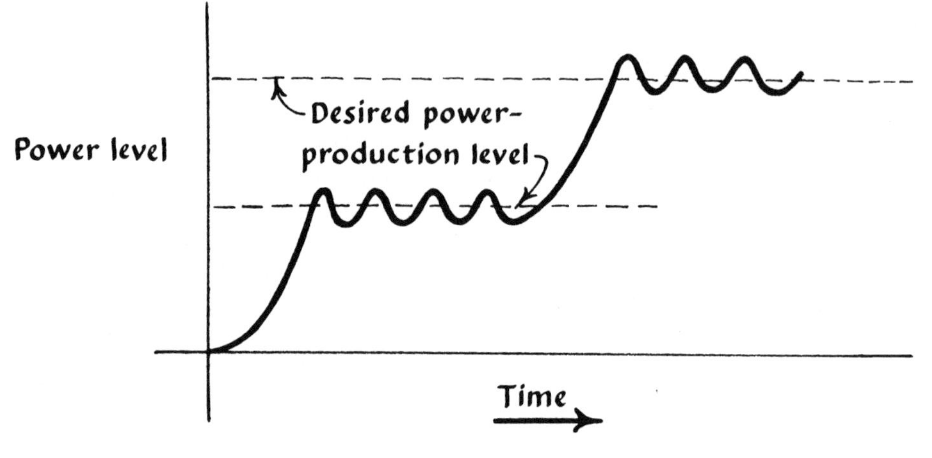

Fig. 17-12 *Behavior of the pile is shown. It starts up, reaches a desired power-production level and is held there by negative-feedback. If more power is called for, the level may be raised by further withdrawal of control rods.*

(power is energy per unit time) reaches a desired level, rods are inserted until k becomes a little less than 1. The control may be achieved either by introducing more rods, or by inserting them further into the pile. As the level falls below the desired point, rods are withdrawn a little, to bring k back to 1. The best operation is to hold k as close to 1 as possible, at whatever level of power the reactor is being run. The reactor can be set, in this way, at a desired level of power production (Figure 17-12).

We remarked that the reaction is extremely fast. If all neutrons appeared with the same promptness, control of the reactor would be a ticklish business, and probably beyond manual control. It turns out, however, that some neutrons among those produced by fission products come out after a little delay. With careful manipulation, then, a pile can be made subcritical to *prompt* neutrons but critical to the sum of the *delayed-plus-prompt* neutrons. (Some "0.4% are delayed at least 0.1 sec. and . . . 0.01% are delayed about a minute.") This makes possible manual control of the reactor if it should be necessary. Information about the rate of change of output is fed back to the controller and so governs the movement of the control rods. The detectors measure neutron flux and temperature, among many other behaviors of the reactor, and this information is used to control the insertion or withdrawal of control rods, the rate of pumping of heat-exchanging agent, and so forth. Figure 17-7, suitably labeled, epitomizes the atomic power plant and its operation.

In cybernetic terms, a bomb is an effector that produces a tendency effect by positive feedback. That it can be made depends on the following considerations. Imagine a piece of U^{235} so large that virtually all neutrons that are produced in it are absorbed and produce fission; that is, few would leak out. Then k would be about 2.5. Each neutron that is absorbed would lead to the production, on the average, of 2.5 neutrons. These neutrons are produced within the volume of the body. They may escape through the surface of the body. For bodies of simple shapes—that is, not objects with highly convoluted surfaces—the volume is proportional to the cube, and the surface to the square, of the dimensions. For example, the volume of a sphere of radius r is $4/3\pi r^3$, while the area is $4\pi r^2$. To prevent a piece of U^{235} from fissioning, then, it should be of such a size that $k < 1$. This is accomplished by having the properly shaped piece small enough that about 2/3 of the neutrons produced escape through the surface. Such a piece of metal would be subcritical.

Gamow has illustrated the principle of one of the types of nuclear bomb. This is shown in section in Figure 17-13. A subcritical thick-walled tube of U^{235} is arranged on an axle. The other end of the axle consists of a subcritical cylinder of U^{235} that just fits inside the tube when the latter, propelled by an explosive charge, is shot down the axle. When the two subcritical masses are brought together, a super-

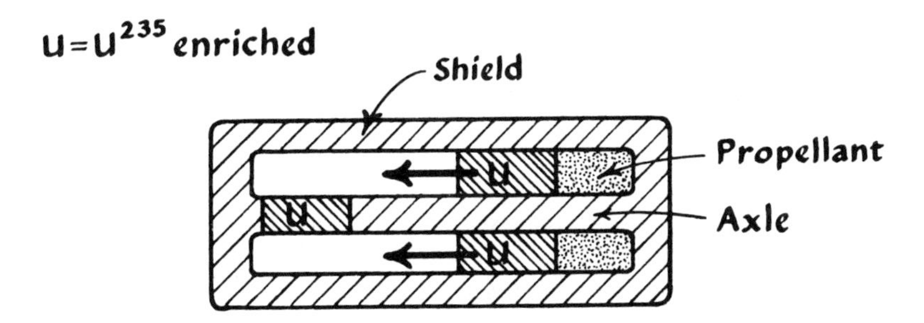

Fig. 17-13 *Scheme of an atomic bomb.* [*From Gamow,* Matter, Earth and Sky, *Ed. 2, Prentice-Hall, Englewood Cliffs, N.J., 1965, Fig. 14-26, p. 369.*] *A subcritical cylinder of uranium enriched in* U^{235}, *mounted on an axle* (*shown in section on the right*) *may be fired along the axis to surround another subcritical cylinder on the left. The whole then becomes critical, and blows up. The radioactive materials are enclosed in a shield.*

critical mass results, with consequent explosion. We can see how this arrangement takes advantage of a change in the ratio of volume to surface. The sudden bringing together of the two parts, each with $k < 1$ and producing a mass with $k > 1$, leads to the explosive chain reaction. Suppose that on the average, in such a bomb, $k = 2$, and that the nuclear fission reaction is very fast, then suppose that one fission occurs in a supercritical mass. (This might be due to inherent or built-in radioactivity or to a cosmic ray.) Then, the multiplication factor doubles the number of neutrons each fraction of a second. If n is the number of the steps, the number of neutrons produced is 2^n, and the amount of energy, at the rate of 200 MeV that results at each step, is then $200 \times 2^{n-1}$ MeV. Thus if, as a guess, the reaction occurred in 10^{-5} sec. per step, after a little over 1/100 sec. one initial neutron would have yielded some 200 million MeV, by this crude calculation. Of course, the speed of the reaction must be such that the massive nuclei do not move very far before they have fissioned in the heavy neutron flux that is produced (Figure 17-14), for the mass must remain supercritical for a sufficient percent of the reaction to take place.

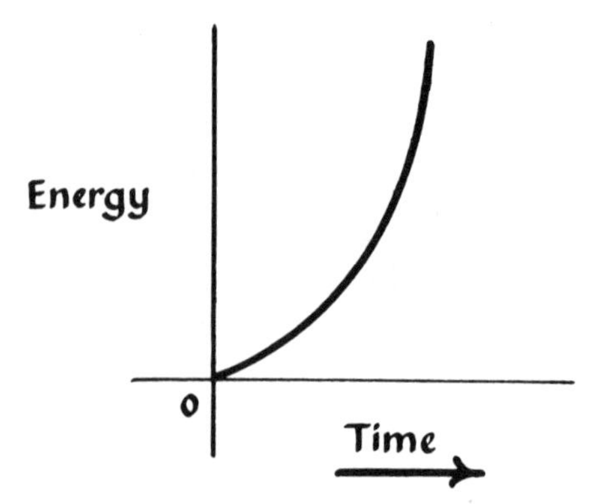

Fig. 17-14 *Plot of energy production of a bomb.*

Here everything is programmed; no overt detectors or governor function. The feedback is in this sense vestigial: a work-factor is part of the output, and this is fed back to actuate other work-factors according to the analysis of the reaction given above. The same might be said of an epidemic or a forest fire. In all these cases control involves interrupting the feedback, and this invokes detection and governing. The natural process usually terminates through exhaustion of the work-factors, or through opposing factors. In the bomb, when nuclides move away from each other the mass becomes subcritical. In a forest fire, the forest terminates, or rains come—possibly as a response to atmospheric disturbance. In an epidemic immunities build up. In the tendency effect of a riot, there are likely to be *agents provocateurs*, hidden or overt, and we have the presence, then, of detectors and governors.

Conceptualization 17.9

We have shown in Section 17.2 that the behavior of an effector as it changes from state to state may be written as a transformation. A partial classification of transformations is necessary at this point. A state (say p_1) goes to another state (say p_2).

$$T: \downarrow \begin{matrix} p_0 & p_1 & p_2 \ldots & p_5 \text{ (operand)} \\ p_1 & p_2 & p_3 \ldots & p_6 \text{ (transform)} \end{matrix} \quad \text{Transition} \qquad (17\text{-}4)$$

In this process each transform becomes the operand for the next step. Moreover, each operand is converted to one definite transform: it is 'single-valued.' It is also 'one-to-one': not only does each operand yield but one transform, but each transform arises only from, and implies, a single definite operand.

A transformation may be 'many-one' if it is single-valued and if a single operand cannot be inferred from a transform. For example, a transformation R:

$$R: \downarrow \begin{matrix} A & B & C & D & E \\ E & A & A & E & A \end{matrix}$$

is single-valued, but the transform 'A,' if it were observed at a certain time, does not indicate what the operand *must* have been; it could have been B, C, or E. Examples of many-one relations are all around us: starting from many different locations on the campus, the members of a class arrive at the same location; the location does not imply a unique path leading to it. However, one's presence in a room with a single door bears a one-one relation to passage through that door; install several doors and the relation becomes many-one, but is still single-valued.

In a *many-valued* relationship an operand may lead to one of several possible transforms, as in the tossing of a coin or the throwing of a die (Chapter 16):

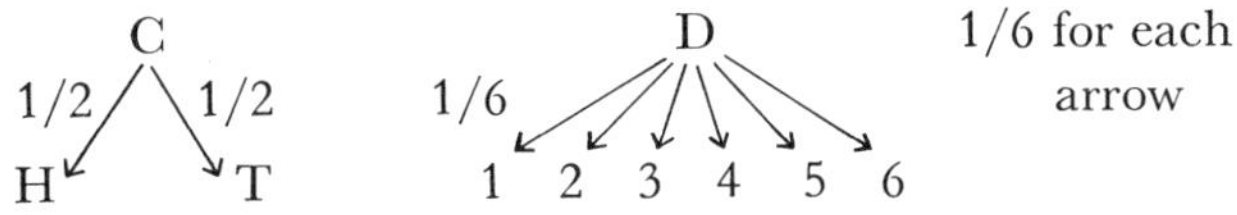

```
      C                          D               1/6 for each
 1/2 / \ 1/2         1/6   / / / \ \ \                arrow        (17-6b)
   H     T                 1 2 3 4 5 6
```

Here each transition is one-one. The states 'C' for the coin and 'D' for the die are not well defined, but the transformations are. For example, the arrows indicate that when the coin is flipped, *either* an 'H' or a 'T' may come up, and the numbers indicate the probability of each.

The transformation T (formula 17-3) is 'not closed' because there is among the transforms an element (the last p, p_6) which is not in the operands. The movement of a particle in a circle, or the second hand around the face of a clock (transformation S), is 'closed'; no new element appears in the transform that is not in the set of operands:

$$\text{S:} \downarrow \begin{array}{cccccc} 0° & 6° & 12° \ldots\ldots & 348° & 354° \\ 6° & 12° & 18° \ldots\ldots & 354° & 0° \end{array}$$

(If the transformation were not one to one, single-valued, and closed, the effector would be no good as a clock.)

There is a transformation in which the operand remains unchanged over a period of time. This is an 'identical' transformation, the formulation and kinematic graph of which is:

$$\begin{array}{l} \text{Identical} \\ \text{Transformation} \end{array} \downarrow \begin{array}{c} \text{A} \\ \text{A} \end{array} \qquad \text{A} \circlearrowleft$$

If the clock of transformation S stopped with the second hand at 6°, this would illustrate the identical transformation, with A = 6°.

As with transformation T (formulation 17-4), the transformation S can be written in the form of an algebraic summary. The anatomy of this process (which is quite general) is shown below. Let the state vector be 'p' and, omitting t, which is implicit in any process, and in this case progresses in units of seconds, write the behavior in degrees of angular deviation:

$$\begin{aligned} p' &= p + 6 \\ p'' &= p' + 6 \\ p''' &= p'' + 6 \\ &\vdots \end{aligned}$$

We can obtain the value of p'' by solving the first two equations to eliminate p'. Thus $p'' = p' + 6 = (p + 6) + 6$. Then, $p''' = [(p + 6) + 6] + 6$, and so forth. By induction, $(m)(x) = x + m \cdot 6$, where m is the number of times the transition occurs. There would have to be a statement that when $(p + 6) = 360°$ the next transition is to $(p + 6)$, otherwise the transformation would not be closed.

17.10 Effector with different possible transitions

Assume an effector with three buttons, and a set of transformations 'U.' When button number one is depressed it goes through the transformation U_1; when 2 is depressed it goes through U_2, and when 3 is depressed, U_3. Only one button may be depressed at any time for after all each part of the effector can do only one thing at a time. The transformations are single-valued and closed:

$$U_1: \downarrow \begin{array}{cccc} a & b & c & d \\ a & b & d & c \end{array} \qquad U_2: \downarrow \begin{array}{cccc} a & b & c & d \\ b & c & d & a \end{array} \qquad U_3: \downarrow \begin{array}{cccc} a & b & c & d \\ b & a & a & c \end{array}$$

The three transformations have common operands but differ in transitions. They are properties of the effector which may be invoked by an outside agency, *i.e.*, by depressing the buttons. The subscripts are 'parameters,' possible inputs to the effector. The inputs are imposed from outside of the effector. U_1, U_2, and U_3 are represented by a matrix of transformations.

↓	*a*	*b*	*c*	*d*
U_1	*a*	*b*	*d*	*c*
U_2	*b*	*c*	*d*	*a*
U_3	*b*	*a*	*a*	*c*

Feedback, II 17.11

Following Ashby, we can now see how feedback may be conceptualized. Suppose we have another effector with transformations V_1 and V_2, and a matrix representing its behavior:

↓	*e*	*f*	*g*
V_1	*g*	*g*	*e*
V_2	*f*	*e*	*f*

The two effectors may be retroactively coupled, meaning that the output of U affects the input of V, and the output of V affects the input of U. The coupling is itself a transformation.

Let V be coupled to T by a relationship 'X' in which the state of V affects the value of U's parameter. For convenience we refer to U's parameter as α, and to V's parameter as β:

$$X: \begin{cases} \text{state of V} \\ \text{value of } \alpha \end{cases} \downarrow \begin{array}{ccc} e & f & g \\ 1 & 3 & 2 \end{array}$$

U is coupled to V by a relationship 'Y' so that the state of U controls the value of V's parameter:

$$Y: \begin{cases} \text{state of U} \\ \text{value of } \beta \end{cases} \downarrow \begin{array}{cccc} a & b & c & d \\ 1 & 2 & 1 & 1 \end{array}$$

The two coupled effectors comprise a new, larger effector, the states of which are specified by the values of U and V. Suppose the effector starts at the state (a, f). In the first transition the state of V, '*f*', will invoke, by relation X, transformation U_3, and *a* will change to b. The state of U, 'a', will invoke, by relation Y, the transformation V_1 and f will go to g. The first transition will be $(a, f) \rightarrow (b, g)$. At the next step, g, invoking U_2, $b \rightarrow c$; and b, invoking V_2, $g \rightarrow f$; thus $(b, g) \rightarrow (c, f)$. Now we can plot the behavior of the larger effector. It corresponds with the new transformation 'R';

$$\text{R:} \downarrow \begin{array}{cccl} (a, f) & (b, g) & (c, f) & \\ (b, g) & (c, f) & (a, g) & \text{etc.} \end{array}$$

The trajectory of the machine, its kinematic graph, is:

$$\begin{array}{ccccccc} (a, f) & \rightarrow & (b, g) & \rightarrow & (c, f) & \rightrightarrows & (a, g) \\ & & & & & & \downarrow \\ & & (a, e) & \leftarrow & (b, f) & \leftarrow & (b, e) \end{array}$$

The effector goes into a cycle which may repeat indefinitely. The output of one effector feeds back to the input of the other: they are retroactively coupled.

Consider the behavior of a particular effector, a household thermostatted heating system. We shall formulate its behavior precisely yet non-quantitatively, based on our observations of what it does. We can be quite precise about its states because we can always recognize a state when it occurs. If we have a state a, then the negative of that state will be symbolized $\sim a$, which is logical notation for 'not a.' We label the states of the work-producing effector:

- f: stand by (do nothing) no heat produced
- a: turn on fuel pump and ignite fuel
- $\sim a$: turn off fuel pump; flame dies out
- b: delay further action until furnace is heated
- $\sim b$: delay further action until furnace has cooled
- c: turn on blower or heat-exchanger pump
- $\sim c$: turn off blower or heat-exchanger pump
- g: furnace in full operation, heat is produced

And we label the behavior of the control effector:

- i: switch (circuit) is closed, electricity flows
- j: switch (circuit) is open, electricity cut off.

The two effectors are described by their transition matrices; 'P' for the work-producing effector and 'Q' for the controller, the thermostat:

↓	f	a	b	c	g	$\sim a$	$\sim b$	$\sim c$
P_1	a	b	c	g	g	.	.	.
P_2	f	.	.	.	$\sim a$	$\sim b$	$\sim c$	f

↓	i	j
Q_1	j	j
Q_2	i	i

The two are now retroactively coupled. From our examination of the effector itself we learn that Q is linked to P's parameter α by 'V':

$$V: \begin{cases} \text{state of Q:} \\ \text{value of } \alpha\text{:} \end{cases} \Big\downarrow \begin{matrix} i & j \\ 1 & 2 \end{matrix}$$

The state of P is linked to the value of Q's parameter, β, by a relation 'W': that when the furnace has raised the temperature of the room to a set value, say 69°, the parameter 2 of Q is invoked; and when the temperature falls below the set 65°, parameter 1 of Q is invoked:

$$W: \begin{cases} \text{state of P:} \\ \\ \text{value of } \beta\text{:} \end{cases} \Bigg\downarrow \begin{matrix} g \to g \text{ until the} & f \to f \text{ until} \\ \text{temperature T} \geqslant 69° & \text{T} \leqslant 65° \\ 1 & 2 \end{matrix}$$

This statement allows for contingencies that could lead to loss of heat. The formulations may seem strange with the letters 'Q', 'P', 'W', 'V', 'α', 'β'. But the entire formulation could be made verbally. The letters mean exactly the same as the sentences they stand for. For example, 'Q' means "the transformation undergone by the thermostat under the parameter 'one' which input, if the furnace is running, causes the furnace to shut itself off, keeps it off."

When the temperature falls to or below 65°F the furnace starts up and runs through the trajectory $f \to a \to b \to c \to g \circlearrowleft$. When the temperature rises to 69°F, $i \to j$ and the parameter 2 of P is invoked. The trajectory is now $g \to \sim a \to \sim b \to \sim c \to f \circlearrowleft$. At 65°F $j \to i$ and the procedure is repeated. The behavior of the whole machine may be symbolized:

$$\begin{array}{l} \qquad (i, f) \to (i, \quad a) \to (i, \quad b) \to (i, \quad c) \to (i, g) \geqslant 69° \\ \longrightarrow \uparrow \qquad\qquad\qquad\qquad\qquad\qquad\qquad\qquad \downarrow \longleftarrow \\ \leqslant 65°\ (j, f) \leftarrow (j, \sim c) \leftarrow (j, \sim b) \leftarrow (j, \sim a) \leftarrow (j, g) \end{array}$$

Essentially, the function of feedback is to nullify or minimize the effects of the contingent factors.

Interim remarks 17.12

These formulations are quite general, as the illustration implies. They have several advantages. An important one is that they require that the system be analyzed, and that specific questions be asked about what kind of effect is present. Is the transformation one-one or one-many, or many-one; is it closed; and so on in detail about input, components, effect, and feedback where present. Another advantage is that if a state can be recognized when it occurs again, then the trajectory may be stated precisely. Also, if the steps can be quantified, the transitions and the whole transformation may be epitomized in a mathematical functional relation. Moreover the formulation is applicable to any process whatsoever. We have dealt here largely with determinate effectors where the transitions are 1:1; however we suggested that the formulation is applicable to effectors that are not determinate, that is to say (in effect), not causal. It is now recognized that once admitting indeterminacy (Section 10.8) we must allow for it everywhere. This is why Chapter 16 is present in this text. It is apparent that when the transition is determinate, the probability is '1.' There are many ways in which the system may not be determinate. For example a transformation may have to be written (on the basis of a large number of observations):

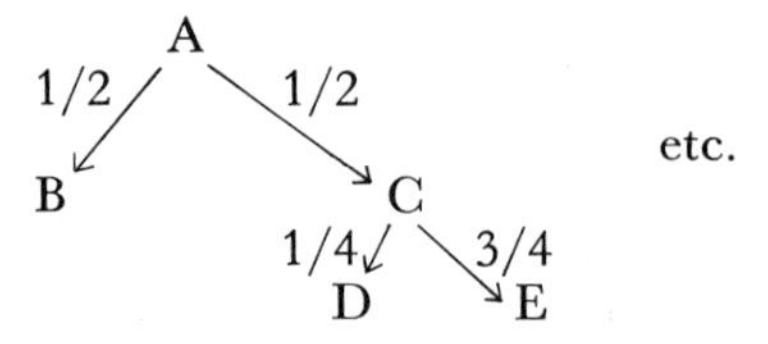

This transformation generates a set of trajectories which differ according to the paths taken by the systems at the branch points (decision points, in human affairs) where possibilities exist. The set of trajectories is spoken of as 'stochastic.' Stochastic processes are important to us because they always suggest that perhaps we are ignorant of some factors. Are the behaviors truly random? Are we overlooking some variables? Whether this question is asked may depend on the philosophical attitude of the questioner. Which does he believe to be fundamental: deterministic or stochastic processes?

17.13 The goal

In the diagrams 17-7 and 17-8 'goal' has been present, in parentheses, but we have avoided the use of that word in the text until now. This is because if there can be said to be a goal in a system with feedback it is certainly not a product of the system but something transcending it as a metalanguage transcends that about which it speaks. One can certainly agree that the goal of the designers of an atomic power plant is to produce power (under certain constraints of economic, political, and other kinds). But it is not proper to say that this is the goal of the plant. Systems with feedback may operate so subtly that it is natural to speak admiringly of them in anthropomorphic terms. This may lead the unwary astray. We shall have to deal with the subject in a later chapter.

17.14 Summary

Cybernetics is the science of processes. It is presented in this text not only because all intelligent people should know something about it, but because being applicable throughout science it makes a connection between all sciences. Motion and change prevail in our world and can be studied in terms of factors and the effects they produce, whether the processes are determinate or stochastic. Feedback is the linkage that operates against chance to make the system determinate, whether as a constancy or as a tendency effector or as a combination of these. The cybernetic analysis even of the most complicated systems gives a basis for hypotheses and so for further study, because it enables relevant questions to be asked: questions about the important inputs; the outputs and goals; the feedbacks and the linkages between subsystems. Cybernetics emphasizes an holistic approach to any problem that is studied by its techniques. It enables the construction of conceptual models even of systems that are too complicated to study in any other way.

Exercises

17.1 Write out the following transformations:

a) $n \rightarrow n + 3$ for n = prime integers between 0 and 10
b) $n \rightarrow n^2$ for n = 1 to 6
c) $n \rightarrow n^2 + 2n + 1$ for n = 1 to 4

17.2 Given the following transformations, convert them to the form in Question 17-1.

$$\text{Ta:} \downarrow \begin{matrix} 1 & 2 & 3 \\ 8 & 9 & 10; \end{matrix} \quad \text{Tb:} \downarrow \begin{matrix} 2 & 3 & 4 \\ 10 & 15 & 20; \end{matrix} \quad \text{Tc:} \downarrow \begin{matrix} 1 & 2 & 3 \\ 1 & 4 & 9 \end{matrix}$$

$$\text{Td:} \downarrow \begin{matrix} 2 & 4 & 6 \\ 1/2 & 1/4 & 1/6; \end{matrix} \quad \text{Te:} \downarrow \begin{matrix} 2 & 3 & 4 \\ 12 & 17 & 22; \end{matrix} \quad \text{Tf:} \downarrow \begin{matrix} 4 & 5 & 6 \\ 1 & 1 & 1 \end{matrix}$$

$$\text{Tg:} \downarrow \begin{matrix} 4 & 5 & 6 \\ 4 & 5 & 6; \end{matrix} \quad \text{Th:} \downarrow \begin{matrix} 2 & 4 & 6 \\ 5 & 10 & 15; \end{matrix} \quad \text{Ti:} \downarrow \begin{matrix} 1 & 2 & 3 \\ 11 & 10 & 9 \end{matrix}$$

17.3 An atomic energy-driven machine is designed in such a way that at each millionth of a second the energy produced by the machine (call the energy E) is increased to E′ by as many units as the previous value of the energy exceeds 100 units. That is, $E' = E + (E - 100)$.

a) Calculate the values of E′ at each millionth of a second for 8 consecutive periods, beginning with E = 110 units.
b) What is the value of E for subsequent millionths of a second when E starts at 100 units?
c) What happens in the case b) should some fluctuation in the energy bring it to 100.01 units?
d) What kind of an effector is this machine?

17.4 Suppose the design of the machine in Exercise 17-3 is changed so that the effector behaves in the following way:

$$E' = 1/2\ (E + F)$$
$$F' = 1/2\ (E - F) + 100$$

a) Calculate what will happen if it should be started at E = 110 and F = 110. Make the calculations for six periods of one-millionth of a second each, starting at $t_0 = 0$.
b) What kind of effector is this?

17.5 Suppose that the 'machine' in Exercise 17-3 were economic-political, with W being an average wage, and *I* being a cost-of-living index. A law pegs the value of W′ at $W' = W + (I - 100)$, and the law states that *I* at the end of each year, when adjustments are made, takes the value of W at the beginning of that year, *e.g.*, $I' = W$. Suppose that the year begins with E = 110, and *I* = 110. What common terms describe the behavior of the economy under this law? (The idea is from Ashby.)

17.6 Suppose that in Exercise 17-5 the law states instead that

$$W' = 1/2[W + (I - 100)]$$
$$I' = 1/2[W - (I - 100)] + 100$$

What kind of an economy could this be called?

17.7 A college faculty's behavior with respect to student attendance undergoes the following changes: "strict attendance requirement" → "a limited number of cuts allowed" → "a cut honor system" → "no attendance requirement" → "strict attendance requirement." Give each a letter designation and show the transformation and the kinematic graph (see next Exercise).

17.8 Let the four behaviors in Exercise 17-7 be A, B, C, D respectively, and let this particular response be S_1; also let other responses S_2, S_3 be possible according to the transformation

↓	A	B	C	D
S_1	B	C	D	A
S_2	C	D	A	B
S_3	D	D	C	C

For example, if the system is in state A, then by S_2 it goes to C. Now let students respond in certain ways according to the scheme:

	↓	*e*	*f*	*g*	*h*
T:	T_1	*f*	*g*	*h*	*e*
	T_2	*g*	*g*	*h*	*h*

Now couple these systems to make a complex system of relations. We will say that the output of subsystem S is coupled to T by the relationship X:

$$X: \begin{cases} \text{value of S:} \\ \text{value of } \alpha: \end{cases} \downarrow \begin{matrix} A & B & C & D \\ 1 & 1 & 2 & 1 \end{matrix}$$

Here α is the subscript of subsystem T. Then if S were at A, its effect on T would be to invoke T_1 behavior; while if S were at state C, it would call forth from T the behavior T_2. In a similar way we couple the output of T to the input of S by the relations Y:

$$Y: \begin{cases} \text{value of T:} \\ \text{value of } \beta: \end{cases} \downarrow \begin{matrix} e & f & g & h \\ 1 & 1 & 2 & 2 \end{matrix}$$

Here β is the subscript which shows what behavior of S is called forth. Thus, if T's output were in state e, it would call forth behavior S_1; if it were in h, S_2, and so forth.
Suppose that the whole (coupled) system is in the state (A, e). What will it do next? By X, A invokes T_1, so e will change to f; and by Y state e invokes S_1, so A goes to B. The system then undergoes the transition (A, e) → (B, f). Write the next 12 transitions of the system. What happens to it, as shown by its kinematic graph?

Answers to Exercises

17.1

$$Ta: \downarrow \begin{matrix} 1 & 2 & 3 \ldots & 7 \\ 4 & 5 & 6 \ldots & 10 \end{matrix} \qquad Tb: \downarrow \begin{matrix} 1 & 2 & 3 \ldots & 6 \\ 1 & 4 & 6 \ldots & 36 \end{matrix}$$

$$Tc: \downarrow \begin{matrix} 1 & 2 & 3 & 4 \\ 4 & 9 & 16 & 25 \end{matrix}$$

17.2 a) $n \rightarrow n + 7$ (n 1, 2, 3); b) $n \rightarrow n \cdot 5$ (n 2, 3, 4);
c) $n \rightarrow n^2$ (n 1, 2, 3); d) $n \rightarrow 1/n$ (n 2, 4, 6);
e) $n \rightarrow 5n + 2$ (n 2, 3, 4); f) $n \rightarrow 1$ (n 4, 5, 6);
g) $n \rightarrow n$ (n 4, 5, 6); h) $n \rightarrow 2n + n/2$ (n 2, 4, 6);
i) $n \rightarrow 12 - n$ (n 1, 2, 3)

17.3 a) 110 → 120 → 140 → 180 → 260 → 420 → 740 → 1380, units
b) 100 units
c) It would 'take off' on an accelerating course.
d) A tendency effector.

17.4 a)

t	E	F	E′	F′
0	110	110	110	100
1	110	100	105	105
2	105	105	105	100
3	105	100	102.5	102.5
4	102.5	102.5	102.5	100
5	102.5	100	101.75	101.75
6	101.75	101.75	—	—

b) A constancy effector.

17.5 Inflation

W	I
110	110
120	110
130	120

17.6 Deflationary to stable

W	I
110	110
60	150
55	110
32.5	122.5

17.7 $A \rightarrow B \rightarrow C \rightarrow D$ (with an arrow from D back to A)

17.8 (A, e) → (B, f) → (C, g) → (A, h) → (C, e) → (D, g) → (B, h) and after six steps the system moves back to (D, g) and thereafter swings around the cycle. (This, of course, a very artificial situation!)

Bibliography

W. R. Ashby, *An Introduction to Cybernetics*, Wiley, New York. (Get the latest edition.) This is the best introduction that I know of. It is well written, explicit, and mathematical. But the mathematics requires nothing beyond ordinary algebra, except in a few sections that may be omitted. There are many exercises by means of which the reader can test his comprehension.

Stafford Beer, *Cybernatics and Management*, Wiley, New York, 1959. This is a fairly comprehensive, non-mathematical treatment. It is deceptive in that a very great deal of subtle material is presented in a pleasant, discursive style. It must be read carefully.

Pierre de Latil, *Thinking by Machine. A Study of Cybernetics*. Houghton Mifflin, Boston, 1957. This book is completely non-mathematical. It is delightfully written with flair and a flamboyant Gallic style. It is essentially philosophical and speculative. It repays reading and rereading. It may be hard to come by this book as it is now out of print.

N. Wiener, *Cybernetics. Or Control and Communication in the Animal and the Machine*. McGraw-Hill, New York, revised 1961. (Get the latest edition.) This is the classic work in the field. The first and last parts of the book are non-mathematical and well worth reading. The middle parts are highly abstruse.

N. Wiener, *The Human Use of Human Beings*. Doubleday, 1954. Discursive, inimitable, worth reading, non-mathematical.

G. Paloczi Horvath, *The Facts Rebel*. Secker and Warburg, London, 1964. The Russian scene from the cybernetic point of view, well documented and worth reading. Comments also on the effects of cybernetics on American culture and on the relations between America and Russia.

Karl W. Deutsch, *The Nerves of Government. Models of Political Communication and Control*. Free Press of Glencoe, Macmillan, New York, 1963. Excellently written illuminating book. The cybernetic approach is applied to government.

Edward F. Haskell, "Mathematical Systematization of 'Environment,' 'Organism,' and 'Habitat.' " *Ecology 21* (1940).

Chapter 18

18.1 Introduction
18.2 Particles
18.3 Fields
18.4 Field properties
18.5 General field characteristics
18.6 Summary

Bodies, Particles, and Fields ~ 18

Introduction 18.1

Modern scientific theory exists in tension between concepts of particles and fields: a tension as old as scientific philosophy. Is a particle a local knot in the otherwise smooth fabric of a field, or is a field particulate but so fine-grained as not to be noticeably discontinuous? Everyone has an intuitive feeling for particles and fields. Particles have boundaries, and inertia; they obey Newton's laws of motion if classical, and Einstein's and Planck's if not. They are tangible things—at least in principle—material, having mass, localizable, in other words small bodies; and everyone experiences bodies. Fields are often unbounded and hard to visualize—indeed a field is invisible and immaterial. But is this entirely so? What is the boundary of a charged particle, for that matter? If we try to find the size of an atom in the usual way by bouncing another particle off it, we find that we get different values depending on the velocity of the probing particle: it comes closer the faster it goes. And what of the field; does it not have mass in view of $\boldsymbol{E} = mc^2$ if it displays an energy-producing property at any point in it?

Fields and particles are distinguished in a sense by their relations to force. A particle may experience a force due to impact with another particle, and also by mechanical connection if it is a larger object. But if a particle not in physical contact with another particle or material object is observed to accelerate, or if a force must be applied to it to keep it from accelerating, we ascribe to the region in which this occurs a force field. We say that the particle or the object experiences a field. Field and particle concepts are convenient ways of dealing with certain phenomena. It is the purpose of this chapter to review and extend these concepts; to unify some of the subjects we have studied.

18.2 Particles

Particle is a multiordinal word—it changes meaning with the context. Newton's apple was certainly a physical body that could be seen, touched, smelled, tasted. In terms of our earlier analysis we would locate it close to the P-plane. Moving into the field of constructs there is the construct *mass*, a quantity defined by Newton and experimentally measurable, which is a property of every material body. Indeed, we saw that a material body is defined by its property of inertia, the quantitative measure of which is mass. For many purposes its mass may be conceived as concentrated at a point, the center of gravity of the body. The mass point is a useful construct because it provides a definite, located point in a reference system from which measurements can be made. It avoids the need to worry about extension of the object: the object may be treated as a point. In mechanics, the term 'particle' has this implication: it has no geometrical extension, but a definite location in space and time, and a definite mass. In some theoretical treatments, bulk matter is taken to be an aggregate of particles, each being, so to say, a condensation from a small local region. This type of 'particle' is clearly purely conceptual. There could not be an actual existent point-mass, for then the geometrical extension being zero, the mass-density (and hence the energy-density) would be infinite. We do not pursue this use of the term 'particle' further, except to recognize that since the space of geometry may be treated as continuous (analogously, an area or a line) and as the aggregate of an infinity of geometrical points, one can see in this a reconciliation of points and continua.

We have studied, in Chapters 8 to 15, another kind of 'particle' where the term now implies a small body with inertia. This small body does have extension, which can be examined by collision experiments, for example (Sections 8.7, 9.2, 10.1). Its inertia can be measured in, say, a mass spectroscope (Section 12.4). Individual particles can be dealt with, as in the Millikan oil-drop experiment (Section 3.13).

These particles are of two kinds: fundamental, or elementary; and complex. Clearly, molecular particles are complex, since they are composed of less complex atoms. Atomic particles are also complex, since they are composed of nuclei and electrons. Nuclei (except for H^1) are complex since they are composed of nucleons. At the present time it is conventional to think of electrons, protons, and neutrons as elementary particles except that, as we saw in Section 9.2, in the nucleus protons and neutrons seem to lose their identity in facile interconversion. The hundred or so particles produced in atom-smashers leave one with the impression that no elementary particle (in the classical sense, as one which is not further divisible) has been reached.

Fields 18.3

The construct 'field' is an invention pure and simple. It is a construct, a concept, designed to represent, or explain, how bodies at a distance can affect each other.

The problem of such interaction is an ancient one, and explanations have included occult 'influences' (invoked every day in our newspaper astrology columns); invisible threads which connect the bodies and draw them together, or make them knowable (as in ancient theories of sight); celestial spheres in mechanical contact which regulated the heavens in ancient cosmologies; lines of force; mathematical equations which describe relationships without being given any physical interpretation. These few examples show the two kinds of approach that may be taken to the explanation of interaction-at-a-distance. In the one, actual mechanical connections of some kind are imagined, or observed, between the interacting bodies. In the other the concept is that of a field. There are two categories of fields. In one class are fields that clearly involve some tangible medium between the interacting bodies, and in the other class fields which do not.

About the middle of the eighteenth century Leonhard Euler developed the theory of the hydrodynamic field, an example of the first type of field. Euler was concerned with the transmission of energy through a fluid: a flowing liquid propagates energy. At any point in the stream, the motion of matter at that point can be described in terms of mass, velocity, and time. The medium is considered to be continuous, at the bulk level with which the theory is concerned. The propagation of energy with or through such a medium, as by starting up wave-action, or by bringing about a flow, takes time and is a function of the physical properties of the medium. The medium occupies space between the source of the energy and the point where it is absorbed. For example, a pebble dropped in a pool of water starts a ripple which transmits energy to the leaf floating some distance away, making it bob up and down. Or, a piece of wood, held below the surface of water, on release bobs to the surface. Held below the surface, it is in a hydrostatic field directed upward. There is a force on it directed upward and countered by the force required to keep it submerged. On its release, its acceleration to the surface is objective evidence for the force and helps to define it.

The other type of field—the type we are chiefly interested in—does not display a medium between interacting bodies. Typical examples are a gravitational field and a magnetic field. With Euler's theory at hand it seemed reasonable to think of these fields in analogous terms, and to imagine "lines of force" (the concept is from Faraday) connecting parts of the field. But it was realized that these lines of force are conceptual models and not actual physical things, with the property of mass, for example. It was the concept of a field, requiring a medium, which led to the hypothesis of the Aether; to the idea of a medium which would support electromagnetic *waves*. If there is a wave, something must "wave," it was supposed.

18.4 Field properties

A field is a region of space in which at any given point some phenomenon can be demonstrated. In the case of an electrostatic field, we have a region of space in which a test *charge* experiences a force. The field has two properties, a force-giving property and an energy-giving property. We must observe that the *electrostatic* field is detected by a *charged* particle, our test charge, or potentially by any *charged* or chargeable object. The response of the test charge depends on the strength oι the field as well as on the quantity of charge on the test object. This is displayed in Coulomb's law, where (as we saw) if we designate the charge on an object that originates the field as q_1, and the charge on our test object as q_2, and the distance apart of the centers of charge s:

$$\mathbf{F} = Kq_1q_2/s^2 \tag{3-3}$$

The field-intensity will be designated $\boldsymbol{E}$. It is defined as the force acting on a point test charge at a certain point, s meters from the center of charge that produces the field, divided by the magnitude of the test charge:

$$\boldsymbol{E} = \boldsymbol{F}/q_2 \tag{3-4}$$

Of course, actual charged objects are often irregular in shape, and often of large size—bodies extended in space. In such cases the simple form of Coulomb's law does not hold. We therefore tend to simplify our problems by imagining the test object and its charge as very small (avoiding thereby the obligation to use special mathematics) and dealing, as we said, with small, spherical charged objects.

From the definition of field strength, or field intensity, E, we can write:

$$\boldsymbol{E} = \mathbf{F}/q_2 = Cq_1/s^2 \tag{3-5}$$

This equation applies to the intensity at any point $\boldsymbol{P}$ in the field of the charged object q_2. It allows calculation of the force-giving property of the field. In Figure 3-28 the field intensity at P is $\boldsymbol{E} = \mathbf{F}/q_2 = Kq_1/s^2$. If q_1 and q_2 are of like sign, the force per unit of test charge will be one of repulsion, and be directed along the line joining the centers of the spheres but away from them. It may be represented as a vector. If the two bodies are of opposite sign, the force will be one of attraction.

General field characteristics 18.5

It is important to note (for the insight it gives to scientific methods) two features of the kinds of fields we have been discussing. One is that the field is a device for grasping something unseen with conceptual hands. A book is held above your desk: the force-giving property of the gravitational field at that point is *mg*. You let go of the book, and it falls. Why? Why didn't it remain where it was? There is nothing that you can see pulling it, or pushing it, toward the desk-top. So you conceive of a field of force, with the force-giving property *mg* at this point, and an energy-giving property *mgs* with respect to your desk-top (Section 3.11).

The second feature of the kind of field we have been discussing is that there is a definite specificity to the field. To detect the field you need an entity with the same kind of field. For example, a pith-ball responds to a gravitational field by hanging vertically downward (Figure 3-3). Its gravitational response depends on its mass and is unaffected by its charge. But all entities which have any mass have gravitational fields of their own, indissolubly associated through Newton's and Coulomb's equations with their field strengths. Thus a gravitational field is responded to by a mass. Similarly, an electrostatic field is responded to by a charged particle, and a magnetic field by a magnet. One might thus say, in summary, that a field is a construct devised and used by scientists for dealing with things unseen; for measuring and manipulating things unseen in terms of things seen that have the appropriate property.

Summary 18.6

Fields are recognized through the behavior of particles that respond to the force- and energy-giving properties of the fields. Fields originate in particles.

Several attitudes compete with each other with respect to the Particle and Field question. One is that in order to have any kind of interaction which produces an effect one must have at least two interacting entities. That is to say, a completely isolated entity is meaningless: it could only be discovered (given meaning) through interaction with something other than itself. Another attitude might be characterized as considering that the field is the fundamental reality. All is field, so to speak, and particles are merely knots or dense places in the ubiquitous field. Still a third position is that neither particle nor field is fundamental, but both are linked through some as yet unknown, hidden factor. This might well be a set of uninterpreted equations, whose mathematical austerity fits them to represent something ultimate.

A fourth attitude rests on the discoveries made in Physics during this century: the duality of light. Look for particle behavior of an X-ray, and you find it in the Compton effect (Section 8.7); look for wave behavior and you find interference phenomena. Or look for particle behavior of an electron, and you find it in the Millikan experiments, yet on testing for wave behavior there are the results of Davisson and Germer (Section 10.6). One must therefore conceive of a property which is immanently particle and wave, and out of which the one manifestation or the other may be evoked with suitable instruments. One may think of a continuum of some kind in which singularities, dense spots, particle-like behaviors may appear, statistically distributed according to the laws of probability, when suitable instruments are used to search. Perhaps the fundamental particles, or the artifacts of nuclear degradation are of this kind. . . .

These speculations have to do with sub-atomic particles, and may not with impunity be extrapolated to the macroscopic level of phenomena. Discrete objects ("large particles") clearly exist, and obey Newton's laws with high predictability. Our daily round rests upon the firm rock of these laws—though it may be interrupted at any time by the consequences of some sub-atomic process.

Exercise

18.1 How could the field-particle interaction be formulated cybernetically? Give some examples.

Bibliography

The books listed below are by no means all that are available on this subject.

David Bohm, *Causality and Chance in Modern Physics.* Harper Torchbooks. Harper & Brothers, New York, 1961 (Van Nostrand, 1957). This is a philosophical book, well written and full of ideas. Some ideas run counter to certain modern schools of quantum theory.

Louis de Broglie, *New Perspectives in Physics*, Basic Books, New York, 1962. A collection of essays that may be read piecemeal. Delightfully written, by one of the pioneers of modern Physics.

Norwood Russell Hanson, *The Concept of the Positron*, Cambridge University Press, London, 1963. This philosophical work is fascinating, for it shows how scientists work. Parts are quite hard going, but the majority of the book is understandable at the present level of knowledge. Rewarding reading for the historian.

Mary B. Hesse, *Forces and Fields. The Concept of Action at a Distance in the History of Physics*. Philosophical Library, New York, 1961. Highly recommended. Parts are mathematically involved, but can be skipped. This book is really a kind of intellectual history which should be enjoyed by any student who has grasped this text so far.

Max Jammer, *Concepts of Force* (1957), *Concepts of Space* (1957), *Concepts of Mass* (1961), Harvard University Press, Cambridge, Massachusetts; *The Conceptual Development of Quantum Mechanics*, McGraw-Hill, New York, 1966. All these books bear on particles and fields. They are scholarly works, heavily documented and so excellent sources of references to the literature. They are somewhat hard going. The last one, read in conjunction with Hanson's book, is most interesting to an historian.

Henry Margenau, *The Nature of Physical Reality. A Philosophy of Modern Physics*, McGraw-Hill, New York, 1950. This highly recommended book, sprightly and authoritative, has been referred to before, as has Polanyi. Michael Polanyi, *Personal Knowledge*, Harper Torchbooks, Harper & Row, New York, 1964.

Chapter 19

19.1 Laws
19.2 Heat and energy
19.3 The Second Law of Thermodynamics, continued
19.4 Entropy
19.5 Some consequences
19.6 Summary

Directional Arrows and Universal Laws ~ 19

Laws 19.1

The laws of science are convenient and concise statements of invariant-relations. One might put these into two classes: those that do not specify a particular direction in time, and those that do: in other words, those that rely upon a certain kind of symmetry in nature, and those that call for dissymmetry. Newton's laws belong to the former class. They say nothing about the direction of time. If one had a sequence of pictures of an object either at rest or in uniform motion in a straight line, a sequence labeled 1, 2, 3 . . . n, for example (Figure 19-1), there would be no clue to whether the sequence began at 1 or at n: whether it was taken as 1, 2, 3 . . . l, m, n, or as n, m, l . . . 1. Again, Newton's second law says that the net unbalanced force acting on an object is directly proportional to, and in the same direction as, the acceleration of the object. A sequence of photographs say, 1, 2, 3 . . . n, of an object moving under the accelerating force of gravity, gives no clue about the direction of the sequence: the object might as well have been rising or falling. Newton's third law, too, contains no clue to a time direction.

This kind of symmetry is basic to classical mechanics. Similarly, the first law of thermodynamics, a law which states the conservation of energy, says nothing about any particular direction of energy flow: if a hot body and a cold body were put in thermal contact in a perfectly insulated container, so that no energy could escape, the law says that the total energy of the system would remain constant. There is nothing in the first law that says that energy may not flow out of the colder into

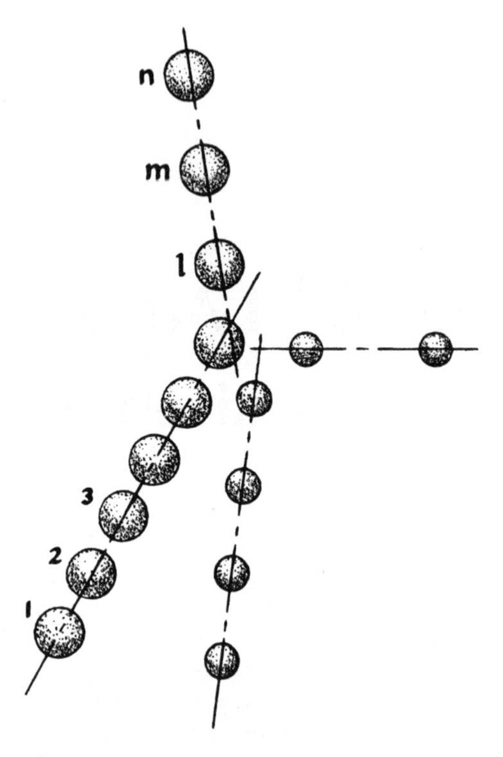

Fig. 19-1 *Multiple flash picture of the collision of a large, massive ball and a small less massive ball, traced from the photographs.*

the hotter body, so that the former would become colder and the latter hotter. However, the second law of thermodynamics does specify direction, as we saw in Section 16.5. There we developed a statistical basis for this law.

19.2 Heat and energy

Thermodynamics is the science of heat energy. Heat energy is a kind of energy that is measurable with a thermometer. When a system goes from one energy state $\boldsymbol{E}_1$ to another $\boldsymbol{E}_2$, measurable by means of a thermometer, the system has undergone a change in heat energy, $\Delta \boldsymbol{E} = \boldsymbol{E}_2 - \boldsymbol{E}_1$. It has absorbed a *quantity* of heat Q. This quantity of heat is equal to the energy change (it is stated in the same units) provided that nothing else occurred to add or remove energy. But we do not write ΔQ for this quantity of heat since we do not imply that the 'heat' has 'changed.'

The usual method of measuring heat is to observe the change in temperature that occurs when heat is absorbed or given off by something the capacity of which to absorb heat is known.

In common with most definitions of properties in the natural sciences the property is given numerical value in relation to the behavior of a chemically pure substance. In this case, the substance chosen is water, which can be purified by careful distillation from glass, or platinum apparatus. The gram-calorie (cal, or g-cal) is the quantity of heat energy that is absorbed by one gram of water when its temperature goes up one degree Centigrade, from 15° to 16°C. This definition is readily applied (Exercises 19-5, 6). Given a certain substance, the quantity of heat energy Q that must be absorbed, or withdrawn, to change the temperature Δt° is directly proportional both to the mass of the object and to the change in the temperature:

$$Q \propto m(t_2^\circ - t_1^\circ), \text{ or } Q \propto m\Delta t^\circ$$

The constant of proportionality $\boldsymbol{c}$ is the *specific heat capacity* of the substance:

$$Q = \boldsymbol{c} m \Delta t^\circ \tag{19-1}$$

Calories are stated per gram per degree Centigrade. 'Kcal,' or kilocalories, are per kilogram per degree Centigrade. Actually $\boldsymbol{c}$ for water is not a constant. At −10° water is a solid with a specific heat capacity of 0.48. The liquid at 0°C has a specific heat capacity of 1.00874, based on 1.00000 at 15°C. At 50° the specific heat capacity is 0.99829, and at 100°C it is 1.00645. At 100°C and constant pressure the specific heat capacity of the gas is 0.4836. Obviously, $\boldsymbol{c}$ is not a constant for a substance; and clearly it is influenced by the state of the substance, solid, liquid, or gas. This is why the temperature, 15°C, is specified, and why the calorie is defined for pure liquid water.

In general, energy may be classified into heat energy and work. Heat energy is associated with kinetic motion of atoms and molecules. Work is usually associated with change in energy of a system operating against some external force. Work may be further classified into electrical work, gravitation work, chemical work, and pressure-volume work. In general, if a system increases in energy by $\Delta \boldsymbol{E}$ joules, this increase is equal to the quantity Q of energy *absorbed* by the system as heat, less any work W *done by* the system upon its surroundings. This is stated as:

$$\Delta \boldsymbol{E} = Q - W \tag{19-2}$$

This is another way of putting the first law. $\Delta \boldsymbol{E}$ is an intrinsic property of the system and depends only on the value of $\boldsymbol{E}$ before and after the process of energy change has occurred. We shall see, however, that the values of Q and W are not properties of this kind; they are sensitive to the path along which the system moves.

We might recall at this point that energy may be defined as that which is transferred between systems that interact. The systems that are meant, here, are those *objects*, with their associated properties, that are singled out for study. Everything else is surroundings. Of course, we recognize that the experimenter must couple himself to the system in some way if he is to observe it, and the energy transferred in this interaction must be taken into account. That two systems interact is certified by changes in their properties.

19.3 The Second Law of Thermodynamics, continued

With this preparatory excursion into some aspects of energy, we return to the second law. We would like to look at the experimental basis for the principle that heat may not be completely converted into work in a complete cyclic process. This principle was first established logically by Sadi Carnot. Another way of putting it is that "you can't get something, like work, for nothing," *i.e.*, going through a complete cycle and back to where you started. It says that perpetual motion machines are not possible. We proceed, following Carnot, by means of a thought-experiment.

Suppose (Figure 19-2) we have a thermostatted block, set at a temperature T_1, or a block large enough that a finite amount of heat may be withdrawn from it without significant change in its temperature. Likewise we have a reservoir at a lower temperature, T_2, to which heat

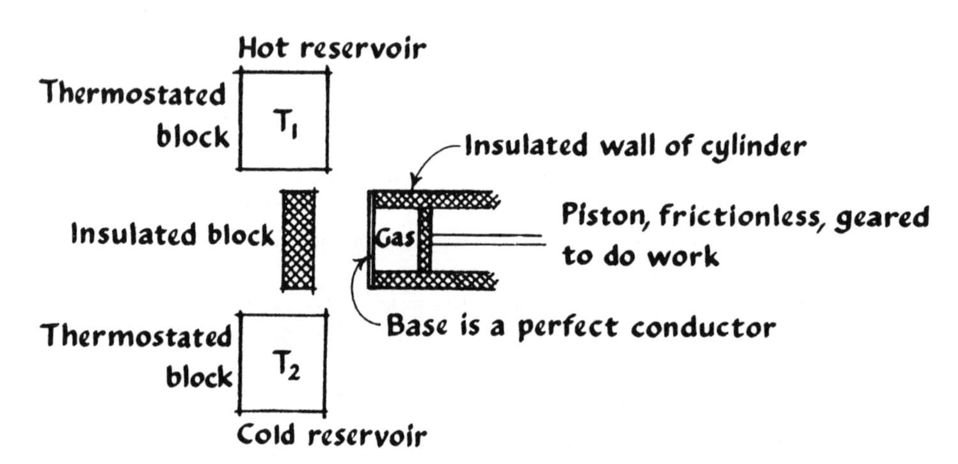

Fig. 19-2 *The machinery for carrying out a Carnot cycle process. Everything operates in an ideal manner.* [*N. L. Sadi Carnot,* Reflections on the motive power of heat. *Ed. R. H. Thurston. John Wiley, New York, 1897. See essay by Sir William Thomson, pp. 127 ff.*]

in finite amount may be added without increasing its temperature significantly. We have a sample of an ideal gas enclosed in a cylinder the walls and piston of which have negligible heat capacity, and are perfectly insulated, so that heat may escape or enter only through the base, which conducts perfectly. We also have an insulated block which we may use to cover the base of the cylinder so that the gas in the cylinder under these conditions may be thermally isolated. We shall carry this sample of gas through a cycle, letting it do work while its temperature and pressure decrease and its volume increases, and bringing it back to its original starting state, by coupling ourselves to the system (conceptually) and moving the parts of the apparatus as required. We imagine that the piston is attached to, say, a flywheel, so that it may do work on the wheel. Thus energy may be stored in the flywheel. We imagine that we have means for making various adjustments to the apparatus, such as arranging for the piston always to work against a force just smaller, but so little smaller that a slight increase would reverse the direction of motion. This kind of arrangement conspires to make the work done 'reversible' and maximum in magnitude. All of these are idealizations, and we conceive that though coupled in this way to the system *we* do not transmit to or receive from it any energy, but that energy coming from the surroundings or given up to them is reversibly transmitted.

To begin, we place the base of the cylinder into contact with the thermostat, or reservoir, at T_1. This holds the temperature at T_1. We now allow the gas to expand: it does a certain amount of work against the piston, pushing it out. At the end of this step the volume has increased from V_1 to V_2 (see Figure 19-3); the pressure has fallen, but the temperature remains the same. What has happened is that the energy to do the work was absorbed from the reservoir held at T_1; thus the work (energy) done is equal to the heat (energy) absorbed from the reservoir. Since T remained constant, the *heat* energy of the gas has the same value at the end as at the beginning of this step. For an ideal gas this work can be shown to be $nRT \ln V_2/V_1$ joules, where n is the number of moles of gas in the cylinder, R is the 'gas constant,' and ln means the natural logarithm, to the base e. (Ignore that we have not derived this formula. It is given in texts referred to at the end of the chapter.)

Now the thermostat is removed and the base of the cylinder is insulated with the block of insulation so that no *heat* energy can enter or leave the system. The gas is then allowed further to expand reversibly and adiabatically to a larger volume V_3, a lower pressure, and a lower temperature T_2. The word 'adiabatic' comes from the Greek and means 'not able to go through.' It means that the system is insulated so that there is no transfer of heat energy from or to the system. Thus the work done is at the expense of the energy stored in the system. In each of

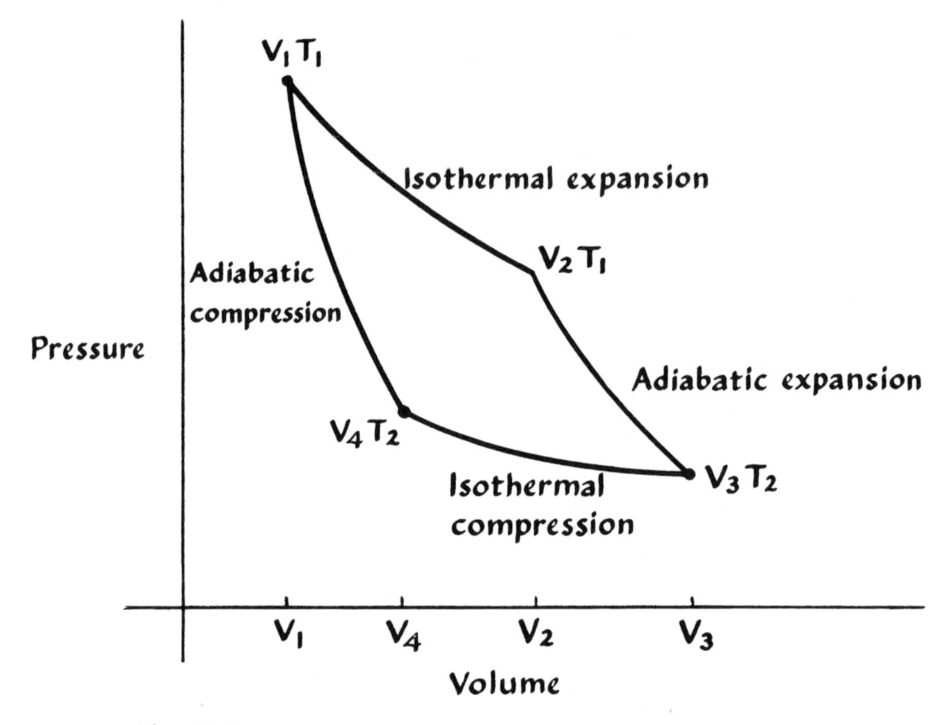

Fig. 19-3 *Pressure-volume diagram of a Carnot cycle.*

these reversible expansions, the physical situation is such that the piston works against a force that is larger at first, when the pressure is larger, and decreases with decrease in pressure, always being, as we said, just a little smaller than the force on the piston. Pressure, P, $= \boldsymbol{F}/A$, where A is the area over which the force $\boldsymbol{F}$ is applied. Since work $W = \boldsymbol{F}\Delta s$, and the volume change is $\Delta V = A\Delta s$, it follows that $W = P\Delta V$. It is possible to calculate the amount of this work (both P and V change); however, we can evaluate it by noticing that since the gas has been allowed to absorb no heat, the work done (according to the first law) must be exactly equal to the change in energy, which by an earlier relation (Equation 19-1, rewritten) must be $Q = n\boldsymbol{c}_V(T_2 - T_1)$, where $\boldsymbol{c}$, the molar heat capacity at constant volume, is assumed constant. Here $n\boldsymbol{c}_V$ is equivalent to $\boldsymbol{c}m$ in Equation 19-1. The gas has done a certain amount of work, along V_2T_1 to V_3T_2 at the expense of its internal energy. The gas molecules, initially moving with higher average speeds at T_1, and striking the piston more frequently, are now moving at lower average speeds at T_2, and are striking the piston less frequently.

We wish to complete the cycle by restoring the system (which is really an idealized gas or steam engine) to its original state, V_1T_1, so that it may continue to operate cyclically. We therefore remove the insulation at the base of the cylinder and bring up the cold reservoir set at T_2. We compress the gas reversibly and isothermally to V_4. As we begin to compress the gas (for the reversible process is carried out slowly) the pressure goes up, and so would the temperature since we are doing work *on* the gas. However, the reservoir at T_2 accepts this increased heat, and the gas temperature remains at T_2. In this way, the compression is carried out isothermally and reversibly to a point V_4T_2. Here the work of compression done on the system comes, one might imagine, from some of the work done in the first two steps. In a real machine it might have been stored in the flywheel. This work is exactly equal to the energy discarded to the reservoir at T_2, and has the value $nRT_2 \ln V_4/V_3$. (We give the results of calculation based on the definition of work as $\Delta P \Delta V$, in the circumstances that both P and V are changing.)

As a last step the gas is compressed adiabatically and reversibly after removing the thermostat at T_2 and replacing the insulation. In this last step the gas has clearly not absorbed any heat energy from the surroundings, but work has been done on it to decrease its volume and to raise its pressure and temperature back to the initial state, V_1T_1. This work is equal to the change in energy, which must be $nc_v(T_1 - T_2)$.

We now sum over the whole cycle: the net work done in the cycle is given by the area inside the quadrilateral in Figure 19-3. (This is equal to the area under the curves from V_1T_1 to V_2T_1 to V_3T_2 less the area under the curves from V_3T_2 to V_4T_2 to V_1T_1.) The work done in step V_2T_1 to V_3T_2, namely $nc_v(T_2 - T_1)$, is clearly equal to and opposite in sign from that done in the step from V_4T_2 to V_1T_1. The two are therefore equal, and cancel when the second is added to the first. The sum of the energies for the other two steps, then, is:

$$W = nR(T_1 \ln V_2/V_1 + T_2 \ln V_4/V_3)$$

Note that $T_1 \ln V_2/V_1$ is a positive number, for V_2 is larger than V_1, and that the other term is negative. By use of the gas laws and algebraic manipulation this equation reduces to

$$W = nR(T_1 - T_2) \ln V_2/V_1$$

from which, by dividing by the heat absorbed at the higher temperature, $Q = nRT_1 \ln V_2/V_1$, one obtains the very fundamental relation

$$W/Q = (T_1 - T_2)/T_1 \text{ (reversible process)} \qquad (19\text{-}3)$$

This relation says that the maximum work (W) that could possibly be obtained from an idealized heat engine working reversibly between T_1 and T_2 is related to the amount of heat absorbed, Q, in the way shown. For example, an engine operating between $T_1 = 373°K$ and $T_2 = 273°K$ can convert a maximum of $(373\text{-}273)/373 \cong 0.27$, that is, 27%

of the heat absorbed into work. This sets a limitation upon all heat engines. It explains why engineers expend a great deal of effort in developing high-temperature boilers. For the only practical way to improve the theoretical operation, and thus the efficiency of an actual heat engine (an efficiency which is always less than ideal) is to increase the numerator by obtaining a larger difference between T_1 and T_2. Of course, the denominator also increases when T_1 is increased. One might thus suppose that a decrease in T_2 presents an alternative, and this it does, up to a point. But usually, economic considerations require that water, the cheapest efficient cooling agent, be used. This sets a lower limit to T_2. The net result of the cycle is that some heat energy always must be 'discarded' into the cold reservoir even in an ideal heat engine. In actual engines which must work with imperfect gases (steam or combustion products of gasoline, for example) and which must contend with frictional forces of many kinds, the efficiency is less than the theoretical—often very much less.

What happens when the process is not carried out reversibly? We noted that the values of Q and W in a process are dependent on the path along which the process occurs. To see this we might look at the kind of process that led Count Rumford to suggest a relation between heat and mechanical energy, and that was utilized by Joule to demonstrate this relation quantitatively. Rumford had noticed that in the boring of cannon, heat was generated. The process was not reversible; the mechanical energy expended in turning the boring tool was dissipated as heat. We might now look at certain analogies between Q and W.

We saw that work of a certain type, namely the force-per-unit-area-over-a-distance type, or PV work, could be calculated either from $P\Delta V$ or from the energy expended if that energy were known, and all in the form of work. P and V are properties of the system, and they differ in an interesting way. The magnitude of V depends upon the geometrical extent of the system. Thus ΔV depends on the area of the piston and the distance it has moved in the machine shown in Figure 19-2. Such a property is said to be *extensive.* Its magnitude is proportional to geometrical extent, or the number of identical entities present, or to the amount of matter. Mass, volume, charge, heat capacity are examples of extensive properties, for their values in a given system are the sums of the properties of the parts.

Pressure belongs in a different category. It is sometimes said that properties of this kind, which are named *intensive* properties, reflect qualities, or potentials, while extensive properties reflect quantities, but this distinction requires careful handling. For example, pressure on a piston is really the summation over a relatively large area and over a relatively long time of the extraordinarily numerous impacts of extremely small particles. The large area and long time, large and long relative to the atomic areas and periods of impact, smooth out the fluctuations in configurations that are characteristic of the atomic-level scene. The intensive property is independent of the extent of the system because of the behavior of statistical aggregates. Examples of intensive properties are pressure, temperature, electrical potential, gravitational potential.

Entropy 19.4

Returning to Q and W, work is quite generally classified in terms of the intensive and the extensive property used to define it. For example, $P\Delta V$ for pressure-volume work; $gh\Delta s$ for gravitational work, where Δs is the distance that a mass is moved against the gravitational potential gh. When we consider Q, we find that T plays the role of P in $P\Delta V$ work, and that a quantity designated S, introduced by Rudolf Clausius (1822–1888) and named by him 'entropy, plays a role analogous to that of V, so that Q in a reversible process may be defined as equal to $T\Delta S$. S is a characteristic of a system. Its value depends only on the states of the system before and after a process and not on the path along which the process occurred. It is an extensive property and its value depends on the amount of substance in the system. We introduce the concept of entropy because it carries a heavy philosophical burden. It is quite possible that, as with relativity, more has been written on the meaning and implications of entropy by philosophers than has been written about actual measurements of the quantity in physical systems. We have introduced the concept in this way to emphasize that we are concerned with a measurable quantity. The dimensions of entropy are those also of heat capacity, namely calories per degree.

Lewis and Randall have pointed out that it is the existence of irreversible properties that necessitates the entropy concept. In a reversible process such as the Carnot cycle in idealized form, the total entropy change is zero for the system. Returning to the example plotted in Figure 19-3, in the first step, the reservoir at T_1 gave up energy to the extent of $nRT_1 \ln V_2/V_1 = Q_1$. Thus Q_1/T_1 has the value $nR \ln V_2/V_1$. But the reservoir at T_1 lost heat in the same amount. Thus the *net* entropy change was zero, though the gas increased in entropy. In the second step (and in the fourth also) no heat was absorbed, and the gas underwent no entropy change. In step number 3, step number one was retraced in terms of the energy, $Q_2 = nRT_2 \ln V_4/V_3$; then Q_2/T_2 has the value $nR \ln V_4/V_3$, which is negative and has the same value as Q_1/T_1. Thus the sum of these steps is zero total entropy change for the gas.

To go to an extreme case which shows how dependent the values of Q and W are upon the path, consider two flasks of equal volume, one evacuated and the other containing a perfect gas, connected together through a stopcock (Figure 19-4). They are in an insulated box so that

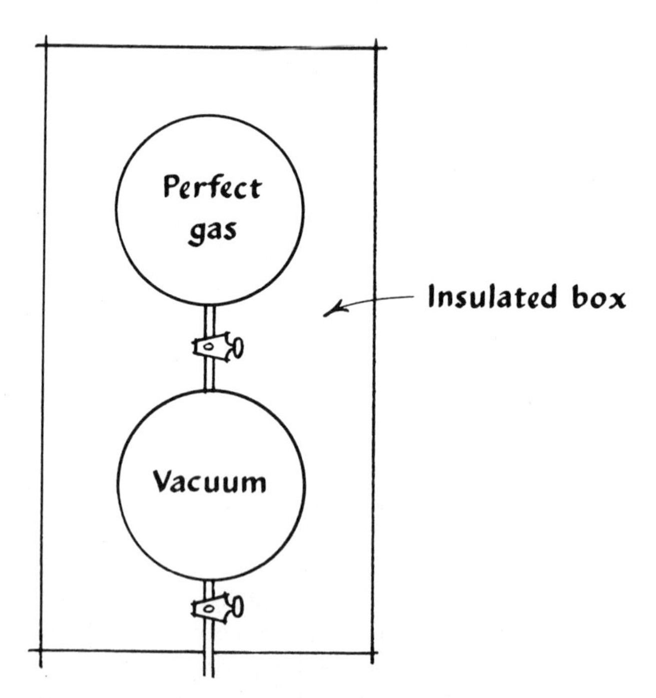

Fig. 19-4 *Two flasks of equal volume are connected by a stopcock. The upper contains a gas. The lower has an additional stopcock through which mercury may be admitted to compress the gas. In the figure both stopcocks are shown closed.*

the two and the contained gas are isolated from all sinks or sources of energy. The stopcock is opened, and the gas distributes uniformly between them: we assume that opening the stopcock was done essentially without contributing energy to the system. If the gas is perfect, it does no work (its molecules do not have to be pulled apart), and the final temperature T remains the same after the expansion as it was before. To find out how much the entropy of the system has changed, the procedure is to restore the original state of the system by a *reversible* process which causes no further dissipation and thus must measure the change in entropy of the irreversible step. Here this might be done by bringing the system into thermal contact with a reservoir at the same temperature T as the gas, and compressing the gas back into the original flask in an isothermal, reversible manner. (Such as by placing the flasks in a vertical position, and then forcing mercury in through the second stopcock.) The work done on the gas *by* the compression, is $W = nRT \ln (V \text{ total}/V \text{ original}) = Q$, which gives us the energy increase of the reservoir. The entropy is Q/T, and this is the value of the entropy change of the gas in this process in which it did no work.

In summary, if a system undergoes a process irreversibly its total entropy increases. If two or more systems interact in an irreversible manner their total entropy is increased. If one or more systems undergo a process reversibly, the total entropy increase is zero. The increase in the entropy of any part of the system, or of one of the linked systems, is equal to the heat which it absorbs divided by its absolute temperature. Thus the construct 'entropy' is necessary because there exist irreversible processes.

Some consequences 19.5

The consequences of Carnot's discovery were stated in grand global form by Clausius. "The energy of the world is constant; the entropy tends to a maximum." Lord Kelvin phrased it as the degradation of energy. This means that while in a closed system reversible reactions do not change the entropy, and thus do not change the 'grade' of energy, irreversible reactions do bring about a loss in entropy and this is associated with a loss in the ability to do work. Thus entropy either remains constant or increases. The Carnot engine irretrievably discards heat energy to the lower-temperature reservoir. Another way of putting the matter is that 'the universe is running down,' which is some kind of an ultimate in extrapolation.

Since in a process such as that described in connection with Figure 19-4, and in general in irreversible processes, the system goes from a less probable to a more probable state in terms of configurations it is sometimes said that overall the universe is going from a less probable to a more probable state. It seems reasonable, however, to look with a jaundiced eye upon statements which purport to tell us about 'the universe.' We have enough difficulty being certain about local situations.

Returning to the discussion of atoms and quanta, entropy has been defined for a system at this level as a function of the number of microstates belonging to some particular configuration in which the system is. The definition is:

$S' = k \ln$ (number of microstates belonging to the particular configuration). (k Is the 'Boltzmann constant.') We already saw that a system tends to go to the configuration with the largest number of microstates, thus spontaneously this S' tends to increase. This is a 'directional arrow' that was discussed from another point of view in Sections 16.6 and 16.7.

There is another 'arrow' that is of ultimate significance to us. This is a directional arrow implied by the evolutionary history of living organisms. Interestingly enough, men had held "a static vision of Nature and Society" until about the first quarter of the eighteenth century and, indeed, the conceptual approaches we now call evolutionary and historical did not begin to gather force until the very beginning of the nineteenth century. One of the powerful driving forces arose from the work of pioneer geologists who were able to interpret fossil remains in time sequence, and to show the extraordinarily long period that has been available for evolutionary development. These geological interpretations were validated and made more precise by studies in radioactivity dating (Section 13.11). Many correlative intellectual and experimental developments, like so many wavelets traveling with different wavelengths, peaked up toward the middle of the nineteenth century in the great theory of organic evolution of Charles Robert Darwin. This theory rested on three propositions, say Toulmin and Goodfield (*loc. cit.*): the doctrine of 'progression,' namely that more complex forms of life appeared on the Earth after the simpler ones; the doctrine of 'transformation,' that the later forms descended from the earlier; and the doctrine that natural selection among variations in the earlier species led to the later ones.

The arrow of time here points in the same direction as that of Carnot and Clausius, but with a different significance. For it certainly seems that the course of evolution must be characterized as moving along the path towards less probable configurations, namely living organisms which develop increased complexity with attendant organization, and with the appearance of intelligence.

We must confirm that while such a course of evolution, as well as the itemizable steps by which it has advanced, implies the progressive decrease in entropy, it has been the deliverance of countless experiments that the second law of thermodynamics is not violated. The local decrease in entropy belonging to the living organism is paid for by the habitat, and more than paid for. The entropy of the larger system increases. Nevertheless *that* there is this directional arrow would seem to many persons to be a matter of the greatest significance.

19.6 Summary

In overall view, we must marvel at the remarkable invariants that have been conceptually dissected out of the flux of experience by physical scientists. The second law of thermodynamics was formulated in the last century and the consequences of it are still being explored. In mundane processes when work is done there is invariably an overall loss. Not all the atoms and molecules involved in the process can be harnessed to move in unison at the technological command. We see this phenomenon throughout our experience.

Exercises

19.1 Give some examples of a) electrical work; b) gravitational work; c) chemical work.

19.2 What is the energy situation when the experimenter couples himself conceptually to a thought experiment?

19.3 In the equation $Q = c\Delta T$, can degrees Centigrade be substituted for degrees Kelvin? That is, is it correct to write $Q = c\Delta T = c\Delta t$? Explain.

19.4 Given the following specific heat capacities (c_v in joules/gram degree Kelvin)

Argon	0.312	Hydrogen chloride	0.572
Carbon monoxide	0.743	Mercury gas	0.0626
Helium	3.13	Nitrogen	0.735
Hydrogen	10.1	Oxygen	0.654

a) Re-code the data in more meaningful terms. Try, for example, the effect of calculating to the molar basis. b) Is the alphabetical listing rational? c) Classify the data further, and see if any invariant(s) is(are) present.

19.5 Assuming a constant specific heat capacity of water over the temperature range involved, calculate the final temperature t of the mixture when 1 kg of water at 12.5°C is mixed with 2.5 kg of water at 30°C.

19.6 The specific heat capacity of milk is 0.900; of butter 0.302 cal/g; of coffee, close to that of water. a) By how much is your coffee cooled when you put cream in it? b) The specific heat capacity of sugar is 0.299 cal/g. By how much does sugar further cool it?

19.7 Draw a diagram to show the successive photographs of a ball or other object moving under the force of gravity, and show why it is impossible to tell which end of the sequence came first in time.

19.8 a) Calculate the maximum efficiency of a steam engine, with superheated steam at 327°C and the condenser at 27°C. b) What factors would decrease this maximum for an actual engine?

19.9 The equation $W/Q = (T_1 - T_2)/T_1$ may be used for calculating the work that must be done to cool a system. In a refrigerator the process is operated in the reverse direction: the gas is allowed to expand at the lower temperature (in coils in the refrigerator).

It absorbs heat, and is then compressed outside of the refrigerator by a motor that does work on it. The heat so produced is released to the atmosphere. In effect, the engine has work done upon it, and it absorbs heat at a lower temperature, giving it off at a higher temperature. Calculate the work needed to freeze 1 kg of water (heat of fusion 80 cal/g) at 0° if T_2 is 25°C. (Note that $Q = 80{,}000$ cal.) In an actual refrigerator would the work be greater or less than this? Why?

19.10 Referring to Figure 19-4, when the stopcock connecting the two flasks is opened, the swarm of gas molecules in the first flask distributes spontaneously between both flasks. Write a brief essay on the difference between this behavior and that of a swarm of bees, migrating, or a flock of birds wheeling in the sky as they gather to migrate. What of the entropy?

19.11 Give a free translation of this passage from the work of Sadi Carnot (*loc. cit.*, p. 10).

> C'est à la chaleur que doivent être attribués les grands mouvements qui frappent nos regards sur la terre; c'est à elle que sont dues les agitations de l'atmosphère, l'ascension des nuages, la chute des pluies et des autres météores, les courants d'eau qui sillonnent la surface du globe et dont l'homme est parvenue à employer pour son usage une faible partie; enfin les tremblements de terre, les éruptions volcaniques reconnaissent aussi pour cause la chaleur.

19.12 Give a free translation of this passage from the work of Sadi Carnot. (*loc. cit.*, p. 11)

> On peut donc poser en thèse générale que la puissance motrice est en quantité, invariable dans la Nature; qu'elle n'est jamais, à proprement parler, ni produite, ni détruite. A la vérité, elle change de forme, c'est a dire qu'elle produit tantôt un genre de mouvement, tantôt un autre; mais elle n'est jamais anéantie.

Answers to Exercises

19.1 For example: a) a current may turn a motor; b) a waterfall may run a mill; c) a chemical reaction may produce electric current, as in a working battery, or may generate heat that may be utilized for doing work.

19.3 Yes. The degrees have the same magnitude, and a *difference* is in question, not an absolute magnitude.

19.4 a) A, 12.5; CO, 20.8; He, 12.5; H_2, 20.4; HCl, 20.9; Hg, 12.6; N_2, 20.7; O_2, 20.9. b) Obviously not. c) The data divide the gases into 2 classes: those with $c_V \sim 12.5$, and those with $c_V \sim 20.5$. The former are monatomic, the latter diatomic.

19.5 1 kg of water changes from 12° to t. The quantity of heat that it absorbs is equal to that lost when 2.5 kg of water change from 30°C to t. Thus:

$$1 \cdot \mathbf{c}\ (t - 12.5°) = 2.5\ \mathbf{c}\ (30° - t). \text{ Therefore, } t = 25°.$$

19.6 Rounding off is reasonable in an estimate like this. Say that the coffee starts at 85°C and goes to t; the cream starts at 15°C, *i.e.*, slightly chilled, and goes to t, with c = 0.8; the sugar starts at 25°C, room temperature. One cup of coffee is about 125 ml = 125 g.; one half-ounce of cream is about 15 ml = about 12.5 g.; one lump of sugar, about 5 g.

19.8 $W/Q = (T_1 - T_2)/T_1 = 600 - 300/600 = 0.5.$

19.9 $W = 80{,}000(27)/273 = 7900$ cal.

19.11 It is to heat that the great motions that catch our eye on the earth should be attributed; to it that atmospheric disturbances are due, the rising of clouds, the fall of showers, and of the other meteorological phenomena, the currents of water that furrow the surface of the globe and of which man has come to bend to his uses only a small part; finally, earthquakes and volcanic eruptions acknowledge as their cause, heat.

19.12 One can then make the general proposition that energy is invariable in quantity in Nature; that it is never, correctly speaking, either produced or destroyed. It is true that it changes its form—that is to say, that it produces now one species of movement, now another—but it is never annihilated.

Bibliography

David Bohm, *Causality and Chance in Modern Physics*. Harper Torchbooks. Harper & Brothers, New York, 1961 (Van Nostrand, 1957). Also, the appendix to *The Special Theory of Relativity*, Benjamin, New York, 1965.

Harold G. Cassidy, *The Sciences and The Arts. A New Alliance*. Harper & Brothers, New York, 1962.

Chalmers W. Sherwin, *Basic Concepts of Physics*. Holt, Rinehart & Winston, New York, 1961.

Gilbert Newton Lewis and Merle Randall, *Thermodynamics and The Free Energy of Chemical Substances*. McGraw-Hill, New York, 1923.

Leon Brillouin, *Science and Information Theory*. Edition 2. Academic Press, New York, 1962.

Nicholas Leonard Sadi Carnot, *Reflections on the Motive Power of Heat*. Ed. by R. H. Thurston, Wiley, New York, 1897. See the essay by Sir William Thomson, pp. 127 ff.

John R. Pierce, *Electrons, Waves and Messages*. Hanover House, Garden City, New York, 1956.

Stephen Toulmin and June Goodfield, *The Discovery of Time*. Harper & Row, New York, 1965, *cf*. p. 125.

Harold F. Blum. *Time's Arrow and Evolution*. Harper Torchbooks. Harper & Brothers, New York, 1962 (Princeton University Press, 1951).

Eugene P. Wigner, "Violations of Symmetry in Physics," *Scientific American, 213* (No. 6, December) p. 28 (1965).

P.S.S.C. Committee, *Physics*. Heath, Boston, 1960.

C. P. Snow, *The Two Cultures and the Scientific Revolution*. (The Rede Lecture, 1959) Cambridge University Press, 1959.

Chapter 20

Toward a Natural Philosophy ~ 20

In our first chapter we stated the objectives of this text: to help non-specialists become literate in science; to show the meaning that modern physical science has for our culture; to further that goal of liberal education which is to enable a person to develop for himself a personal philosophy. This philosophy, an approach to a 'life view,' should be constructed, we said, on that which knowledge of the past, experience, and his own enlightened observation tell him of the fit between facts, theories, hopes, and ideals and the way things are or might be. The contribution that physical science might make to this is to show in substantive ways how scientists have gone about solving problems, and how the knowledge so gained has influenced our thinking through philosophy and technology, and to show how a person may use his mind.

The student who has conscientiously worked his way through this text must have gained some knowledge of modern physical science. What is more important, if he came to this course uncertain of his powers he should by now have developed a degree of confidence. He must surely have become "less timid in the face of life's possibilities." He could be a scientist if he wanted to, and had the temperament to be, because the qualities of mind and attitude that are required for competence in this area are not different from those required in any other intellectual enterprise. If the student has developed increased confidence in his own intellectual powers then our objectives have been met in large part. It remains only to suggest (for the ultimate goal is unreachable and the paths toward it as many as there are students and teachers) what we might hope for in assisting the student to a life view.

We live on a 'Space ship Earth,' in Kenneth Boulding's phrase. We know a great deal about this Earth and are quite certain that once upon a time the Earth was barren of life of any kind. But what is once upon a time when there is no sentient human being to be aware of time? We do not know the most fundamental things about our origins, though we have many speculations. We know a good deal about our recent history—some six thousand years of it. During the course of time there have been discovered certain invariant-relations, chiefly derived from human behavior. These are taught in our greatest literature and are actually what it is about, what makes it great. Among them are the "essential counsels and truths" that are "valid at all times and places" which Toynbee found in his study of the great world religions. The Golden Rule is an example of one of these invariant-relations. Among them, too, but of more recent discovery are the great counsels and truths, valid at all times and places, that have been formulated as scientific laws: Newton's laws of motion that describe in minute detail the behaviors of mundane objects and of the Solar System's orbits; the laws of the conservation of mass-energy, of momentum, of charge; the invariants implied by these laws and the great relativistic invariant-relation known as the Lorentz transformation. All these World Laws that describe the behavior of Nature and Human Nature are part of the heritage that the literate person may rely upon in constructing his own life view. They are laws that cannot be legislated out of existence, though many of them can be ignored, especially but not exclusively in the realm of human nature. (For example, few people would sponsor laws against the teaching of physical science, but laws against teaching Darwinian evolution were once common. In our own time we have the example of Lysenko.) But this does not change the consequences of natural laws, it defers payment to others and other generations. The person who would be a responsible steward of this Space ship Earth cannot ignore these laws in his life view.

Today's "intelligentsia" are on the whole illiterate with respect to science. This statement applies to a generation of elders, and a younger generation who will become the elders of the next twenty or thirty years. It is a condition not easy to correct, and at the same time one of considerable danger to our Nation. Many vocal members of this class of persons know that they are ignorant of science, and are defensive, unsympathetic, and with raised hackles; others withdraw. Still others—the saving remnant—attempt to correct their college deficiencies. Unfortunately their voices are not loud at present. Our hope for a more rational and balanced culture rests largely with the present generation of students, and that is what has motivated the writing of this book.

Broadly speaking, this is what I have tried to show: scientific activity arises out of the specifically human will-to-meaning implemented by the pattern-forming ability of our brains. The things we perceive in whatever way are named and compared; they are put into classes. Relationships are sought between the things (this helps to classify them) and the classes of things. These relationships are themselves related wherever possible. Thus we build a marvelous network of relationships which supports what I have called our life view. These mental operations are possible because the original perceptions—our empirical contacts with the World—are transformed into symbolic form, and because the symbols can be manipulated by logical rules including the spare language of mathematics. This interconnected field of constructs, connected at innumerable places to the empirical world, even though it is probably in error here and there, and though it is and will forever be incomplete, confers meaning on its parts, and on the scientists who are weaving it. It is a powerful bearer of meaning to the humanist who unites his own endeavors to this grand intellectual enterprise. For he, too, can weave into the pattern in his own ways his own meaning.

I have tried to show substantively as well as philosophically how far-reaching is the impact of science on each life and on any society. I have tried to indicate how a scientist thinks, or may think, in a given situation, and to do this I have necessarily included actual examples of an appropriate kind. What I would like to have conveyed is the conviction that this kind of thinking, in its systematic application, which has a lot to do with common sense—but applied as a policy—should be fruitful in dealing with other than solely conventional scientific problems.

The scientist does not use scientific methods and thinking in all areas of his life. This is because many aspects of life, to be specific those treated in the Humanities, are not amenable to this approach. Also, our cultural milieu makes it virtually impossible to be consistently scientific: when he leaves his laboratory for the street he has re-entry problems. Nevertheless, he has to keep trying.

One point we have tried in various ways to make is that science rests upon—or at least implies—a body of beliefs: for example, that we *can* learn something significant about the World outside ourselves, and about how others see it; that there are orderly relations to be found in the welter of appearances.

So we start with an hypothesis: man lives in three dimensions, the physical, the mental, and the spiritual. Let us review what the book has taught about these three dimensions of life view.

The physical world is ruled by physical, chemical, and biological laws. It is the dimension of matter in motion; of matter in processes of transformation; of material form and function. At a sub-atomic level, our theories and experiments tell us, all matter comprises relatively massive nuclei and shells of electrons. We know that the nuclei have internal structure. The mass-energy that they are is subject to inherent constraints that imply order, though as yet we have only dim intimations of what this order may be. And at this level we feel certain—perhaps with a tinge of apprehension—that surprises await us. Minute size, rapid motions, powerful interactions at this nuclear level are properties that cannot be grasped by our macroscopic visual abilities, but can only be comprehended through mathematics.

At the next level of complexity, the atomic level, we are still below the range of adequate visualization though electron and other subtle microscopes give us pictures that help to satisfy our visual senses. We know more about patterns and constraints here: the fantastic variety of the world arises from the functioning of relatively few kinds of atomic structures. The atomic structures show a remarkable pattern of nucleus and electron shells, and we have a simple numerology which describes the detailed orbital structure of these shells. In the outermost orbitals reside electrons that give the atom its characteristic chemical properties. Through their functioning they produce the forms of molecules and bulk matter.

The molecular level comprises structures composed of atoms. Here we glimpsed a small fraction of the elegance and variety of structures that delight imaginative scientists. The molecular forms depend on atomic functions as atomic forms depend on nuclear functions. There is an holistic relationship here. It is at the atomic-molecular level that we come upon a phenomenon of vast importance to us. We exist on this Earth by the gift of sunlight. These solar quanta, streaming through space, eight minutes from the boiling surface of the Sun, fall upon green leaves, are absorbed, and by the molecular machines in these leaves are changed in size. The photon entering the molecule excites, and enables, chemical transformations that keep us in oxygen to breathe and food to support our minds and spirits. We have reason for giving light a central role in the scheme of our philosophy. The aggregation of molecules and giant molecules called polymers produce the variety of bulk matter.

We continually meet the interplay of variety and constraint. We emphasize that although probability rules the orbitals of atoms, and the behaviors of atoms in gases, and many behaviors of molecules and bulk matter, yet we can live securely with physical laws that predict accurately the behavior of macroscopic objects. The reason for this as we saw is that atoms and molecules are so small relative to our world of tangible objects, and so numerous, that statistics are in our favor. Fluctuations tend to cancel: never will all the air molecules in this room spontaneously move up to the ceiling and leave us gasping or dead below. Moreover, Planck's quantum of action has so small a value that the graininess of energy at this level is quite beyond our perception in normal mundane affairs.

In Science we have been equally concerned with the mental

dimension of our existence. If we were to reside solely at the plane of perceptions, we would be largely creatures of chance, surprised at every turn of fate and incapable of learning, because everything new would be unique. But our minds are not built that way. From the earliest age, it seems, we begin to seek patterns and invariant-relations between the occurrences in our experience. Some of these we discover. Probably the majority are taught us in accord with and to accord with our cultural milieu. Many of these must later be unlearned as we become educated and accultured in broader realms of knowledge and experience.

The lessons taught by Science are spectacular ones. By creative travail great scientists have brought out of their imaginations constructs and invariant-relations based on publicly verifiable experience of the most far-reaching and powerful kind. There is no question about the seminal roles that Newton, Boyle, Dalton, Faraday, Maxwell, Mendeleev, Rutherford, Planck, Einstein played in their days, about the vast changes their work and ideas brought about in succeeding generations.

The interplay of perceptions and constructs mediated along operational pathways gives us a metaphor of how phenomena sensed as being 'out there' in the external world undergo symbolic transformation into something new and often strange 'in here' in our minds. The vast network of constructs, interlocked by invariant-relations, mathematical rules, and correlation schemes, carries an internal logic that is most convincing. And when this is connected to perceptions, capable of being checked along operational pathways by anyone who can understand what is involved, the whole structure of physical reality receives the strongest validation: that conferred by the way things are.

Meaning inheres in Science for the very reason of this vast validated network of theory and experiment. No wonder that many scientists have anchored their life view exclusively to this rock. Such a single-minded life view is, we have suggested, partial, therefore at fault.

The spiritual dimension, the specifically human dimension, is strongly supported by Science. This is the dimension of the will-to-meaning, the dimension of the actualization of values. Mind and spirit are seldom mentioned in modern physical textbooks. Perhaps they have been exorcised by the (still prevalent) positivism of the last century and the early part of this one. This may be why, or at least part of the reason why, many perceptive students have been repelled by their experiences with 'Science.' We hold that Science is only a part of the knowledge, experience, and action that comprise a liberal education, and that to avoid science courses is a serious error. This is because the student may easily become maimed in that sensitive will-to-meaning that is part of his humanity. We see evidence of such maiming all about us: a 'feeling-is-all' attitude that begins as an exploration but soon develops into anti-intellectualism. (For if one does not cultivate certain powers of the mind they inevitably atrophy.) The mind has a self-protective ability to rationalize which makes healing under these conditions more and more difficult with the passage of time. One result of any anti-intellectualism is that it destroys freedom to think. Those who spawn it may be its first victims. We see evidence of this maiming of the will-to-meaning in the nihilistic philosophies that suckle on truly deep ignorance, and on an egocentric concept of 'absurdity' and similar slogans that is highly damaging to its devotees, to judge by their cries of anguish, many of which have a hollow sound, resonant with neologisms of doubtful reference. Science students appear to be almost immune to the "identity crises" that seem to plague some students. One reason is that they are busy *creating* their identity as partners in the enterprise of science: they are connected to a rational enterprise that makes connections between physical, mental, and spiritual aspects of the World.

Science is a radical conservative force, the motive power of much that is of the highest value in our culture. Radical because of the policy to attack anomalies in theory, and to correct areas of misfit between theory and the way Nature is. Conservative because unwilling to give up something imperfect just because it is clearly imperfect until something better comes along. Unwilling, that is, to relinquish one jot of the victories won by reason over unreason, by the rational mind over irrational impulse. Valuable for this recognition of a 'directional arrow' in the Universe: that it takes rationality to recognize irrationality, never the other way around.

We inveigh equally against the one-sided humanist and the one-sided scientist. Each may easily think he has the exclusive 'word' and if he gains power in any context he becomes dangerous.

What we offer the student as a gift of Science toward his life view are some of the values that he himself has actualized in his studies, as well as the object lessons that are to be inferred. Science is a communal effort that values individual enterprise and insight. Being the work of people who range over the gamut of nobilities and failings, it is not surprising that individual enterprise and insight often have a hard time when up against prevalent conventions. There is a kind of cultural inertia that resists new ideas unless the culture itself is changing. In that case, ideas have to at least appear new to be accepted. We claim no superior perfection or superhuman practitioners for science. We do, however, insist that the great and noble be honored. Here, then, are some of the values that Science offers the intelligent student.

Honesty, cooperation, high seriousness. The philosopher Bertrand Russell considered the kernel of the scientific attitude to be "the refusal to regard our own desires, tastes, and interests as affording a key to the understanding of the world." This attitude makes it possible for people to communicate together even in matters that might otherwise arouse partisanship. It is part of the objectivity of the judge or umpire that permits the facts to be weighed. It is particularly important when uncertainty and inconclusive evidence cloud the issue.

Development of meaning. Science and its associated technology have worked so many miracles that we almost can't see them. How did these come about if not through the fantastic interconnections between data and constructs that enable each part of a science to be connected to other parts, thus giving meaning to both? The scientist himself grows in meaning as he helps to build this great physical and mental structure. Science is an embodiment of our collective will-to-meaning in its area of application. Here the values that are actualized are specifically human ones.

Existential and essential values. Every student who has worked in a laboratory has experienced the intransigence of inanimate matter. Science is built out of Nature, and a scientist would surely not derogate the existential aspects of life. But his life view equally comprises the essential aspects; the invariant-relations that serve as touchstones for actual observations. Thus as a person who exists, suffers, is responsible, the scientist agrees with the existentialist: he can oppose the limitations of Nature and assert his freedom in the face of them. What one wants to teach is man's freedom to be himself, and to belong to a community of responsible persons; for freedom, as wise men have always insisted, is contingent upon, but not dependent upon, constraints. Our very existence is dependent on our physical substrate. Science shows examples of how men have transcended these constraints either by subduing them or by consciously submitting to them.

Moral and ethical stance. Human actions have moral dimensions from the examination of which ethical presuppositions may be inferred. Science is action, in part. It involves choice and the responsibility that goes with choice. It must therefore be evident that Science has moral dimensions. The 'objectivity' that is a grand feature of scientific methods is that of the judge. It arises from and implies an ethical position based on principles which we have just suggested: honesty, cooperation, high seriousness, disinterestedness of an important kind, rational attitudes, responsible search for truth, and the development of meaning through the balanced actualization of essential and existential values. The tremendous successes of Science and its associated technologies would seem to validate its principles in its realm of action. This must surely be taken into account in an educated person's life view.

Each person's life view is a matter for himself. We do not attempt to prescribe. We have attempted only to provide some of the means for realization as they are offered in physical Science, so that each person may choose and emphasize his own. In this way we have hoped to show the relevance of Science to the modern literate person, and to show why many highly intelligent, creative, and compassionate people have devoted their lives to Science.

Appendixes

Appendix 1. Trigonometry
Appendix 2. Some Physical Constants
Appendix 3. Some Relative Nuclidic Masses
Appendix 4. Alphabetical List of the Elements

Appendix ~ 1

Trigonometry

Trigonometry has to do with the properties and measurement of triangles. This Appendix is designed to recall to the student who studied trigonometry in secondary school the essentials of what he learned at that time, and to introduce these essentials to anyone who has not studied trigonometry.

Directed segments, or 'lines.' A positive and a negative direction may be assigned to any line, one the opposite in direction to the other. The convention that is universally accepted is that a line is positive in the direction 0 toward ^+X or ^+Y or ^+Z in the Cartesian frame of reference, and negative in the opposite directions. Given a line $PQ \underset{P \quad Q}{\longrightarrow}$ the direction P to Q is positive, that from Q to P is negative, thus $PQ = -QP$; $-PQ = QP$. The radius of a circle is always positive when measured from the center outward.

Angles and their magnitude. Suppose the line segment PQ be placed on the Cartesian axes so that P coincides with the origin O, and Q lies along the X axis. Now let the line OQ rotate in a counter-clockwise direction (Figure 1). When it lies along the line OY, it will have described or generated an angle of 90°; when it lies along $O(-X)$, 180°; when along $O(-Y)$, 270°; and when it returns to OX, 360°. One full circle is 360°. By continuing to rotate in the counter-clockwise direction, which by convention is the *positive* direction, angles of any magnitude may be described. In this illustration the *vertex* of the angle is at O. Angles may be added and subtracted. For example, in Figure 2(a) two positive angles, α and β with vertices at O, are added; in Figure 2(b) a positive and a negative angle are added. In this text we shall be concerned almost entirely with positive angles.

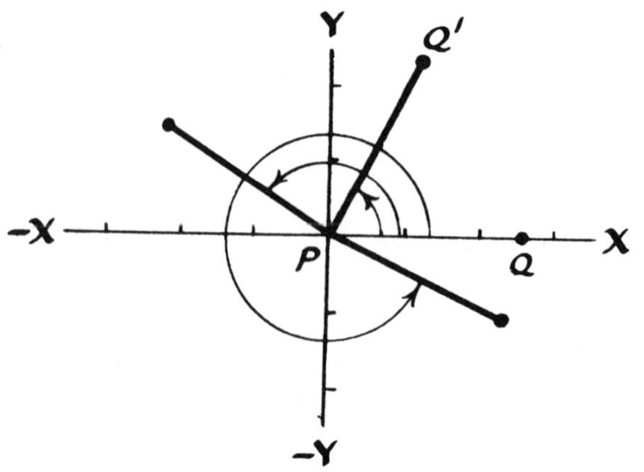

Fig. 1

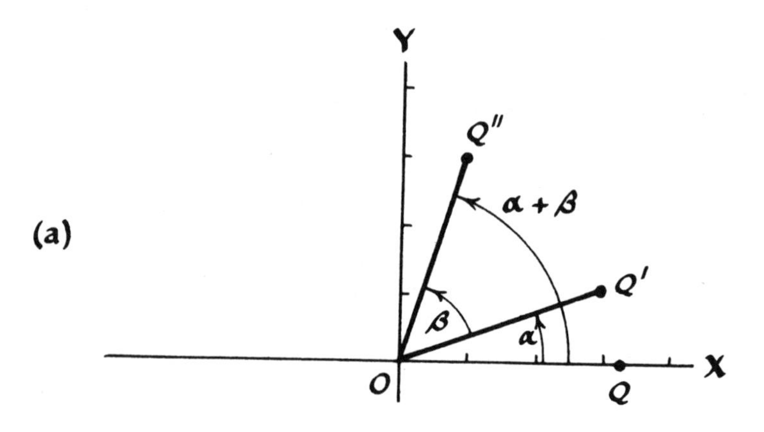

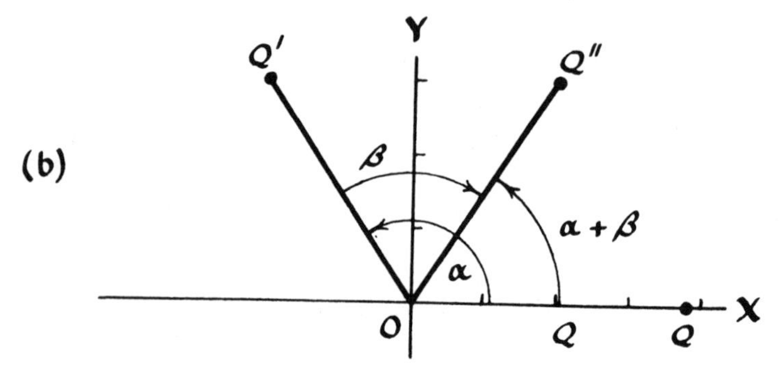

Fig. 2

Radian. Angles may be measured in degrees, of which there are 360 in a full circle, or in radians, of which there are 2π in a full circle. A radian is an angle such that if its vertex is placed at the center of a circle with radius r, it will intercept an arc equal in length to r. Thus in Figure 3, the radius of the circle is OQ, then if $OQ = QQ'$,

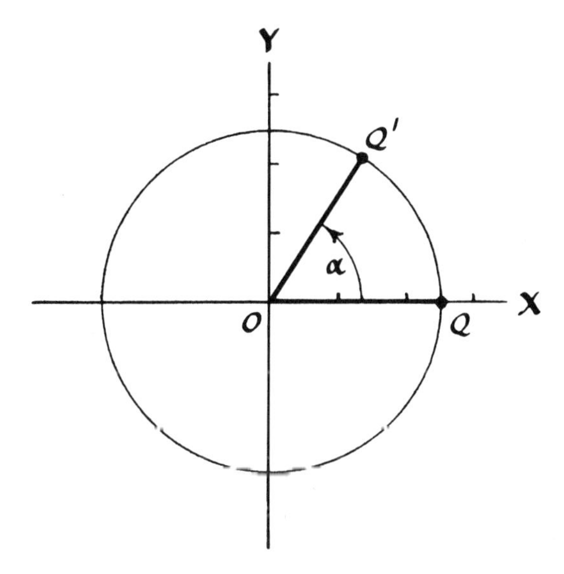

Fig. 3

the angle α has a magnitude of one radian. With this definition, $90° = 2/\pi$ radians, $180° = \pi$ radians, and so on. One radian = $57°17'44''$. The convenience of radian measure can be guessed from the relation that if r is the length of the radius of a circle and s the length of an arc of the circle then the angle α subtended by the arc for any circle whatsoever is, in radians:

$$\alpha = s/r$$

Trigonometric relations. If a line is dropped from Q in any of the Figures, perpendicular to the X axis and reaching it, there is constructed a right-angle triangle. As the line OQ of length r is rotated about O, it is apparent that the distance x from O to the point where the perpendicular meets the X axis changes with the angle rotated. Thus in Figure 4, it starts with a length r, when OQ lies on the X axis, and decreases in length to zero as OQ is rotated through the first quadrant to lie on the Y axis. It would then increase to r at 180°, decrease to zero at 270°, and return to r at 360°. At the same time a perpendicular dropped from Q to the Y axis marks off a distance to

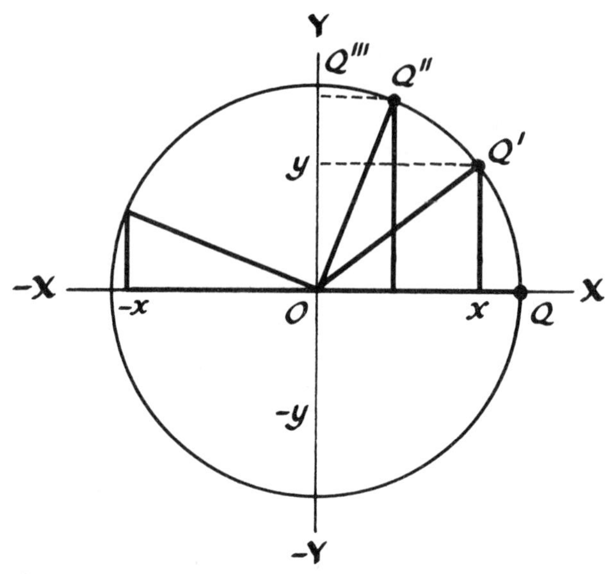

Fig. 4

O, y, which increases from zero to r at 90°, decreases to zero at 180°, increases to r at 270°, and returns to zero at 360°. Trigonometry is concerned with the relations between these distances and the associated angles. There are many of these relations. We are interested, however, in three only: the *sine* (sin), *cosine* (cos), and *tangent* (tan) of the angle.

These are defined as follows. We draw a right triangle, as in Figure 5, and label the angles α, β, and γ, letting α represent the angle with vertex at O, γ the right angle, and β the remaining angle. The distance OQ is labeled r. In every case we are dealing with a directed line. The distance OM is labeled x and is directed along the X axis; the distance ON is labeled y and is directed along (parallel to) the Y axis. The distance x is sometimes called the *abscissa* of the point Q, and the distance y the *ordinate* of the point Q. The radial distance r is then usually just the *distance*, O to Q, and is plus when measured in this direction.

$\sin \alpha = y/r =$ ordinate/distance = side opposite/hypotenuse
$\cos \alpha = x/r =$ abscissa/distance = side adjacent/hypotenuse
$\tan \alpha = y/x =$ ordinate/abscissa = side opposite/side adjacent

Notice that each of these three trigonometric relations, or functions of the angle, is defined as a *ratio* between two line segments: each is therefore a dimensionless number. This is very convenient because it follows that they apply to any right triangle whatsoever. These ratios can be calculated by geometry, and the use of the Pythagorean rela-

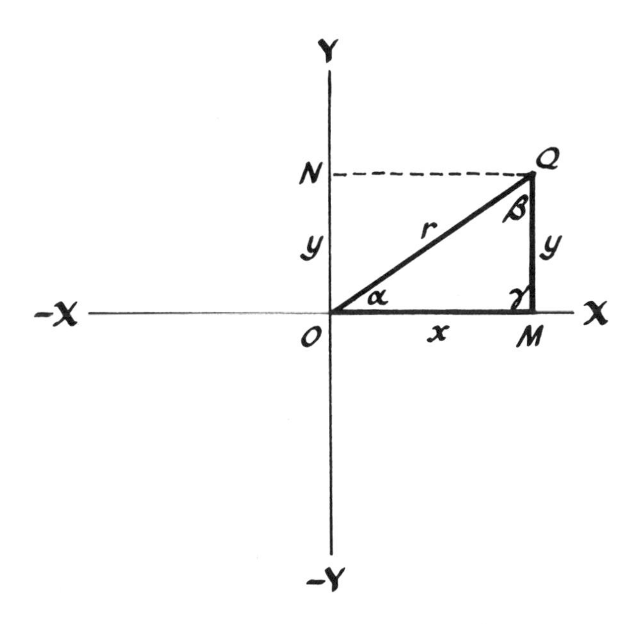

Fig. 5

tion that $r^2 = x^2 + y^2$. For example, in Figure 6(a) the construction shows an equilateral triangle OQR with vertex at O, and with the X axis bisecting the angle QOR. The three angles of any Euclidean triangle sum to 180°; thus the angles of the equilateral triangle are 60° each, whence the angle α must be 30°. Then

$$\sin\alpha = \sin 30° = 1/2 = 0.500$$
$$\cos\alpha = \cos 30° = \sqrt{3}/2 = 0.866$$
$$\tan\alpha = \tan 30° = 1/\sqrt{3} = 0.577$$

Similarly, in Figure 6(b):

$$\sin 45° = 1/\sqrt{2} = (1/2)\sqrt{2} = 0.707$$
$$\cos 45° = 1/\sqrt{2} = (1/2)\sqrt{2} = 0.707$$
$$\tan 45° = 1/1 = 1$$

And in Figure 6(c):

$$\sin 60° = \sqrt{3}/2 = 0.866$$
$$\cos 60° = 1/2 = 0.500$$
$$\tan 60° = \sqrt{3}/1 = 1.732$$

When the angles are greater than 90°, the same kinds of calculations may be used, only now one must pay close attention to signs. Thus in Figure 7(a) the $\cos\alpha$, $-\sqrt{3}/2$, is negative, as is the tangent; and so on as in 7(b) and 7(c). These relations are summarized for the four quadrants:

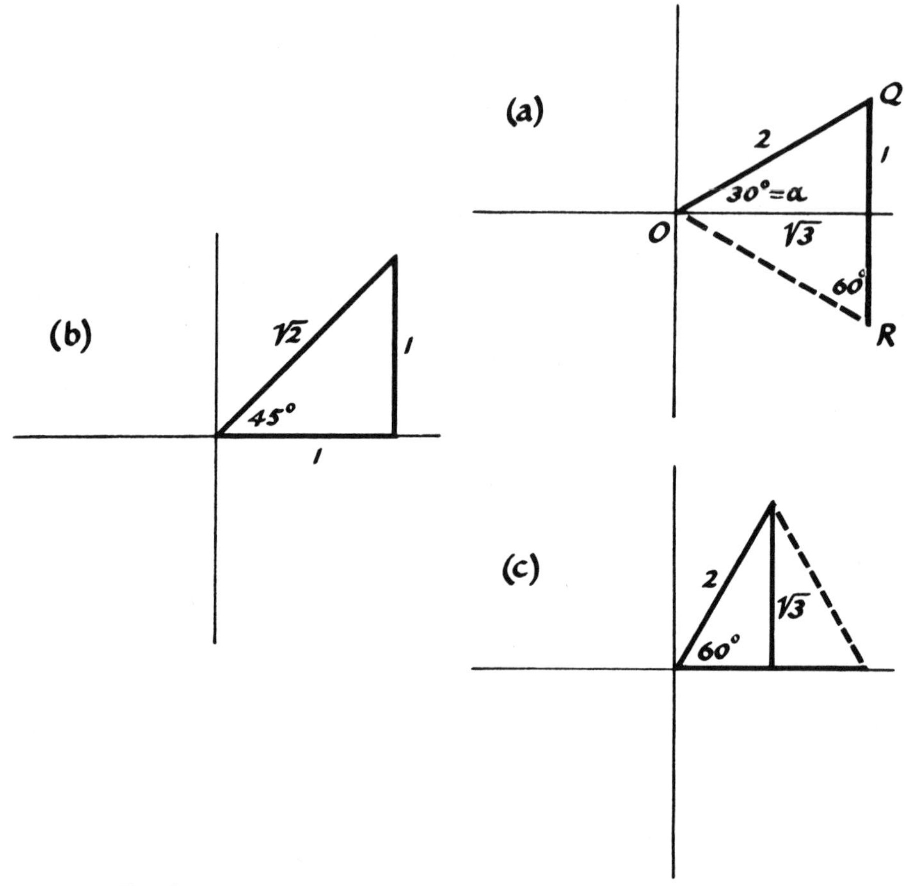

Fig. 6

Function	Quadrant I	Quadrant II	Quadrant III	Quadrant IV
sin α	$+/+=+$	$+/+=+$	$-/+=-$	$-/+=-$
cos α	$+/+=+$	$-/+=-$	$-/+=-$	$+/+=+$
tan α	$+/+=+$	$+/-=-$	$-/-=+$	$-/+=-$

If we were to center our attention upon the angle β, then by our definitions, its sine would be side opposite/hypotenuse, *i.e.*, x/r in Figure 5.

Examples of use

The elegance and usefulness of these relations can be recognized from a simple example. More forceful are the applications made in the text.

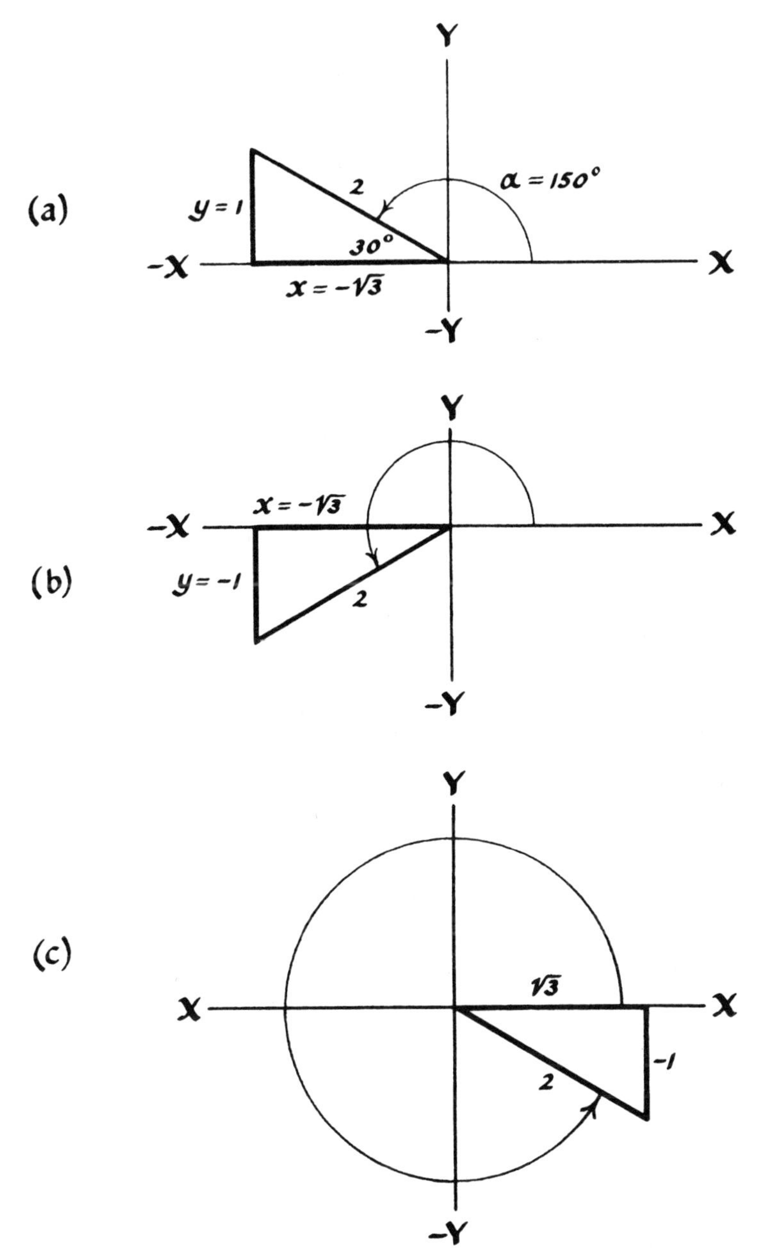

Fig. 7

Example: A large tree is at a certain distance from a house. If it should blow over will it reach to the house? Letting the base of the house be at O and the distance to the tree be x, and the height of the

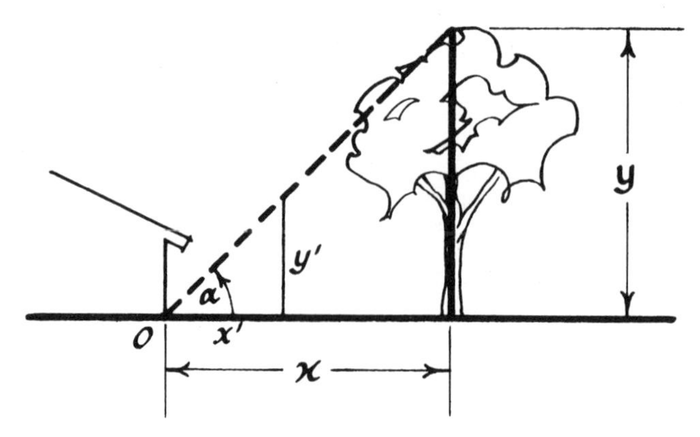

Fig. 8

tree be y, that is, erecting a Cartesian frame of reference in the most convenient way, we can see that $\tan \alpha = y/x$. We can measure x along the ground, and if we could measure α, then we could calculate y by the use of the table of natural tangents. α could be measured with a surveyor's instrument. Another way is to have someone hold a pole of length y' so that we can sight along the line from close to the ground at O to the tree-top just touching the top of the pole. Now the pole is at a distance x' from the house. Then:

$$y'/x' = \tan \alpha = y/x$$

Since we know x', x and y', we can calculate y.

In this text we arrange problems and discussions that involve trigonometry so as to deal as much as possible with right triangles. Sometimes, however, it is necessary to employ triangles in which no angle is 90°. There are simple rules which are useful on such occasions. These may be derived by constructing appropriate right triangles on the oblique triangle with which one is concerned, as may be seen from the following example of the derivation of the law of sines. This law says that in a plane triangle any two sides are to each other as the sines of the opposite angles. In Figure 9(a) and (b) are two triangles with sides a, b, c and angles α, β, γ respectively opposite these sides. We draw d from the vertex of γ, perpendicular to the side c, or to an extension of c. In each figure,

$$\sin \alpha = d/b$$
$$\sin \beta = d/a$$

When we divide the first equation by the second, we obtain:

$$a/b = \sin \alpha / \sin \beta$$

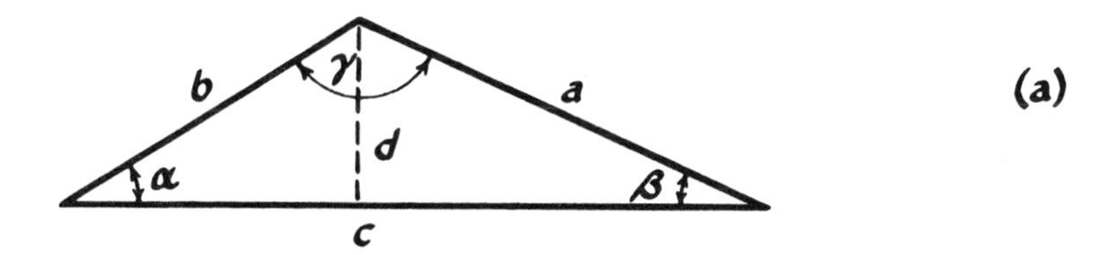

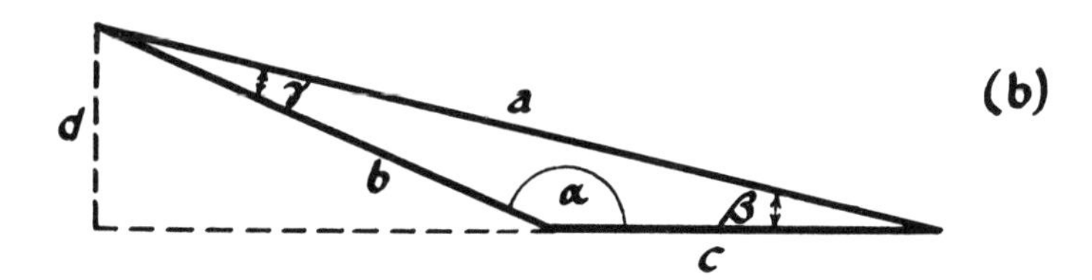

Fig. 9

Similarly, dropping a perpendicular to a from the vertex of α, we obtain

$$b/c = \sin \beta / \sin \gamma$$

These equations may be rearranged to yield the *law of sines:*

$$\frac{a}{\sin \alpha} = \frac{b}{\sin \beta} = \frac{c}{\sin \gamma}$$

By similar reasoning one may obtain other useful relations:

$$\sin^2 \alpha + \cos^2 \alpha = 1$$

$$x^2 + y^2 = r^2$$

$$\frac{x^2}{r^2} = \sin^2 \alpha; \; y^2/r^2 = \cos^2 \alpha.$$

Appendix ~ 2

Some Physical Constants*

Adjusted Values of Constants

Constant	Symbol	Value	Unit
Speed of light in vacuum	c	2.997925×10^{8}	meter/sec; ms^{-1}
Elementary charge	e	1.60210×10^{-19}	coulomb; C
Avogadro constant	N_A	6.02252×10^{23}	per mole; mol^{-1}
Electron rest mass	m_e	9.1091×10^{-31}	kg
		5.48597×10^{-4}	atomic mass unit; u
Proton rest mass	m_p	1.67252×10^{-27}	kg
		1.00727663	u
Neutron rest mass	m_n	1.67482×10^{-27}	kg
		1.0086654	u
Faraday constant	F	9.64870×10^{4}	$C\ mol^{-1}$
Planck constant	h	6.6256×10^{-34}	joule/sec; Js
Rydberg constant	R_∞	1.0973731×10^{7}	m^{-1}
Bohr radius	a_0	5.29167×10^{-11}	m
Electron radius	r_e	2.81777×10^{-15}	m
Wien displacement constant	b	2.8978×10^{-3}	m °K
Stefan-Boltzmann constant	σ	5.6697×10^{-8}	$W\ m^{-2}\ °K^{-4}$
Gravitational constant	G	6.670×10^{-11}	$N\ m^{2}\ kg^{-2}$
		Defiined values and equivalents	
Standard acceleration of free fall	g	9.80665	$m\ s^{-2}$
Coulomb law constant	C‡	8.98759×10^{9}	$N\ m^{2}C^{-2}$

* As recommended by the National Academy of Sciences—National Research Council. *Physics Today 17*, No. 2, February 1964, pp. 48–49.

‡ In the text, K is used in place of C.

Notes: The numbers given are subject to some uncertainty in the last one or two figures. The student, in using them for calculation should round off to the first two or three significant numbers. For example, the speed of light, 3.0×10^8 m/sec; or Avogadro's number, 6.02×10^{23} per mole. Rounding off does *not* apply, obviously, to nuclear calculations. The units are in the international system. The symbols used are: meter, m; kilogram, kg; second, s; degree Kelvin, °K; unified atomic mass unit, u; mole, mol (which is defined as the amount of substance containing the same number of atoms as *12g* of pure C^{12}. This is the chemical mole, and note that the mass is taken in grams). Coulomb, C; joule, J; watt, W; newton, N; volt, V.

Appendix

Some Relative Nuclidic Masses

Z	A	Symbol	Mass (u)
0	1	n	1.008665
1	1	H	1.007825
1	2	D	2.014102
1	3	T	3.016049
2	4	He	4.002603
3	7	Li	7.016005
4	9	Be	9.012186
5	11	B	11.009305
6	12	C	12.000000
7	14	N	14.003074
8	16	O	15.994915
9	19	F	18.998405
10	20	Ne	19.992440
82	206	Pb	205.974459
86	222	Em	222.017530
88	226	Ra	226.025360
90	234	Th	234.043570
91	234	Pa	234.043370
92	235	U	235.043933
92	238	U	238.050760

Selected from F. Everling, L. A. König, J. H. E. Mattauch, and A. H. Wapstra, *Nuclear Physics*, *18*, 529–569 (1960).

Appendix ~ 4

*Alphabetical List of the Elements**

Element	Symbol	Atomic number, Z	Atomic weight	Element	Symbol	Atomic number, Z	Atomic weight
Actinium	Ac	89	[227]	Erbium	Er	68	167.26
Aluminum	Al	13	26.9815	Europium	Eu	63	151.96
Americium	Am	95	[243]	Fermium	Fm	100	[255]
Antimony	Sb	51	121.75	Fluorine	F	9	18.9984
Argon	A	18	39.948	Francium	Fr	87	[223]
Arsenic	As	33	74.9216	Gadolinium	Gd	64	157.25
Astatine	At	85	[210]	Gallium	Ga	31	69.72
Barium	Ba	56	137.34	Germanium	Ge	32	72.59
Berkelium	Bk	97	[249]	Gold	Au	79	196.967
Beryllium	Be	4	9.0122	Hafnium	Hf	72	178.49
Bismuth	Bi	83	208.980	Helium	He	2	4.0026
Boron	B	5	10.811	Holmium	Ho	67	164.930
Bromine	Br	35	79.909	Hydrogen	H	1	1.00797
Cadmium	Cd	48	112.40	Indium	In	49	114.82
Calcium	Ca	20	40.08	Iodine	I	53	126.9044
Californium	Cf	98	[249]	Iridium	Ir	77	192.2
Carbon	C	6	12.01115	Iron	Fe	26	55.847
Cerium	Ce	58	140.12	Krypton	Kr	36	83.80
Cesium	Cs	55	132.905	Lanthanum	La	57	138.91
Chlorine	Cl	17	35.453	Lead	Pb	82	207.19
Chromium	Cr	24	51.996	Lithium	Li	3	6.939
Cobalt	Co	27	58.9332	Lutetium	Lu	71	174.97
Copper	Cu	29	63.54	Magnesium	Mg	12	24.312
Curium	Cm	96	[245]	Manganese	Mn	25	54.9380
Dysprosium	Dy	66	162.50	Mendelevium	Md	101	[256]
Einsteinium	Es	99	[253]	Mercury	Hg	80	200.59

Alphabetical List of the Elements (Continued)

Element	Symbol	Atomic number, Z	Atomic weight	Element	Symbol	Atomic number, Z	Atomic weight
Molybdenum	Mo	42	95.94	Scandium	Sc	21	44.956
Neodymium	Nd	60	144.24	Selenium	Se	34	78.96
Neon	Ne	10	20.183	Silicon	Si	14	28.086
Neptunium	Np	93	[237]	Silver	Ag	47	107.870
Nickel	Ni	28	58.71	Sodium	Na	11	22.9898
Niobium	Nb	41	92.906	Strontium	Sr	38	87.62
Nitrogen	N	7	14.0067	Sulfur	S	16	32.064
Nobelium	No	102		Tantalum	Ta	73	180.948
Osmium	Os	76	190.2	Technetium	Tc	43	[99]
Oxygen	O	8	15.9994	Tellurium	Te	52	127.60
Palladium	Pd	46	106.4	Terbium	Tb	65	158.924
Phosphorus	P	15	30.9738	Thallium	Tl	81	204.37
Platinum	Pt	78	195.09	Thorium	Th	90	232.038
Plutonium	Pu	94	[242]	Thulium	Tm	69	168.934
Polonium	Po	84	[210]	Tin	Sn	50	118.69
Potassium	K	19	39.102	Titanium	Ti	22	47.90
Praseodymium	Pr	59	140.907	Tungsten (Wolfram)	W	74	183.85
Promethium	Pm	61	[147]	Uranium	U	92	238.03
Protactinium	Pa	91	[231]	Vanadium	V	23	50.942
Radium	Ra	88	226.05	Xenon	Xe	54	131.30
Radon	Rn	86	[222]	Ytterbium	Yb	70	173.04
Rhenium	Re	75	186.22	Yttrium	Y	39	88.905
Rhodium	Rh	45	102.905	Zinc	Zn	30	65.37
Rubidium	Rb	37	85.47	Zirconium	Zr	40	91.22
Ruthenium	Ru	44	101.07				
Samarium	Sm	62	150.35				

* The atomic weights are based on ${}_6C^{12}$ taken as 12.000, and are those adopted in 1961 by the International Union of Pure & Applied Chemistry. The weights in brackets are approximate. The atomic weights are for the naturally occurring mixture of isotopes, as used by the chemist.

Index

*n refers to a footnote

Acceleration, average, definition, 50
  centripetal, conceptualization, 108
    definition, 109
Accelerator, definition, 322
  linear, 322
Acid, definition, 371
Action, definition, 7
Adiabatic, definition, 481
Aether, 472
Alpha particle, definition, 310
Alpha-decay, nuclear effect of, 336
Annihilation, 329
Aromatic compound, definition, 397
Ashby, W. R., 439
Asymmetric molecule, definition, 407
Atkinson, Robert, 350
Atom, definition, 258
Atomic mass unit, definition, 322
  relation to electron volts, 341
Avogadro's number, 366
Axes, Cartesian, 38

Barn, definition, 347
Barrie, J. M., 193
Base, definition, 371
Becquerel, Henri, 309
Beta-decay, nuclear effect of, 336
Beta-particle, definition, 310
Bethe, Hans, 350
Binding energy, calculation of, 342
  nuclear, 342
'Black body', 222
Bohr, Niels, 239, 269
Bohr atom, 270
Boltzmann, Ludwig, 431
Bolyai, John, 192
Bond, coordinate covalent, 360, 365
  covalent, 260, 360
  double, 391
  hydrogen, 374
  ionic, definition, 260
  metallic, definition, 360, 365
  percent ionic character, 364
  shared-electron, definition, 260
  triple, 391
Boulding, Kenneth, 495

Canonical representation, 443
Carbon atom, structure of, 384
Carnot, Sadi, 480
Carnot cycle, 480
Cartesian axes, 38
C-field, definition, 16
Cockcroft, J. D., 115
Chance, prevalence of, 423
Chemical symbols for elements, 257
Chemistry, definition, 359
Cis, definition, 395*n*
Clausius, Rudolf, 485
Coin, outcomes of tossing, 424
Compound, definition, 359
Compton, Arthur H., 240
Configuration, absolute, 411
  in statistical mechanics, 430
Conformation, definition, 389
Constancy effect, definition, 450
Constant, equilibrium, 371
Constraint, measure of, 256

Construct, definition, 16
  fundamental, 138
Copernicus, 19
Correspondence, one-to-one, 443
Couffignal, Louis, 439
Coulomb, Charles, 74, 473
Coulomb law, 75
Critchfield, Charles, 352
Critical mass, nuclear, 453
Crookes tube, 70, 291
Cross-section, nuclear, 347
Cubists, 192
Current, alternating, 90
  direct, 90
  electrical, definition, 90
  induced, 119
Cybernetics, definition, 439
Cyclotron, description, 115

Darwin, Robert Charles, 488
Dating, radioactive, 352
Davisson, C. J., 281
De Broglie, Louis, 277
De Latil, Pierre, 445
Decay constant, nuclear, 338
Decay, nuclear, chance in, 338
Decay series, nuclear, 336
Descartes, René, 38
Deuterium, 257
Dextro, definition, 409
Dimensions, of existence, mental, 497
  physical, 497
  spiritual, 497
Dimensions, of units, use of, 79
Distance, relativity of, 184
Du Fay, Charles, 61

Eddington, Sir Arthur, 432
Effector, definition, 440
Einstein, Albert, 51
Electricity, relation to magnetism, 205, 210
  resinous, 62
  static, field intensity, 84
    potential, 85
    resinous, 62
    separation of charge, 65
    sign of charge, 62
    vitreous, 62
  vitreous, 62
Electron, charge on, 91
  definition, 281
  wave nature of, 281
Electron volt, definition, 118
  relation to joules, 341*n*
    to kilocalories, 341*n*
    to kinetic temperature, 349*n*
    to mass units, 341*n*
Electroscope, 62
Electrophilic, definition, 393*n*
Element, chemical, definition, 257, 359
Eliade, Mircea, 128
Emission of radiation, by gas, mechanism of, 292
  by solid, mechanism of, 291
Enantiomer, definition, 407
Enentiomorph, definition, 407
Energy, 273
  binding, of nuclide, definition, 258
  kinetic, definition, 88
    in cyclotron, 118
  nuclear, comparison to other sources, 341
  potential, definition, 88
  relativity of, 187
  stellar, 350
  threshold value of a metal, 230
Energy barrier, 325
Energy flow, direction of, 431
Energy level, definition, 273
Entropy, decrease of, 485
  increase of, 485
Equations, chemical, calculations with, 367
  writing of, 366
Equilibrium constant, 371
Ethyl alcohol, 400
Euler, Leonhard, 471
Existential frustration, 12
Experience, definition, 7
Experiment, definition, 19
Explosion, nuclear, 345
Exponents, discussion of, 15

Feedback, definition, 433, 459
  negative, 450
  positive, 450
Feller, William, 425
Fermi, Enrico, 345
Field, definition, 469
  electrical, definition, 73
Fisher, Emil, 411
Fission, nuclear, 335

process of, 346
FitzGerald, G. F., 155
Focillon, Henry, 18
Force, gravitational, 88
lines of, 71
magnetic, 103
lines of, 99
Form and function, 37
Formic acid, 406
Formula, empirical, definition, 385
molecular, definition, 385
structural, definition, 385
Foucault, Jean, 129
Frame of reference, 37, 45
Einsteinean, 175
Galilean, 171
Frankl, Viktor E., 12
Free fall, discussion of, 46
Freezing, heat of, 374
Frisch, Otto R., 345
Frustration, existential, 12
Fuel oil, 389
Function, definition, 42
Function and form, 37
Fusion, process of, 348

Galileo, 20, 51, 171
Galvanometer, use of, 120
Gamma ray, definition, 310
Gamow, George, 285
Gasoline, 389
Gauss, Carl Friedrich, 192
Geiger-Müller tube, 315
Geometry, non-euclidean, 192
Germer, L. H., 281
Gibbs, J. Willard, 431
Goethe, Johann Wolfgang von, 127
Goodfield, June, 488
Grimaldi, F. M., 129
Grissom, I., 10
Ground state, 297
Group, organic, definition, 299, 387
naming of, 387

Haber process, 376
Half-life, definition, 336
Hanson, N. R., 129
Hardware, embodying logic, 29
Heat, quantity of, 478
Heat capacity, 479
Heat-exchanger, use of, 348
Heisenberg, Werner, 284, 286
Hertz, Heinrich Rudolph, 204
Hertz, experiments of, 212
Hogben, Lancelot, 154
Houtermans, Fritz, 350
Humanities, definition, 7, 8
Huygens, Christiaan, 140
Hybridization, 385
Hydration, of ions, 370
Hydrocarbons, paraffin, naming, 386
Hydrogen bond, 374
energy of, 374
length of, 374

Impulse, 46
Indeterminacy, 284
Inside straight, 426
Instrumentation, problems of, 119, 234
Interactions, strong, in nucleus, 325
Interference, constructive, 138
destructive, 138
Interferometer, description of, 144
Michelson, 144
Ion, definition, 68
Isotope, 257
detection of, 320

Jammer, Max, 20
Jeans, James, 226
Joule, relation to kilocalorie, 341*n*

Kelvin, Lord, 487
Kerosene, 389
Kilocalories, conversion to electron volts, 341*n*
conversion to joules, 341*n*
Kinetic temperature, conversion to electron volts, 349*n*
conversion to joules, 349*n*
Kirchoff, Gustav R., 222
Knowledge, definition, 7
Körner, G., 397

Lavoisier, Antoine Laurent, 368
Lawrence, E. O., 115
Le Chatelier's principle, 376
Lenard, P., 229
Length, non-proper, definition, 184
proper, definition, 184
units of, 36
Levo, definition, 409
Lewis, Gilbert N., 228, 485

Light, quanta of, 228
Limit, definition, 111
Livingston, M. Stanley, 115
Lobachevski, Nikolai I., 192
Lorentz, H. A., 154
Lorentz equation, 169
 derivation, 176
LP gas, 389

Machine, definition, 446
Magnet, behavior of, 97
 lines of force of, 100
Magnetic flux density, 102
Magnetic bottle, for plasma, 350
Magnetism, relation to electricity, 207, 210
Margenau, Henry, 16
Mass, relativity of, 185
Mass action, law of, 369
Mass deficit, calculation of, 342
 definition, 342
Mass-energy, 188
Mass spectrograph, 317
Mass units, relation to electron volts, 341*n*
Maxwell, James Clerk, 204
Meaning, 11
Meitner, Lise, 345
Melting, heat of, 374
Mendeleev, Dmitri, 298
Meta-, definition, 397
Metal, definition, 301
Metalanguage, 18, 45
Meyer, Lothar, 298
Michelson, A. A., 143
Michelson-Morley experiment, 149
Microstate, in statistical mechanics, 428
Millikan, Robert A., 91
Millikan oil-drop experiment, 91
Minkowski, H., 190
Mixture, definition, 360
Molar solution, definition, 368
Molecule, definition, 260
Molecular orbital, 360
Momentum, conservation of, 187
Morley, E. W., 150
Motion, conceptualization, 38
 uniform circular, 107
Multiplication factor, in fission, 451

N, magnetic, definition, 99
Naming of unsaturated hydrocarbons, 391
Names, trivial, 387
Natural gas, 389
Newton, Sir Isaac, 46, 128, 470, 477
Newton (unit), 46
Newton's first law of motion, 38, 134
 second law of motion, 46
Neutron, definition, 257
 delayed, 454
 prompt, 454
 thermal, 347
Nicolson, Marjorie, 20
Nomenclature, international rules of, 388
Non-metal, definition, 302
North, geographic, definition, 99
Northrop, F. S. C., 16
Nucleon, definition, 256
Nucleophilic, definition, 393*n*
Nucleus, binding energy of, 324
 energy levels in, 324, 327
 definition of, 256
Nuclide, definition, 256

Operand, definition, 441
Orbital, definition, 294
Orbital, *pi*, 392
Ortho-, definition, 397
Oxidation, definition, 372

Pair-production, 329
Para-, definition, 397
Paraffin, 389
 burning of, 389
Parameter, definition, 459
Particle, definition, 469
Pauli, Wolfgang, 296
Pauli principle, 296
Pearson, Julius, 234
Perception, discussion of, 13
Period, chemical, definition, 299
Periodic classification of the elements, 298
Person, literate, characterization of, 5
Phase, definition, 263, 373
 gas, properties of, 264
 liquid, properties of, 263
 solid, properties of, 263
Philosophies, 8
Photoelectric effect, 229
Photoelectric equation, 230, 231

Photoelectron, definition, 230
Photomultiplier tube, 313
Photon, definition, 228
particle behavior of, 239
Pile, nuclear, 451
Plasma, definition, 349
temperature of, 350
Planck, Max, 226, 431
Polanyi, Michael, 153
Polymer, definition, 261
P-plane, definition, 16
Positron, definition, 328
Possibility, logical, 27
Potential, electrical, 90
ionization, of elements, 302
Power, definition, 91
Powers of ten, prefixes for, 36
Priestly, J. B., 193
Principal quantum number, 274
Probability thinking, 432
Projectile, nuclear, production of, 322
Property, extensive, 484
intensive, 484
Proposition, definition, 24
Proton, definition, 257
Ptolemy, 19

q/m of electron, determination of, 112
Quantum, definition, 228
Quantum number, 273
definition, 294
Quantum mechanics, definition, 277

Radical, definition, 390
Radian, definition, 107
Radiation, electromagnetic, source, 204
ionizing ability of, 311
Radioactive series, 339
Radioactive substance, definition, 310
Radioactivity, detection, 309
induced, 342
Randall, Merle, 485
Ratio, nature of, 111
Ray, of light, definition, 130
Rayleigh, Lord, 226
Reaction, addition, mechanism of, 393
chain, mechanism of, 390
chemical, effect of heat on, 375
heat of, 375
proof of mechanism of, 412
rate of, 368
heterogeneous, 371
homogeneous, 371
Reactor, nuclear, 348
Reduction, definition, 372
Reference, frame of, 149
Reflection of light, geometry of, 130
Refraction of light, law of, 131
Relationship, many-one, 457
many-valued, 457
single-valued, 457
Relativity, special theory, 175
Retroaction, definition, 443
Retroactive coupling, 459
Riemann, Georg F. B., 192
Right-hand convention for magnetic force, 101
Rings, nomenclature, 395
Rowland, Henry, 204
Rule, Golden, 496
Rutherford, Lord Ernest, 115, 322

S, magnetic, definition, 99
Scalar, definition, 80
Scattering, of light, mechanism, 130
Schrödinger, Erwin, 278
Science, contribution to life-view, 500
essential aspects of, 501
existential aspects of, 501
meaning in, 499
moral aspects of, 501
Sciences, definition, 7, 8
Scintillation counter, 313
Set, definition, 21
Sherwin, C. W., 181
SHM, simple harmonic motion, 155
Simple harmonic motion, 155
Soddy, Frederick, 322
Solid state, 374
Solution, molar, definition, 368
South, geographic, definition, 99
Space, absolute, 147
Spaceship Earth, 6, 495
Specialization, dilemma of, 5
Spectrum, definition, 131
electromagnetic, 203, 217
Sphere of knowledge, experience and action, 9
Spin, electron, 362
Statement, compound, 25
open, 28
Stefan-Boltzmann law, 223

Structures, equivalent, 388
  organic, proof of, 397, 400
Symbols, logical, for propositions, 25
  for sets, 22

Table, periodic, of the elements, 298
Technologies, 8
Temperature, kinetic, definition, 349*n*
Tendency effect, definition, 450
Thermodynamics, definition, 478
  second law of, 427
Thomson, J. J., 291, 317
Time, absolute, 147
  relativity of, 178
    test of, 183
Time interval, non-proper, definition, 183
  proper, definition, 183
Time's arrow, 432
Toulmin, Stephen, 488
Toynbee, Arnold, 496
Trans, definition, 395*n*
Transform, definition, 441
Transformation, identical, 458
Transformations, coupling of, 459
Transmutation, 336
Tritium, 257
Truth table, 25
Truth value, 25
Tunnel effect, 326

Universal gravitation constant, 88
Unsaturation, definition, 391

Vaporization, heat of, 375
Variable, dependent, 44
  independent, 44
Variety, measure of, 256
Vector, addition, definition, 82
  definition, 80
  multiplication, 104
  subtraction, definition, 83
Velocity, average, definition, 48
  definition, 42
Venn diagram, definition, 22
Vinegar, 402
Von Weizsacher, Carl, 350

Walton, E. T. S., 115
Watt, James, 443
Wave, amplitude of, 137
  energy of, 158
  formation of, 132
  frequency of, 138
  length, definition, 136
  mathematical description, 154
  period, 137
  velocity, definition, 136
Wavelength, measurement, 138, 143
Wave mechanics, definition, 277
Waves, electromagnetic, 211
  interference phenomena in, 138
  radio, 212
Well, potential, in nucleus, 325
Wiener, Norbert, 439
Will-to-meaning, 12, 499
Wien displacement law, 225
Wilson, C. T. R., 316
Wilson cloud chamber, 316
Wood alcohol, 402
Work, definition, 86
  electrical, 87
  gravitational, 88
  reversible, 481
World line, 190
World point, 190

Young, Thomas, 129